Computer Aided Virtual Manufacturing Using Creo Parametric

Computer Aided Virtual Manufacturing Using Creo Parametric

Paul Obiora Kanife

Computer Aided Virtual Manufacturing Using Creo Parametric

Easy to Learn Step by Step Guide

Applicable to both Creo Parametric 2.0 & 3.0

Springer

Paul Obiora Kanife
Notting Hill
UK

ISBN 978-3-319-23358-1   ISBN 978-3-319-23359-8 (eBook)
DOI 10.1007/978-3-319-23359-8

Library of Congress Control Number: 2015949504

Springer Cham Heidelberg New York Dordrecht London

Printed on acid-free paper

Springer International Publishing AG Switzerland is part of Springer Science+Business Media
(www.springer.com)

This book is dedicated to the greatest designer and manufacturer man has ever known. The giver of wisdom, knowledge and free will. The Omniscience, Omnipresence GOD, to whom I give all the glory. Also to my amazing mother, late Mrs. Monica Anayochukwu, Vincent Chidiebere and Dorothy Nnenna Kanife.

This book is dedicated to the greatest designer and manufacturer man has ever known. The giver of wisdom, knowledge and free will. The Omniscience, Omnipresence GOD, to whom I give all the glory. Also to my amazing mother, late Mrs. Monica Anayochukwu, Vincent Chidiebere and Dorothy Nnenna Kanife.

Preface

My drive for writing this tutorial book is based on the fact that during my M.Sc. programme in Advanced Manufacturing Systems Engineering at Brunel University London, I found that there were fewer tutorial/textbooks out there in the market that covered all areas of computer virtual manufacturing using the Creo Parametric software. Since this book focuses more on Computer Aided Manufacturing (CAM) and less on 3D Modelling using Creo Parametric 2.0 software formerly known as Pro Engineer, new users to this software will be introduced briefly to 3D modelling using a step-by-step guide to illustrate the process.

This book is aimed at undergraduate and master's level manufacturing engineering students and also mechanical engineering and design students who have CAD and CAM as an elective option. This book can also be used by technicians, technologists and engineers who want to learn CAM using the Creo Parametric software programme, which is a commercially available product and a registered trademark of PTC Inc. in the United States and other countries.

This book is written in such a way so as to make learning CAM using this software program fun and easy to follow and understand. The step-by-step guide illustrated in this tutorial will be better understood if you take time to read the text carefully, thinking and observing what happens. Therefore, just clicking through the command sequence without observing and paying attention to what happens is not enough. You will also learn more by exploring and experimenting with the different commands on your own.

The first tutorial chapter covers a brief introduction to Creo Parametric. Chapter 2 gives a brief introduction to 3D modelling, adding material to Part and the Engraving process. Face milling, Profile milling, Surface milling, and Volume Rough milling using Mill Volume and Window milling processes are covered in the subsequent chapters (Chaps. 3–7). Chapters 8 and 9 covers Expert Machinist and Electric Discharge Machining, which are 2½ and 2 axis machining processes respectively. Chapters 10–11 cover CNC Lathe Area Turning (2 axis machining process), drilling and boring. Five axis machining, drilling and tapping operations are covered in Chaps. 12–13.

In the practical manufacturing application/situation, it is highly advisable to consult the Machine Tool manual and other relevant books and tooling handbook manuals, if the user wants to use relevant formulas to calculate the cutting parameters, and to use the accurate Tool for each different operation. All dimensions of the Cutting Tools, Tooling Materials and the Cutting Parameters used in this tutorial are for illustration purposes only.

Any relevant suggestions/advice on how to improve subsequent editions of this book are highly welcome.

The 3D models of Parts used in this tutorial can be emailed to the user on request. All requests should be sent to the following email address: paul.kp123@yahoo.co.uk.

Acknowledgments

To my dad Mr. Sylvester Uchechukwu and my late mum Mrs. Monica Anayochukwu Kanife, I say a huge thank you for laying the foundation on which I'm standing and for all your love, words of wisdom and encouragement throughout my educational pursuits. Not forgetting all my family members for their valuable words of support and prayers.

A huge thank you to Barrister Cordelia Kanife for all the wonderful words of encouragement and moral support throughout the ups and downs of writing this book.

To all the lecturers and professors who I have learnt from, I say a huge thank you.

To Anthony Doyle and the editorial team at Springer publishing UK, I say a huge thank you for having faith in my work and publishing it.

A huge thank you to PTC Inc. for granting me the copyright permission, without which this book would not have gone to print. And also to Pennie Stone, the permission and compliance team at Parametric Technology UK (PTC Inc.).

To those who have contributed in one way or the other whom I did not mention here, I say a huge thank you.

Contents

About the Author

Paul Obiora Kanife is a lecturer in the Department of Mechanical Engineering at Coventry University College UK. He currently teaches Mechanical Engineering, Engineering Science, Design and Engineering Graphics using Creo Parametric design software. He is a Fellow of the Institute of Manufacturing (F.I.Manf.) and also a Member of the Institution of Engineering and Technology (MIET) in UK. He holds an MSc in Advanced Manufacturing Systems Engineering from Brunel University London and has many years of combined practical industrial engineering and teaching experience.

List of Figures

Chapter 1
Introduction

1.1 What Is Creo Parametric

Creo Parametric (Previously knows as Pro Engineer and Wildfire) is one of the most sophisticated and advanced solid modelling programmes available in the market. It is the registered trade mark property of PTC Inc. (Parametric Technology Inc.) in the United States and other countries. It drives its power from the extremely rich command set that understandably requires constant practice to master effectively and efficiently.

Creo Parametric contains powerful suite of programmes that are used in the design, analysis and manufacturing of virtual unlimited range of products. Creo Parametric is one of other software packages of PTC Inc, like Creo Simulate, Creo Direct, Creo Layout, Creo Schematics, etc. Creo Parametric is used to carryout 3D Modelling of Part and Assembly, and production of engineering drawings. Optional modules are available to handle more designs and application task like, sheet metal operation, mould, wiring harness and piping design, NC machining, expert machining, etc. Structural analysis (static, deformation, buckling, fatigue and vibration), Thermal and Dynamic motion analysis of mechanism can also be carried out because they are integrated unto the Creo Parametric module application.

Creo Parametric uses the context sensitive menu method/approach, this implies that more options windows and selection menus will only be available when they are applicable to the current task.

When designing with Creo Parametric, the final 3D model of Part that is created forms the basis for all engineering, assembly and design functions. The design of Parts and Assemblies, and the creation of related drawings, forms the foundation of engineering graphics.

Creo is a parametric, feature based solid modelling software programme.

Parametric definition implies that you can relate the attributes of one feature to the other using numeric formula. The physical shape of the part is driven by the values which are assigned to the primary dimensions of its features. Modifying a

P.O. Kanife, *Computer Aided Virtual Manufacturing Using Creo Parametric*,
DOI 10.1007/978-3-319-23359-8_1

feature dimensions or other attributes at any time, will instigate changes through the model.

By Feature based, this implies that 3D model and assembly are created by defining high level and meaningful features like holes, slots, sweeps, extrusions, etc. As features are modified, the entire object automatically updates after regeneration. The idea behind feature-based modelling is that the designer constructs an object so that it is composed of individual features that describe the way the geometry is supposed to behave if its dimensions change. Feature can either create or remove material from the model.

By solid modelling, it means that the 3D model created contains all the information that a real 3D solid object would contain. The created part has volume; which implies that the mass and inertial of the part is known once a value is provided for the density of the material.

Features are the basic foundation blocks use to create an object. Features can be categorised into base features, sketched features, datum features and referenced features.

Base Feature
A set of datum planes makes up the base features which references the default coordinate system on Creo Parametric GUI environment. The base feature is important in Creo parametric modeling because all future model geometry will reference this feature indirectly or directly and any changes to the base feature will affect the geometry of the entire part.

Sketched Feature
Sketched features are created by revolving, extruding, sweeping or blending a sketched cross section. Material can either be created or removed by protruding or cutting the feature from the existing model.

Datum Feature
They do not have mass or volume and can be hidden without affecting solid geometry. They are mostly used to provide sketching planes. Example of datum features are axes, lines, curves and points. There are three primary types of datum Features: (a) datum planes, (b) datum axes and (c) datum points (there are also datum curves and datum coordinate Systems).

Referenced Feature
They do not need to be sketched because a reference feature example rounds, shell, holes, uses the exiting geometry for positioning also uses an inherent form.

1.2 Starting Creo Parametric

This section introduces a step-by-step guide on how to start Creo Parametric CAD software for new users.

Different methods of starting Creo Parametric CAD software are:

Method 1

Click on the Start menu icon on your computer to activate its panel content ≫ Now click on Creo Parametric on the menu list. This action will automatically activate the Creo Parametric software installed in your computer hard drive.

Method 2

Click on the Start menu to activate its panel content ≫ Click on All Programs ≫ Now click on PTC Creo folder on the menu list to activate its drop down menu list ≫ Click on Creo Parametric. This action will automatically activate the software.

Method 3

Click on Creo Parametric icon on your desktop and this action will automatically activate the Creo Parametric version of the software installed in your hard drive as shown in Fig. 1.1.

Note: If you encountered any difficulty activating/starting Creo Parametric, please consult you technical instructor. This tutorial in this text assumes that you are using the default setting on Creo Parametric software configuration

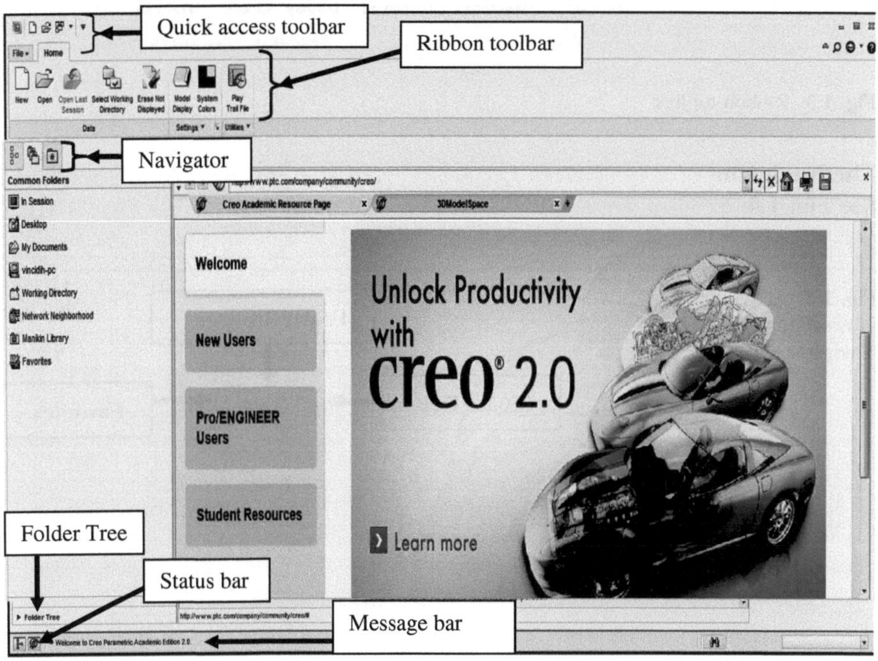

Fig. 1.1 Welcome to Creo Parametric 2.0 window in concise form

The Quick access toolbar

The Quick Access toolbar allows user easy and quick access to commands used frequently. The toolbar can be customised by adding and removing sets of options or individual commands (Fig. 1.2).

Fig. 1.2 Quick Access toolbar

The Ribbon toolbar

The Ribbon toolbar contains commands/operations that can be used for all modes of the system. The Ribbon toolbar contains command buttons organised as a set of tabs. The tabs can be customised by adding, removing or moving buttons sets of options or individual commands (Fig. 1.3).

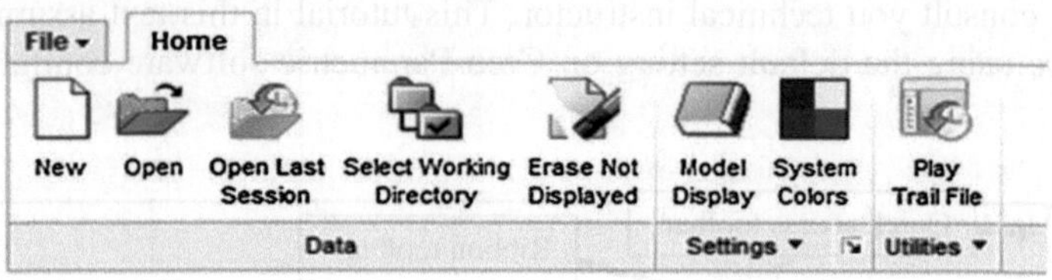

Fig. 1.3 Ribbon toolbar

The Navigator

See Fig. 1.4.

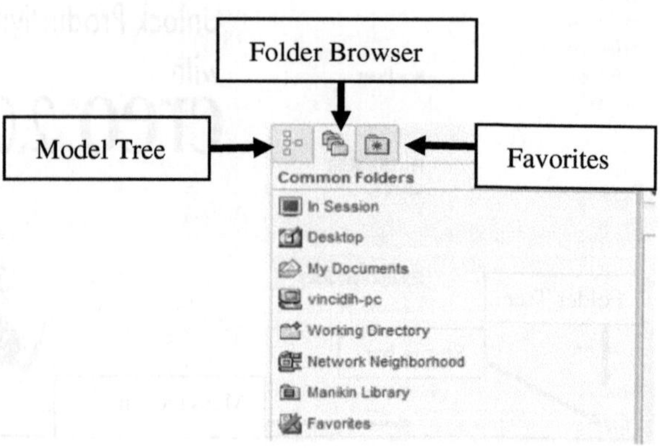

Fig. 1.4 Active Folder browser contents on the Navigator

The Web Browser

The web browser allows the user to quickly access model information and online documentation and help. It can also be used for general web browsing. The web browser will close automatically when you open a model (Fig. 1.5).

Fig. 1.5 Web browser in concise form

The Navigator and Browser Display Control

The web browser can be opened or closed by clicking on the Web Navigator icon as indicated by the arrow above (Fig. 1.6).

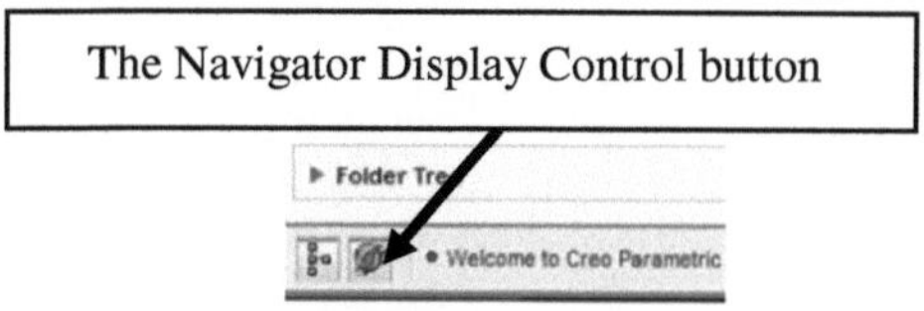

Fig. 1.6 The Navigator display control button

The Mouse Function

A three-button mouse is highly recommended when using Creo Parametric as shown in Fig. 1.7.

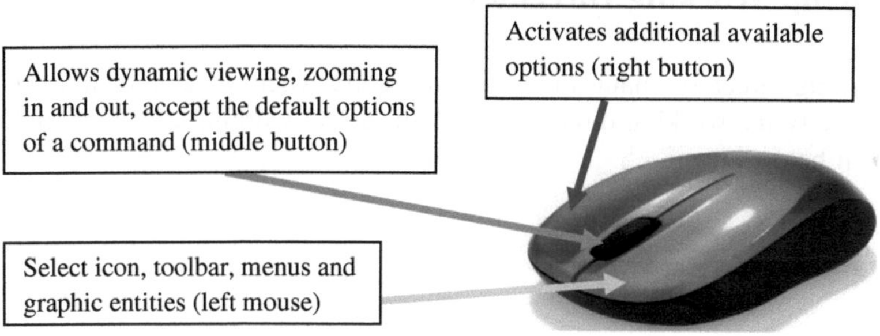

Fig. 1.7 The Mouse functions

The Left Mouse Button

The left mouse button is used to carryout most of the operations and activities in Creo parametric like picking graphic entities and items, clicking on icon and menus.

The Middle Mouse Button

It is used to accept the default settings of a prompt. It can also be used as a shortcut to carryout dynamic viewing functions like Pan, Rotate and Zoom in or out.

The Right Mouse Button

This is used to query a selection and also to activate additional available options.

The Keyboard Shortcuts in Creo Parametric

Ctrl C = Copy
Del = Delete
Ctrl F = Find
Ctrl O = Open file
Ctrl V = Paste
Ctrl N = New file
Ctrl V = Paste
Ctrl Y = Redo
Ctrl G = Regenerate
Ctrl R = Repaint
Ctrl S = Save file
Ctrl D = Standard view
Ctrl Z = Undo

1.3 Setup Working Directory

The working directory enables the user to specify where to save their work.

To specify the working directory, click on Select working directory folder icon as shown below.

The select working directory dialogue window is activated on the main window as shown in Fig. 1.8.

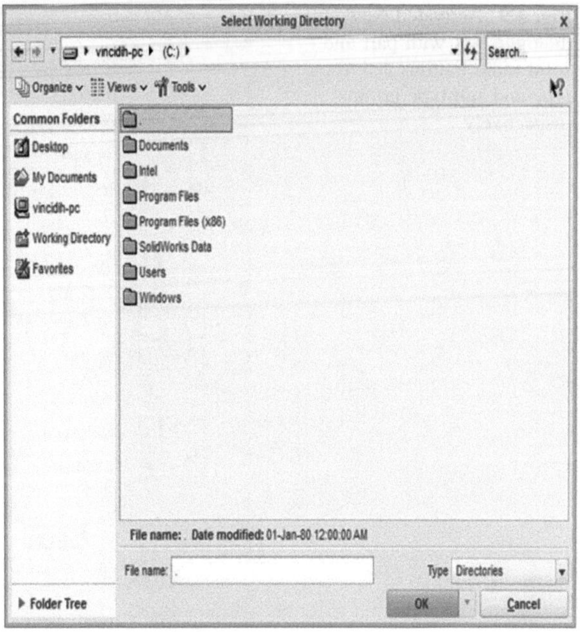

Fig. 1.8 Activated select working directory window

Now click on the folder or drive that you want to use as your Working Directory. Click on OK tab to exit the select working directory dialogue window.

1.4 Starting Creo Parametric Manufacturing Application

Click on New icon as shown below.

The New dialogue box is activated on the main window as shown in Fig. 1.9.

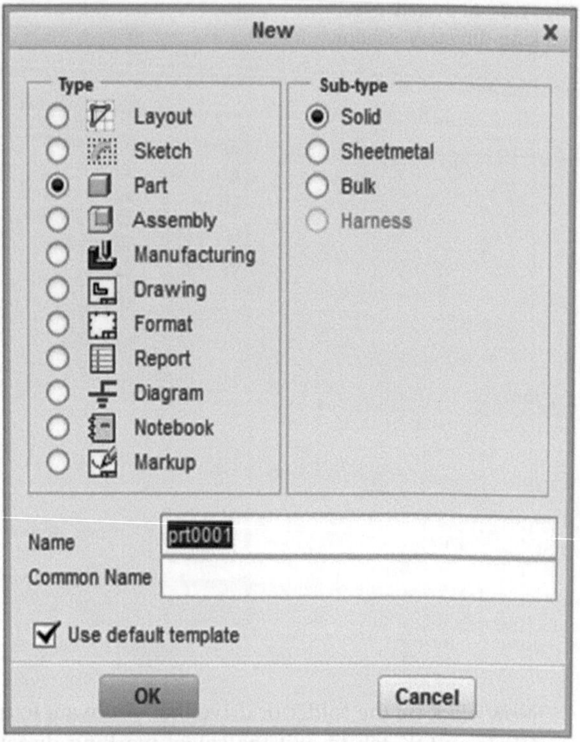

Fig. 1.9 Activated New dialogue box with part and solid radio buttons active on type and sub-type groups respectively

Note: If the user wants to create 3D Model Part, just click in the Part radio button as shown in Fig. 1.9. It is highly advisable for new users to click and open a different radio button to see for themselves each application GUI window. The graphic user interface (GUI) window for each module/application will only open with the full menus, if they are supported by the appropriate license.

The new dialogue box as shown in Fig. 1.9, acts as a gateway to which other module or application you want can be started in Creo Parametric software programme.

In this tutorial section, our main interest for now is to activate the manufacturing application.

To activate the manufacturing application module ≫ Click in the manufacturing radio button on the Type group in Fig. 1.9 ≫ Click in the NC Assembly radio button on the sub-type group in Fig. 1.9 ≫ Change the name or leave the default name generated by system ≫ Click in the square box of Use default template to clear the Check Mark icon.

The new dialogue box will now look as shown in Fig. 1.10.

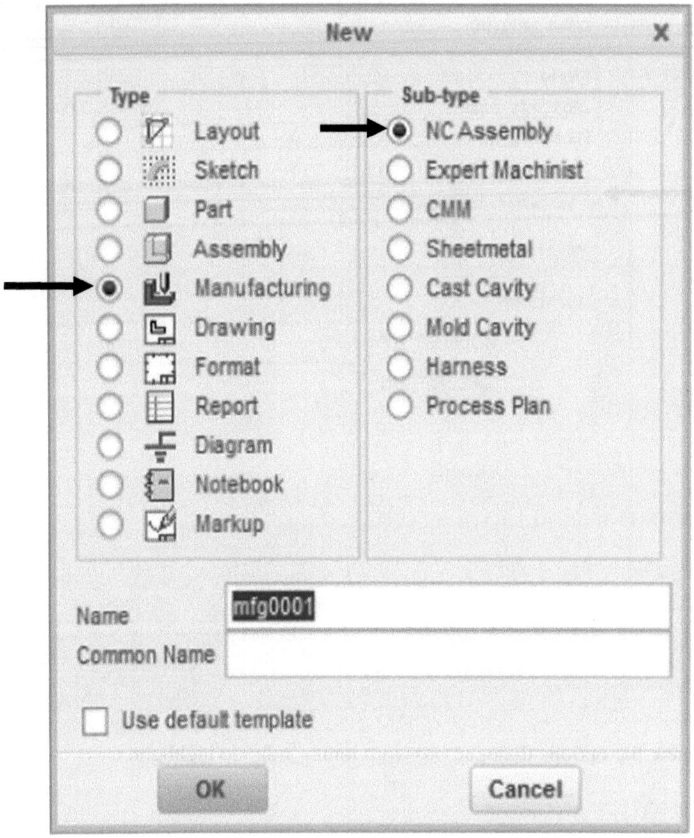

Fig. 1.10 New dialogue box with manufacturing and NC Assembly radio buttons active on Type and Sub-type groups

Click on OK tab to exit the New dialogue box.

The new file options dialogue window is activated on the main window ≫ On the Template group, click on "mmns_mfg_nc" for this tutorial. On the Parameters group, type whatever you want in the MODELED_BY and DESCRIPTION section boxes as shown Fig. 1.11.

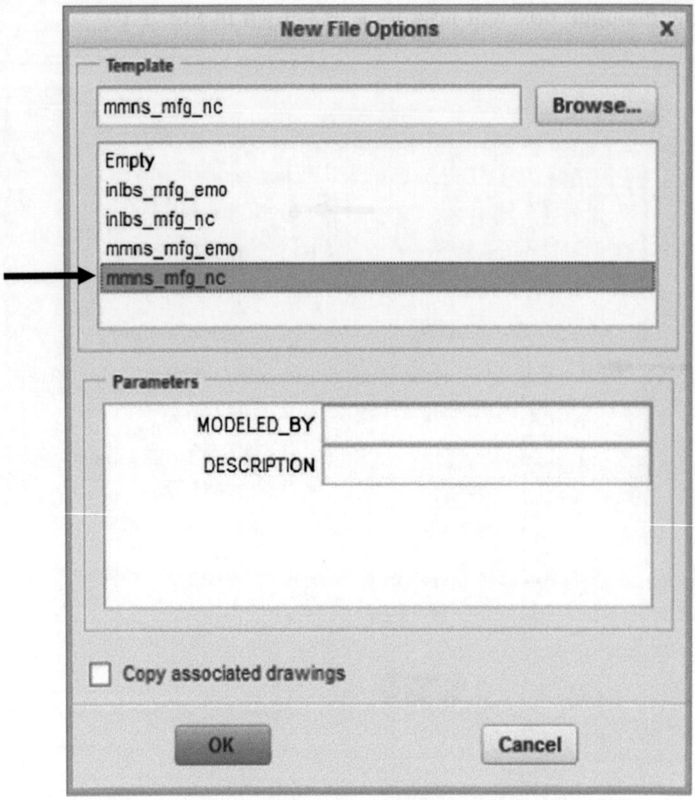

Fig. 1.11 New file options dialogue box with mmns_mfg_nc highlighted

Click on OK tab to exit the New File Options window.

Creo Parametric 2.0 (Your version of Creo) Graphic User Interface (GUI) window for manufacturing application/module is activated as shown in Figs. 1.12 and 1.13.

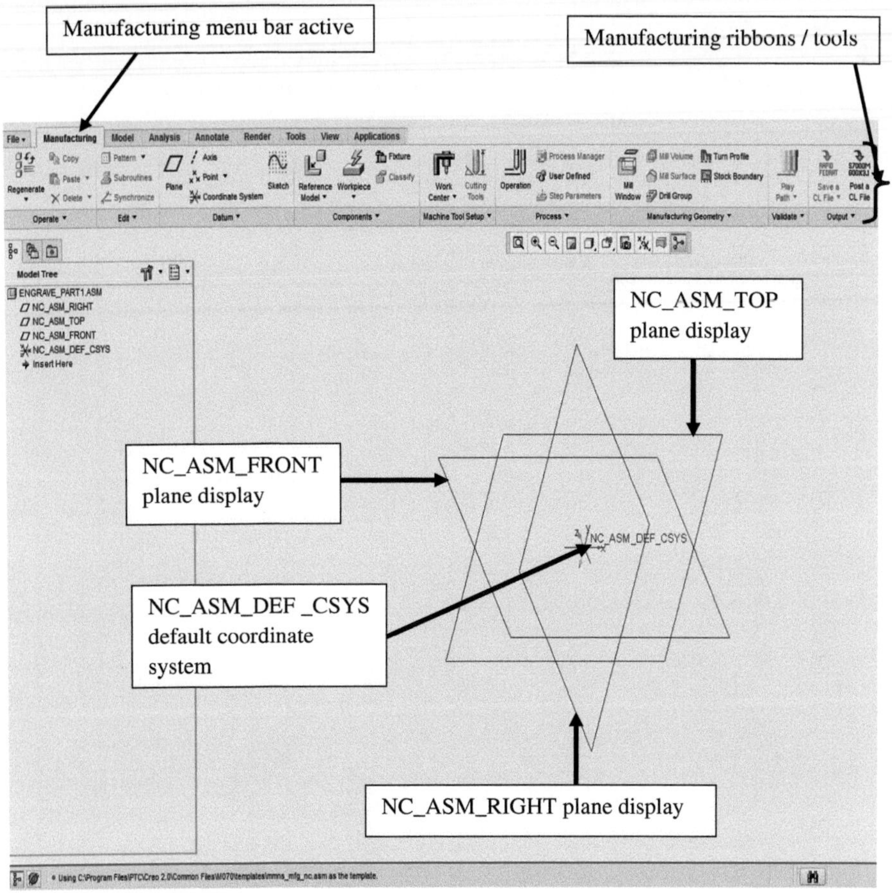

Fig. 1.12 Activated manufacturing main graphic user interface (GUI) window

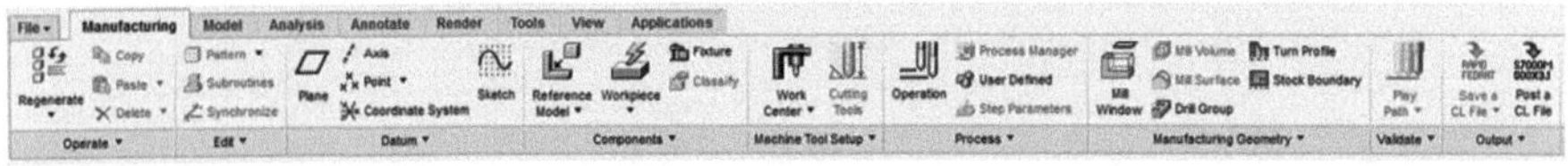

Fig. 1.13 Activated manufacturing ribbon in concise form

Chapter 2
Engraving Tutorial

Engraving is used to inscribe (a design, writing, character, etc.) onto a block, plate or other surface by carving, etching with acid or other process.

The Engraving process is a three axes milling machining process.

Engraving is used when tool follows a specified trajectory.

The following step-by-step guide will be covered in this Engraving tutorial.

- Introduction to 3D Modelling of Part
- Adding Colour Visualisation to the Part
- Adding Material to the Part
- Activating the Manufacturing application
- Importing and constraining the 3D Part
- Machine Coordinate (Programme Zero) creation
- Adding Cutting Tool
- Add Engraving Parameters (example feed rate, spindle speed, depth of cut, etc.)
- Generate Cutter Location (CL) Data
- G-codes generation

2.1 Step-by-Step Guide on How to Create a 3D Model Part (REC_PRT1)

This step-by-step guide on how to create an Engraved 3D Part is to aid new users to Creo Parametric.

2.1.1 How to Activate the Modelling Application

Start Creo Parametric either from your Computer desktop or from Programme.

P.O. Kanife, *Computer Aided Virtual Manufacturing Using Creo Parametric*,
DOI 10.1007/978-3-319-23359-8_2

Click on New icon as shown below.

Alternatively, Click on File to activate its drop-down menu list ≫ Now click on New icon on the drop-down list as indicated by the arrow below.

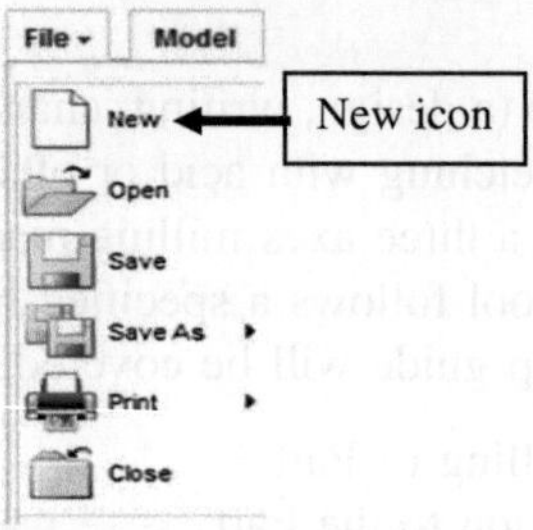

The New dialogue box is activated ≫ Click in the Part radio button on the Type group ≫ Click in the Solid radio button on the Sub-type group ≫ Type REC_PRT1 in the Name section box ≫ Clear the Check Mark icon in the Use default template square box and the New dialogue box will look as shown in Fig. 2.1.

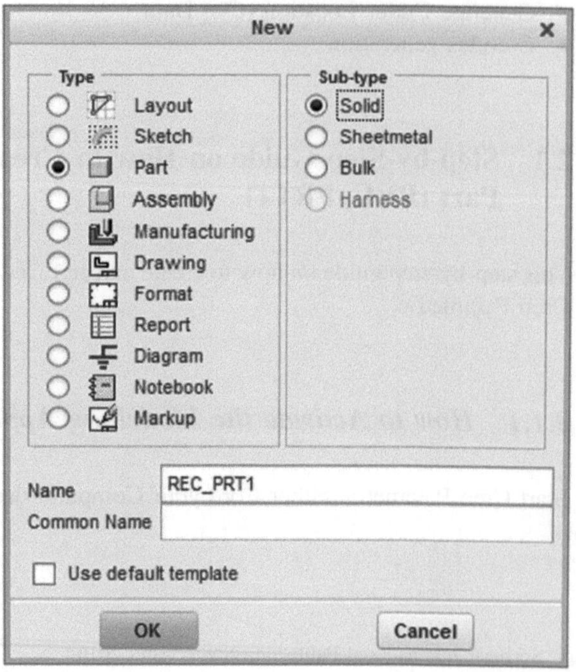

Fig. 2.1 New dialogue box with Part and Solid highlighted

Click on OK tab to exit.

The New File Options dialogue window is activated ≫ On the Template group select "solid_part_mmks" as the unit to be used in this tutorial ≫ Clear the Copy associate drawings Check Mark icon in the square box. The New File Options dialogue window is as shown in Fig. 2.2.

Note: You can also select either "mmns_part_solid" or "solid_part_mmks" if you are working in SI units.

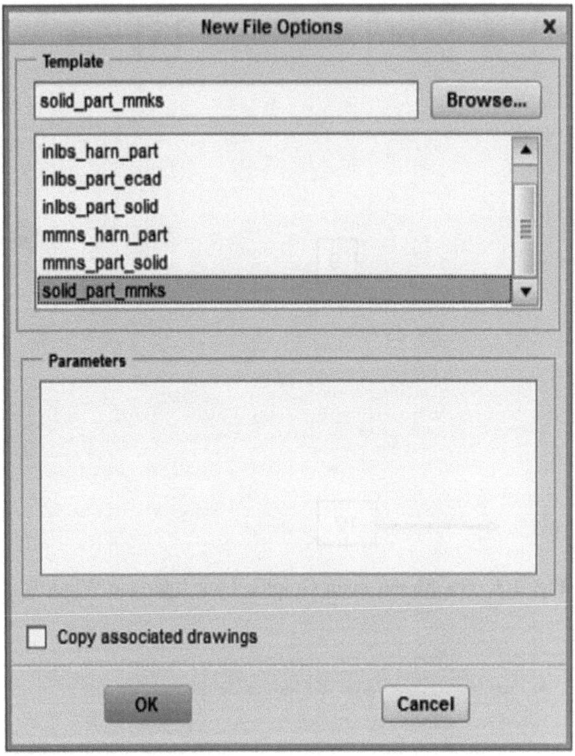

Fig. 2.2 New File Options dialogue box with solid_part_mmks highlighted

Now click on OK tab to exit.

Note: Any of the standard units of measurement above can be selected based on the unit you are working with.

The Model Graphic User Interface (GUI) window is activated as shown (See Figs. 2.3).

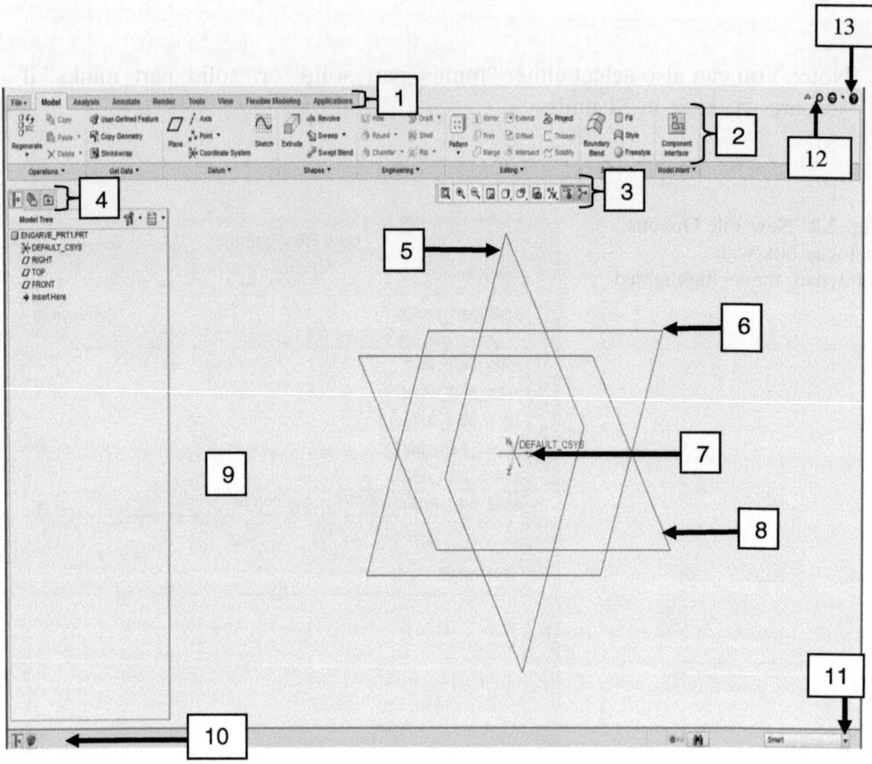

Fig. 2.3 The Model main Graphic User Interface Window

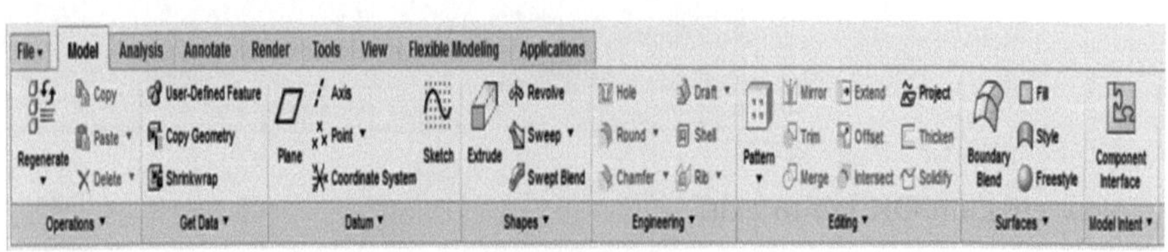

Fig. 2.4 Activated Model ribbon/tools in concise form

where

- 1 represents the Menu Bar (Main Tab)
- 2 represents the Model Ribbons (Model menu list)
- 3 represents the Default Graphics Toolbar (View and Display Tools)

- 4 represents the Navigator (Contains the Model Tree, Folders and Favourites)
- 5 represents the Right Datum Plane
- 6 represents the Top Datum Plane
- 7 represents the Default Coordinate System
- 8 represents the Front Datum Plane
- 9 represents the Main Graphic Window (Graphic User Interface Window (GUI))
- 10 represent the Message Bar
- 11 represent the Filter Selection drop-down list (Active filter)
- 12 represent the Command Search
- 13 represent the Creo parametric Help

2.1.2 *Activate 2D Sketch Application*

Click on the Sketch icon on the Model menu list as shown below.

The Sketch dialogue box is activated on the main graphic window as shown in Fig. 2.5.

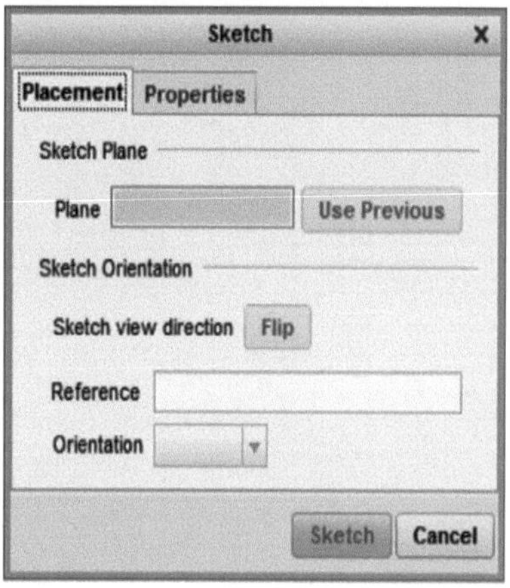

Fig. 2.5 Activated Sketch dialogue box

Click on the Placement tab ≫ On the Sketch Plane group, click in the Plane section box ≫ Now click on the Top Datum Plane on the Model Tree and the Sketch dialogue box is automatically updated. Make sure that the Orientation is set to Right as shown. See Fig. 2.6.

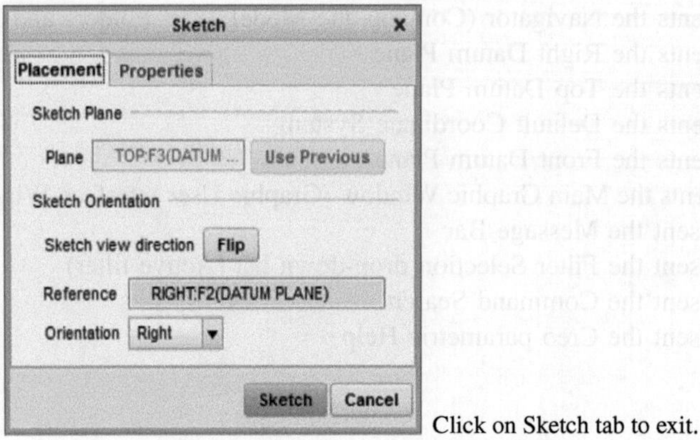

Fig. 2.6 Updated Sketch dialogue box

The Sketch ribbon tools are activated as shown in Fig. 2.7.

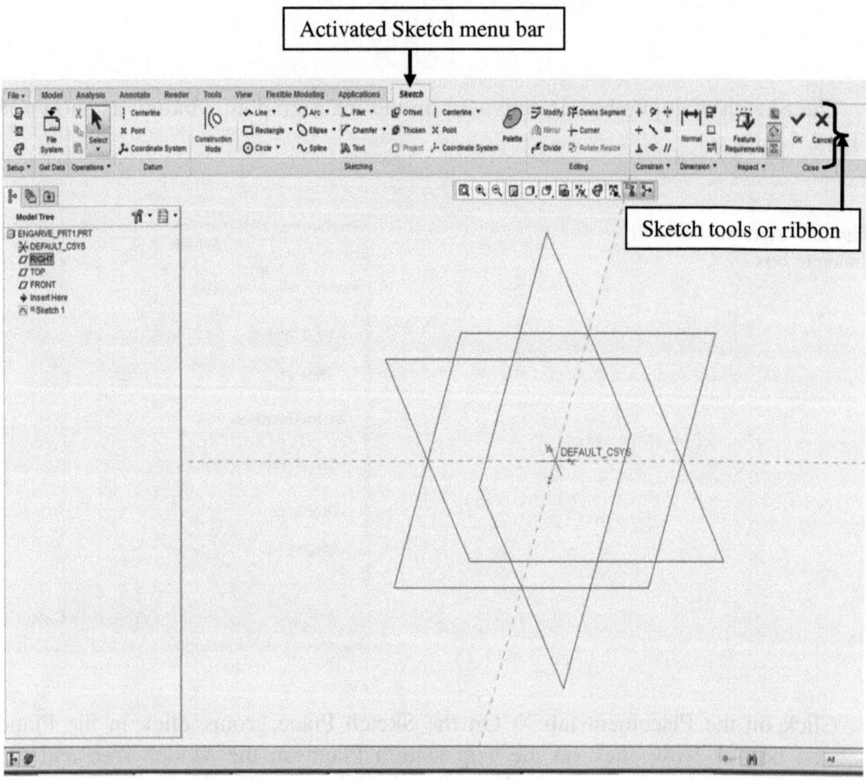

Fig. 2.7 Activated Sketch tools and datum plane display on the main Graphic User Interface (GUI) window

The concise Sketch ribbon tools are activated. See Fig. 2.8.

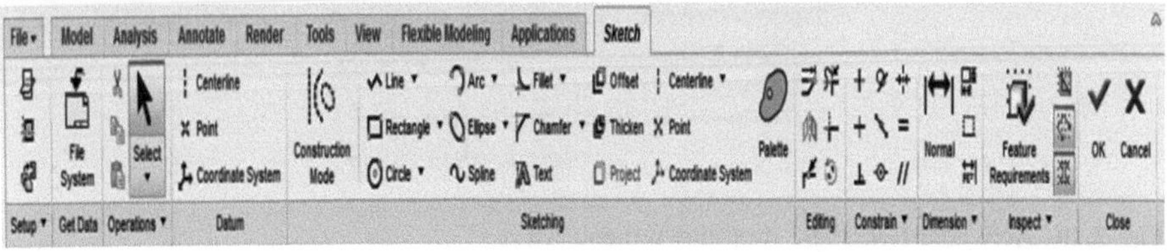

Fig. 2.8 Concise Sketch tools in concise form

Now click on the Sketch view icon  to position the Top Datum Plane as the sketch plane (orient the sketch plane parallel to the screen) as shown in Fig. 2.9.

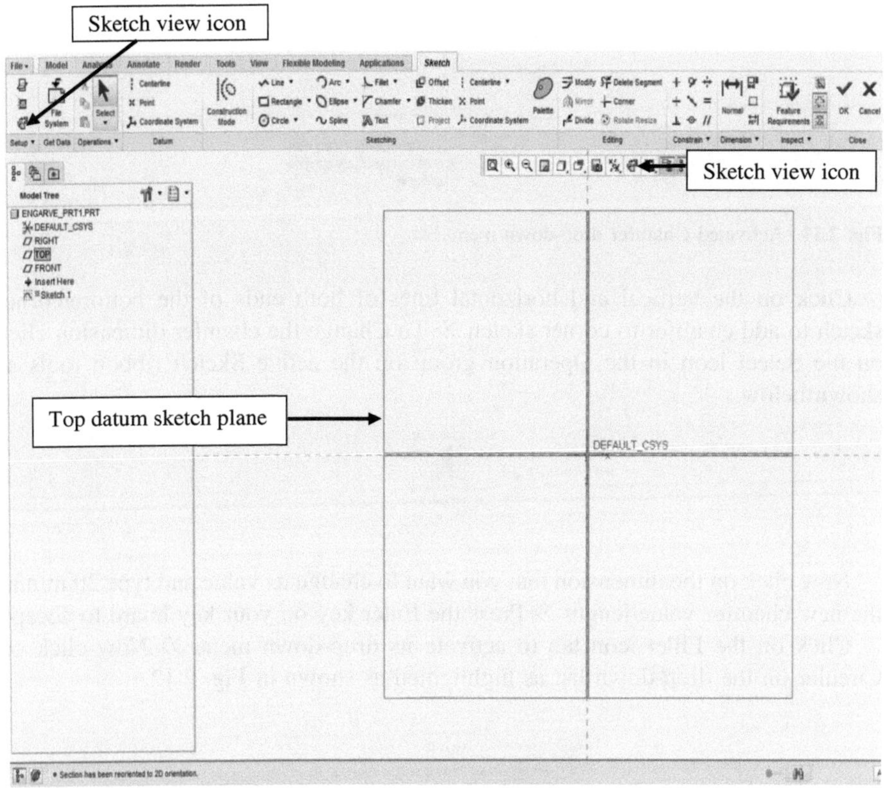

Fig. 2.9 Orientation of Sketch view on the main graphic window

Click on the Rectangle icon downward arrow to activate its drop-down menu >> Now click on Centre Rectangle on the drop-down list as shown in Fig. 2.10.

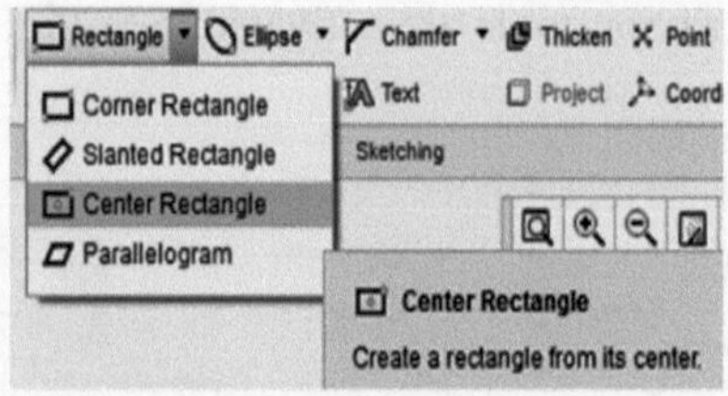

Fig. 2.10 Activated Rectangle drop-down menu list

The mouse pointer changes to an X symbol at its tip.

Place the arrow at the intersection point between the vertical and horizontal dotted line on the sketch plane and draw/sketch a rectangle of 400 × 250 mm.

Click on the Chamfer icon tab to activate its drop-down menu list ≫ Now click on Chamfer on the drop-down list as highlighted and shown in Fig. 2.11.

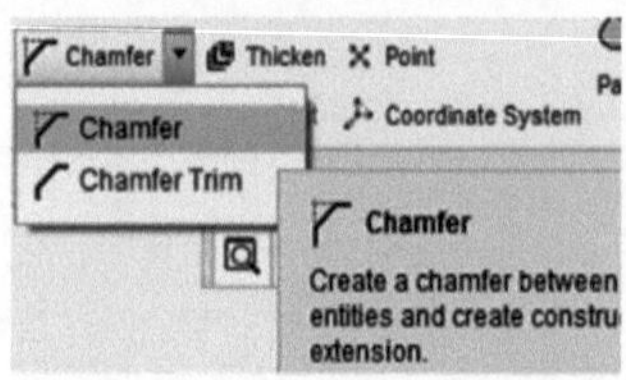

Fig. 2.11 Activated Chamfer drop-down menu list

Click on the vertical and horizontal lines of both ends of the bottom/corner sketch to add chamfer to corner sketch ≫ To Change the chamfer dimension click on the Select icon in the Operation group on the active Sketch ribbon tools as shown below.

Now click on the dimension that you want to change its value and type 20 mm as the new chamfer value/length ≫ Press the Enter key on your key board to accept.

Click on the Fillet icon tab to activate its drop-down menu ≫ Now click on Circular on the drop-down list as highlighted as shown in Fig. 2.12.

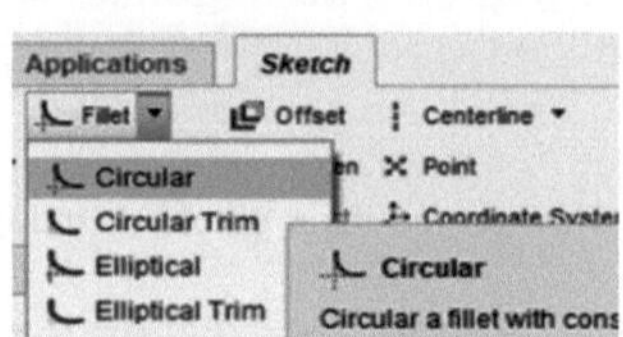

Fig. 2.12 Activated Fillet drop-down menu list

Click on both the vertical and horizontal lines of the upper sketch to add the fillet to both top end corners/edges ≫ To change the fillet dimension, click on the Select icon as shown below.

Now click on the dimension that you want to change its value and type 20 mm as the new fillet radius ≫ Press the Enter key on your key board to accept.

The changes made to the sketch are as shown in Fig. 2.13.

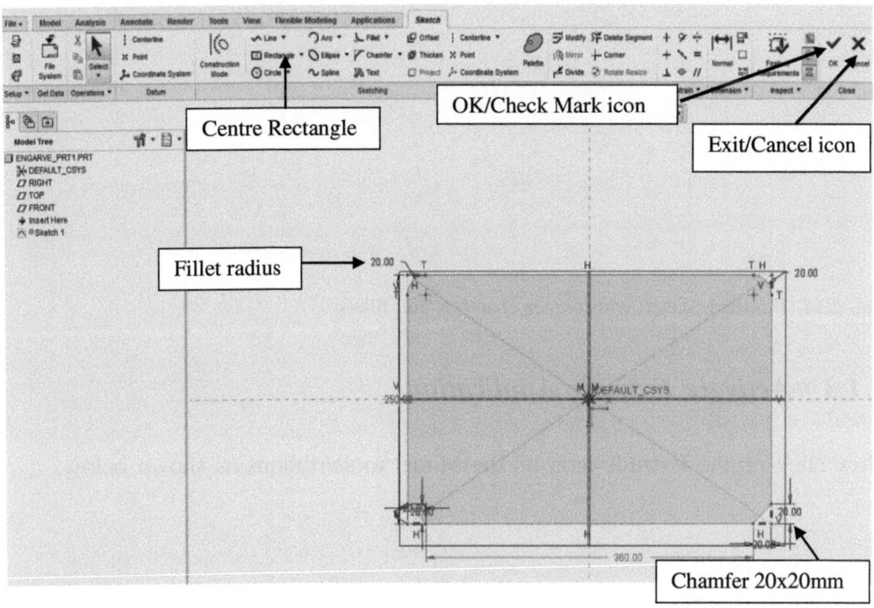

Fig. 2.13 Sketching of the rectangle on the main graphic window

Click on the OK icon as indicated by the arrow shown in Fig. 2.13; if you are satisfied with your sketch to exit the Sketch application.

The main GUI window changes as shown in Fig. 2.14.

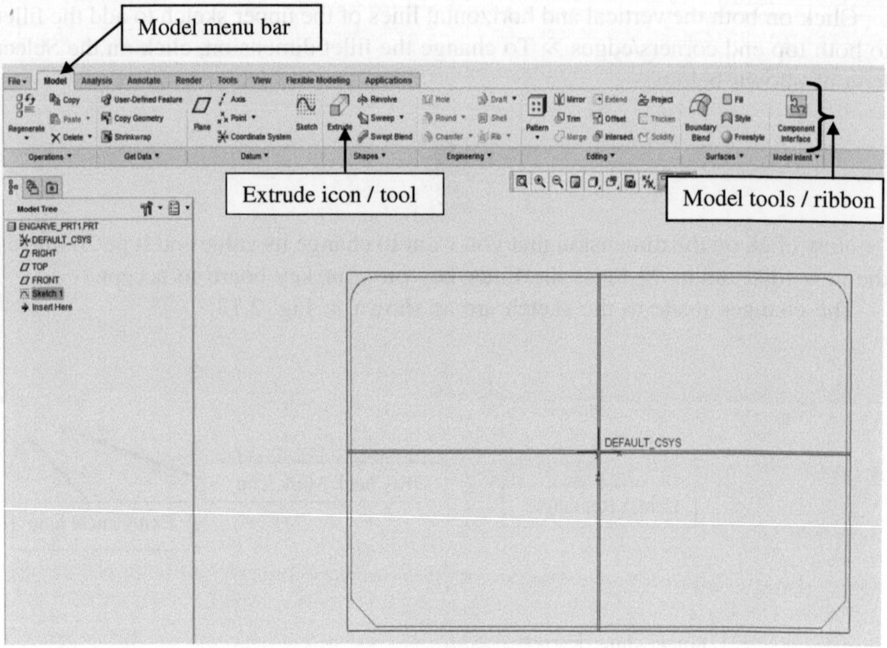

Fig. 2.14 Finished Sketch with corner chamfers and fillets

2.1.3 Activate Extrude Application

Now click on the Extrude icon on the Model tools/ribbons as shown below.

The Extrude dashboard tools are activated as shown in concise form (Fig. 2.15).

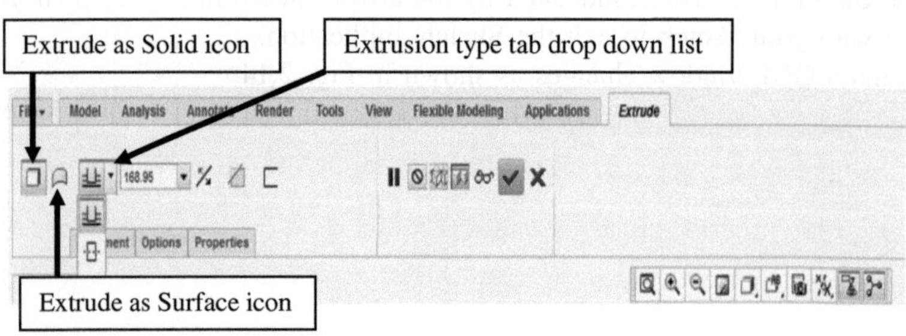

Fig. 2.15 Concise Extrude dashboard

Make sure that the Extrude as Solid icon tab is activated as indicated in Fig. 2.15 ≫ Click on the Extrusion type tab downward arrow to activate its drop-down menu, now click on "Extrude from sketch plane by a specified depth value" on the drop-down menu list. ≫ In the "Extrude from sketch plane by a specified depth value" value section box, type 50 mm to be depth of extrusion as illustrated in Fig. 2.16.

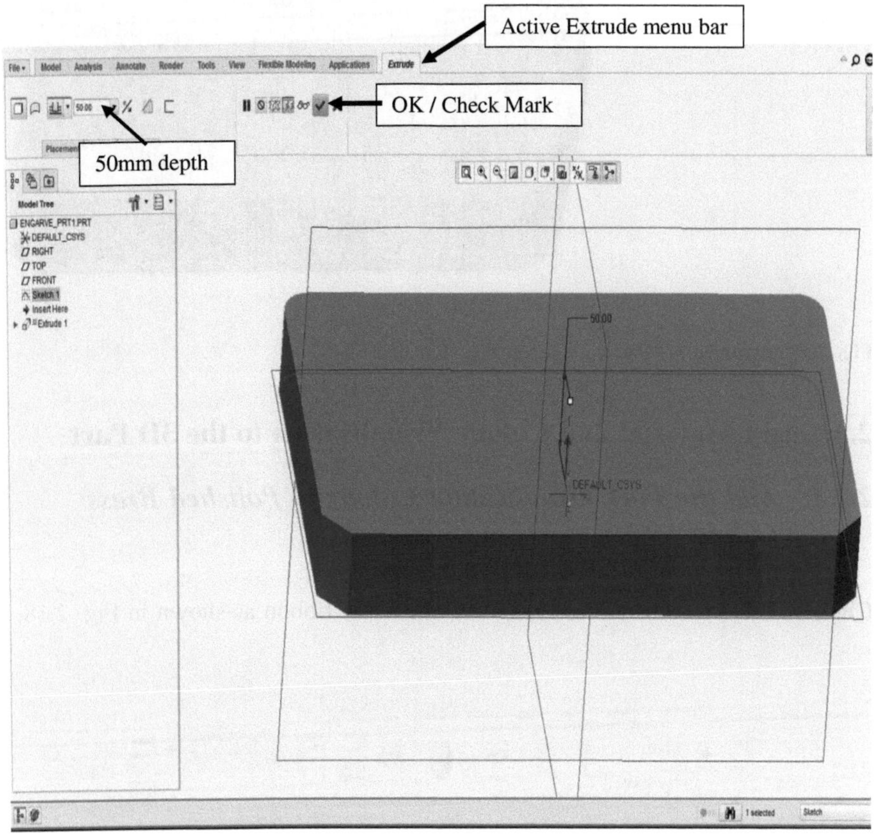

Fig. 2.16 Extruded sketched Part

Click on the OK/Check Mark icon.

After clicking on Check Mark, the Part changes on the main GUI window as shown in Fig. 2.17.

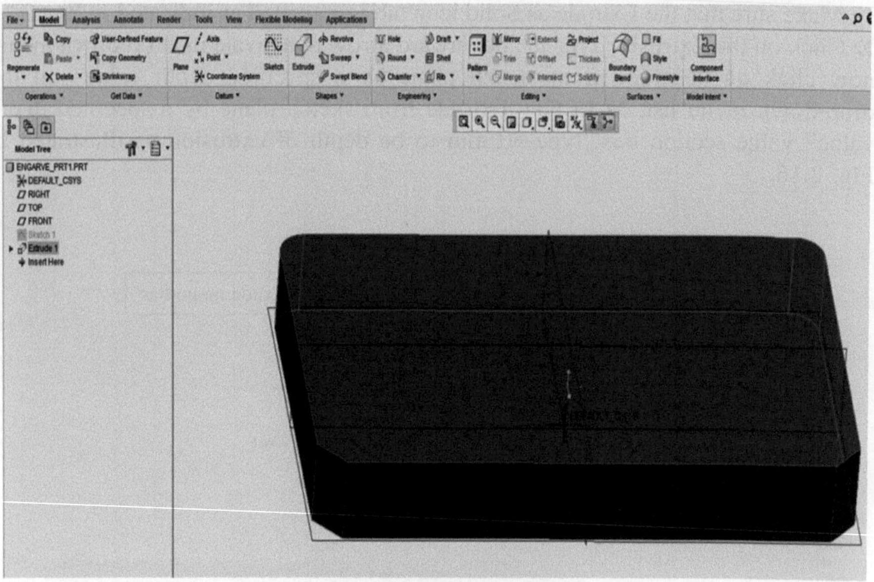

Fig. 2.17 Extruded 3D Part with corner chamfers and fillets

2.2 Add Material and Colour Visualisation to the 3D Part

2.2.1 Add the Part Visualisation Colour to Polished Brass Colour

Click on the View menu bar to activate the View ribbon as shown in Fig. 2.18.

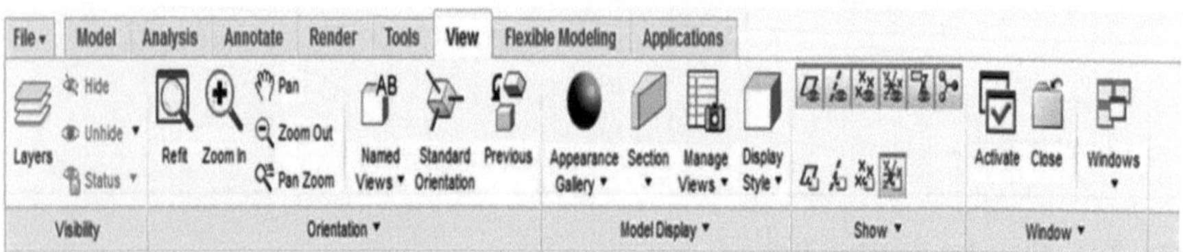

Fig. 2.18 Activated View ribbon in concise form

Click on the Appearance Gallery icon 

to activate its content.

Go to Library, and click on the "std-metals.dmt" tab to activate its content as illustrated in Fig. 2.19.

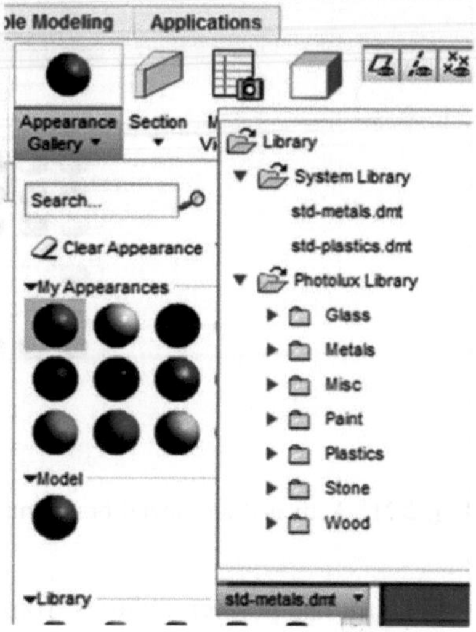

Fig. 2.19 Activated std-metals.dmt group

On the drop-down menu list, click on the Metals folder arrow to expand its contents ≫ Now click on the "adv-metal-brass.dmt" on the list as indicated by the arrow shown in Fig. 2.20.

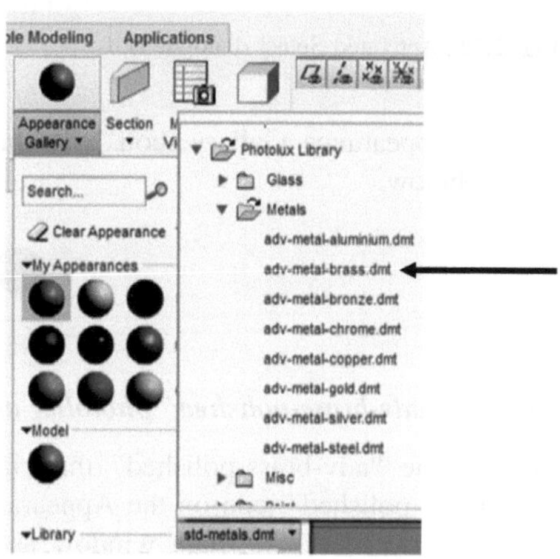

Fig. 2.20 Activated Metals folder drop-down list

On the "adv-metal-brass.dmt" list, now click on the "adv-brass-polished" photolux appearance as indicated by the arrow as shown in Fig. 2.21.

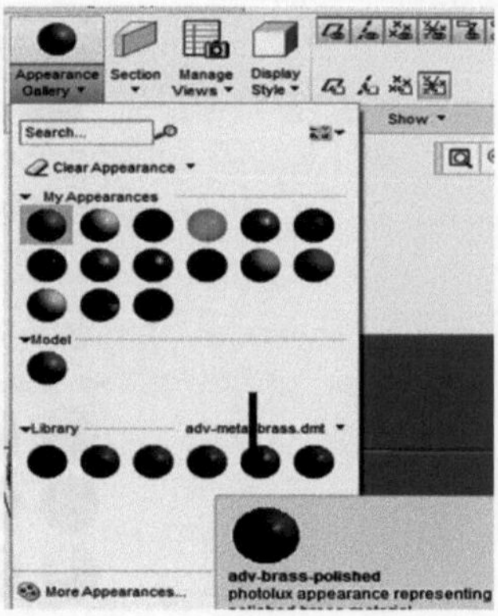

Fig. 2.21 Activated adv-metal-brass.dmt group

The mouse pointer changes from an arrow to a blush icon ≫ The Select dialogue box is activated on the main window as shown in Fig. 2.22.

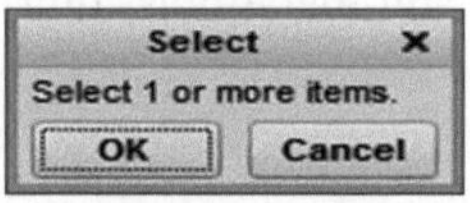

Fig. 2.22 Activated Select dialogue box

The Appearance Gallery icon will change to "adv-brass-polished" colour as shown below.

Add the "adv-brass-polished" photolux appearance

To add the "adv-brass-polished" material colour to the Part ≫ Click on the "adv-brass-polished" icon on the Appearance Gallery tab, and the Select dialogue box is activated of the main window as shown in Fig. 2.23. If it is not active already.

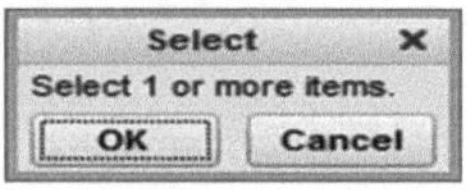

Fig. 2.23 Activated Select dialogue box

Now start clicking on the Part surfaces and edges while holding down the Ctrl key on your key board ≫ Click on the OK tab when you are done to exit the Select dialogue box. The Part changes colour as shown in Fig. 2.24.

Fig. 2.24 The "adv-brass-polished" material colour added to the Part

Note: Remember to release the Ctrl tab while selecting the Part surfaces and edges, by doing so you will be able to rotate the Part.

2.2.2 Add Material to the Part

To add the "adv-brass-polished" material to the Part ≫ Go to File on the menu bar and click on its downward arrow to activate its drop-down menu ≫ Now click on Prepare to activate its content ≫ On the Prepare model for distribution list, click on Model Properties as highlighted in Fig. 2.25.

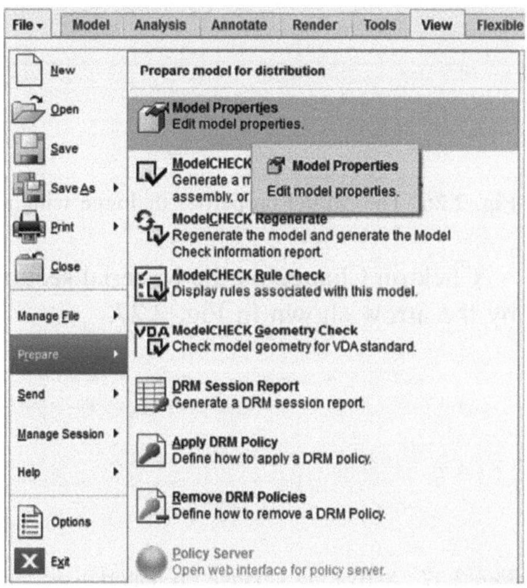

Fig. 2.25 Activated Prepare panel content

The Model Properties dialogue window is activated as shown in Fig. 2.26.

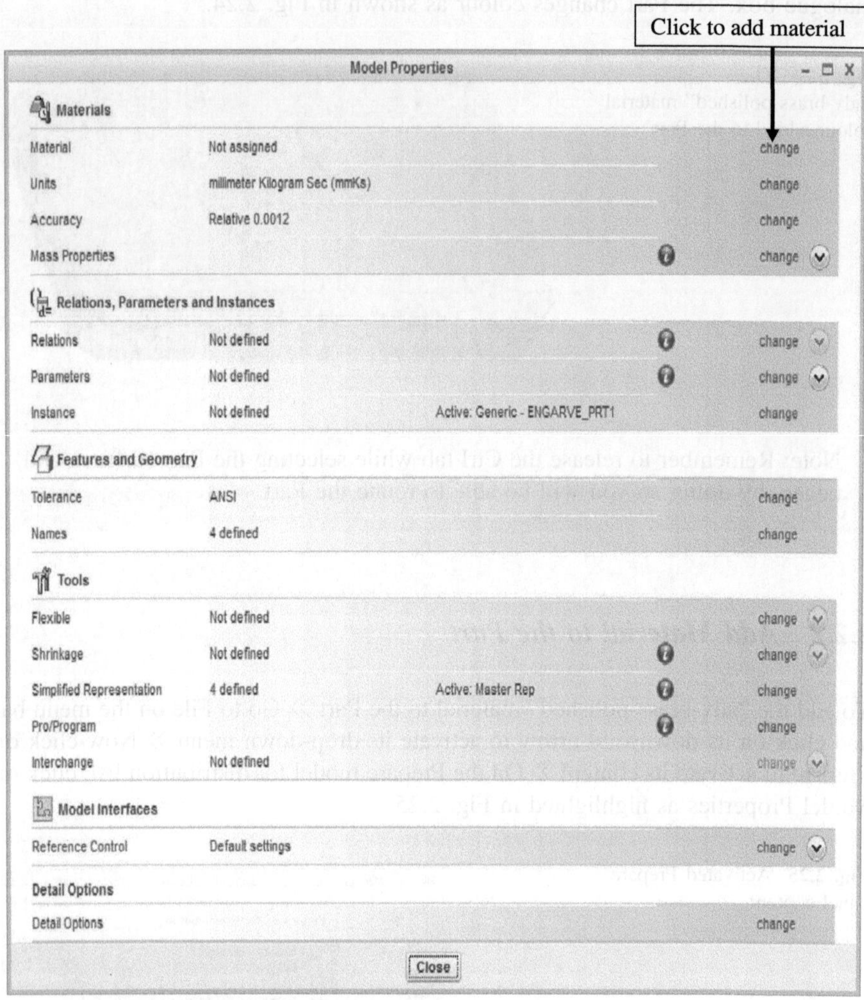

Fig. 2.26 The Model properties dialogue window

Click on Change on the Material section, under the Materials group as indicated by the arrow shown in Fig. 2.27.

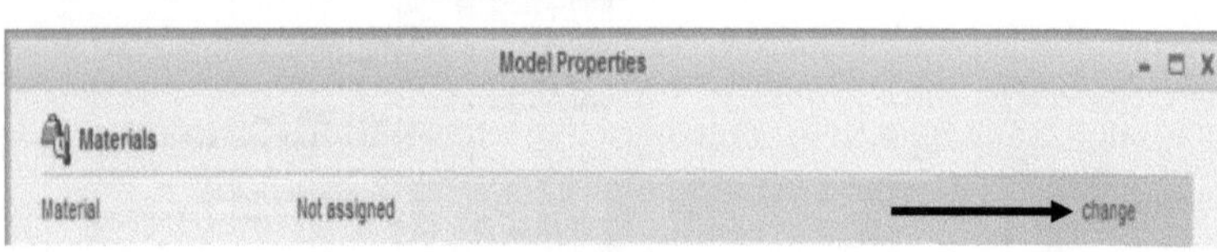

Fig. 2.27 Activating Change on material section

The Materials dialogue window is activated as shown in Fig. 2.28.

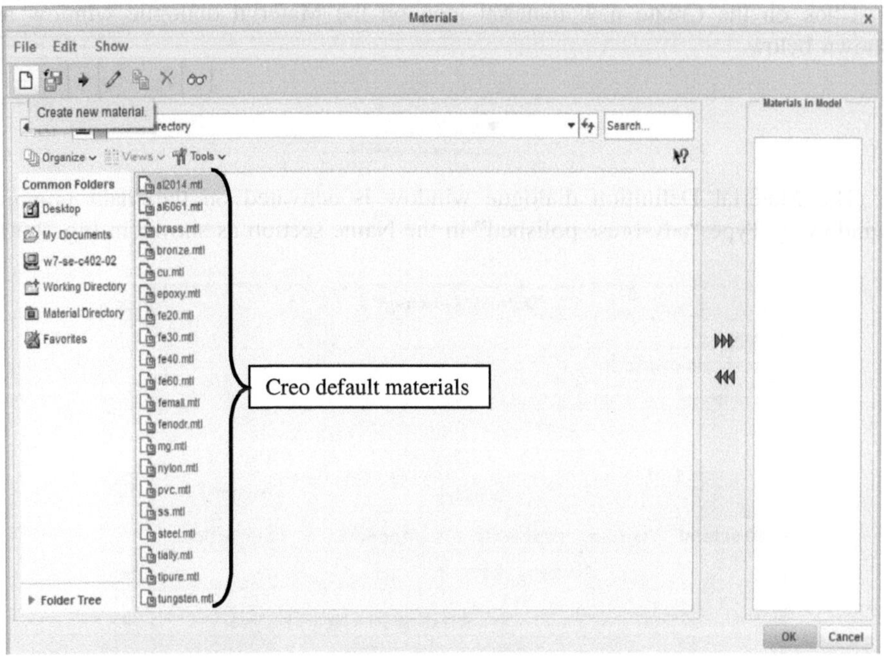

Fig. 2.28 Activated Materials dialogue window

Choose from the lists of default materials if you want to. And click on the red arrow as shown below at the top to assign material to the Part.

The material assigned now appears on the Material in Model section box as indicated by the arrow shown in Fig. 2.29.

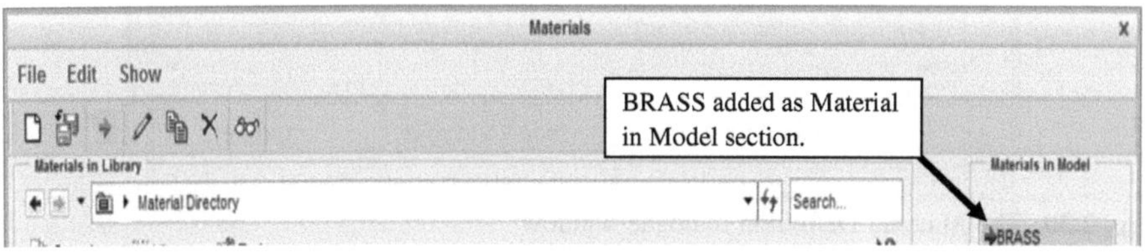

Fig. 2.29 BRASS in Materials in Model section

Now click on the Ok tab to exit the Materials dialogue box.

Alternatively, add material to Part through Create new material as will be illustrated and used in this tutorial.

Click on the Create new material icon on the Material dialogue window as shown below.

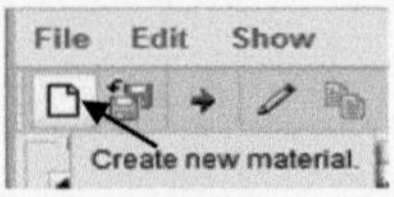

The Material Definition dialogue window is activated on the main graphic window ≫ Type "adv-brass-polished" in the Name section as shown in Fig. 2.30.

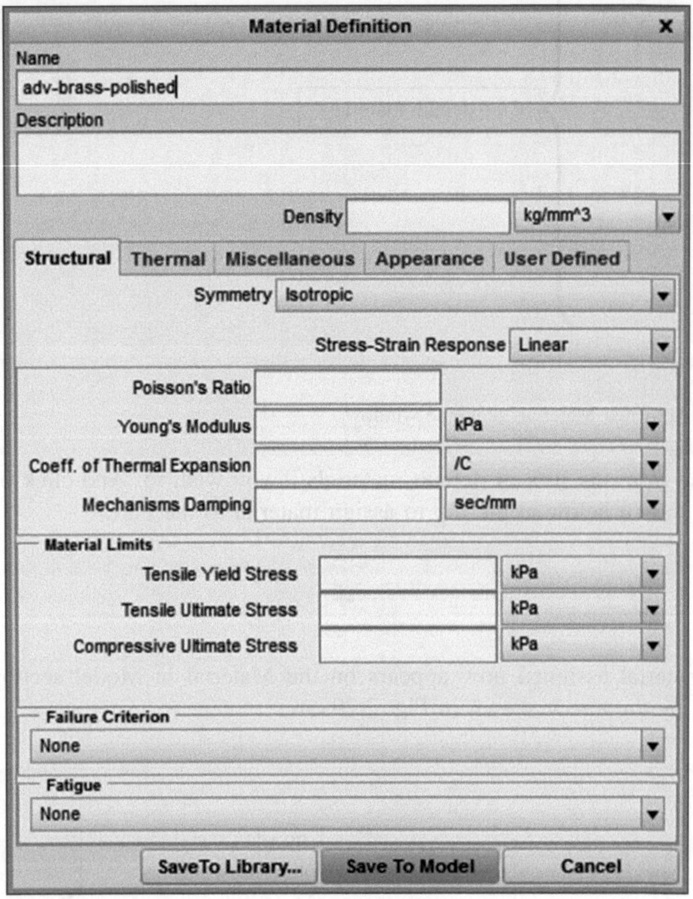

Fig. 2.30 The Material Definition dialogue window

Note: For advanced users, you can create your own new Material properties if you know the material Poisson's Ratio, Young's Modulus etc.

Click on the Appearance tab ≫ Now click on the New tab as indicated by the arrow in Fig. 2.31.

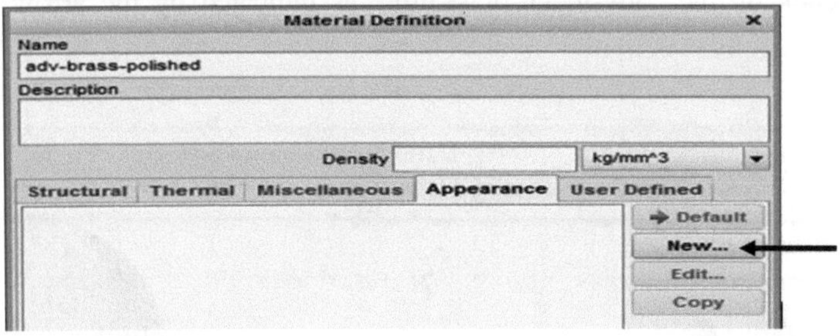

Fig. 2.31 Activating new appearance

The Material Appearance Editor dialogue window is activated as shown in Fig. 2.32.

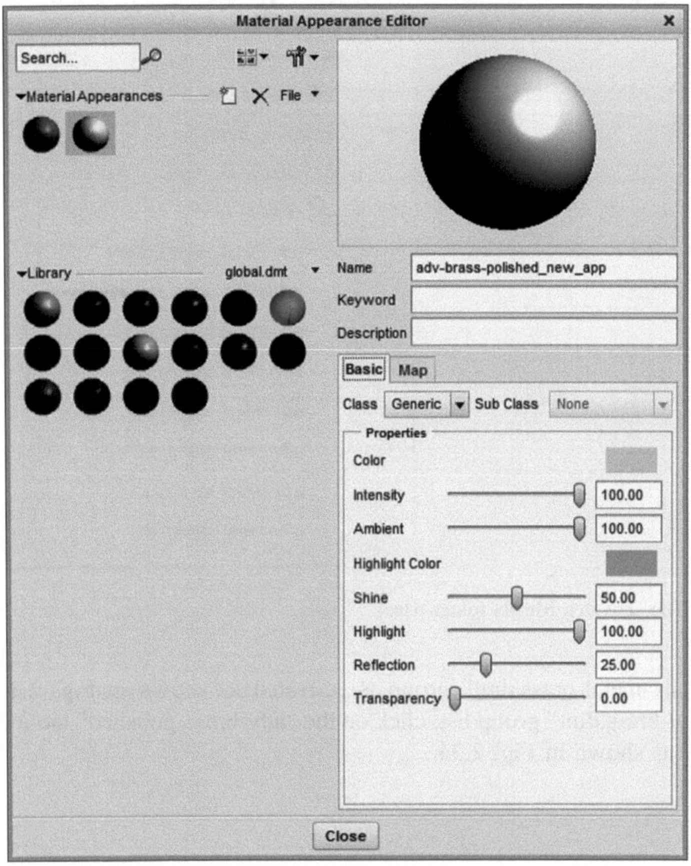

Fig. 2.32 The Material Appearance Editor dialogue window

On the Material Appearance Editor dialogue window ≫ Click on the "global. dmt" downward pointing arrow to activate its content ≫ On the drop-down menu list, click on the Metal folder arrow to activate its content.

Now click on the "adv-metal-brass.dmt" as indicated by the arrow shown in Fig. 2.33.

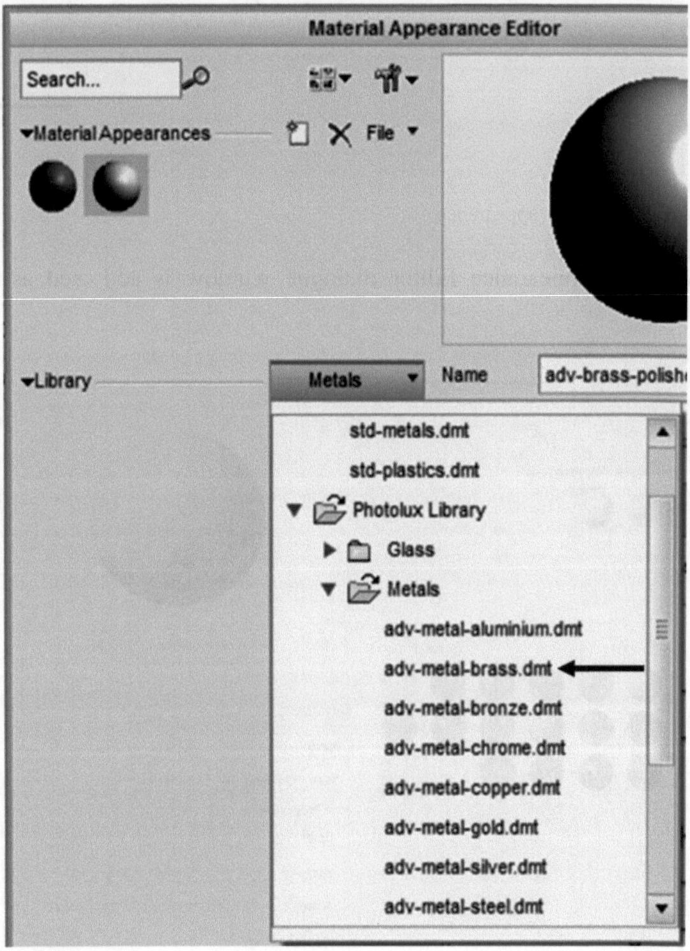

Fig. 2.33 The activated Metals folder panel

The "adv-metal-brass.dmt" group is activated as shown in Fig. 2.34. On the "adv-metal-brass.dmt" group list, click on the "adv-brass-polished" tab indicated by the arrow as shown in Fig. 2.34.

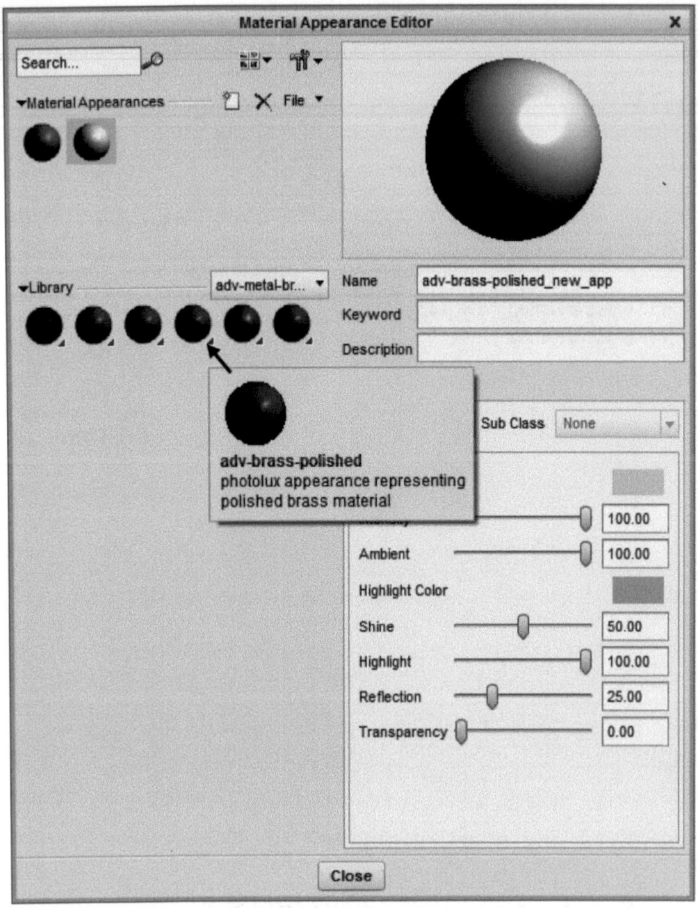

Fig. 2.34 The selected “adv-brass-polished” material to be added to Part

Now click on Close tab to close the Material Appearance Editor dialogue window.

On the Material Definition dialogue window, click on the “adv-brass-polished_new_app” ≫ Now click on the Default Red arrow tab as shown in Fig. 2.35.

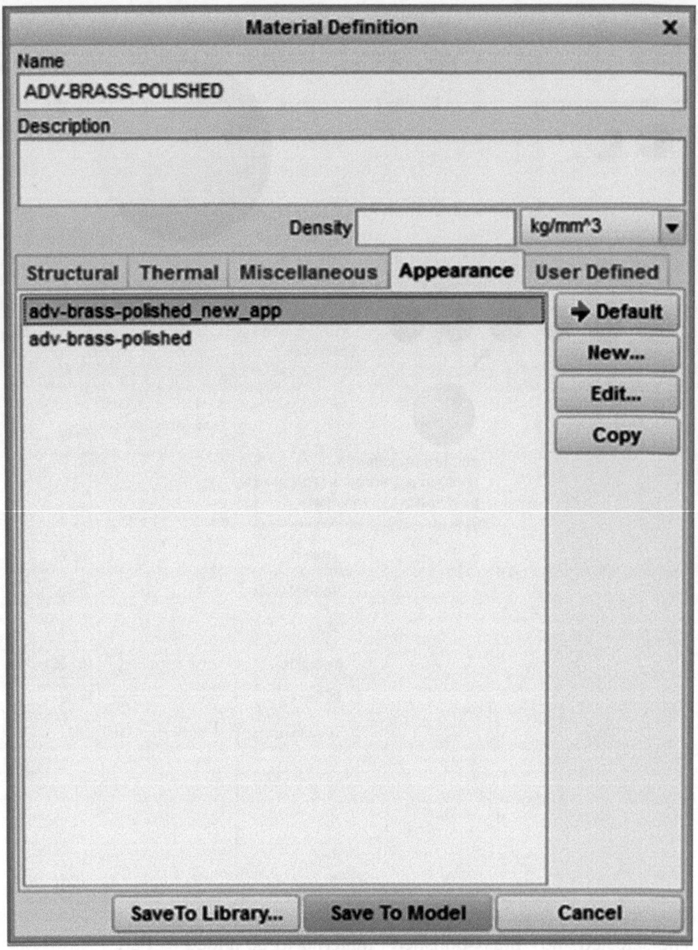

Fig. 2.35 The "adv-brass-polished_new_app" material is now added to Appearance section box

Click on the Save To Model tab as highlighted above to save the created material in the Materials Directory.

Note: When saving to a different material library folder click on Save To Library tab.

The ADV-BRASS-POLISHED is now in the Material in Model section as indicated by the arrow shown in Fig. 2.36.

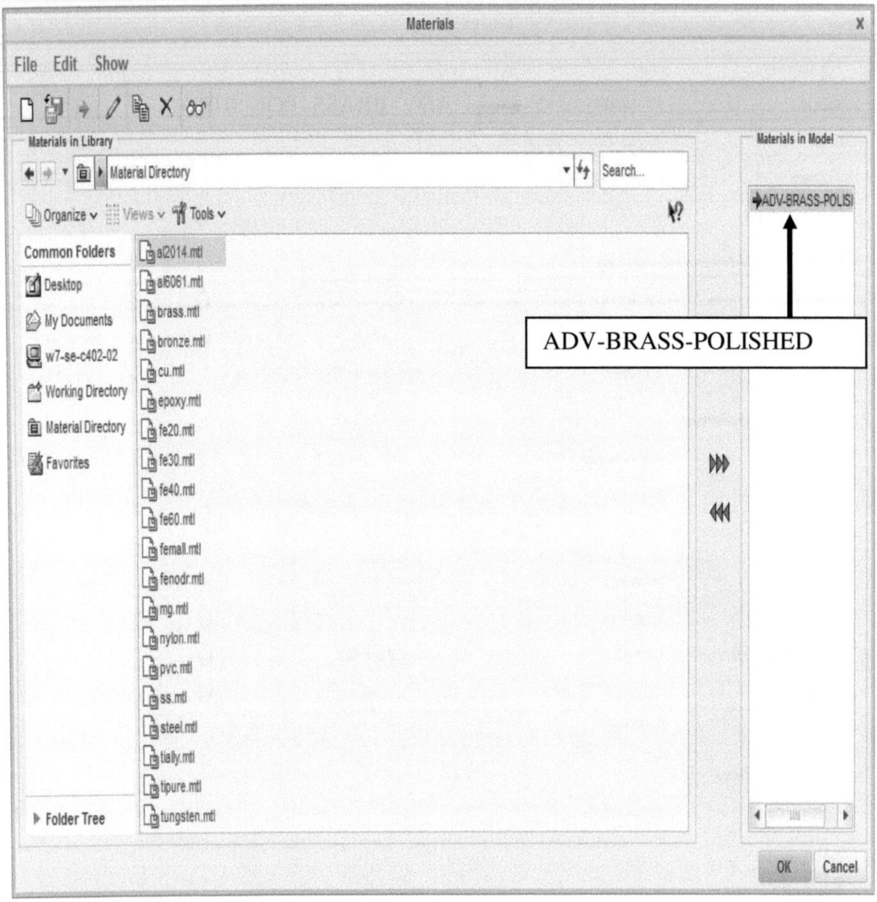

Fig. 2.36 The "adv-brass-polished_new_app" material is now added on the Model section box

Now click on the OK tab to exit.

The Model Properties dialogue window is activated indicating that the "adv-brass-polished" material is now the material added to the 3D Part as indicated by the arrow in Fig. 2.37.

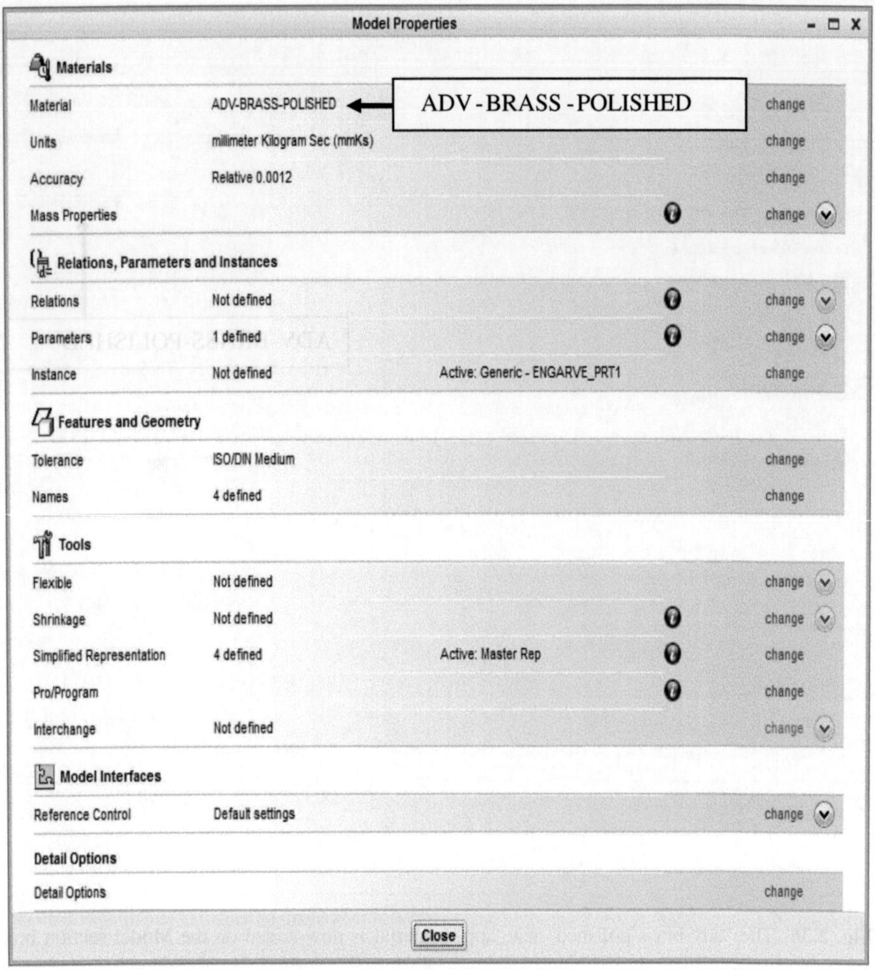

Fig. 2.37 Model Properties dialogue box showing the added material on the Material section

Now click on the Close tab to close the window.
Click on the Save icon to save your work.

2.3 Create the Engraving Imprint on the Part Surface

On the Model menu bar ≫ Click on the Sketch icon as shown below.

The Sketch dialogue box is activated on the main graphic window ≫ Click on the Placement tab ≫ Now click on the Top surface of the Part. The Plane, Reference and Orientation section boxes are updated automatically by the system. Make sure that Orientation is set to Right. See Fig. 2.38.

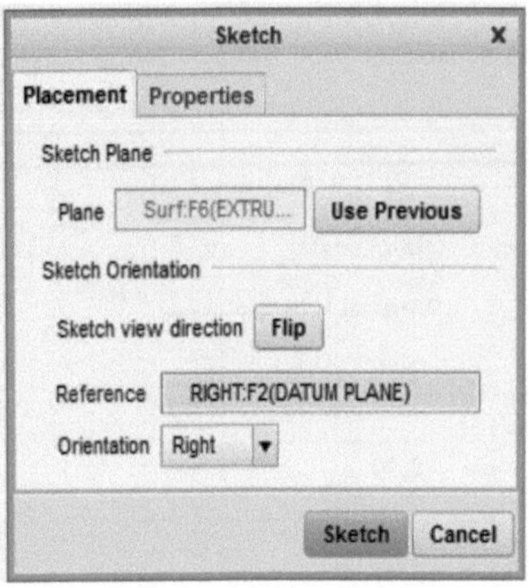

Click on the Sketch tab to exit.

Fig. 2.38 Updated Sketch dialogue

The Sketch tools are activated with all the sketching tools (See Fig. 2.39).

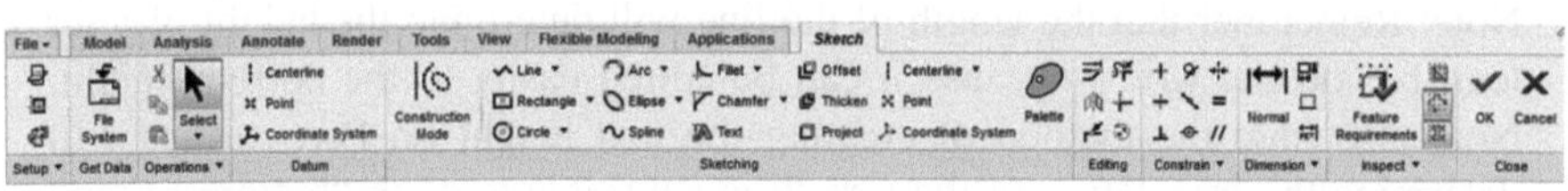

Fig. 2.39 Activated Sketch tools

To imprint “MANUFACTURE” on the Part surface ≫ Click on the Text icon and now click on the Part surface to determine the start point, click again to determine the text height. The Text dialogue box is activated as shown in Fig. 2.40.

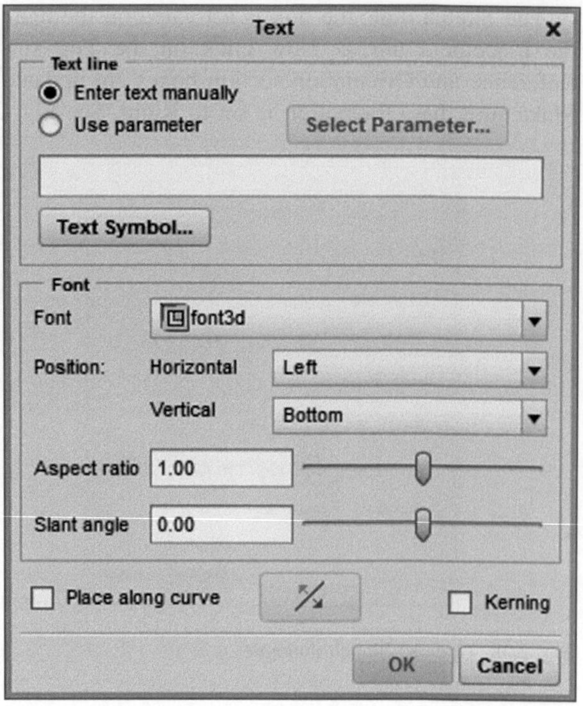

Fig. 2.40 Activated Text dialogue box

Note: Remember that the length of the line will determine the height of your text, but do not worry because it can be edited easily to the correct height. Note also, that the method of creating the text height on Part surface will determine the orientation of the text. It is advisable to click "bottom to top" and not top to bottom on the Part surface.

On the Text dialogue box, click in the "Enter text manually" radio button on the Text line group to make it active ≫ Now type "MANUFACTURE" in the empty section box.

On the Font group, click on the downward pointing arrow on the Font section box ≫ On the activated drop-down list click on cal_alf as shown in Fig. 2.41.

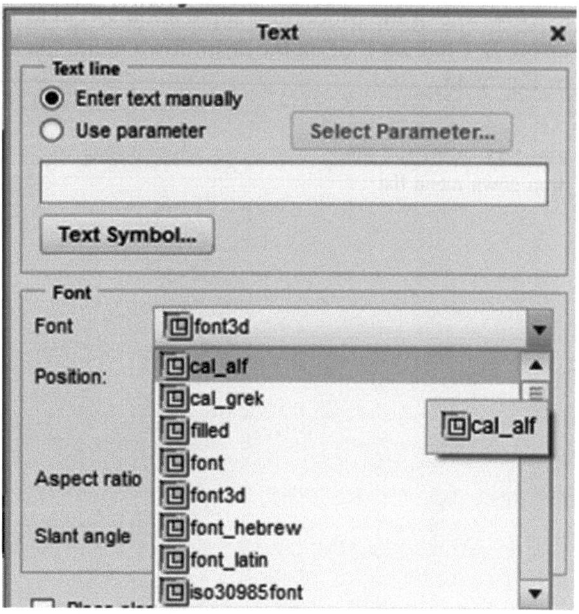

Fig. 2.41 Activated Font drop-down menu list

On the Font group, set the Horizontal, Vertical and Aspect ratio section boxes as illustrated in Fig. 2.42.

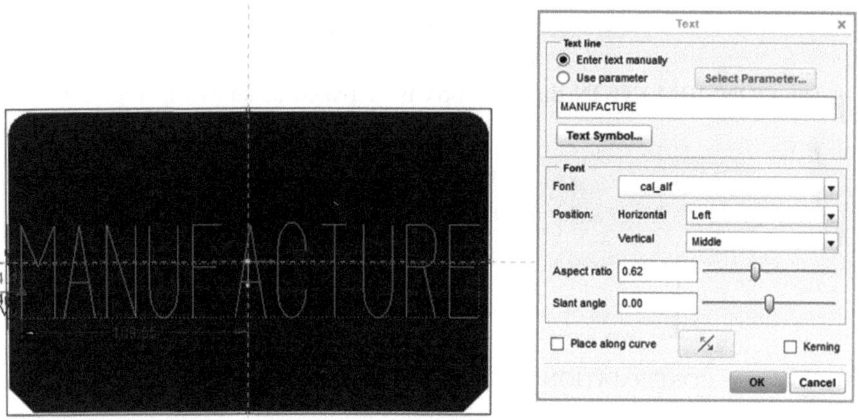

Fig. 2.42 MANUFACTURING imprint on Part surface and Text dialogue box

Click on the OK tab to exit the Text dialogue box.

Now click the Check Mark icon to exit the Sketch application.

Save the Engraved Part in the selected directory, folder or drive as ENGRAVED_PRT1.

Now to exit the Modelling application ≫ Go to File and click on the downward arrow ≫ Click on Exit on the drop-down menu list as indicated by the arrow shown in Fig. 2.43.

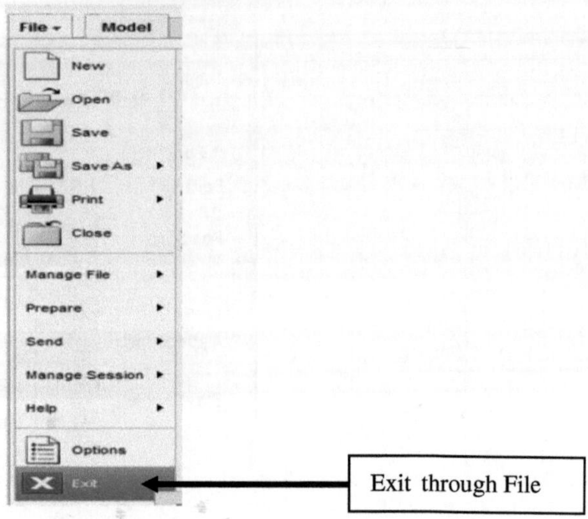

Fig. 2.43 Activated File drop-down menu list

Alternatively, click on the Close icon on the GUI main window as indicated by the arrows shown below.

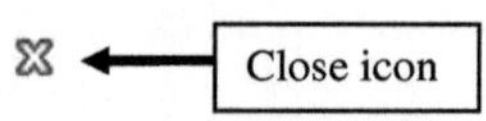

The CONFIRMATION dialogue box is activated as shown in Fig. 2.44.

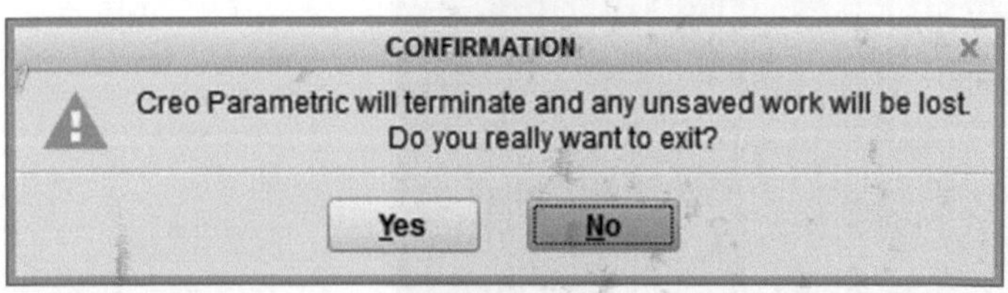

Fig. 2.44 The CONFIRMATION dialogue box

Click on the Yes tab to confirm and exit Creo Parametric completely.

Congratulations for getting to this stage in this tutorial. For a new user, all that is required is patience and to practice as much as you can and play with different application/toolbar icon to get familiar with using them.

2.4 Manufacturing Procedure of the Engraved 3D Part

2.4.1 Activate the Manufacturing Application

Start Creo Parametric either from your Computer desktop or from Programme.

Click on File to activate its drop-down menu list ≫ Now click on New on the drop-down menu list as indicated below by the arrow as shown in Fig. 2.45.

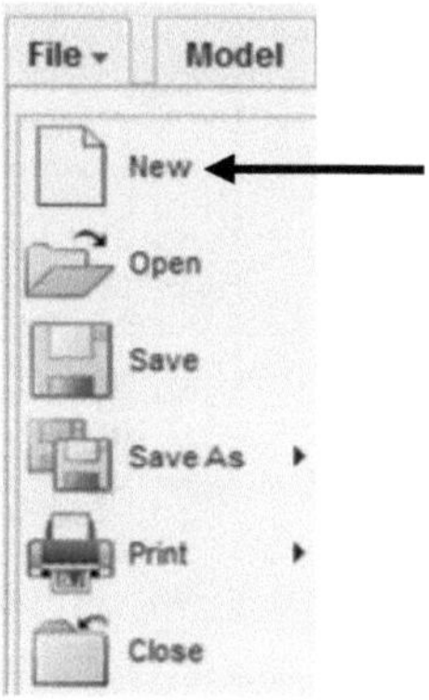

Fig. 2.45 Activated File drop-down menu list

The New dialogue box is activated on the main graphic window as shown in Fig. 2.46.

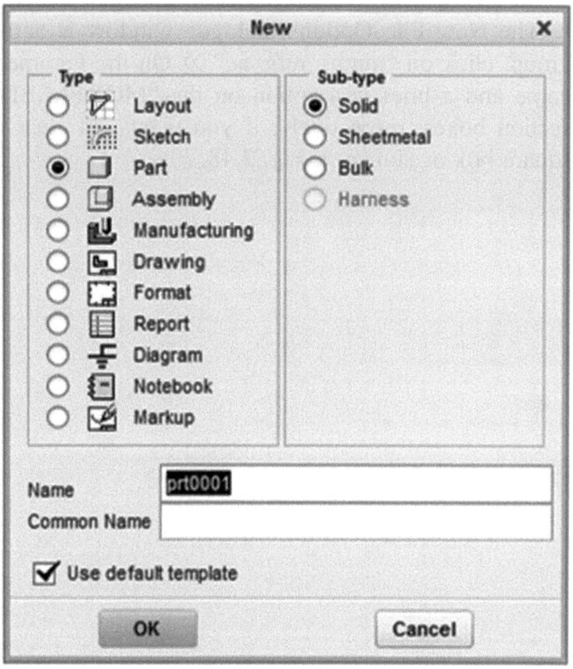

Fig. 2.46 The activated New dialogue box

Change the Radio button on the Type and Sub-type groups on the New dialogue box

On the Type group, click in the Manufacturing radio button to make it active ≫ On the Sub-type group, click in the NC Assembly radio button to make it active.

Type the name (Engave_PRT1) in Name section box. Clear the Use default template as shown in Fig. 2.47.

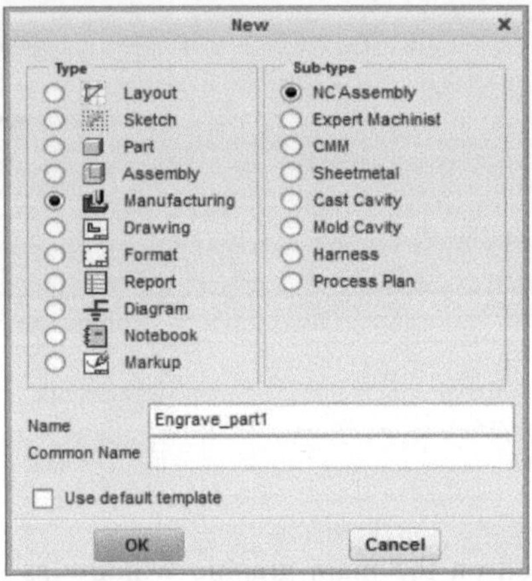

Now click on OK tab to exit.

Fig. 2.47 Manufacturing and NC Assembly radio buttons activated

The New File Options dialogue window is activated ≫ Under the Template group, click on "mmns_mfg_nc" ≫ On the Parameters group, you can type your name and a brief description on the "MODELLED_BY" and "DESCRIPTION" section boxes, respectively, if you want and clear the copy associated drawings square box as shown in Fig. 2.48.

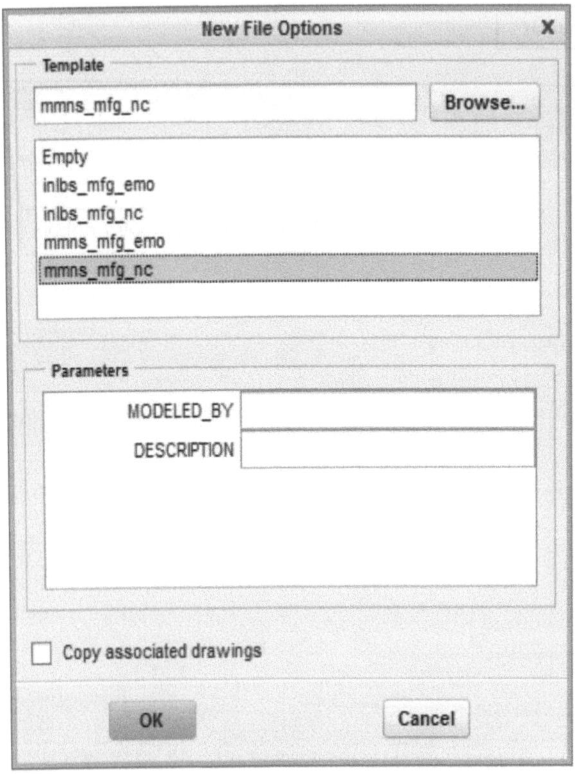

Click on the OK tab to proceed.

Fig. 2.48 The New File Options dialogue box

The Manufacturing main Graphic User Interface (GUI) window is activated as shown in Fig. 2.49.

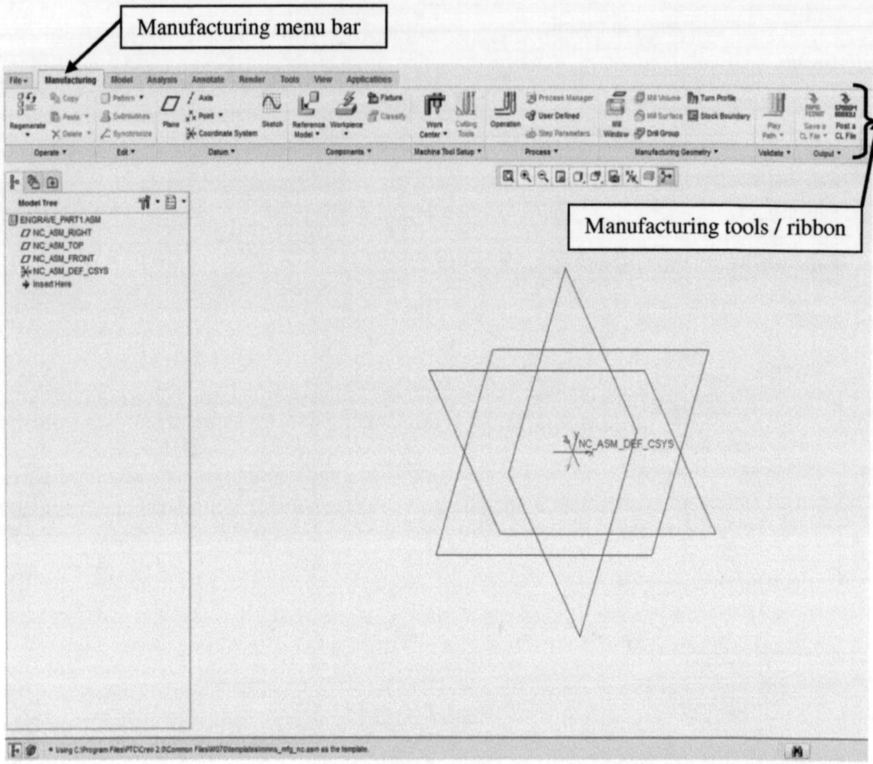

Fig. 2.49 Activated Manufacturing main GUI window

2.4.2 Import and Constrain the Reference Model

Import 3D Part

Click on the Reference Model icon to activate its drop-down list ≫ Now select Assembly Reference Model on the drop-down list as illustrated below.

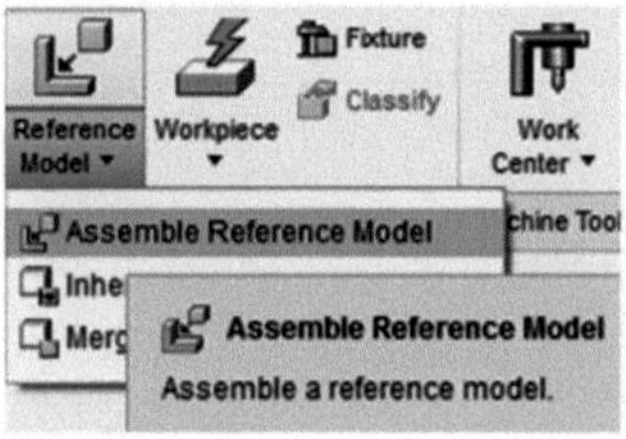

Note: The Assemble Workpiece option as shown below is used to assemble together two Parts, where one is the imported Part and the other is the imported Workpiece.

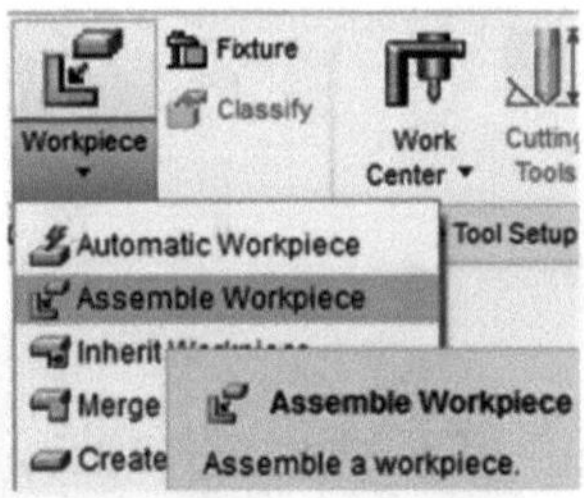

The Open dialogue window is activated. Make sure that the file name you want to open is the correct one on the File name section box as shown in Fig. 2.50.

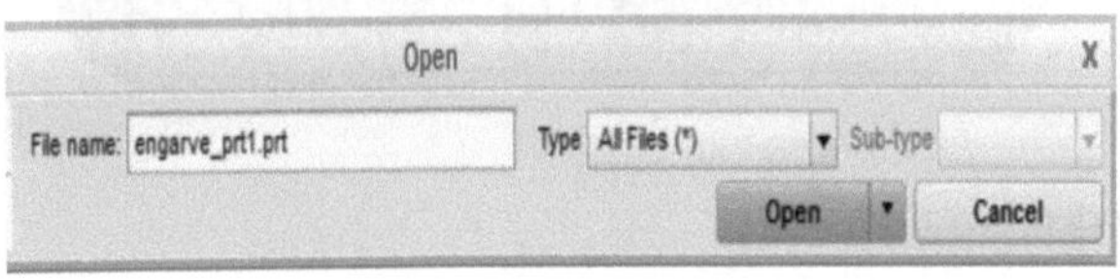

Click on the Open tab.

Fig. 2.50 The activated Open dialogue window in concise form

The 3D Part (engrave_prt1) is now imported on the main graphic window.

Constrain the Imported Part

Constrain the imported Part with the ASM datum plane by right clicking on the right mouse on the main GUI window and then click on Default Constraint on the drop-down menu list as illustrated in Fig. 2.51.

Fig. 2.51 Constraining imported Part method 1

Alternatively, Click on the Automatic section box to activate its drop-down menu list ≫ Now click on Default on the drop-down menu list as indicated by the arrow in Fig. 2.52.

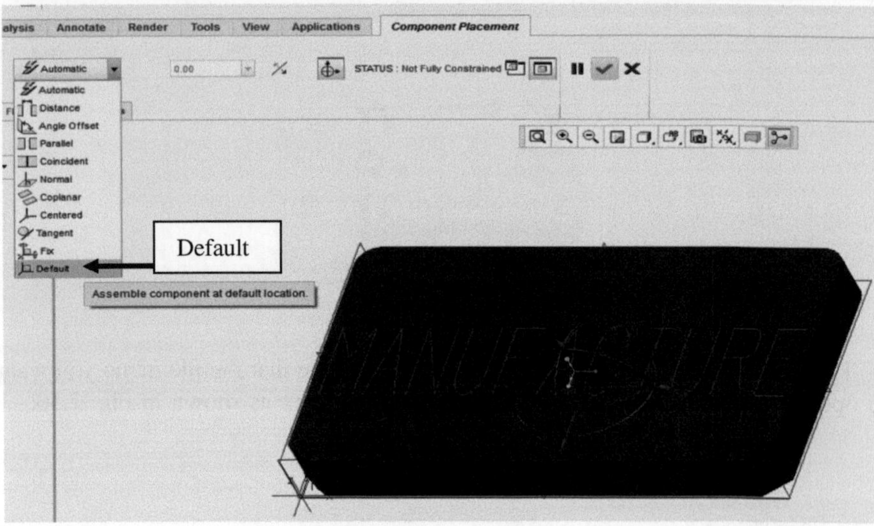

Fig. 2.52 Constraining imported Part method 2

The Part is now fully constrained on the main graphic window as shown in Fig. 2.53. The information on the STATUS section, reads Fully Constrained as indicated by the arrow in Fig. 2.53.

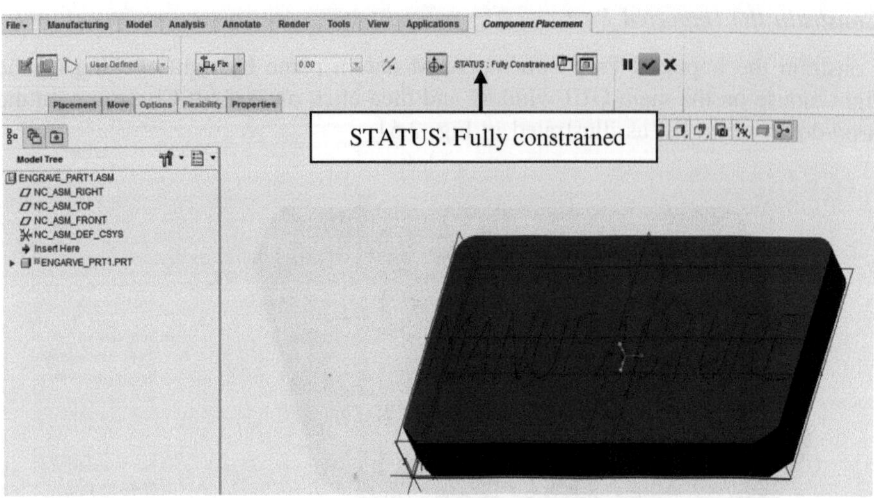

Fig. 2.53 Fully constrained Part on the Manufacturing main GUI window

Click on the Check Mark icon  to exit.

2.4.3 *Adding Stock to the Part*

No Stock will be created/added to Part in this Engraving process, because Reference Part and Stock (Workpiece) have the same dimensions.

2.5 Add Coordinate System (Programme Zero) to the Part

A Coordinate System or Programme Zero will be created on the Workpiece with respect to which all the Machine Coordinates will be referenced/obtained.

Click on the Coordinate System icon tab on the Datum group as shown below.

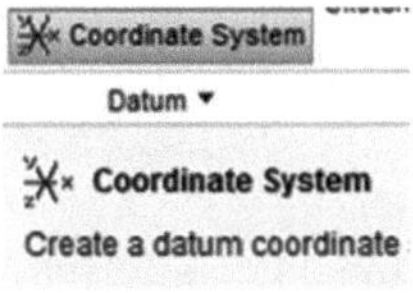

The Coordinate System dialogue box is activated on the main graphic window as shown in Fig. 2.54.

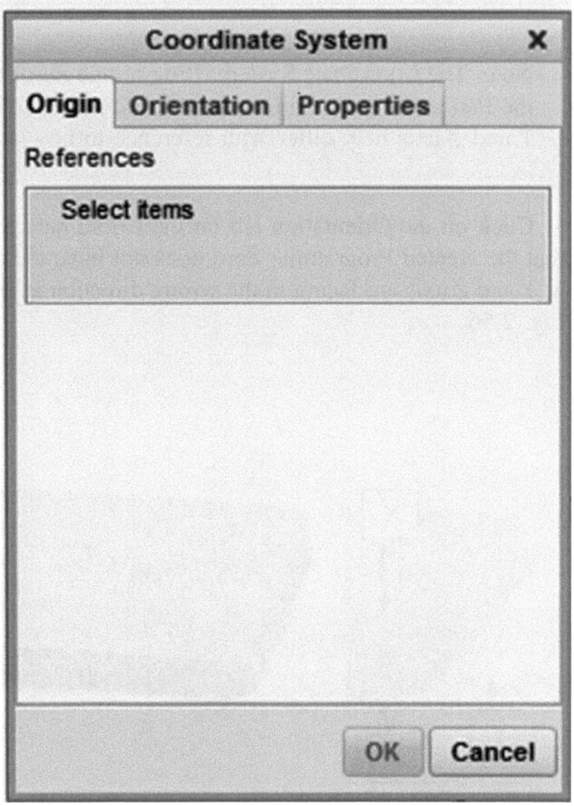

Fig. 2.54 Activated Coordinate System dialogue box

Now click on the Top surface and two Side surfaces of the Workpiece while holding down the Ctrl key when clicking. The selected Top and Sides surfaces will appear on the References section box on the Coordinate dialogue box as illustrated in Fig. 2.55.

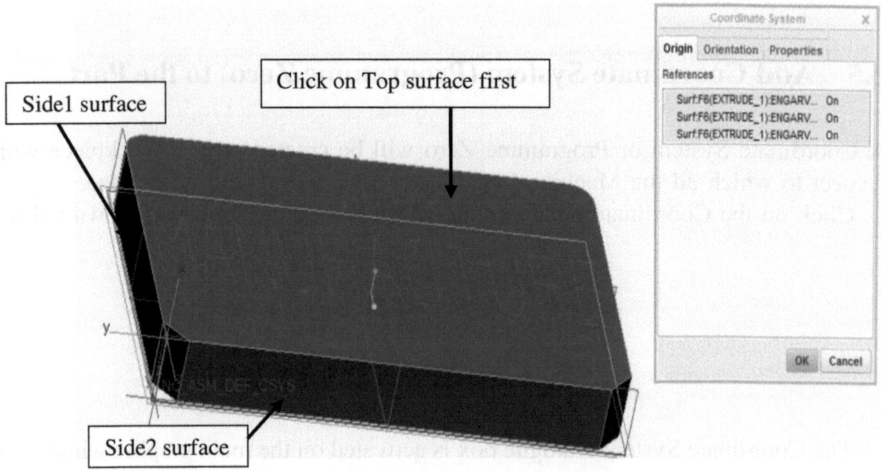

Fig. 2.55 Selected surfaces and the Coordinate Systems dialogue box

Note: The Coordinate System (Programme Zero) can be created anywhere on the Part as long as the correct procedures are followed. Your position of *X*, *Y* and *Z* axis may differ with reference to how you selected those surfaces.

Click on the Orientation tab on the Coordinate System dialogue box ≫ Notice that the created Programme Zero does not have the correct orientation because the *X*, *Y* and *Z* axis are facing in the wrong direction as indicated by the three arrows in Fig. 2.56.

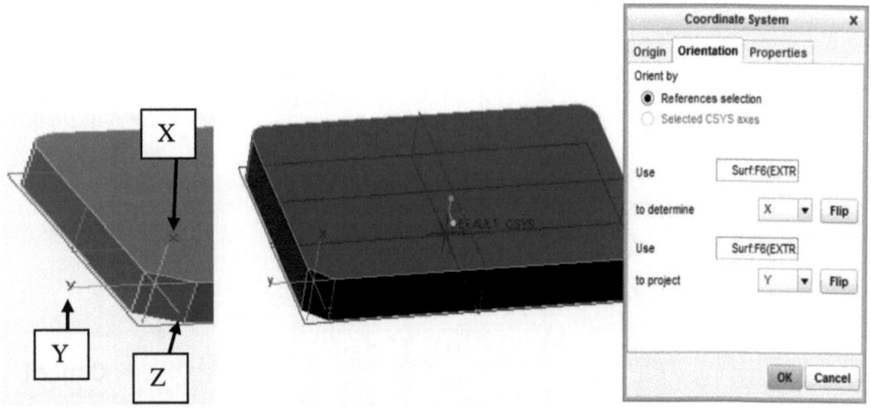

Fig. 2.56 Wrong orientation of *X*, *Y* and *Z*-axes of the Coordinate Systems

Orientate the X, Y and Z coordinate axes to the correct orientation

Click on the Orientation tab ≫ On the Orient by group, click in the "References selection" radio button ≫ Click on "to determine" section box and click on *Z* axis on the drop-down list ≫ Click on "to project" section box and click on *X* axis on the drop-down list ≫ Now click on the Flip tab, to flip *X* axis orientation. The arrows shown in Fig. 2.57 illustrate the above description, and show the correct orientation for the created Programme Zero.

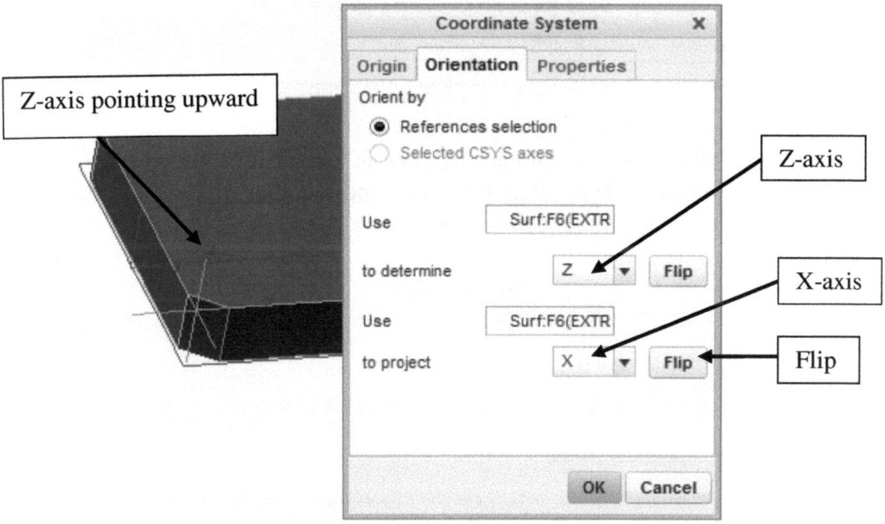

Fig. 2.57 Correct orientations of the Coordinate Systems *X*, *Y* and *Z*-axes

Click on the OK tab to exit the Coordinate System dialogue box.

A new Coordinate System (Programme Zero) called ACS1 is created. The new ACS1 created can be re-named by slow clicking on ACS1 on the Model Tree. Alternatively, right click on ACS1, now click on rename on the drop-down list. In this tutorial, ACS1 will not be re-named as indicated by the arrow on the Model Tree as shown in Fig. 2.58.

Fig. 2.58 Arrow indicating ACS1 on the Model Tree

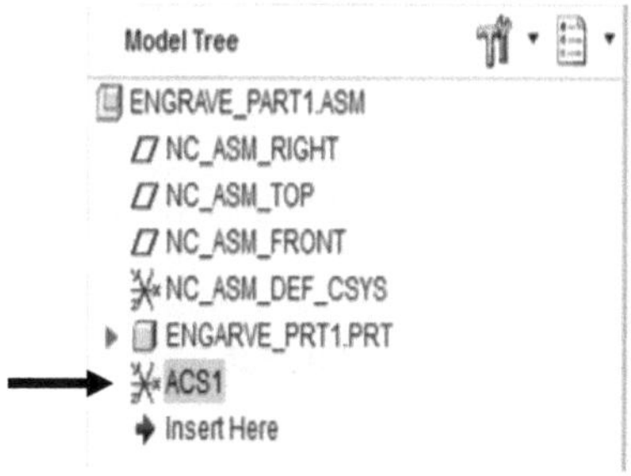

2.6 Set up the Work Centre

Click on the Work Centre icon on the Manufacturing menu bar ≫ On the activated drop-down menu list, click on Mill as shown below.

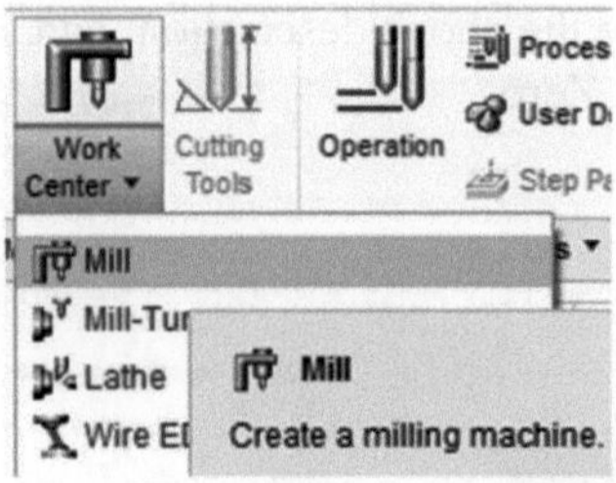

The Milling Work Centre dialogue window is activated as shown in Fig. 2.59. Make note of the Name, Type, Post Processor and Number of Axis section boxes.

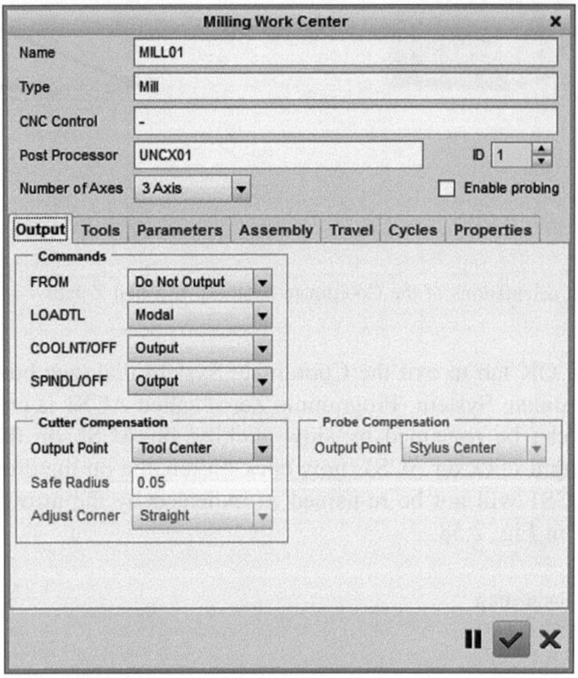

Fig. 2.59 Activated Milling Work Centre dialogue box

Click on the OK tab ✓ to exit the Milling Work Centre dialogue window.

2.7 Set up the Operation

Click on Operation icon as shown below

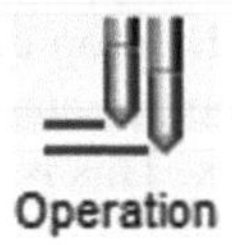

The Operation dashboard tools are activated as shown in Fig. 2.60.

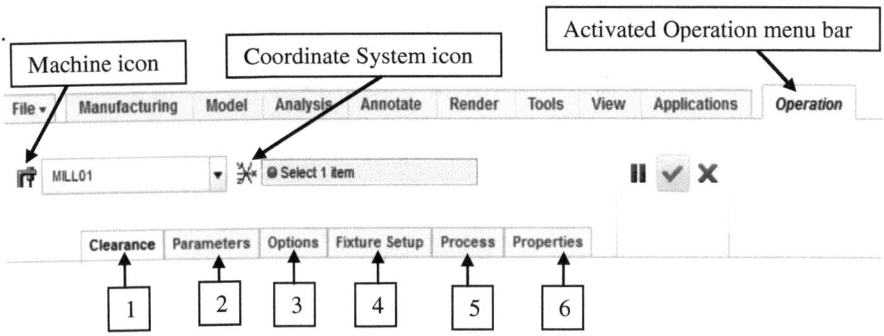

Fig. 2.60 Activated Operation tools in concise form

where

1. Represent Clearance tab
2. Represent Parameters tab
3. Represent Options tab
4. Represent Fixture Setup tab
5. Represent Process tab
6. Represent Properties tab

To add ACS1 (New Coordinate System)

Click in the Coordinate System icon section box ≫ Go to the Model Tree and click on ACS1. ACS1 is automatically added onto the Coordinate System icon section box.

Make sure that MILL01 is on the Machine icon section box. MILL01 is automatically added to the Machine icon section box by the system.

To activate the Clearance tab

Now click on the Clearance tab (that is 1 in Fig. 2.60) to activate the Clearance panel.

On the Retract group, click on the Type section box to activate its drop-down menu, ≫ Now click on Plane on the drop-down menu list ≫ Click on the Part surface and the Reference section box is automatically updated by the system ≫ Type 10 mm in the Value section box. This is now the Retract distance for the Tool as shown in Fig. 2.61.

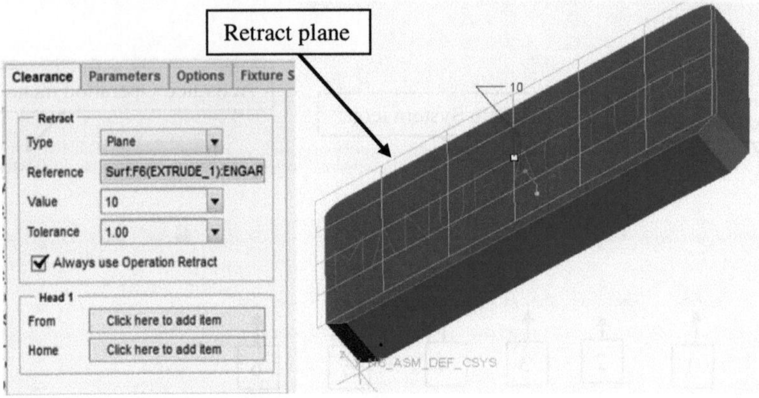

Fig. 2.61 Activated Clearance panel and the Retract plane

Click on the Check Mark icon  to exit the Operation setup.

Note: The Retract surface is the surface that the Tool will move up to after each machining process.

Check the Model Tree to make sure that the ACS1, MILL01 and OP010 [MILL01] as created are on the Model Tree as shown in Fig. 2.62.

Fig. 2.62 Model Tree showing all the steps achieved

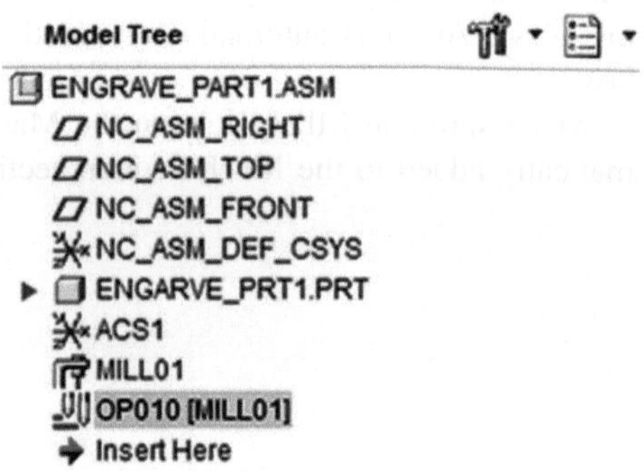

2.8 Set up the Cutting Tool

Click on the Cutting Tools icon

The Tools Setup dialogue window is activated on the main graphic window ≫ Click on the General tab, type the Tool name in the Name box ≫ Click on the Type section box to activate its contents, now click on the Grooving tool on the menu list ≫ Click on the Material section box, type the Tool material name. Add dimensions to the Tool and click on the Apply tab. See Fig. 2.63.

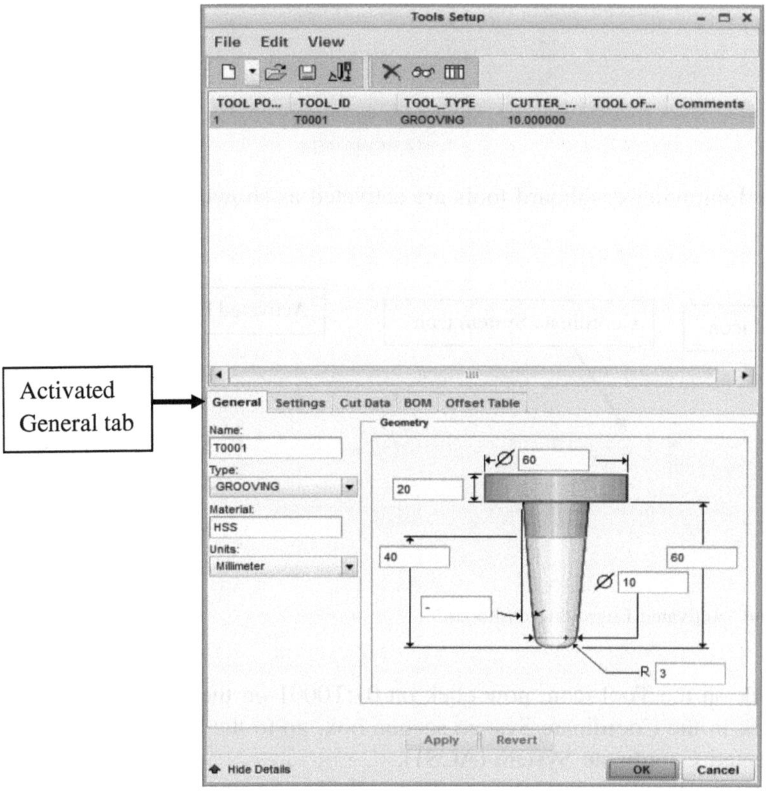

Fig. 2.63 Tools Setup dialogue box indicating the Grooving Tool

Click on OK tab to exit Tools Setup dialogue window.

Note: All dimensions of the cutting Tool are theoretical as shown in Fig. 2.63. They are used in this tutorial for illustration purposes only. Correct cutting Tool material must be selected for each machining applications in practical application.

2.9 Activate the Engraving Process

Click on the Mill menu bar ≫ Now click on Engraving icon tab on the Milling group as shown below.

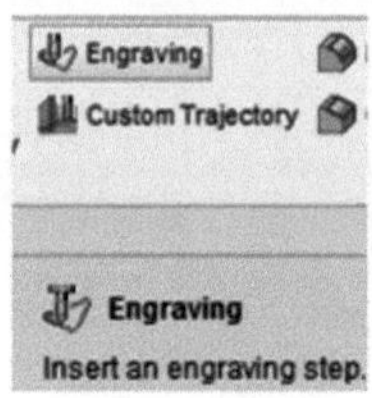

The Engraving dashboard tools are activated as shown in Fig. 2.64.

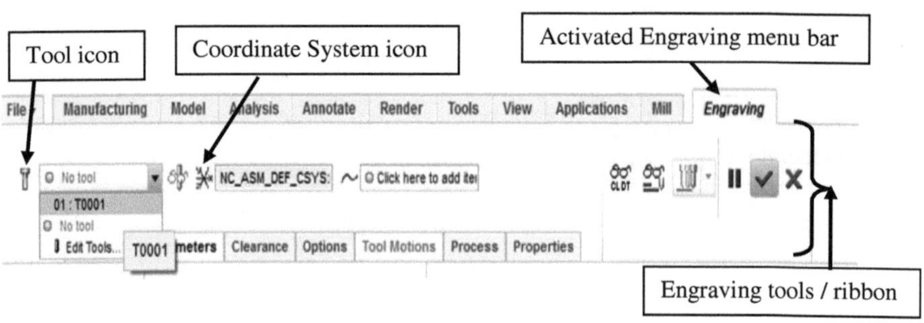

Fig. 2.64 Activated Engraving dashboard

Click on the Tool icon, now click on 01:T0001 on the drop-down list.

Click in the Coordinate System section box, go to the Model Tree and click on the created Coordinate System (ACS1).

Click on the Datum Curve or Curve Profile section box, now click on the "MANUFACTURE" imprinted text on the Part surface as indicated by the arrows in Fig. 2.65.

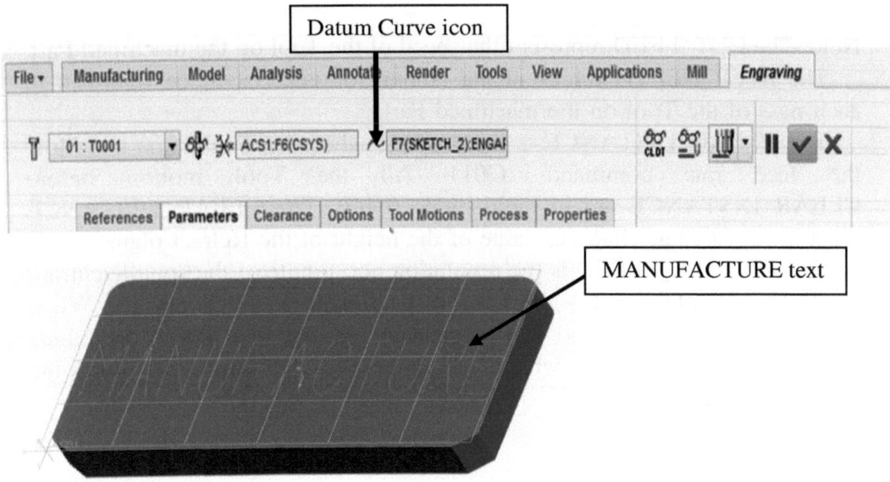

Fig. 2.65 T0001, ACS1, MANUFACTURE text added into their section boxes

Click on the Parameters tab in Fig. 2.65 to activate its panel, now type in the parameter values for the CUT_FEED, STEP_DEPTH, CLEAR_DIST, SPINDLE_SPEED as shown in Fig. 2.66.

Fig. 2.66 Engraving Parameters values

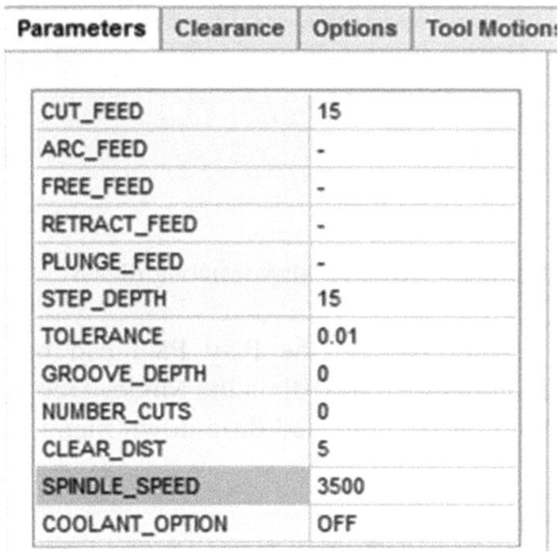

CUT_FEED	15
ARC_FEED	-
FREE_FEED	-
RETRACT_FEED	-
PLUNGE_FEED	-
STEP_DEPTH	15
TOLERANCE	0.01
GROOVE_DEPTH	0
NUMBER_CUTS	0
CLEAR_DIST	5
SPINDLE_SPEED	3500
COOLANT_OPTION	OFF

Click on the Check Mark icon ✓ to exit the Engraving application.

Note: The CUT_FEED represent the speed of the Tool on the machined Part.

The STEP_DEPTH represent the amount of material to be removed in each pass of the Tool on the machined Part.

The CLEAR_DISTANCE is the distance when the Tool starts to follow the feed rate command (G01). All the Tool motion before CLEAR_DISTANCE are in rapid mode (G00). The CLEAR_DISTANCE must always be less than the value of the height of the Retract plane.

The SPINDLE_SPEED is the revolution per minute of the spindle/cutting Tool. The spindle speed is the biggest determiner of the Tools life. When Tool is running/turning too fast, it generates excess heat (there are others ways to generate heat too), which softens the Tool and ultimately allows the edge to dull.

2.9.1 Activate Play Path

Click on the Manufacturing menu bar ≫ Click on the Play Path icon to activate its drop-down list, now click on Play Path as indicated by the arrow in Fig. 2.67.

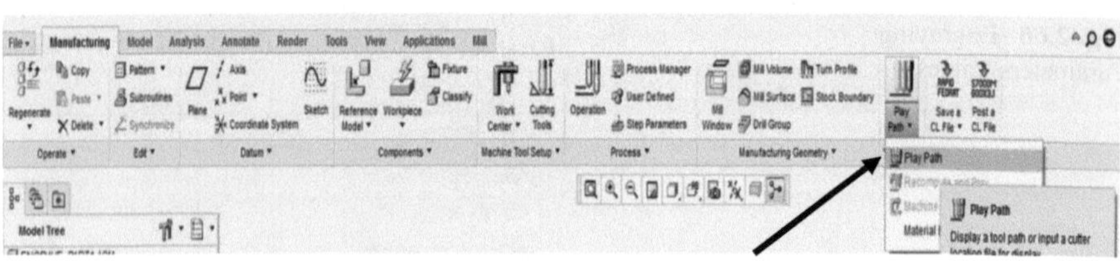

Fig. 2.67 Activated Manufacturing ribbon

Alternatively, the Play Path can be activated by right click the mouse on Engraving 1[OP010] on the Model Tree as shown in Fig. 2.68 ≫ Now click on Play Path on the activated drop-down menu list.

Fig. 2.68 Activating Engraving 1[OP010] on the Model Tree

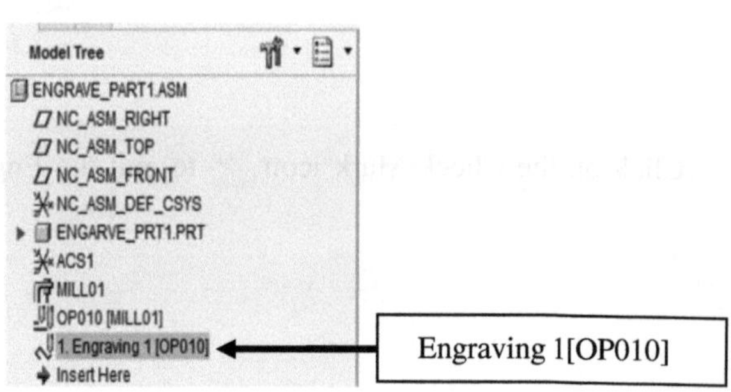

The cutting Tool and the Play Path dialogue box are activated on the Manufacturing main graphic window as shown in Fig. 2.69.

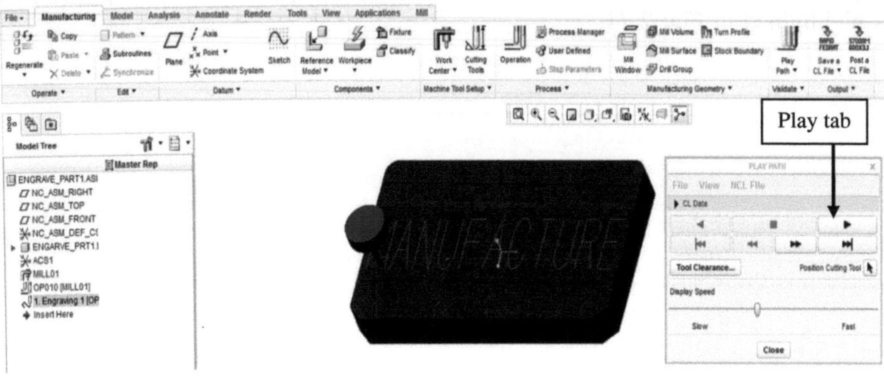

Fig. 2.69 Activated Engraving Tool and the PLAY PATH dialogue box

Click on the Play tab to start the Engraving operation.

Note: The Tool speed can be slowed or increased by adjusting the Display Speed dial/knob on the PLAY PATH dialogue box to enable better view of the Engraving operation.

The end of the Engraving process is as shown in Fig. 2.70.

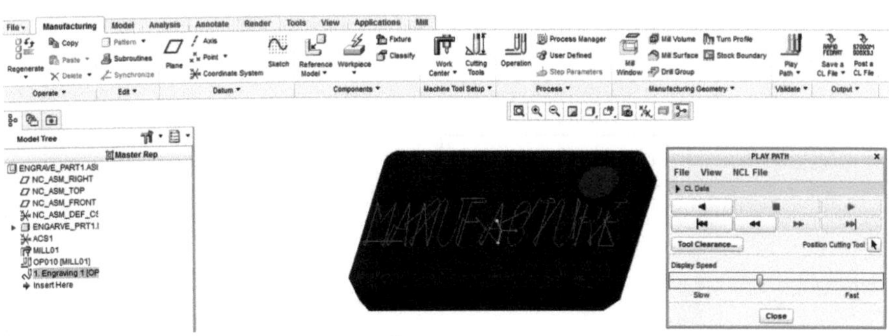

Fig. 2.70 End of on screen Engraving process

Note: When the mouse is right clicked on the MILL01, OP010 [MILL01] or Engraving 1 [OP010] on the Model Tree and Edit Definition is clicked on the drop-down menu list, there parameters can be edited.

2.10 Add Material Removal (nccheck_Type) Application

Click on File to activate its menu list ≫ Now click on Options on the drop-down list as highlighted in Fig. 2.71.

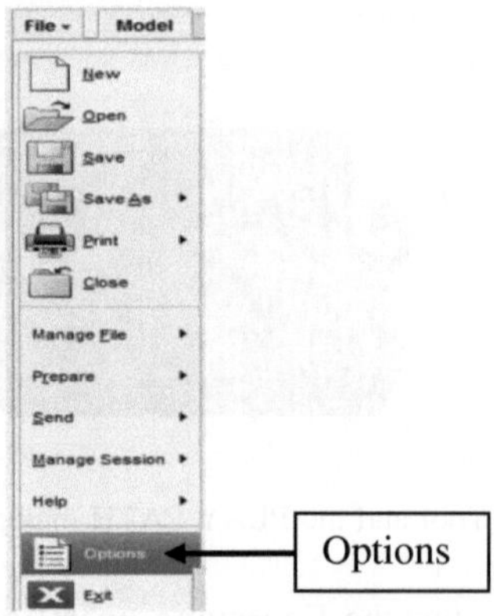

Fig. 2.71 Activated File group

The Creo Parametric Options dialogue window is activated as shown in Fig. 2.72.

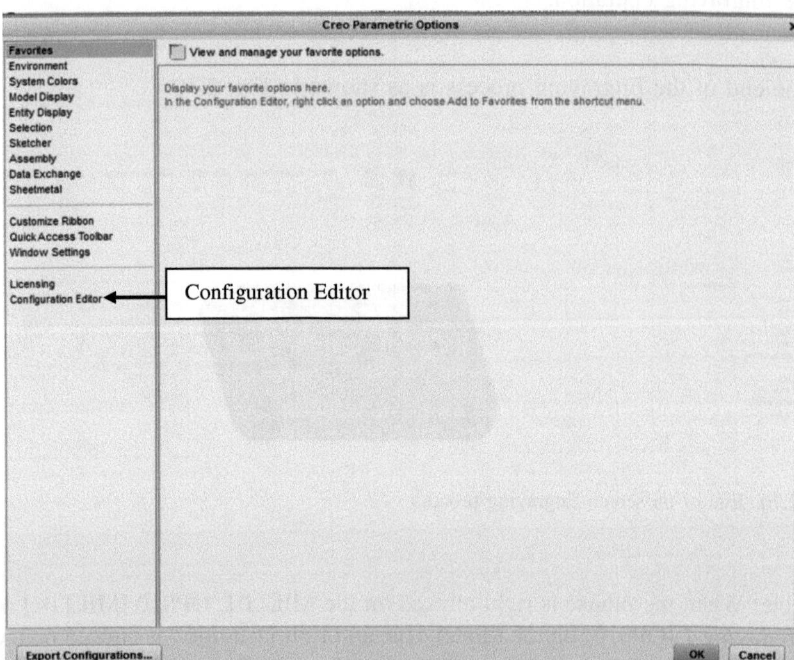

Fig. 2.72 Activated Creo Parametric Options dialogue window

Click on Configuration Editor on the menu list as indicated by the arrow in Fig. 2.72.

The Configuration Editor application tools are activated as shown in Fig. 2.73.

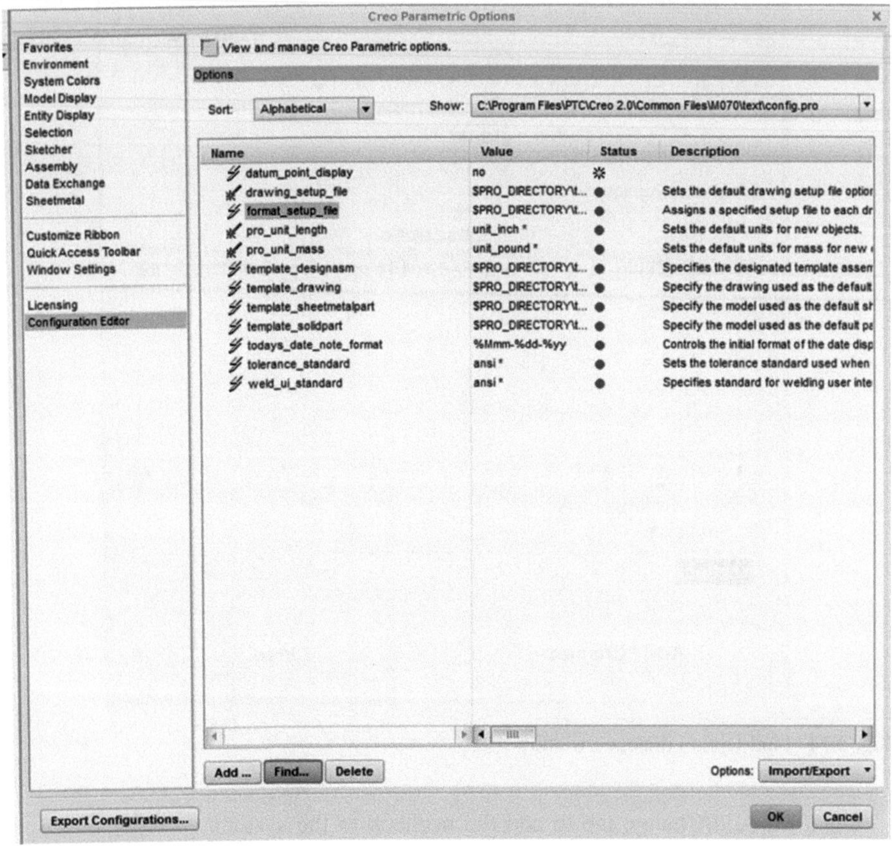

Fig. 2.73 Activated Configuration Editors application tools

Click on Find tab.

The Find Option dialogue window is activated on the main graphic window ≫ Now type "nccheck_type" in the Type keyword section box ≫ Click on Find Now tab ≫ Click on the downward pointing arrow on the Set value section box, now click on nccheck on the drop-down menu list. See Fig. 2.74.

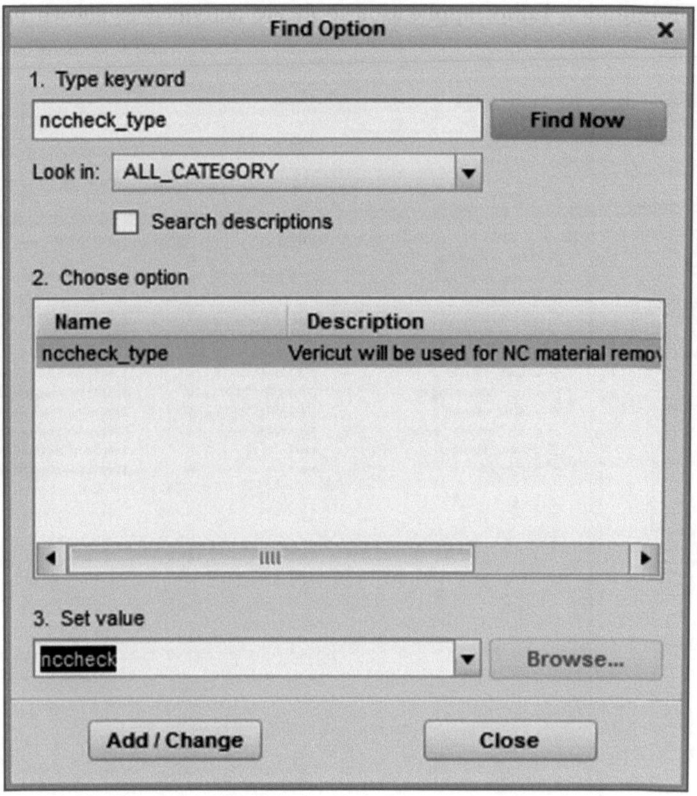

Fig. 2.74 Find Option dialogue window

Click on Add/Change tab to add the nccheck to the system.
Save into the configuration file when prompted to do so.
Click on the Close tab to exit the Find Option dialogue window.

2.11 Material Removal Simulation Activation

To view the material removal simulation after adding the "nccheck_type" to the configuration file.

Got to the Model Tree and click on Engraving 1 [OP010] on the Model Tree to highlight it ≫ Now right click on the mouse to activate Engraving 1 [OP010] menu list, now click on Material Removal Simulation on the drop-down menu list.

The NC CHECK Menu Manager dialogue box is activated on the main graphic window as shown in Fig. 2.75.

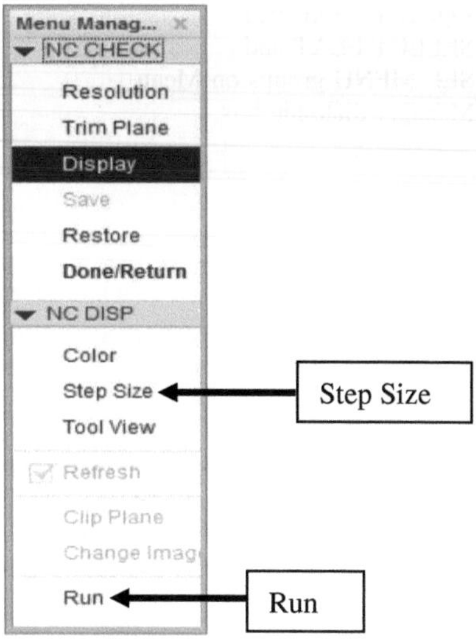

Fig. 2.75 Activated NC CHECK and NC DISP groups on the Menu Manager dialogue box

Click on Run to start the Material Removal Simulation.

To reduce the speed of the material removal process » Click on the Step Size and add a lower step size value » Click on Run again, and the simulation animation will run very slow.

Click on Done/Return on the NC CHECK group to exit the Material Removal Simulation process.

2.12 Creating the Cuter Location (CL) Data

Click on Manufacturing menu tab » Now click on Save a CL File icon on the output group as illustrated below.

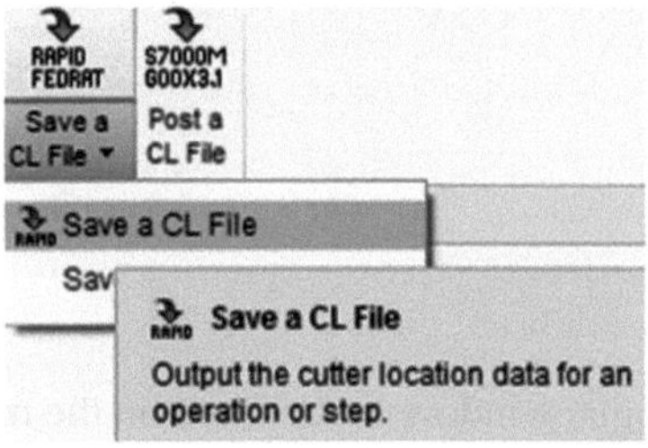

The SELECT FEAT Menu Manager dialogue box is activated » Click on Operation on the SELECT FEAT group and click on OP010 on the SEL MENU group as highlighted in Fig. 2.76.

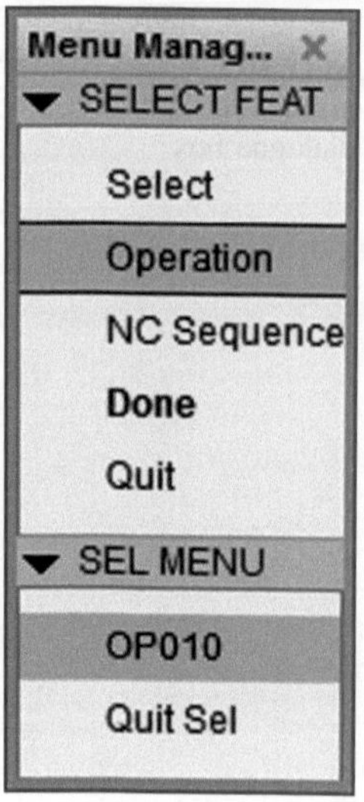

Fig. 2.76 Activated SELECT FEAT and SEL MENU groups on Menu Manager dialogue box

Once OP010 is clicked, the PATH Menu Manager dialogue box is activated on the main graphic window ≫ Click on File on the PATH group and Check Mark the "CL File" and "Interactive" square boxes on the OUTPUT TYPE group if they are not automatically Check Marked by system as shown in Fig. 2.77.

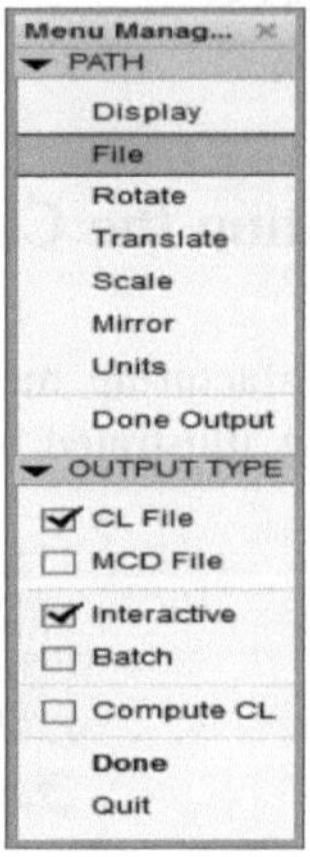

Fig. 2.77 Activated PATH and OUTPUT TYPE groups on Menu Manager dialogue box

Click on Done.

The Save a Copy dialogue window is activated on the main graphic window. See Fig. 2.78.

Fig. 2.78 Activated Save a Copy dialogue window in concise form

Make sure that the name on the New Name section box is correct. Alternatively, Type Engrave_op010 as the new name, if you did not type Engrave_part1 in the New dialogue box at the early stage.

Click on the OK tab to exit.

The CL File(*.ncl) is now saved on the already chosen directory folder/drive.

Click on Done Output on the PATH group as indicated by the arrow shown in Fig. 2.79.

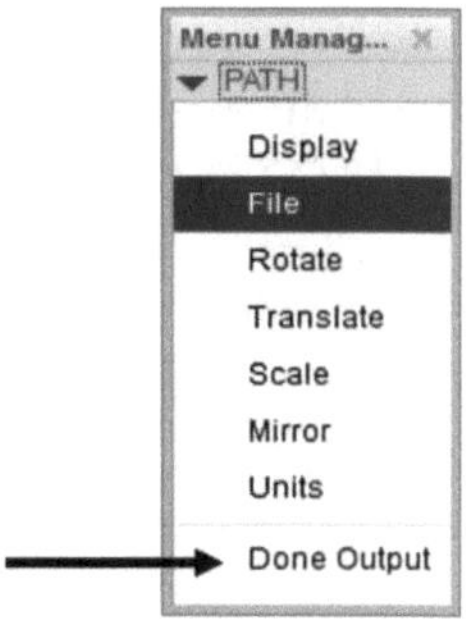

Fig. 2.79 Activated PATH group on the Menu Manager dialogue box

2.13 Create the G-Code Data

Click on Manufacturing toolbar » Click on the Post a CL File icon on the output group as shown below.

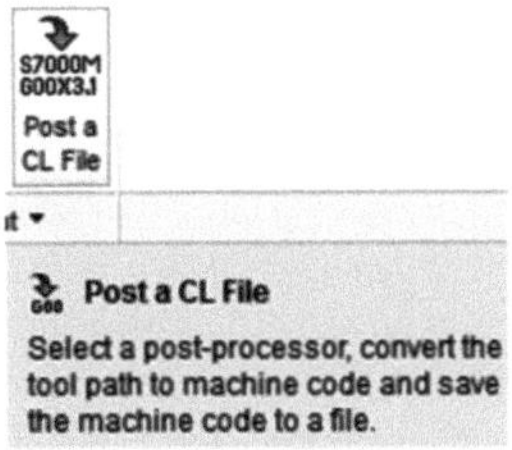

The Open dialogue window is activated on the main graphic window. See Fig. 2.80.

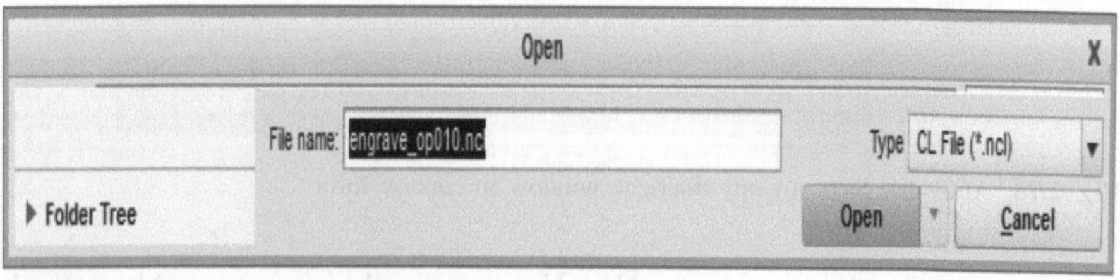

Fig. 2.80 Activated Open dialogue window in concise form

Make sure that the file name on the File Name section box is correct.

Click on the Open tab to exit.

The PP OPTIONS Menu Manager dialogue box is activated on the main graphic window ≫ Check Mark Verbose and Trace square boxes, respectively, if they are not automatically Checked Marked by the system as shown in Fig. 2.81.

Fig. 2.81 Activated PP OPTIONS Menu Manager dialogue box

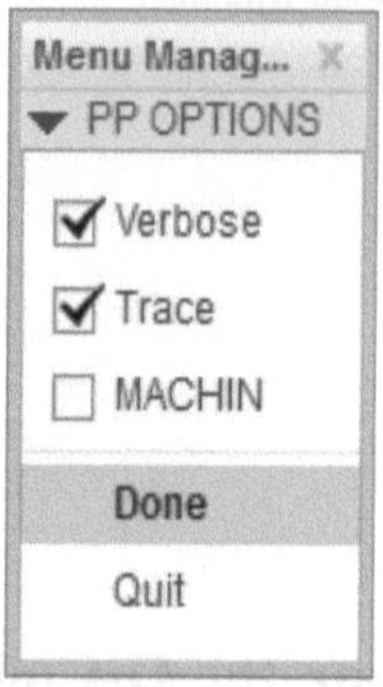

Click on Done.

The PP LIST Menu Manager dialogue box is activated on the main graphic window with all the available Post Processor (PP) list as shown in Fig. 2.82.

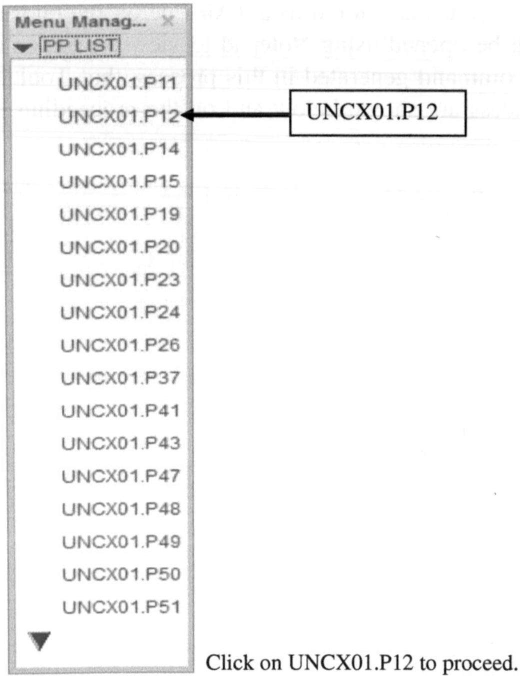

Fig. 2.82 Activated PP LIST Menu Manager dialogue box

> Note: The UNCX01.P12 generates G-codes for four axes Milling Machine but can also generate G-codes for three axes Milling Machine.

The INFORMATION WINDOW opens up on the graphic window as shown in concise form. See Fig. 2.83.

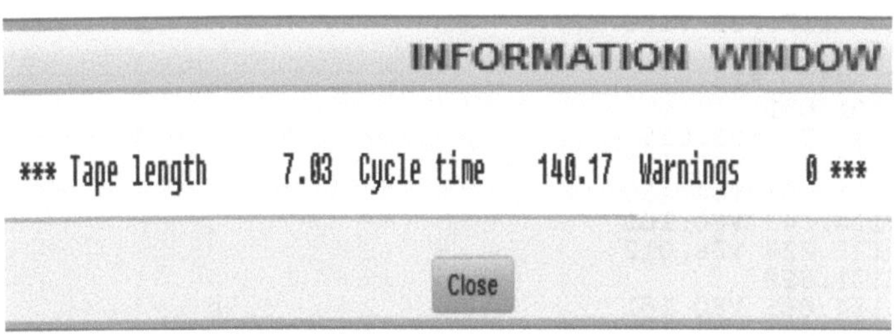

Fig. 2.83 Activated INFORMATION WINDOW in concise form

Check to see if there are any warnings in case the programme has encountered any problem during generating the G-codes.

Click on Close tab on the INFORMATION WINDOW dialogue window to exit.

The generated G-codes are stored as a TAP File on the chosen work directory. The TAP File will be opened using Notepad to view the generated G-codes.

The G-Codes command generated in this process, that Tool will follow during the Engraving process are shown below and on the proceeding pages.

engrave_op010.tap - Notepad

File Edit Format View Help

```
N5 G71
N10 ( / ENGRAVE_PART1)
N15  G0 G17 G99
N20  G90 G94
N25  G0 G49
N30 T1 M06
N35 S3500 M03
N40 G0 G43 Z60. H1
N45 X38.28 Y78.012
N50 Z55.
N55 G1 Z50. F15.
N60 Y155.377
N65 X24.367 Y120.993
N70 X10.454 Y155.377
N75 Y78.012
N80 Z60.
N85 G0 X72.07
N90 Z55.
N95 G1 Z50. F15.
N100 X59.151 Y155.377
N105 X46.231 Y78.012
N110 Z60.
N115 G0 X51.2 Y108.098
N120 Z55.
N125 G1 Z50. F15.
N130 X67.101
N135 Z60.
N140 G0 X105.86 Y155.377
N145 Z55.
N150 G1 Z50. F15.
N155 Y78.012
N160 X82.008 Y155.377
N165 Y78.012
N170 Z60.
N175 G0 X117.786 Y155.377
N180 Z55.
N185 G1 Z50. F15.
N190 Y101.651
N195 X118.78 Y93.055
N200 X119.774 Y88.757
N205 X121.761 Y84.459
N210 X124.743 Y80.161
N215 X127.724 Y78.012
N220 X131.699
N225 X134.681 Y80.161
N230 X137.662 Y84.459
N235 X139.65 Y88.757
N240 X140.644 Y93.055
N245 X141.638 Y101.651
N250 Y155.377
N255 Z60.
N260 G0 X177.415
N265 Z55.
```

```
N270 G1 Z50. F15.
N275 X153.563
N280 Y78.012
N285 Z60.
N290 G0 Y120.993
N295 Z55.
N300 G1 Z50. F15.
N305 X173.44
N310 Z60.
N315 G0 X215.18 Y78.012
N320 Z55.
N325 G1 Z50. F15.
N330 X202.261 Y155.377
N335 X189.341 Y78.012
N340 Z60.
N345 G0 X194.31 Y108.098
N350 Z55.
N355 G1 Z50. F15.
N360 X210.211
N365 Z60.
N370 G0 X248.97 Y133.887
N375 Z55.
N380 G1 Z50. F15.
N385 X247.976 Y142.483
N390 X246.982 Y146.781
N395 X244.995 Y151.079
N400 X242.013 Y155.377
N405 X239.032 Y157.526
N410 X235.057
N415 X232.075 Y155.377
N420 X229.094 Y151.079
N425 X227.106 Y146.781
N430 X226.112 Y142.483
N435 X225.118 Y133.887
N440 Y101.651
N445 X226.112 Y93.055
N450 X227.106 Y88.757
N455 X229.094 Y84.459
N460 X232.075 Y80.161
N465 X235.057 Y78.012
N470 X239.032
N475 X242.013 Y80.161
N480 X244.995 Y84.459
N485 X246.982 Y88.757
N490 X247.976 Y93.055
N495 X248.97 Y101.651
N500 Z60.
N505 G0 X262.884 Y155.377
N510 Z55.
N515 G1 Z50. F15.
N520 X282.76
N525 Z60.
N530 G0 X272.822
N535 Z55.
N540 G1 Z50. F15.
N545 Y78.012
N550 Z60.
N555 G0 X296.673 Y155.377
N560 Z55.
N565 G1 Z50. F15.
N570 Y101.651
```

```
N575 X297.667 Y93.055
N580 X298.661 Y88.757
N585 X300.649 Y84.459
N590 X303.63 Y80.161
N595 X306.612 Y78.012
N600 X310.587
N605 X313.568 Y80.161
N610 X316.55 Y84.459
N615 X318.537 Y88.757
N620 X319.531 Y93.055
N625 X320.525 Y101.651
N630 Y155.377
N635 Z60.
N640 G0 X332.451 Y78.012
N645 Z55.
N650 G1 Z50. F15.
N655 Y155.377
N660 X348.352
N665 X352.327 Y153.228
N670 X354.315 Y151.079
N675 X355.309 Y148.93
N680 X356.303 Y142.483
N685 Y133.887
N690 X355.309 Y127.44
N695 X354.315 Y125.291
N700 X352.327 Y123.142
N705 X348.352 Y120.993
N710 X332.451
N715 Z60.
N720 G0 X346.364
N725 Z55.
N730 G1 Z50. F15.
N735 X356.303 Y78.012
N740 Z60.
N745 G0 X392.08 Y155.377
N750 Z55.
N755 G1 Z50. F15.
N760 X368.228
N765 Y78.012
N770 X392.08
N775 Z60.
N780 G0 X368.228 Y120.993
N785 Z55.
N790 G1 Z50. F15.
N795 X388.105
N800 Z60.
N805 M30
%
```

Chapter 3
Face Milling Operation

Face milling machining process is used to smoothen or clear any excessive material on the part/workpiece surface. It is usually the first NC sequence used in machining a part.

The steps below will be followed when creating the face milling manufacturing application.

- Start the manufacturing application
- Import the saved part into the main manufacturing GUI window
- Constrain the imported part
- Add stock (workpiece)
- Add programme zero (Machine Coordinate System)
- Add machine work centre
- Tool is added to the setup
- Face milling operation sequence is created and parameters are added
- Generate the Cutter Location (CL) data
- Generate the G-codes data for the operation

3.1 Start the Manufacturing Application

Start Creo Parametric 2.0 either from your computer desktop or from the programme.

Click on the New icon as shown below.

P.O. Kanife, *Computer Aided Virtual Manufacturing Using Creo Parametric*,
DOI 10.1007/978-3-319-23359-8_3

The New dialogue box is activated on the main window as shown in Fig. 3.1.

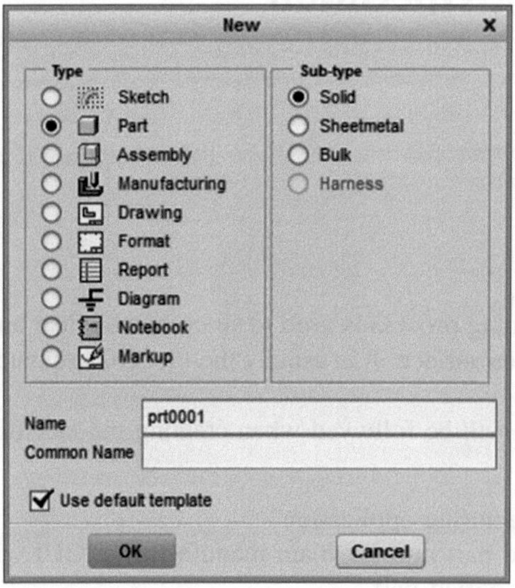

Fig. 3.1 Activated New dialogue box

On the Type group, click in the Manufacturing radio button to make it active ≫ On the Sub-type group, click in the NC Assembly radio button to make it active ≫ Type the name (Mfg_rec_prt1) in the name section box. Clear the Use Default Template ≫ The New dialogue box changes as shown in Fig. 3.2.

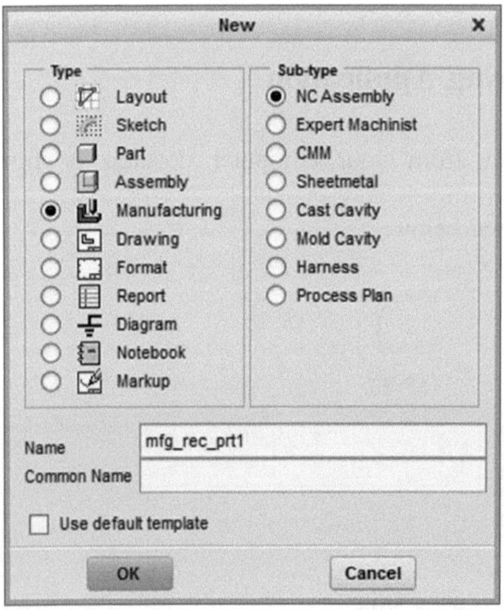

Click on the OK tab to proceed.

Fig. 3.2 New dialogue box with Manufacturing and NC Assembly radio buttons activated

The New File Options dialogue window is activated on the main window ≫ On the Template group, click on mmns_mfg_nc ≫ On the Parameters group, type your name. In this tutorial, we shall type "Paul" in the "MODELLED_BY" section box. In the "DESCRIPTION" section box, type "Part for face milling" if you want ≫ Clear the "Copy associated drawings" option. The New File Options dialogue box will changes as shown in Fig. 3.3.

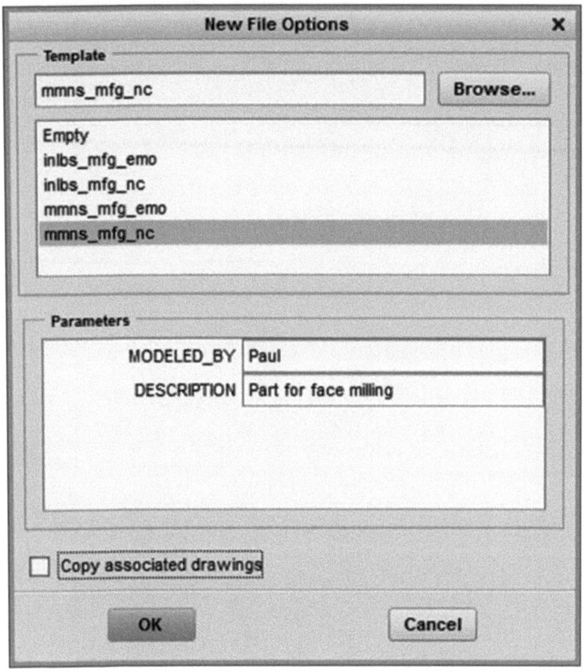

Fig. 3.3 Activated New File Options dialogue box with mmns_mfg_nc highlighted and clicked

Click on the OK tab to exit.

The Manufacturing main GUI window is activated as shown in Fig. 3.4.

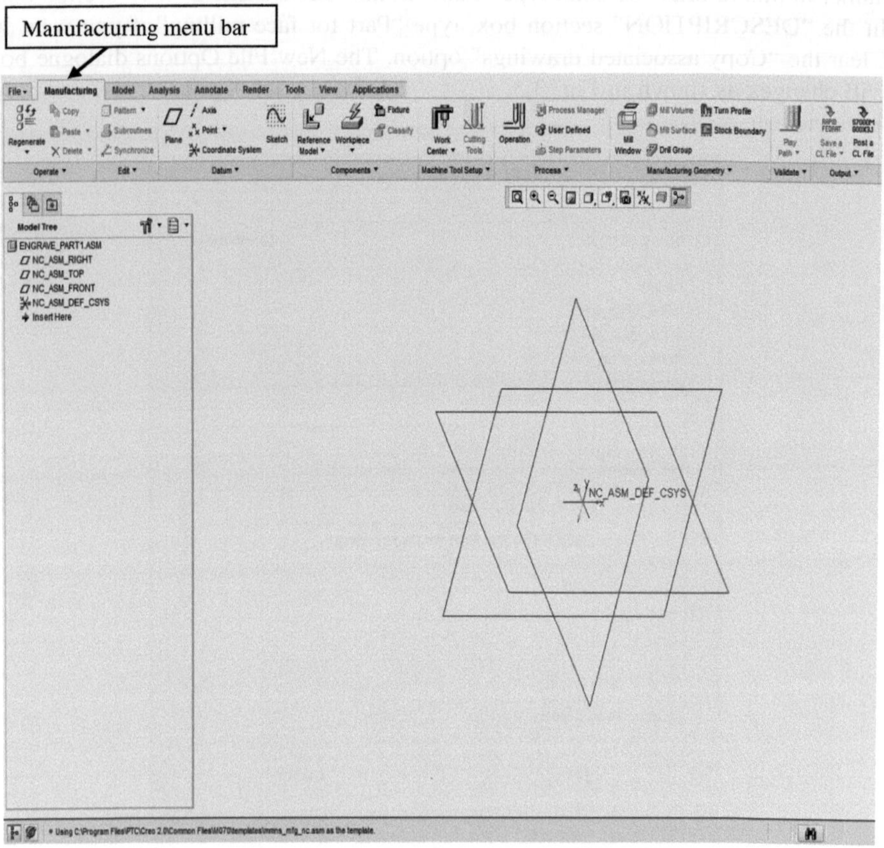

Fig. 3.4 Activated Manufacturing main Graphic User Interface window

3.2 Importing and Constraining the Saved 3D Part

Importing the 3D Part

Click on the Reference Model icon, now click on Assembly Reference Model on the drop-down menu list as illustrated below.

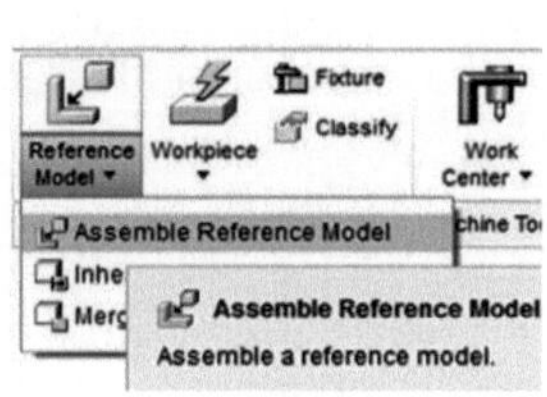

The File Open dialogue window is activated on the main graphic window. See Fig. 3.5. Make sure that the correct part to be opened is on the file name section box.

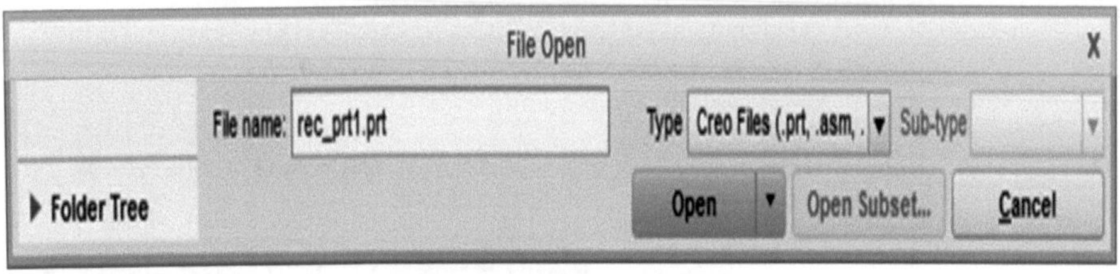

Fig. 3.5 Activated File Open dialogue window in a concise form

Click on the Open tab to exit.

The part is now imported into the Manufacturing main graphic window.

Constrain the Imported Part

To constrain the imported part with the ASM datum plane, right click on the right mouse in the main GUI window and then click on Default Constraint on the drop-down menu list as indicated by the arrow as shown in Fig. 3.6.

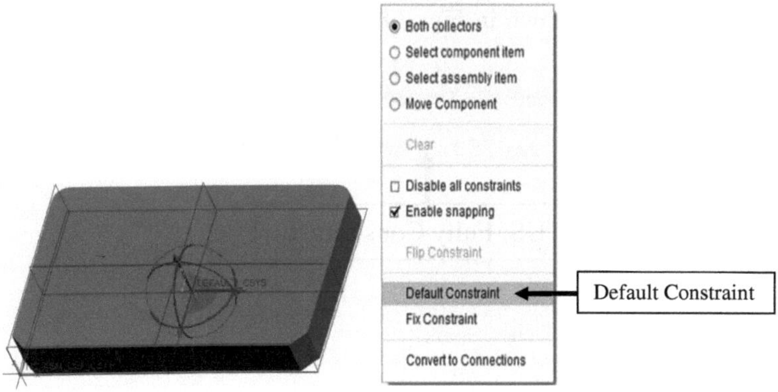

Fig. 3.6 Constraining imported part using method 1

Alternatively, click on the Automatic section box to activate its drop down menu list ≫ Now click on Default on the drop-down menu list as indicated by the arrow in Fig. 3.7.

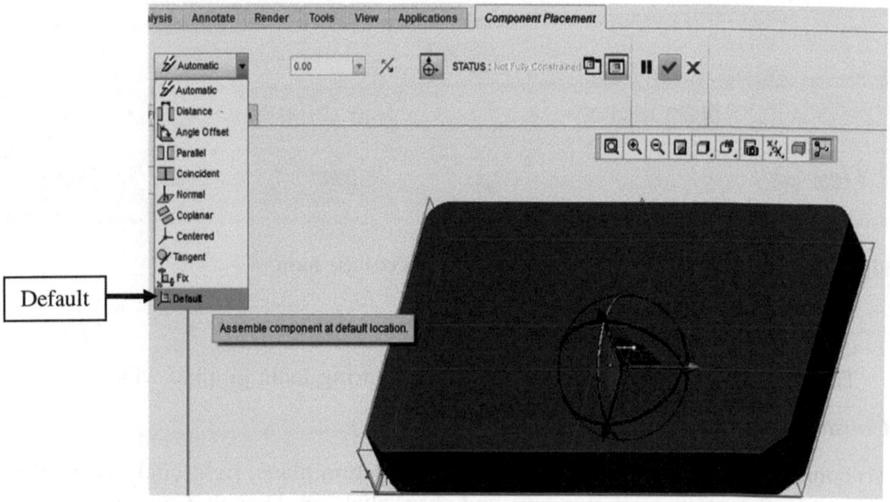

Fig. 3.7 Constraining the part using method 2

The information on the STATUS section, must read Fully Constrained as indicated by the arrow shown in Fig. 3.8.

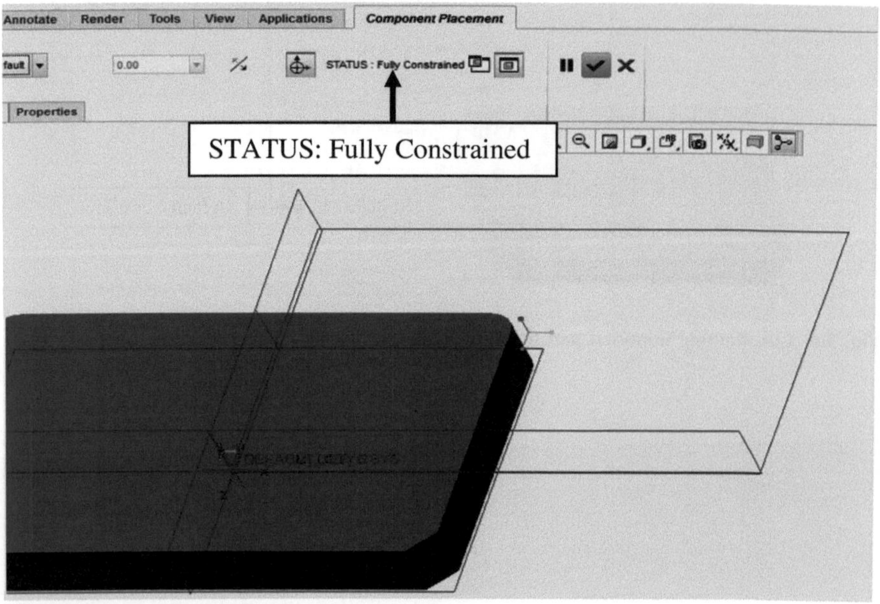

Fig. 3.8 Part fully constrained

Click on the Check Mark icon to accept and exit Component Placement application.

The 3D part is now fully constrained on the main graphic window as shown in Fig. 3.9.

Fig. 3.9 Part after exiting Component Placement application

3.3 Adding Stock to the Part

Click on Workpiece icon to activate its drop-down menu list ≫ Now select Automatic Workpiece on the drop-down menu list as shown below.

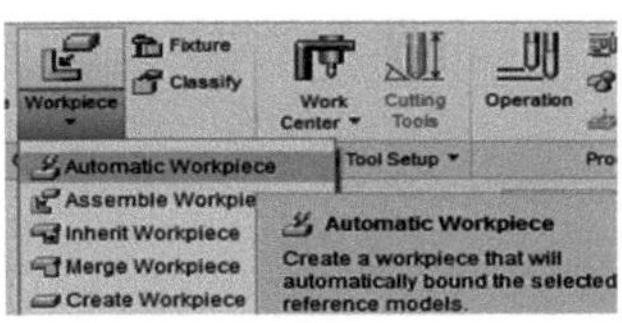

The auto workpiece creation dashboard tools are activated. See Fig. 3.10, shown in concise form.

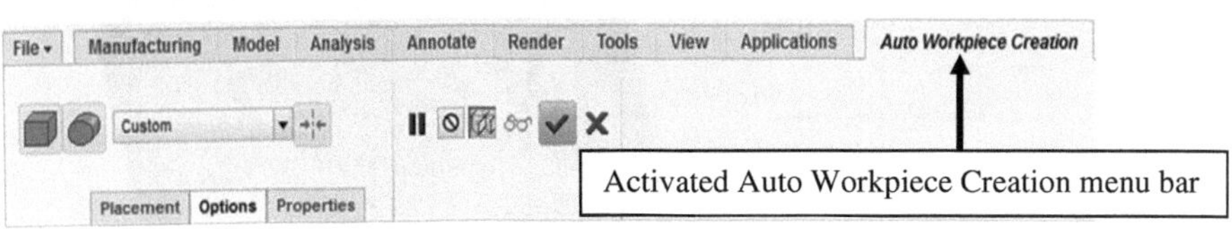

Fig. 3.10 Activated Auto-Workpiece creation application ribbon

Make sure that the icon tab  is active.

Click on the options tab to activate the Options panel.

On the Linear Offsets group, add dimension to the +Y section box indicated by the arrow in Fig. 3.11.

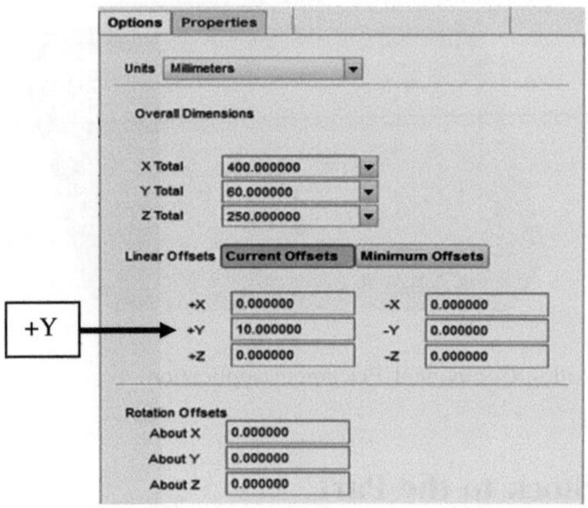

Fig. 3.11 Activated Options panel

10 mm of stock is added to the part in the +Y direction section box as shown in Fig. 3.12.

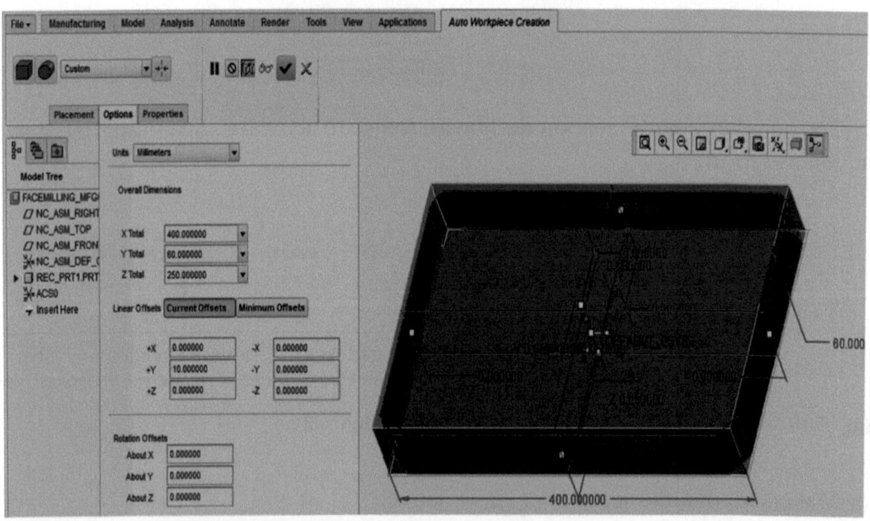

Fig. 3.12 Workpiece dimensions

Click on the Check Mark icon to exit.

Stock is now added into the part and the part changes on the main Manufacturing graphic window as shown in Fig. 3.13.

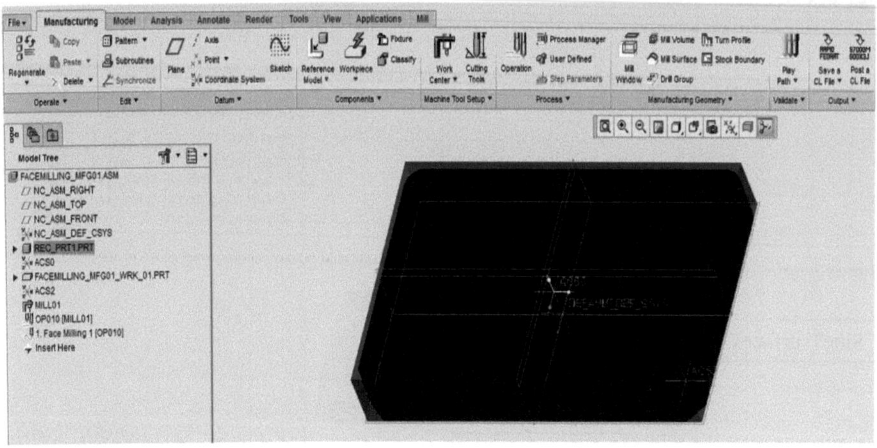

Fig. 3.13 Created stock on part in the main Manufacturing graphic window

3.4 Create Coordinate System (Programme Zero)

Click on the Coordinate System icon tab 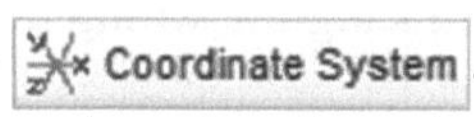

.

The Coordinate System dialogue box is activated on the main graphic window as shown in Fig. 3.14.

Fig. 3.14 Activated Coordinate System dialogue box

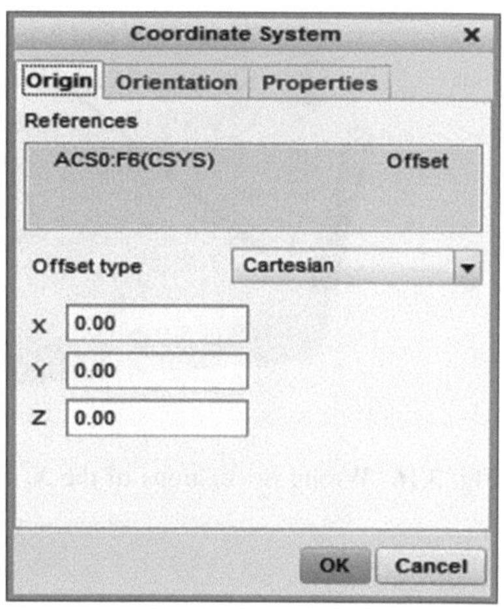

Click on the part top surface and the two side surfaces while holding down the Ctrl key on the key board as shown below. The References section box on the Coordinate System dialogue box is automatically updated by the system. See Fig. 3.15.

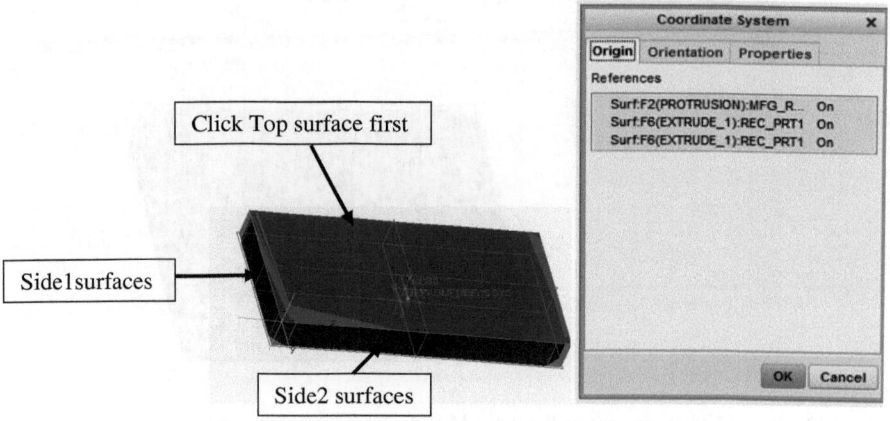

Fig. 3.15 Selection of the part surface used in creating the Coordinate System axes

The created Programme Zero does not have the correct orientation because the *X*, *Y* and *Z*-axes are facing the wrong direction as illustrated in Fig. 3.16.

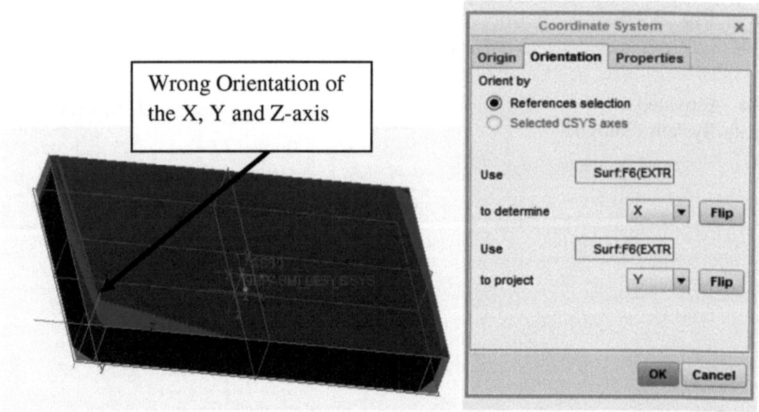

Fig. 3.16 Wrong orientations of the *X*, *Y*, and *Z*-axis of the Coordinate System

Orient X, Y, and Z-axis to the correct position

Click on the Orientation tab on the Coordinate System dialogue box to activate its contents ≫ On the orient by group, click the "References selection" radio button ≫ Click the downward arrow on the "to determine" section box to activate its drop-down menu list, now click the *Z*-axis on the drop-down list ≫ Click on the downward arrow on the "to project" section box, now click the *X* axis on the activated drop-down menu list ≫ Now click on the Flip tab, to flip *X* axis orientation.

The correct orientations for the created Programme Zero, as described above are indicated by the arrows as shown in Fig. 3.17.

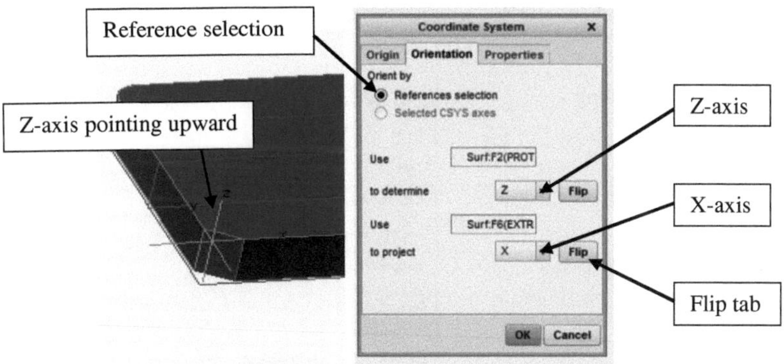

Fig. 3.17 Correct orientations of the *X*, *Y*, and *Z*-axis of the Coordinate System

Click on the OK tab to exit.

A new Coordinate System (Programme Zero) is created, called ACS1.

The new ACS1 created will be renamed by slowly clicking on ACS1 in the Model Tree and then type ACS2. Alternatively, right click on ACS1 to activate its drop-down menu list, now click on Rename on the drop-down list. In this tutorial, ACS1 will be renamed as ACS2 as shown in the Model Tree below.

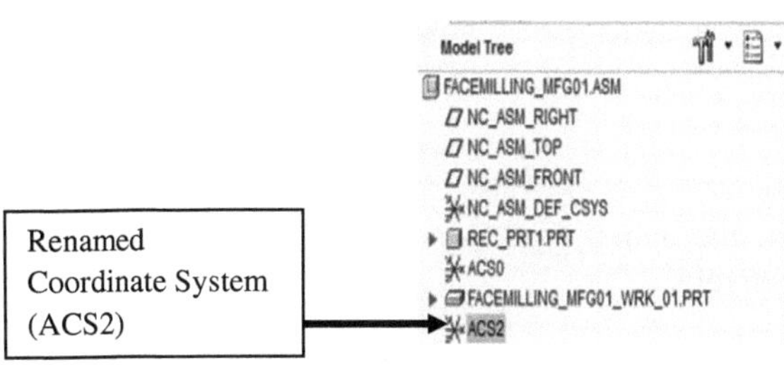

3.5 Setup Work Centre

Click on Work Centre icon to activate its drop-down menu list ≫ Now click on Mill on the drop-down list as shown below.

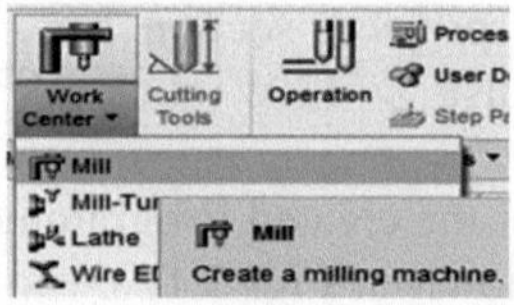

The Milling Work Centre dialogue window is activated on the main graphic window as shown in Fig. 3.18.

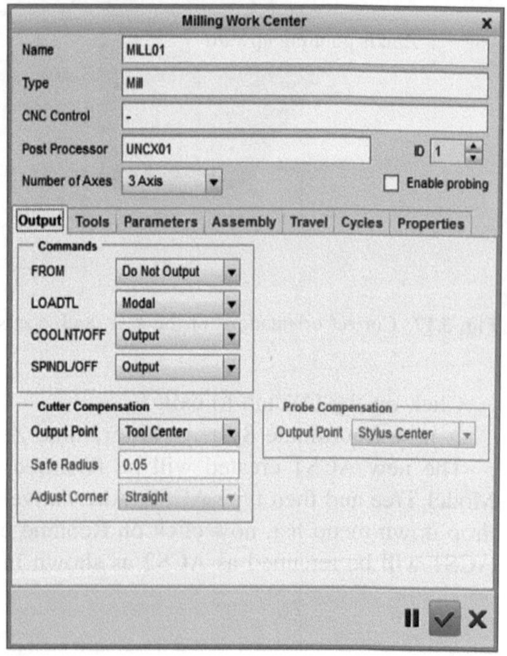

Fig. 3.18 Activated Milling Work Centre dialogue window

> Make note of the name, type, post processor, number of axis, etc, section boxes, respectively.

Click on the Check Mark icon on the Mill Work Centre dialogue window to exit.

3.6 Setup the Operation

Click on the Operation icon as shown below.

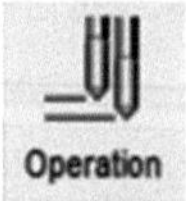

The operation dashboard tools are activated as shown in a concise form in Fig. 3.19.

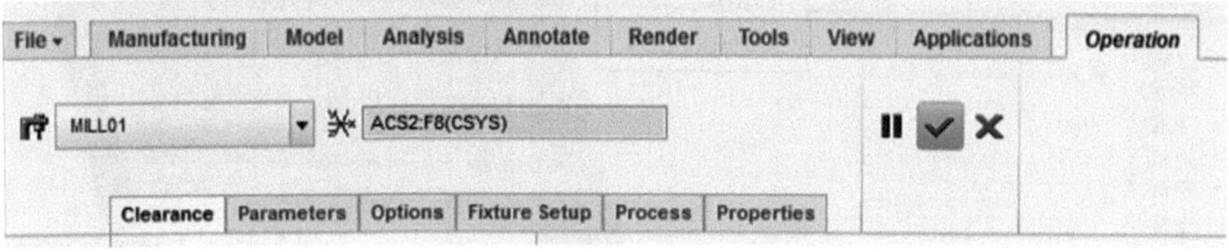

Fig. 3.19 Activated Operation dashboard

To add ACS2, if it is not automatically added by the system >> Click on the Coordinate System icon section box >> Go to the Model Tree and click on the created Coordinate System (ACS2).

Click on the Clearance tab to activate its panel >> On the retract group, click on the Type section box and click on Plane on the activated drop-down list as shown in Fig. 3.20.

Fig. 3.20 The Type drop-down menu list on the Retract group

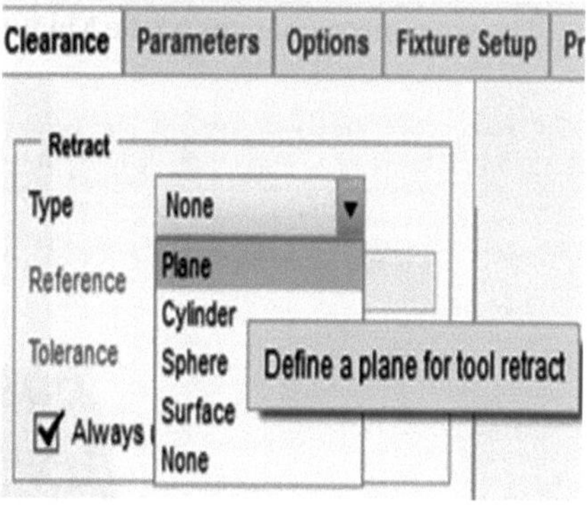

Click on the workpiece top surface and the Reference section box is automatically updated by the system ≫ Type 10 mm in the Value section box, as illustrated by the arrow in Fig. 3.21.

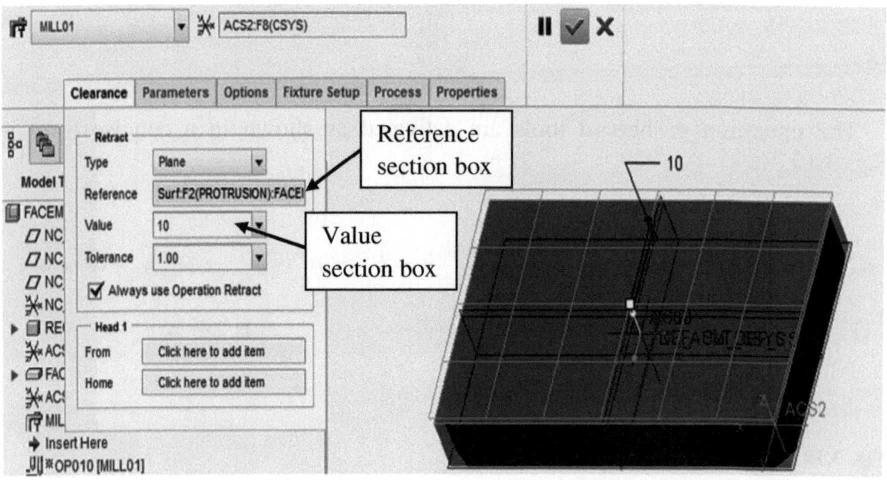

Fig. 3.21 Activated Clearance panel indicating the Type and References updates

Now click on the Check Mark icon ✓ to exit.

Note that the retract surface is the surface that the tool will move up to, after each machining process.

The workpiece changes on the Manufacturing main graphic window. See Fig. 3.22.

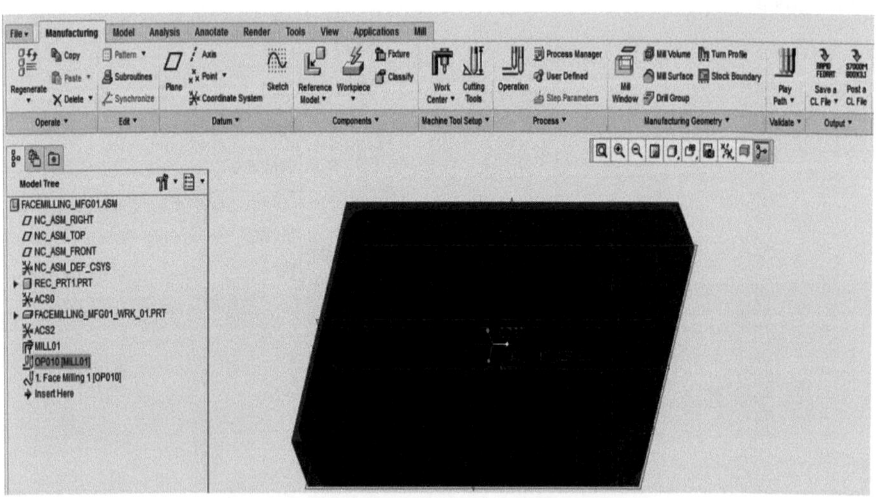

Fig. 3.22 Workpiece after adding the work centre and operation parameters

Check the Model Tree as shown in Fig. 3.23 to make sure that the ACS2, MILL01, and OP010 [MILL01] are in the Model Tree as created.

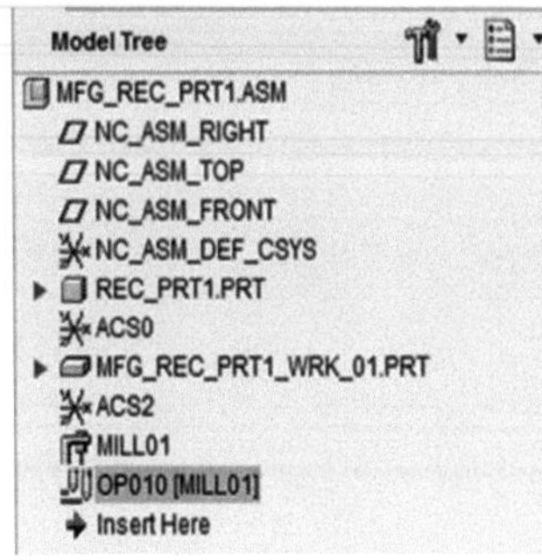

Fig. 3.23 Model Tree after creating Auto-Workpiece, ACS2, work centre, and operation

3.7 Setup the Cutting Tool

Click on the Cutting Tools icon 

.

The Tools Setup dialogue window is activated on the main graphic window ≫ Click on the General tab to activate its content ≫ Type the tool name in the name section box or leave the generated default Tool name ≫ Click on the downward arrow on the Type section box and now click on End Mill Tool on the activated drop-down menu list ≫ Click on the Material section box, and type the tool material. Add dimensions to the tool. See Fig. 3.24.

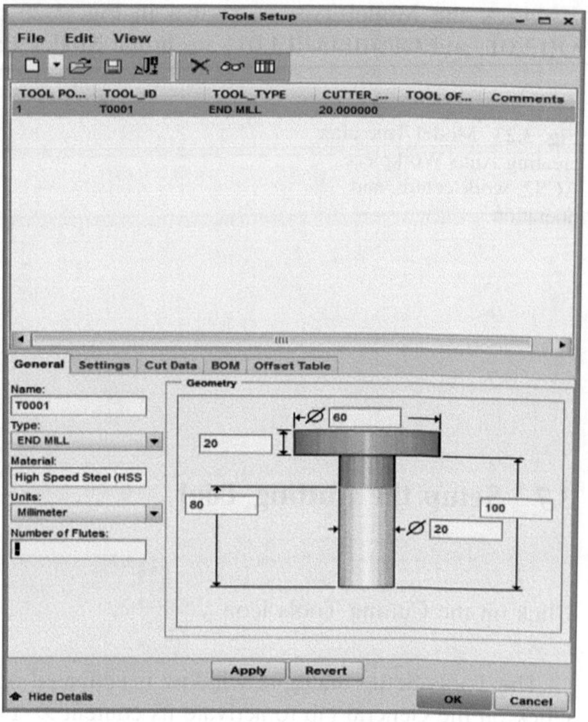

Fig. 3.24 Activated Tools Setup dialogue window with all added parameters on Tool

Click on the Apply tab.
Click on the OK tab to exit the Tools Setup dialogue window.

3.8 Setup the Face Milling Operation

Click on the Mill menu bar to activate the Mill ribbon >> Now click on Face milling icon on the milling group as indicated by the arrow shown below.

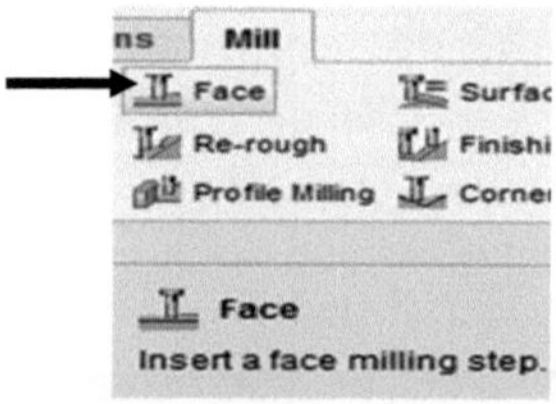

The Face milling dashboard tools are activated. See Fig. 3.25.

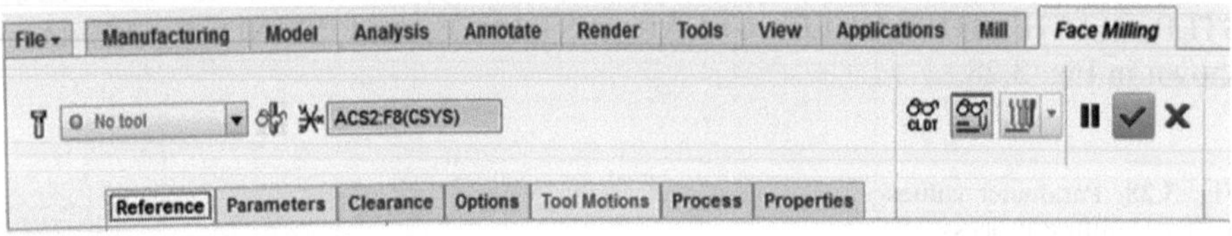

Fig. 3.25 Activated Face Milling dashboard in concise form

To add “01:T0001”, click on the Tool icon section box to activate its drop-down menu list ≫ Now click on “01:T0001” on the activated drop-down menu list.

Click on the Coordinate System section box; now click on the created Coordinate System (ACS2) on the Model Tree, if it is not add automatically by the system.

Click on the Reference tab to activate its panel as shown in Fig. 3.26.

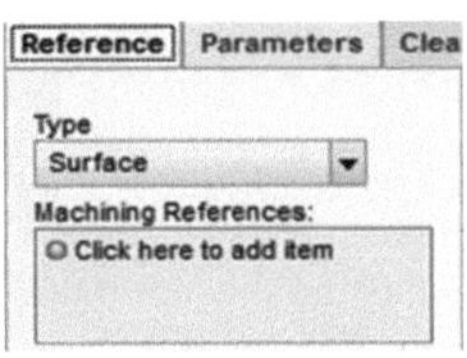

Fig. 3.26 Activated Reference panel

Now click on the “Type” section box; and now click on Surface on the activated drop-down menu list ≫ Click on the workpiece top surface and the Machining References section box is automatically updated by the system.

Note: You can also use the filter selection to select the top surface of the workpiece. Go to the bottom of the main graphic window and click on the downward arrow on the active filter to activate its drop-down menu list. Now, click on reference model surface as highlighted in the drop-down list in Fig. 3.27.

Fig. 3.27 Active Filter selection drop-down list panel

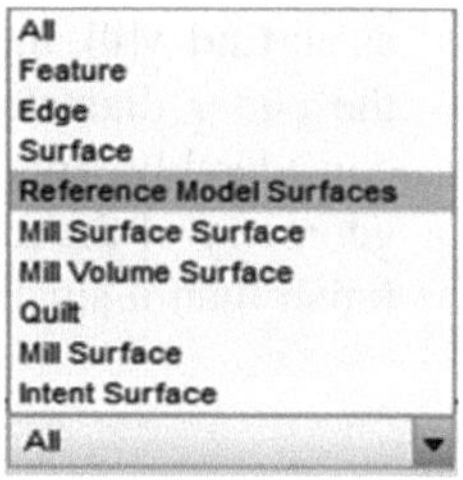

Now click on the top surface of the Workpiece and the Machining References section box is automatically updated/added.

Click on the “Parameters” tab to activate its panel. Type values for the CUT_FEED, STEP_DEPTH, STEP_OVER, CLEAR_DISTANCE, and SPINDLE_SPEED as shown in Fig. 3.28.

Fig. 3.28 Parameter values for face milling process

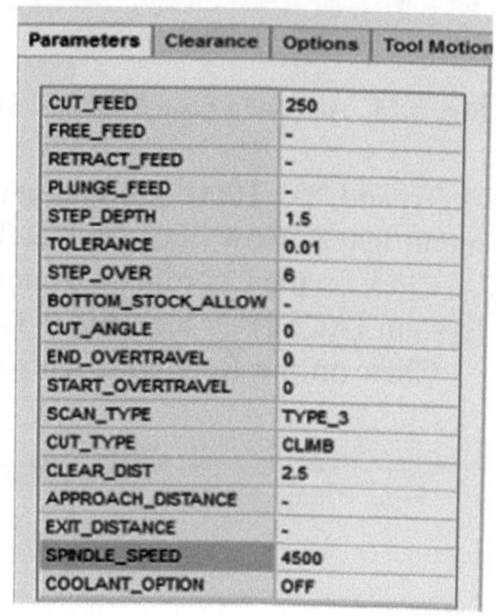

CUT_FEED	250
FREE_FEED	-
RETRACT_FEED	-
PLUNGE_FEED	-
STEP_DEPTH	1.5
TOLERANCE	0.01
STEP_OVER	6
BOTTOM_STOCK_ALLOW	-
CUT_ANGLE	0
END_OVERTRAVEL	0
START_OVERTRAVEL	0
SCAN_TYPE	TYPE_3
CUT_TYPE	CLIMB
CLEAR_DIST	2.5
APPROACH_DISTANCE	-
EXIT_DISTANCE	-
SPINDLE_SPEED	4500
COOLANT_OPTION	OFF

Click on the Check Mark icon ✓ to exit.

Note: The CUT_FEED represents the speed of the tool on the Part surface being machined.

The STEP_DEPTH represent the amount of material to be removed in each pass of the tool on the machined part.

The CLEAR_DISTANCE is the distance when the tool starts to follow the feed rate command (G01). All the tool motions before CLEAR_DISTANCE are in rapid mode (G00). The CLEAR_DISTANCE must always be less than the values of the height of retract plane/distance.

The TOLERANCE controls the accuracy to which the cutter path follows the shape of the component.

The STEP_OVER is the distance the tool moves between adjacent tool path tracks. The distance or Step-over value determines whether the surface finish on a Workpiece is rough or smooth. While using a flat bottom tool such as an End Mill, the Step-over value normally ranges between 70 and 80 % of the cutter diameter. When using a Ball nose cutter, the Step-over will be considerably smaller, when roughing and finishing mainly due to the geometry of the tool. A larger Step-over will ultimately give a rough surface finish than a small Step-over.

The SPINDLE_SPEED is the revolution per minute of the spindle/cutting tool. The spindle speed is the biggest determiner of the tool life. Running too fast generates excess heat (there are others ways to generate heat too), which softens the tool and ultimately allows the edge to dull.

3.8.1 *Activate Face Milling Play Path*

Click on the Manufacturing menu bar ≫ Click on the Play Path icon to activate its drop-down menu list, now click on Play Path on the drop-down menu list as indicated by the arrow in Fig. 3.29.

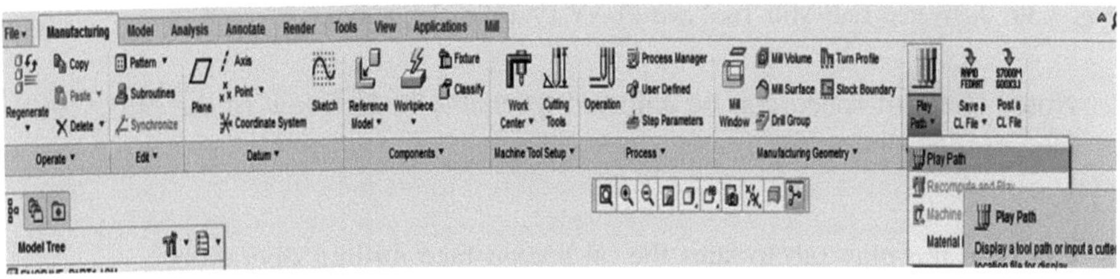

Fig. 3.29 Manufacturing ribbon toolbar with Play Path tab activated

Alternatively, right click the mouse on Face Milling 1[OP010] on the Model Tree indicated by the arrow as shown below.

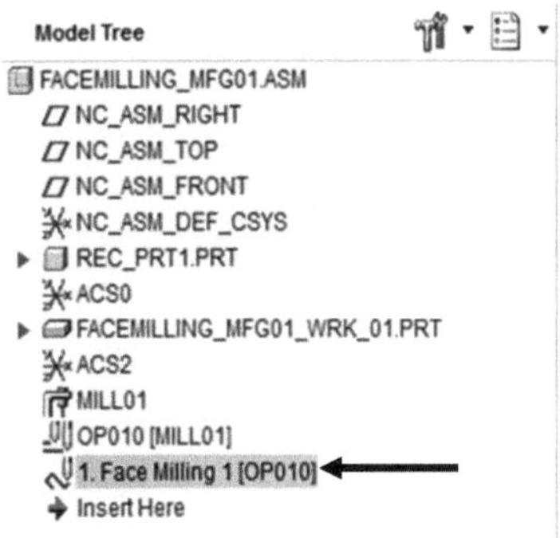

Now click on Play Path on the activated drop-down menu list.

The cutting Tool and the Play Path dialogue box are activated on the Manufacturing main graphic window as shown in Fig. 3.30.

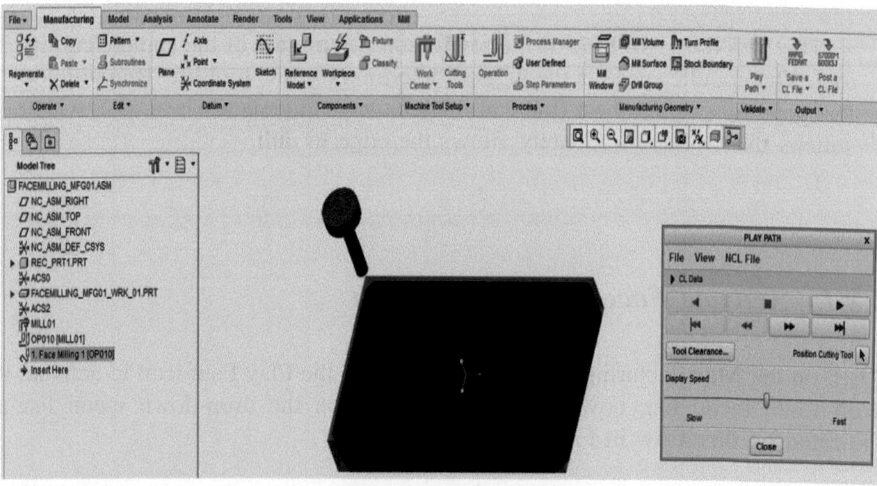

Fig. 3.30 Activated End Mill Tool and PLAY PATH dialogue box

Note: The tool speed can be adjusted to enable better view of the on screen face milling operation.

Click on the play tab to start the on screen face milling operation. The end of the face milling operation is as shown in Fig. 3.31.

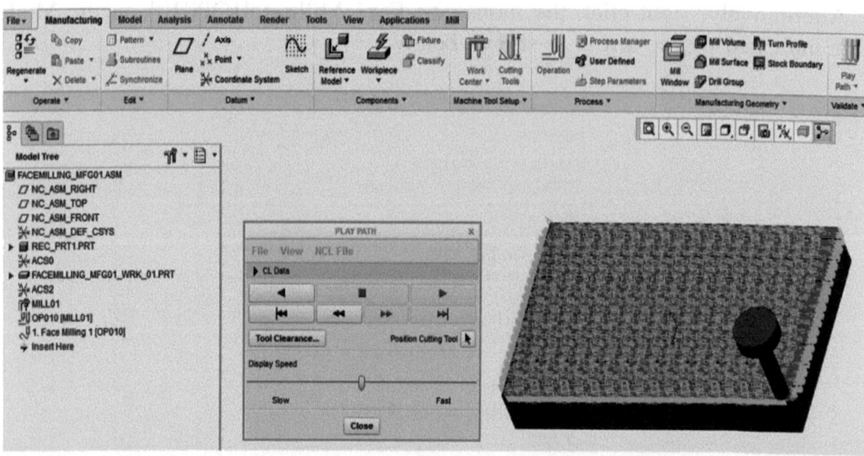

Fig. 3.31 End of on screen Face milling operation Face Milling operation

Click on the Close tab to exit the on screen face milling operation.

Note: To edit the parameters of OP010 [MILL01], face milling 1 [OP010] and MILL01 on the Model Tree, just highlight and right click the mouse on each and then click on Edit Definition on their respective activated drop-down menu list.

3.9 Activate the Material Removal (nccheck_type)

Click on File to activate its drop-down menu list ≫ Now, select Options on the drop-down menu list as indicated by the arrow in Fig. 3.32.

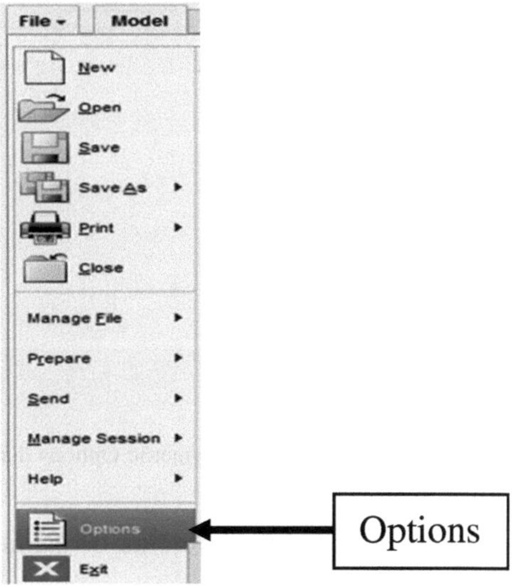

Fig. 3.32 Activated File drop-down menu list

The Creo Parametric Options dialogue window is activated as shown in Fig. 3.33.

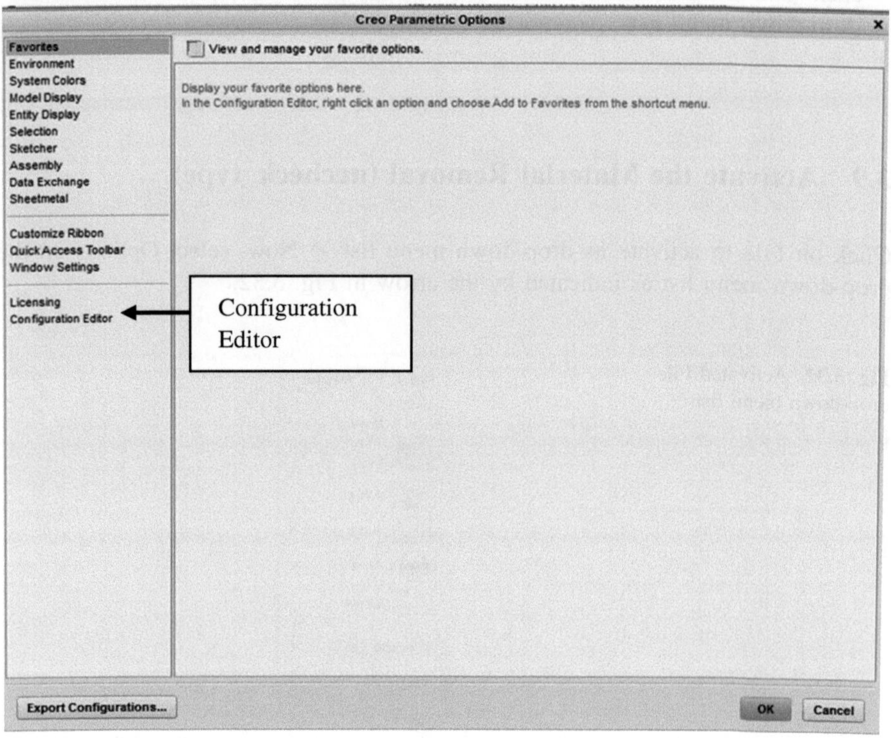

Fig. 3.33 Activated Creo Parametric Options dialogue window

Click on Configuration Editor on the menu list as indicated by the arrow shown in Fig. 3.33.

The Configuration Editor application tools are activated as shown in Fig. 3.34.

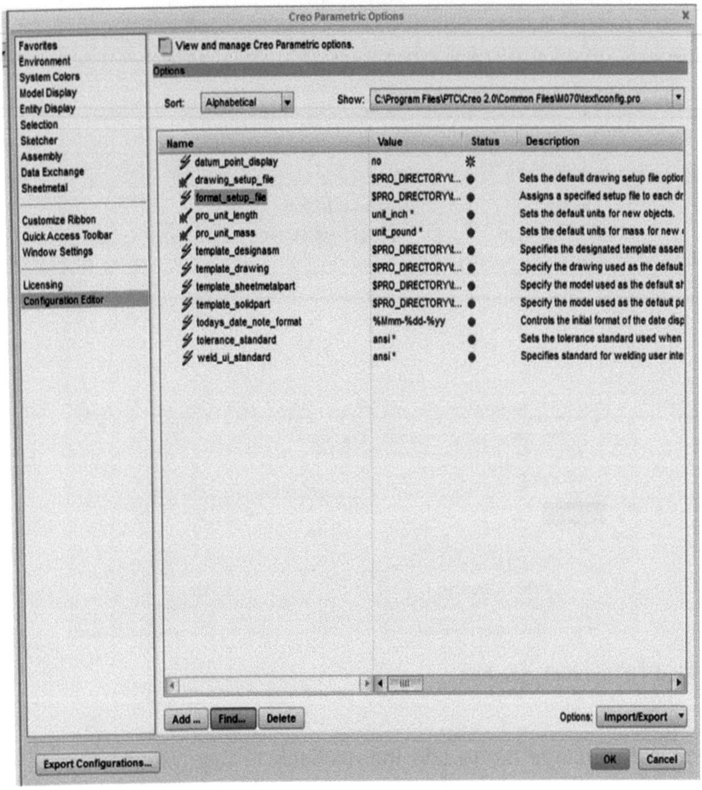

Fig. 3.34 Activated Configuration Editor application tools

Click on the Find tab.

The Find Option dialogue window is activated on the main graphic window ≫ Now, type “nccheck_type” in the Type keyword section box ≫ Click on Find Now tab ≫ Click on the Set value section box, and then click on “nccheck” on the activated drop-down menu list as shown in Fig. 3.35.

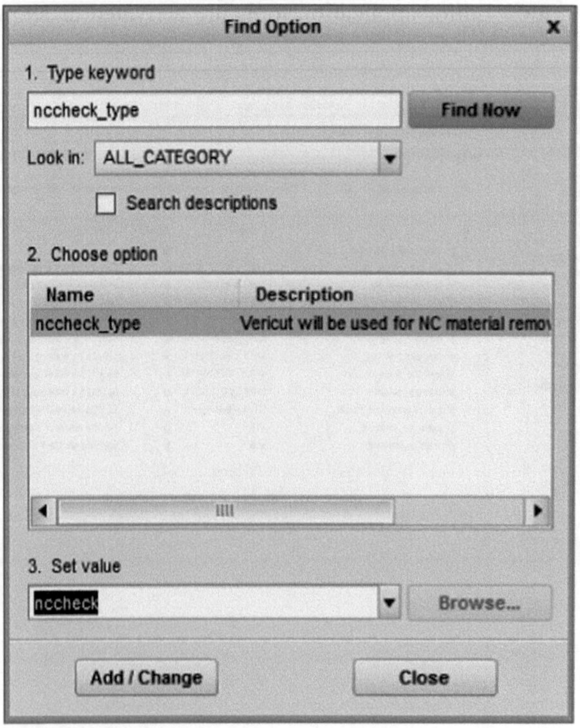

Fig. 3.35 Find Option dialogue box

Click on Add/Change tab to add the nccheck to the system.
Save into the configuration file when prompted to do so by the system.
Click on the Close tab to exit the Find Option dialogue window.

3.10 Material Removal Simulation

To see the material removal simulation after adding the 'nccheck_type' to the configuration file.

Right click on Face Milling1 [OP010] on the Model Tree ≫ Click on Material Removal Simulation on the activated drop-down menu list.

The NC CHECK Menu Manager dialogue box is activated as shown in Fig. 3.36.

Fig. 3.36 Activated NC CHECK Menu Manager dialogue box

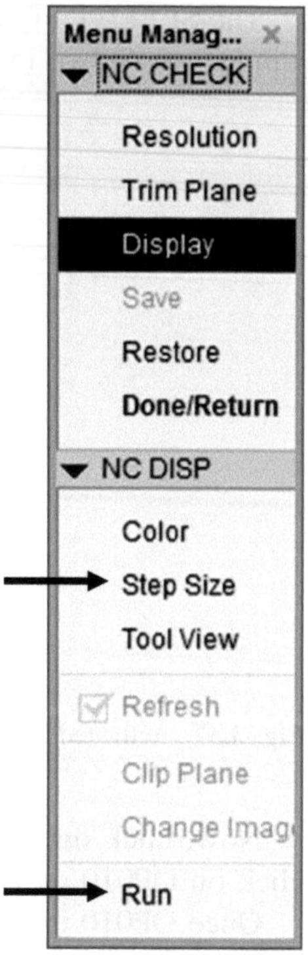

Click on Run to activate the Material Removal Simulation. Now click on the Step Size and change the automatic generated value to a lower value, this will slow down the on screen animation. Click on Done/Return on the NC CHECK group to exit the Material Removal Simulation.

3.11 Create the Cutter Location (CL) Data

Click on the Save a CL File icon on the Manufacturing menu bar ≫ Click on Save a CL File on the activated drop-down menu list as shown below.

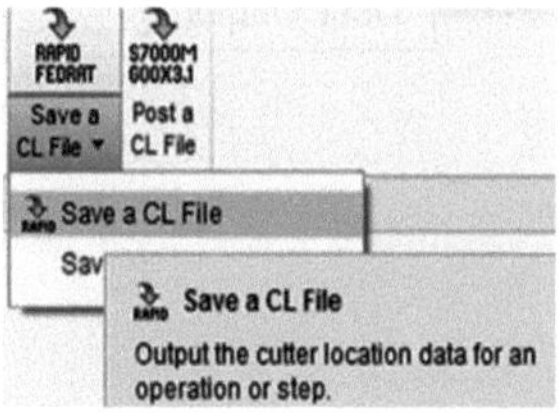

The SELECT FEAT Menu Manager dialogue box opens up on the main graphic window as shown in Fig. 3.37.

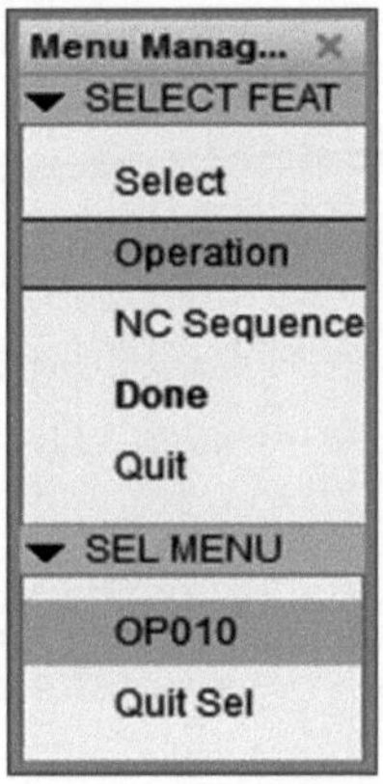

Fig. 3.37 Activated SELECT FEAT Menu Manager dialogue box

Now, click on Operation if it is not automatically selected by the system and click on OP010 on the SEL MENU group as highlighted above.

Once OP010 is clicked on the SEL MENU group, the "PATH" Menu Manager dialogue box opens up on the main graphic window.

On the PATH group, click on File and Check Mark the "CL File" and "Interactive" square boxes on the OUTPUT TYPE group as shown in Fig. 3.38, if they are not automatically selected by the system.

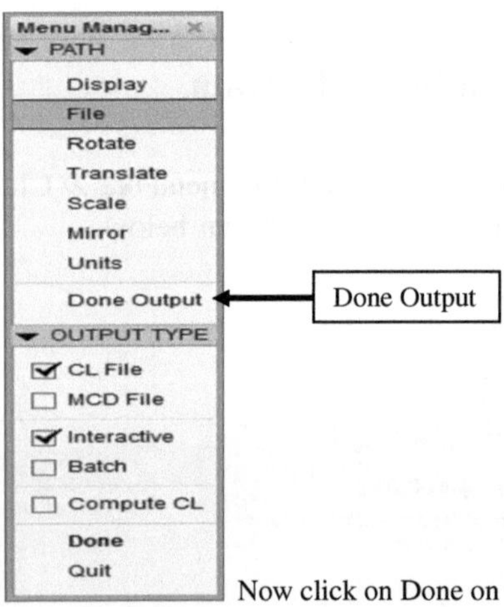

Now click on Done on the OUTPUT TYPE group.

Fig. 3.38 PATH Menu Manager dialogue box

The Save a Copy dialogue window is activated on the main graphic window. See, Fig. 3.39.

Fig. 3.39 Activated Save a Copy dialogue window in concise form

Make sure the File Name and Type are correct.

Click on the OK tab to exit.

The CL File (Data) is now saved to the correct chosen directory.

Click on Done Output on the "PATH" group on the Menu Manager dialogue box as indicated by the arrow in Fig. 3.38.

3.12 Creating the G-Code Data

Click on the Post a CL File icon tab 

.

The Open dialogue window pops up on the main graphic window. See Fig. 3.40.

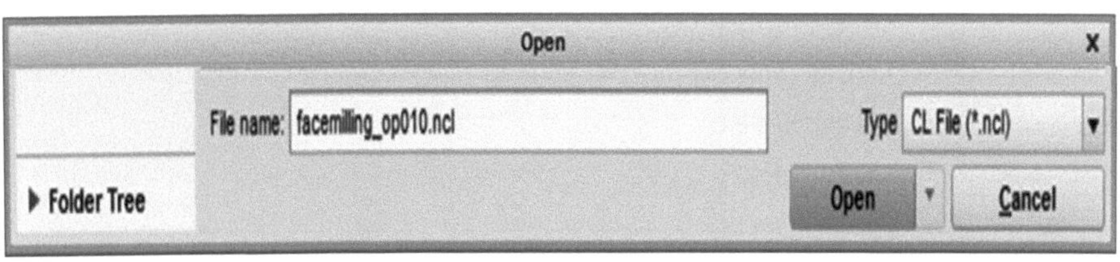

Fig. 3.40 Activated Open dialogue window in a concise form

Click on the saved 

file on the Open dialogue window. Make sure that the correct file name and extension (.ncl) is in the "File name" section box.

Now click on the Open tab to exit.

The PP OPTIONS Menu Manager dialogue box is activated on the main graphic window » Check Mark "Verbose" and "Trace" if they are not automatically Checked Marked by the system as shown in Fig. 3.41.

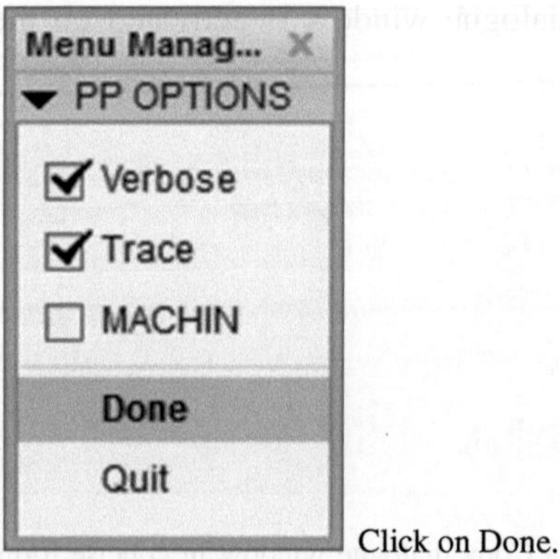

Fig. 3.41 Activated PP Options Menu Manager dialogue box with Verbose and Trace Checked Marked

The system automatically generates the PP (Post Processor) LIST as shown (Fig. 3.42).

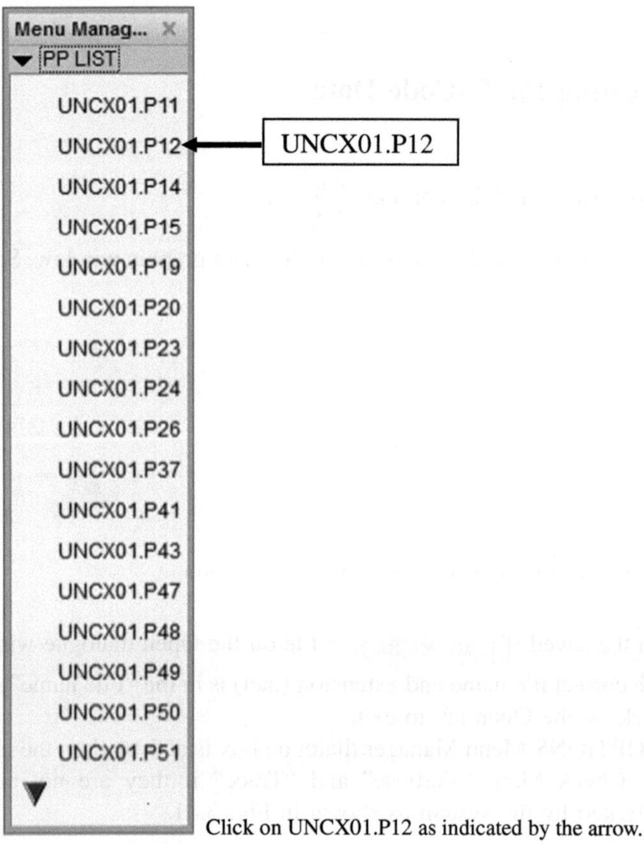

Fig. 3.42 Activated PP LIST Menu Manager dialogue box

The INFORMATION WINDOW dialogue window opens up on the main graphic window as shown in Fig. 3.43 in a concise form. Check to see if there are any warnings in case the programme has encountered any problem during the generation of G-codes.

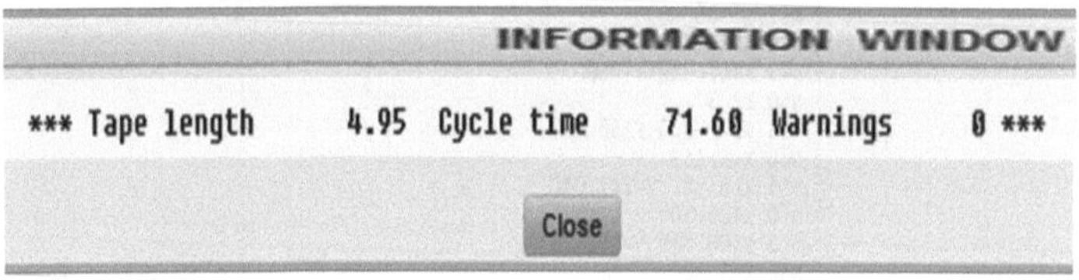

Fig. 3.43 Activated INFORMATION WINDOW in concise form

Click on Close tab on the INFORMATION WINDOW dialogue window to exit.

The generated G-code data is stored as a TAP file in the chosen work directory. The TAP file will be opened using Notepad to view the G-codes.

Some of the G-codes commands generated for the face milling operation are shown in the proceeding pages.

op010.tap - Notepad

File Edit Format View Help

```
N5 G71
N10 ( / FACEMILLING_MFG01)
N15  G0 G17 G99
N20  G90 G94
N25  G0 G49
N30 T1 M06
N35 S4500 M03
N40 G0 G43 Z10. H1
N45 X402.359 Y-250.
N50 Z2.5
N55 G1 Z-1.5 F250.
N60 X-2.359
N65 X-6.507 Y-244.048
N70 X406.507
N75 X408.886 Y-238.095
N80 X-8.886
N85 X-9.921 Y-232.143
N90 X409.921
N95 X410.  Y-226.19
N100 X-10.
N105 Y-220.238
N110 X410.
N115 Y-214.286
N120 X-10.
N125 Y-208.333
N130 X410.
N135 Y-202.381
N140 X-10.
N145 Y-196.429
N150 X410.
N155 Y-190.476
N160 X-10.
N165 Y-184.524
N170 X410.
N175 Y-178.571
N180 X-10.
N185 Y-172.619
N190 X410.
N195 Y-166.667
N200 X-10.
N205 Y-160.714
N210 X410.
N215 Y-154.762
N220 X-10.
N225 Y-148.81
N230 X410.
N235 Y-142.857
N240 X-10.
N245 Y-136.905
N250 X410.
N255 Y-130.952
N260 X-10.
N265 Y-125.
N270 X410.
N275 Y-119.048
N280 X-10.
N285 Y-113.095
N290 X410.
N295 Y-107.143
N300 X-10.
N305 Y-101.19
N310 X410.
N315 Y-95.238
N320 X-10.
N325 Y-89.286
N330 X410.
```

op010.tap - Notepad

File Edit Format View Help

```
N335 Y-83.333
N340 X-10.
N345 Y-77.381
N350 X410.
N355 Y-71.429
N360 X-10.
N365 Y-65.476
N370 X410.
N375 Y-59.524
N380 X-10.
N385 Y-53.571
N390 X410.
N395 Y-47.619
N400 X-10.
N405 Y-41.667
N410 X410.
N415 Y-35.714
N420 X-10.
N425 Y-29.762
N430 X410.
N435 Y-23.81
N440 X-10.
N445 Y-17.857
N450 X410.
N455 X406.047 Y-11.905
N460 X-6.047
N465 X-.095 Y-5.952
N470 X400.095
N475 X394.142 Y0.
N480 X5.858
N485 Z10.
N490 G0 X402.359 Y-250.
N495 Z1.
N500 G1 Z-3. F250.
N505 X-2.359
N510 X-6.507 Y-244.048
N515 X406.507
N520 X408.886 Y-238.095
N525 X-8.886
N530 X-9.921 Y-232.143
N535 X409.921
N540 X410. Y-226.19
N545 X-10.
N550 Y-220.238
N555 X410.
N560 Y-214.286
N565 X-10.
N570 Y-208.333
N575 X410.
N580 Y-202.381
N585 X-10.
N590 Y-196.429
N595 X410.
N600 Y-190.476
N605 X-10.
N610 Y-184.524
N615 X410.
N620 Y-178.571
N625 X-10.
N630 Y-172.619
N635 X410.
N640 Y-166.667
N645 X-10.
N650 Y-160.714
N655 X410.
N660 Y-154.762
```

op010.tap - Notepad

File Edit Format View Help

```
N665 X-10.
N670 Y-148.81
N675 X410.
N680 Y-142.857
N685 X-10.
N690 Y-136.905
N695 X410.
N700 Y-130.952
N705 X-10.
N710 Y-125.
N715 X410.
N720 Y-119.048
N725 X-10.
N730 Y-113.095
N735 X410.
N740 Y-107.143
N745 X-10.
N750 Y-101.19
N755 X410.
N760 Y-95.238
N765 X-10.
N770 Y-89.286
N775 X410.
N780 Y-83.333
N785 X-10.
N790 Y-77.381
N795 X410.
N800 Y-71.429
N805 X-10.
N810 Y-65.476
N815 X410.
N820 Y-59.524
N825 X-10.
N830 Y-53.571
N835 X410.
N840 Y-47.619
N845 X-10.
N850 Y-41.667
N855 X410.
N860 Y-35.714
N865 X-10.
N870 Y-29.762
N875 X410.
N880 Y-23.81
N885 X-10.
N890 Y-17.857
N895 X410.
N900 X406.047 Y-11.905
N905 X-6.047
N910 X-.095 Y-5.952
N915 X400.095
N920 X394.142 Y0.
N925 X5.858
N930 Z10.
N935 G0 X402.359 Y-250.
N940 Z-.5
N945 G1 Z-4.5 F250.
N950 X-2.359
N955 X-6.507 Y-244.048
N960 X406.507
N965 X408.886 Y-238.095
N970 X-8.886
N975 X-9.921 Y-232.143
N980 X409.921
N985 X410. Y-226.19
N990 X-10.
```

op010.tap - Notepad

File Edit Format View Help

```
N995 Y-220.238
N1000 X410.
N1005 Y-214.286
N1010 X-10.
N1015 Y-208.333
N1020 X410.
N1025 Y-202.381
N1030 X-10.
N1035 Y-196.429
N1040 X410.
N1045 Y-190.476
N1050 X-10.
N1055 Y-184.524
N1060 X410.
N1065 Y-178.571
N1070 X-10.
N1075 Y-172.619
N1080 X410.
N1085 Y-166.667
N1090 X-10.
N1095 Y-160.714
N1100 X410.
N1105 Y-154.762
N1110 X-10.
N1115 Y-148.81
N1120 X410.
N1125 Y-142.857
N1130 X-10.
N1135 Y-136.905
N1140 X410.
N1145 Y-130.952
N1150 X-10.
N1155 Y-125.
N1160 X410.
N1165 Y-119.048
N1170 X-10.
N1175 Y-113.095
N1180 X410.
N1185 Y-107.143
N1190 X-10.
N1195 Y-101.19
N1200 X410.
N1205 Y-95.238
N1210 X-10.
N1215 Y-89.286
N1220 X410.
N1225 Y-83.333
N1230 X-10.
N1235 Y-77.381
N1240 X410.
N1245 Y-71.429
N1250 X-10.
N1255 Y-65.476
N1260 X410.
N1265 Y-59.524
N1270 X-10.
N1275 Y-53.571
N1280 X410.
N1285 Y-47.619
N1290 X-10.
N1295 Y-41.667
N1300 X410.
N1305 Y-35.714
N1310 X-10.
N1315 Y-29.762
N1320 X410.
```

op010.tap - Notepad

File Edit Format View Help

```
N1325 Y-23.81
N1330 X-10.
N1335 Y-17.857
N1340 X410.
N1345 X406.047 Y-11.905
N1350 X-6.047
N1355 X-.095 Y-5.952
N1360 X400.095
N1365 X394.142 Y0.
N1370 X5.858
N1375 Z10.
N1380 G0 X402.359 Y-250.
N1385 Z-2.
N1390 G1 Z-6. F250.
N1395 X-2.359
N1400 X-6.507 Y-244.048
N1405 X406.507
N1410 X408.886 Y-238.095
N1415 X-8.886
N1420 X-9.921 Y-232.143
N1425 X409.921
N1430 X410. Y-226.19
N1435 X-10.
N1440 Y-220.238
N1445 X410.
N1450 Y-214.286
N1455 X-10.
N1460 Y-208.333
N1465 X410.
N1470 Y-202.381
N1475 X-10.
N1480 Y-196.429
N1485 X410.
N1490 Y-190.476
N1495 X-10.
N1500 Y-184.524
N1505 X410.
N1510 Y-178.571
N1515 X-10.
N1520 Y-172.619
N1525 X410.
N1530 Y-166.667
N1535 X-10.
N1540 Y-160.714
N1545 X410.
N1550 Y-154.762
N1555 X-10.
N1560 Y-148.81
N1565 X410.
N1570 Y-142.857
N1575 X-10.
N1580 Y-136.905
N1585 X410.
N1590 Y-130.952
N1595 X-10.
N1600 Y-125.
N1605 X410.
N1610 Y-119.048
N1615 X-10.
N1620 Y-113.095
N1625 X410.
N1630 Y-107.143
N1635 X-10.
N1640 Y-101.19
N1645 X410.
N1650 Y-95.238
N1655 X-10.
N1660 Y-89.286
```

op010 - Notepad

File Edit Format View Help

```
N1665 X410.
N1670 Y-83.333
N1675 X-10.
N1680 Y-77.381
N1685 X410.
N1690 Y-71.429
N1695 X-10.
N1700 Y-65.476
N1705 X410.
N1710 Y-59.524
N1715 X-10.
N1720 Y-53.571
N1725 X410.
N1730 Y-47.619
N1735 X-10.
N1740 Y-41.667
N1745 X410.
N1750 Y-35.714
N1755 X-10.
N1760 Y-29.762
N1765 X410.
N1770 Y-23.81
N1775 X-10.
N1780 Y-17.857
N1785 X410.
N1790 X406.047 Y-11.905
N1795 X-6.047
N1800 X-.095 Y-5.952
N1805 X400.095
N1810 X394.142 Y0.
N1815 X5.858
N1820 Z10.
N1825 G0 X402.359 Y-250.
N1830 Z-3.5
N1835 G1 Z-7.5 F250.
N1840 X-2.359
N1845 X-6.507 Y-244.048
N1850 X406.507
N1855 X408.886 Y-238.095
N1860 X-8.886
N1865 X-9.921 Y-232.143
N1870 X409.921
N1875 X410. Y-226.19
N1880 X-10.
N1885 Y-220.238
N1890 X410.
N1895 Y-214.286
N1900 X-10.
N1905 Y-208.333
N1910 X410.
N1915 Y-202.381
N1920 X-10.
N1925 Y-196.429
N1930 X410.
N1935 Y-190.476
N1940 X-10.
N1945 Y-184.524
N1950 X410.
N1955 Y-178.571
N1960 X-10.
N1965 Y-172.619
N1970 X410.
N1975 Y-166.667
N1980 X-10.
N1985 Y-160.714
N1990 X410.
N1995 Y-154.762
N2000 X-10.
N2005 Y-148.81
N2010 X410.
N2015 Y-142.857
N2020 X-10.
N2025 Y-136.905
N2030 X410.
N2035 Y-130.952
N2040 X-10.
N2045 Y-125.
N2050 X410.
N2055 Y-119.048
N2060 X-10.
N2065 Y-113.095
N2070 X410.
N2075 Y-107.143
N2080 X-10.
N2085 Y-101.19
N2090 X410.
N2095 Y-95.238
N2100 X-10.
N2105 Y-89.286
N2110 X410.
N2115 Y-83.333
N2120 X-10.
N2125 Y-77.381
N2130 X410.
N2135 Y-71.429
N2140 X-10.
N2145 Y-65.476
N2150 X410.
```

op010 - Notepad

File Edit Format View Help

```
N2155 Y-59.524
N2160 X-10.
N2165 Y-53.571
N2170 X410.
N2175 Y-47.619
N2180 X-10.
N2185 Y-41.667
N2190 X410.
N2195 Y-35.714
N2200 X-10.
N2205 Y-29.762
N2210 X410.
N2215 Y-23.81
N2220 X-10.
N2225 Y-17.857
N2230 X410.
N2235 X406.047 Y-11.905
N2240 X-6.047
N2245 X-.095 Y-5.952
N2250 X400.095
N2255 X394.142 Y0.
N2260 X5.858
N2265 Z10.
N2270 G0 X402.359 Y-250.
N2275 Z-5.
N2280 G1 Z-9. F250.
N2285 X-2.359
N2290 X-6.507 Y-244.048
N2295 X406.507
N2300 X408.886 Y-238.095
N2305 X-8.886
N2310 X-9.921 Y-232.143
N2315 X409.921
N2320 X410. Y-226.19
N2325 X-10.
N2330 Y-220.238
N2335 X410.
N2340 Y-214.286
N2345 X-10.
N2350 Y-208.333
N2355 X410.
N2360 Y-202.381
N2365 X-10.
N2370 Y-196.429
N2375 X410.
N2380 Y-190.476
N2385 X-10.
N2390 Y-184.524
N2395 X410.
N2400 Y-178.571
N2405 X-10.
N2410 Y-172.619
N2415 X410.
N2420 Y-166.667
N2425 X-10.
N2430 Y-160.714
N2435 X410.
N2440 Y-154.762
N2445 X-10.
N2450 Y-148.81
N2455 X410.
N2460 Y-142.857
N2465 X-10.
N2470 Y-136.905
N2475 X410.
N2480 Y-130.952
N2485 X-10.
N2490 Y-125.
N2495 X410.
N2500 Y-119.048
N2505 X-10.
N2510 Y-113.095
N2515 X410.
N2520 Y-107.143
N2525 X-10.
N2530 Y-101.19
N2535 X410.
N2540 Y-95.238
N2545 X-10.
N2550 Y-89.286
N2555 X410.
N2560 Y-83.333
N2565 X-10.
N2570 Y-77.381
N2575 X410.
N2580 Y-71.429
N2585 X-10.
N2590 Y-65.476
N2595 X410.
N2600 Y-59.524
N2605 X-10.
N2610 Y-53.571
N2615 X410.
N2620 Y-47.619
N2625 X-10.
N2630 Y-41.667
N2635 X410.
N2640 Y-35.714
```

op010 - Notepad

File Edit Format View Help

```
N2645 X-10.
N2650 Y-29.762
N2655 X410.
N2660 Y-23.81
N2665 X-10.
N2670 Y-17.857
N2675 X410.
N2680 X406.047 Y-11.905
N2685 X-6.047
N2690 X-.095 Y-5.952
N2695 X400.095
N2700 X394.142 Y0.
N2705 X5.858
N2710 Z10.
N2715 G0 X402.359 Y-250.
N2720 Z-6.5
N2725 G1 Z-10. F250.
N2730 X-2.359
N2735 X-6.507 Y-244.048
N2740 X406.507
N2745 X408.886 Y-238.095
N2750 X-8.886
N2755 X-9.921 Y-232.143
N2760 X409.921
N2765 X410. Y-226.19
N2770 X-10.
N2775 Y-220.238
N2780 X410.
N2785 Y-214.286
N2790 X-10.
N2795 Y-208.333
N2800 X410.
N2805 Y-202.381
N2810 X-10.
N2815 Y-196.429
N2820 X410.
N2825 Y-190.476
N2830 X-10.
N2835 Y-184.524
N2840 X410.
N2845 Y-178.571
N2850 X-10.
N2855 Y-172.619
N2860 X410.
N2865 Y-166.667
N2870 X-10.
N2875 Y-160.714
N2880 X410.
N2885 Y-154.762
N2900 X410.
N2905 Y-142.857
N2910 X-10.
N2915 Y-136.905
N2920 X410.
N2925 Y-130.952
N2930 X-10.
N2935 Y-125.
N2940 X410.
N2945 Y-119.048
N2950 X-10.
N2955 Y-113.095
N2960 X410.
N2965 Y-107.143
N2970 X-10.
N2975 Y-101.19
N2980 X410.
N2985 Y-95.238
N2990 X-10.
N2995 Y-89.286
N3000 X410.
N3005 Y-83.333
N3010 X-10.
N3015 Y-77.381
N3020 X410.
N3025 Y-71.429
N3030 X-10.
N3035 Y-65.476
N3040 X410.
N3045 Y-59.524
N3050 X-10.
N3055 Y-53.571
N3060 X410.
N3065 Y-47.619
N3070 X-10.
N3075 Y-41.667
N3080 X410.
N3085 Y-35.714
N3090 X-10.
N3095 Y-29.762
N3100 X410.
N3105 Y-23.81
N3110 X-10.
N3115 Y-17.857
N3120 X410.
N3125 X406.047 Y-11.905
N3130 X-6.047
N3135 X-.095 Y-5.952
N3140 X400.095
N3145 X394.142 Y0.
N3150 X5.858
N3155 Z10.
N3160 M30
%
```

Chapter 4
Volume Rough Milling Operation

4.1 Volume Rough Milling Operation Using Mill Volume

The Volume Rough milling operation is used when excess materials have to be removed from the Part/Workpiece surface as will be demonstrated in this Volume Rough milling tutorial.

Volume Rough milling operation is a 2.5 axis Machining process.

Volume Rough milling operation can be achieved by using either the Mill Volume or the Mill Window milling process. The Mill Volume process will be demonstrated in this tutorial.

The steps below will be followed when creating the Volume Rough milling operation.

- Start the Manufacturing Application
- Importing and constraining the saved Part on the Manufacturing GUI window
- Add Stock Material
- Create Programme Zero (Machine Coordinate System)
- Add the Work Centre
- Setup the Operation
- Add Cutting Tool (Create Cutting Tool)
- Create the Pocket Mill Volume on the Workpiece
- Create the Volume Rough milling Operation Sequence (By adding Tools and Parameters of Sequence)
- Create the Material Removal Simulation process

P.O. Kanife, *Computer Aided Virtual Manufacturing Using Creo Parametric*,
DOI 10.1007/978-3-319-23359-8_4

4.2 Activate the Manufacturing Application

Start Creo Parametric either from your computer desktop or from Programme ≫ Click on the New icon as indicated by the arrow below.

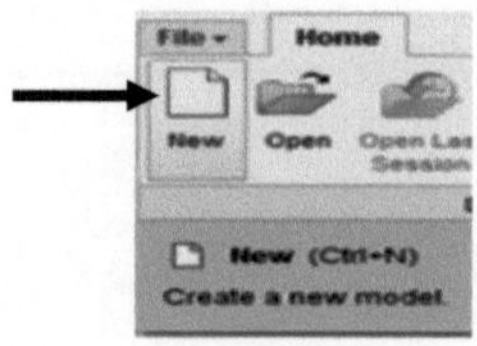

The New dialogue box is activated on main window ≫ Click on the Manufacturing and NC Assembly radio buttons ≫ In the Name section box, type any name. In this tutorial, the "mfg_Vol_Mill" will be typed in the name section box as shown in Fig. 4.1.

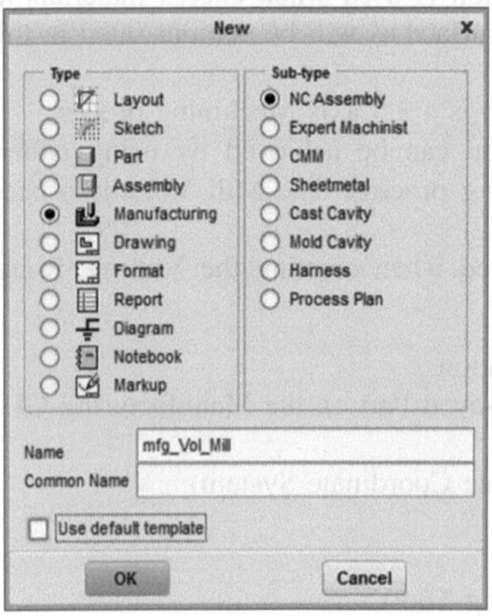

Click on OK tab to proceed.

Fig. 4.1 Manufacturing and NC Assembly radio buttons activated

The New File Options window opens up on the main window ≫ Click on "mmns_mfg_nc" in the Template group ≫ In this tutorial, we type 'Your name' in the MODELLED BY section box and 'Volume Mill_Rough Volume' in the DESCRIPTION section box as shown in Fig. 4.2.

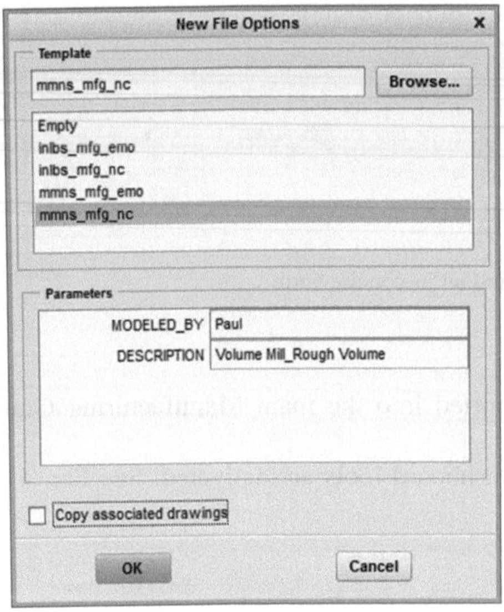

Click on OK tab to exit.

Fig. 4.2 New File Options dialogue box with mmns_mfg_nc highlighted

The Manufacturing application GUI (Graphics User Interface) window is activated.

4.2.1 Import the Saved Part

Importing the 3D Part (vol_mill.prt1)

Click on the Reference Model icon, now click on Assembly Reference Model on the activated drop-down list as shown below.

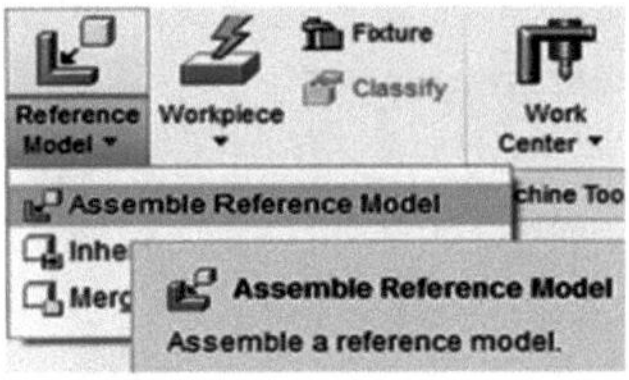

The Open dialogue window is activated on the main graphic window. See in Fig. 4.3. Make sure that the correct Part to be opened is on the File name section box.

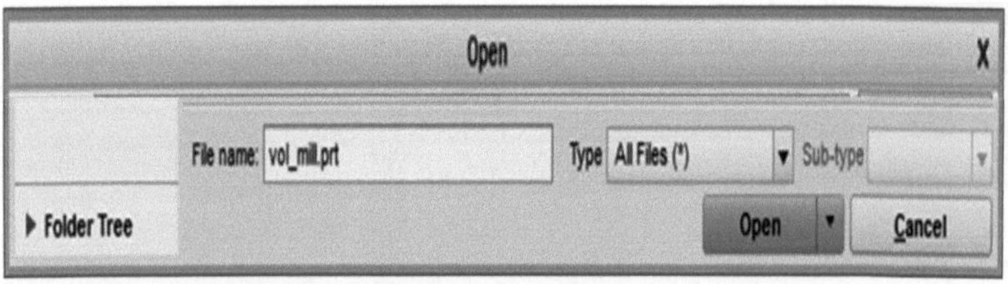

Fig. 4.3 Activated Open dialogue window in concise form

Click on Open tab to exit.

The Reference Model is imported into the main Manufacturing Graphic User Interface (GUI) window.

The Component Placement dashboard tools are activated. See Fig. 4.4.

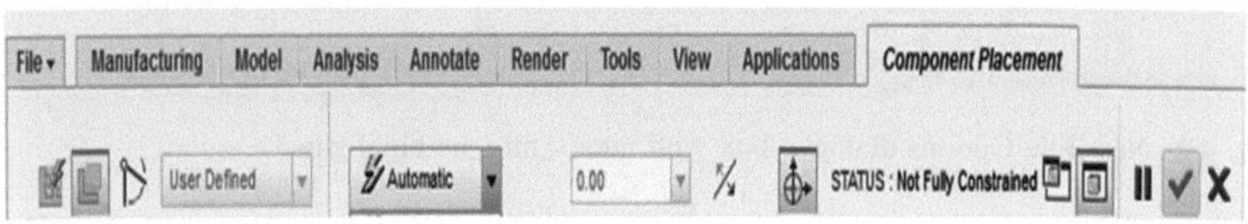

Fig. 4.4 Activated Component Placement dashboard in concise form

4.2.2 *Constraining the Imported Part*

Orient the imported Part with the ASM datum plane, right click on the GUI window and then click on Default Constraint on the drop-down list as indicated by the arrow in Fig. 4.5.

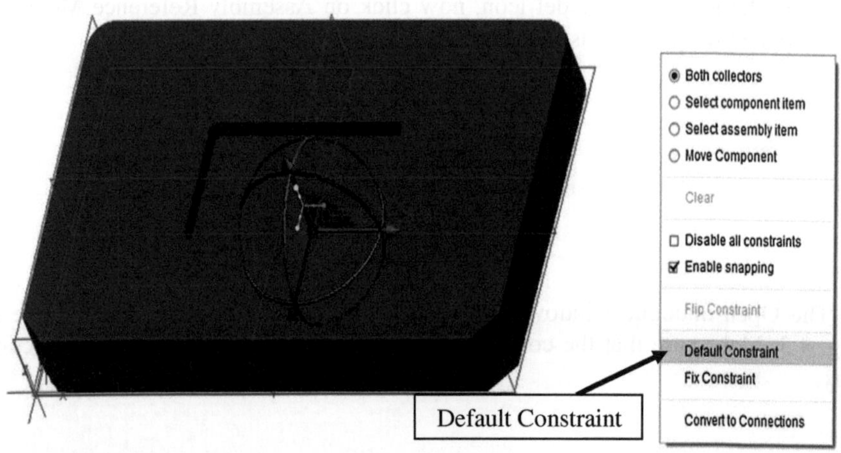

Fig. 4.5 Constraining the imported Part using Method 1

Alternatively, click on the downward arrow in the Automatic section box, now click on Default on the activated drop-down menu list as indicated by the arrow in Fig. 4.6.

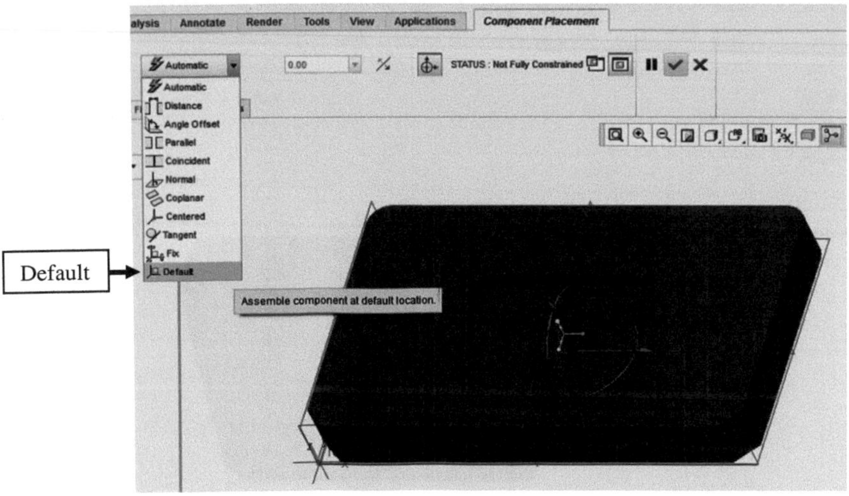

Fig. 4.6 Method 2 of constraining the imported Part

The information on the STATUS bar at the top, should read Fully Constrained as indicated by the arrow in Fig. 4.7.

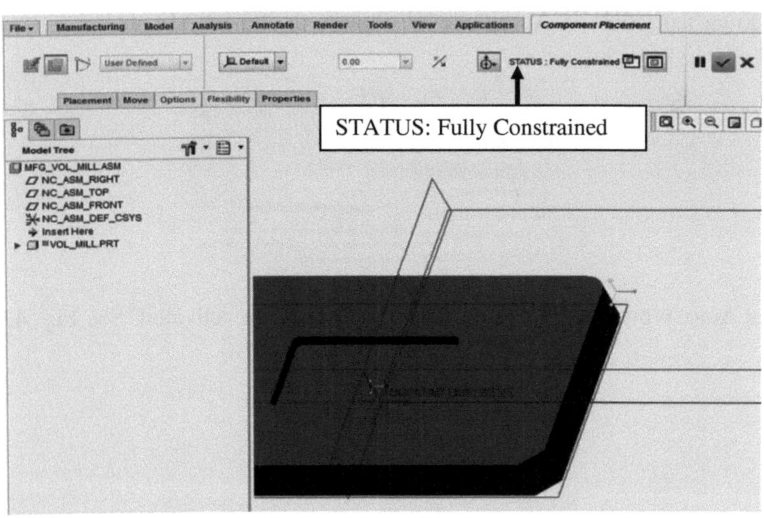

Fig. 4.7 Part is fully constrained on the main graphic window

Click on the Check Mark icon to exit.

After exiting the Component Placement application; the Part appearance changes on the main Manufacturing graphic window as shown in Fig. 4.8.

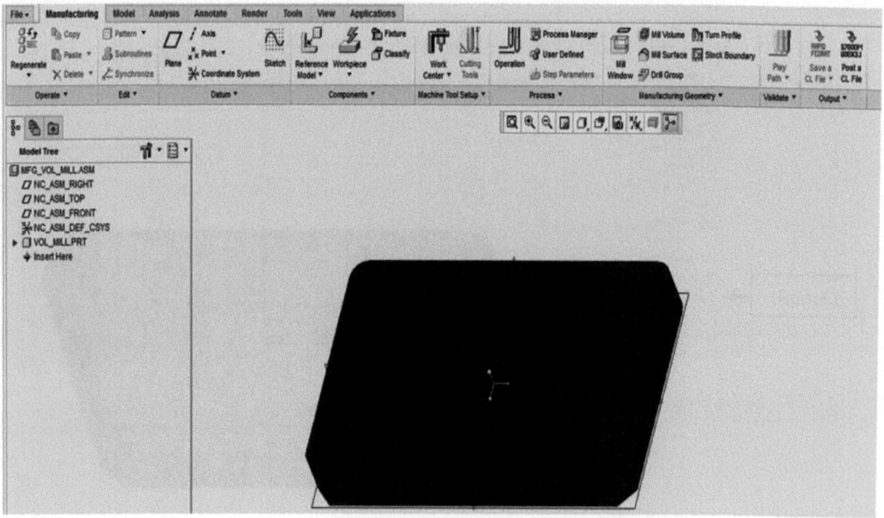

Fig. 4.8 Part appearance after exiting the Component Placement application

4.3 Adding Stock to the Part

Click on the drop-down arrow on the Workpiece icon ≫ On the activated drop-down list, click on Automatic Workpiece as illustrated below.

The Auto Workpiece Creation dashboard tools are activated. See Fig. 4.9.

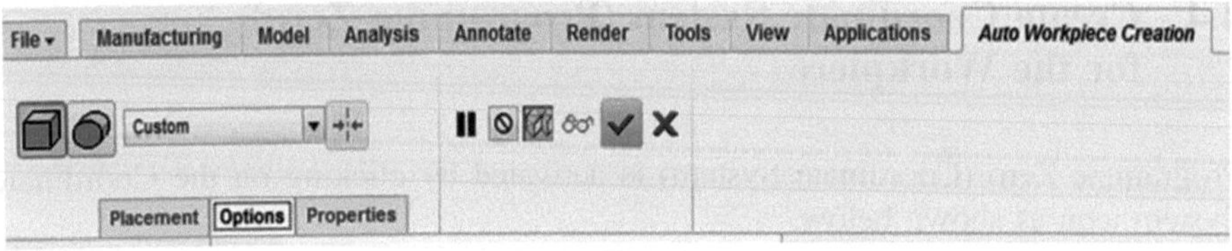

Fig. 4.9 Activated Auto-Workpiece Creation tools in concise form

Click on the Options tab to activate the Options panel. No values are added to the X, Y, and Z section boxes on the Linear Offsets group as shown in Fig. 4.10.

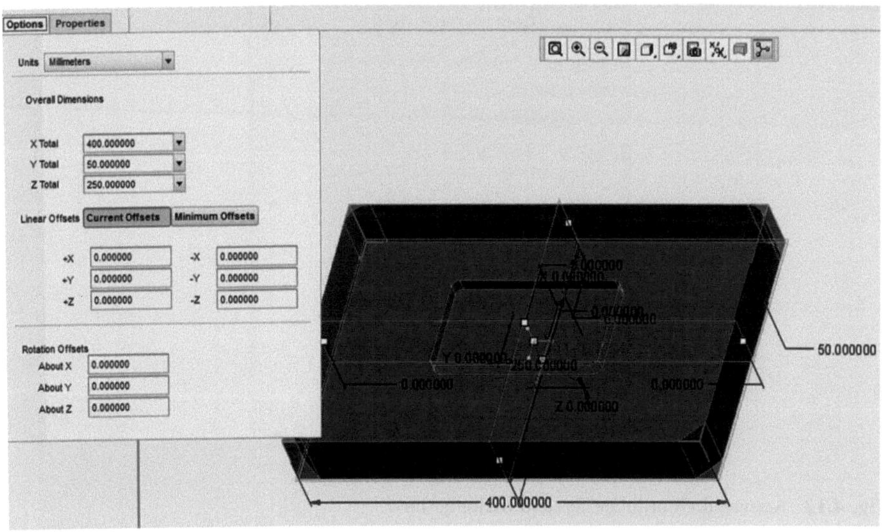

Fig. 4.10 Activated Options panel showing Workpiece dimensions

Click on the Check Mark icon  to exit.
Stock is added to the Part as shown in Fig. 4.11.

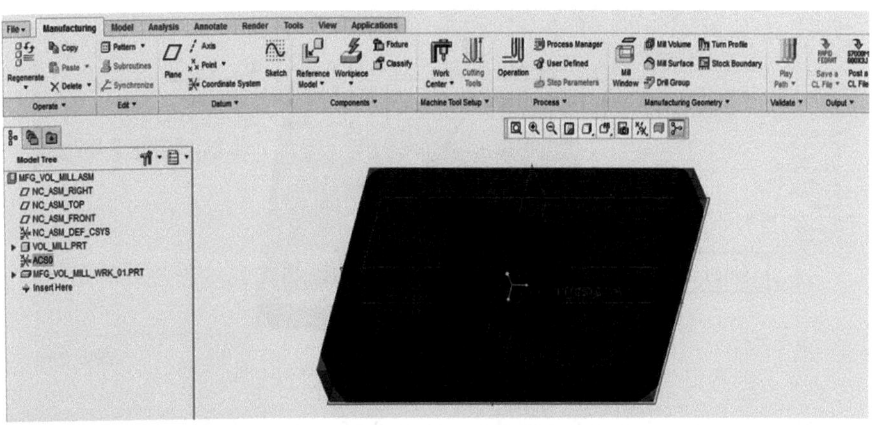

Fig. 4.11 Stock added to imported 3D Part

4.4 Create Coordinate System (Programme Zero) for the Workpiece

Programme Zero (Coordinate System) is activated by clicking on the Coordinate System icon as shown below.

The Coordinate System dialogue box opens up on the main graphic window as shown in Fig. 4.12.

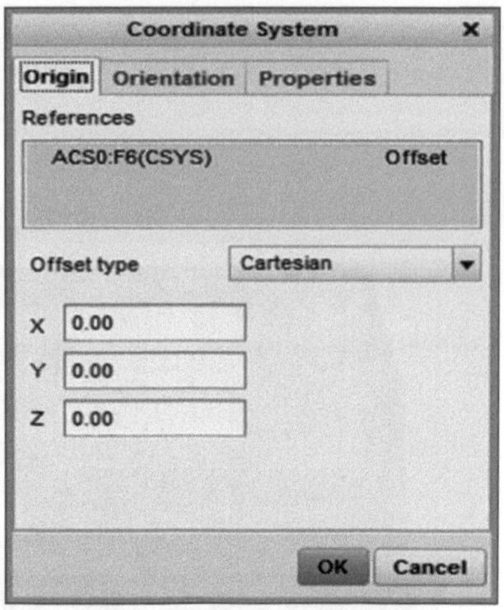

Fig. 4.12 Activated Coordinate System dialogue box

Now, click on the top and two side surfaces of the Workpiece while holding down the Ctrl key when clicking on the surface. The selected top and side surfaces will be added automatically onto the References section box by the system as illustrated in Fig. 4.13.

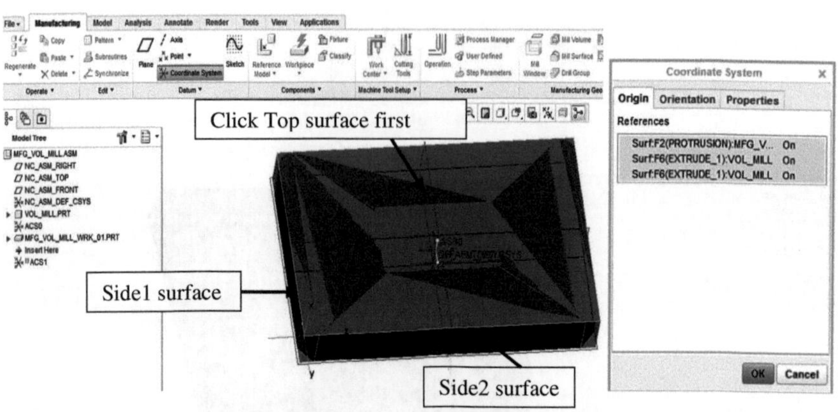

Fig. 4.13 Selecting the Part surface used in creating the Coordinate System

The created Programme Zero is not in the correct orientation because the X, Y, and Z axes are facing the wrong directions as indicated by the arrows in Fig. 4.14.

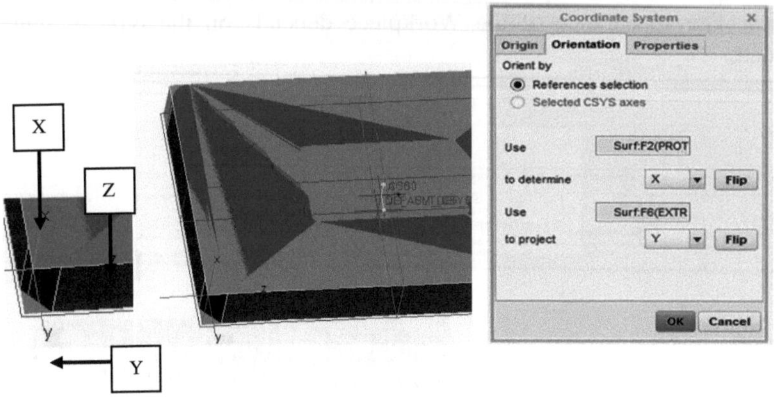

Fig. 4.14 Wrong orientations of the *X*, *Y*, and *Z* axis of the Coordinate System

To orient the X, Y, and Z coordinate axes to the correct orientation

Click on the Orientation tab to activate its contents ≫ On the "Orient by" group, click on the radio button on the "References selection" ≫ Click on the downward pointing arrow on the "to determine" section box and click on the Z axis on the activated drop-down menu list ≫ Click on the downward pointing arrow on the "to project" section box and click on the X axis on the activated drop-down list. Now click on the Flip tab, to flip X axis orientation. The correct orientations for the created Programme Zero axes are indicated by the arrows in Fig. 4.15.

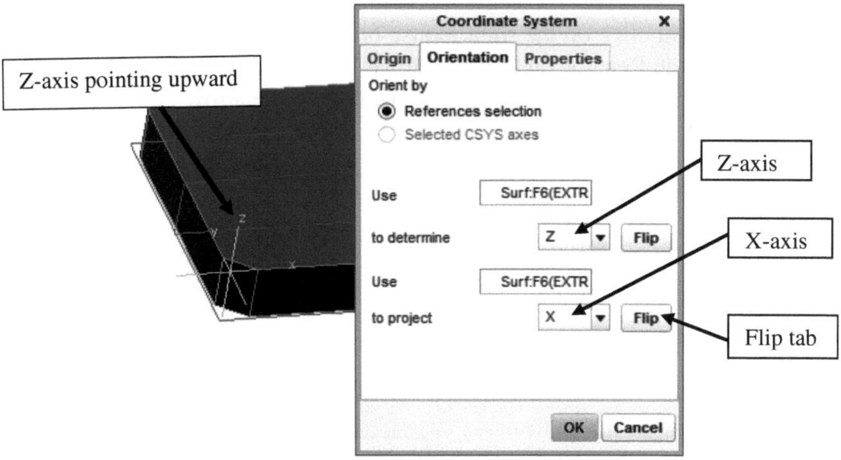

Fig. 4.15 Correct orientations of the *X*, *Y*, and *Z* axis of the Coordinate System

Click on the OK tab to exit. A new Coordinate System (ACS1) is created.

On the Manufacturing main graphic window, the Workpiece changes as shown in Fig. 4.16.

Note: The appearance of the Workpiece depends on the type of material added.

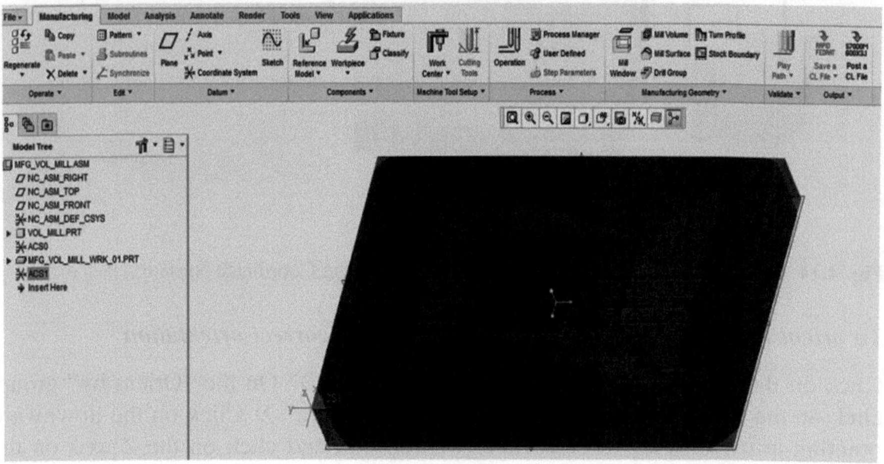

Fig. 4.16 Part on the main window after adding Stock and Coordinate System

Rename the created Coordinate System

To rename ACS1, slowly click on the ACS1 on the Model Tree and then type ACS3. Alternatively, right click on the right mouse button on ACS1 to activate its options menu, now click on Rename on the drop-down list. Now, rename ACS1 to ACS3 by typing it. ACS1 is renamed to ACS3 as indicated by the arrow in Fig. 4.17.

Fig. 4.17 Model Tree indicating the renamed ACS1 to ACS3

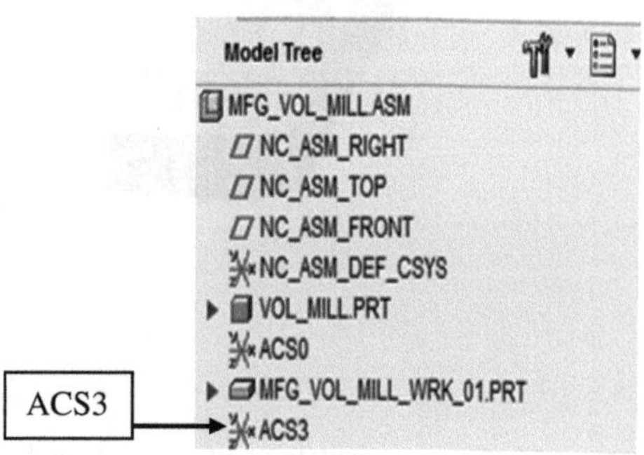

4.5 Set up the Work Centre

Click on the Work Centre icon to activate its drop-down menu list ≫ Now click on Mill in the activated menu list as shown below.

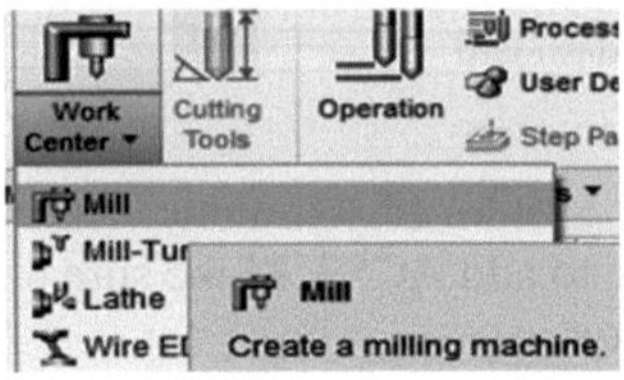

The Milling Work Centre dialogue window is activated on the main graphic window as shown in Fig. 4.18. Make note of the Name, Type, Post Processor, Number of Axis, etc.

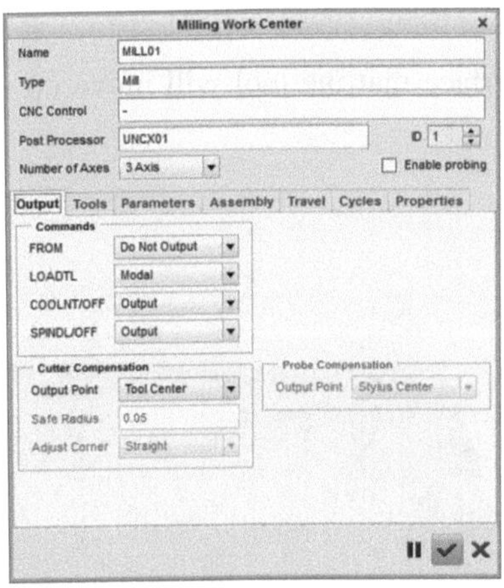

Click on the Check Mark icon.

Fig. 4.18 Acitivated Milling Work Centre dialogue window

4.6 Set up Operation

Click on the Operation icon as shown below.

The Operation dashboard tools are activated as shown in Fig. 4.19.

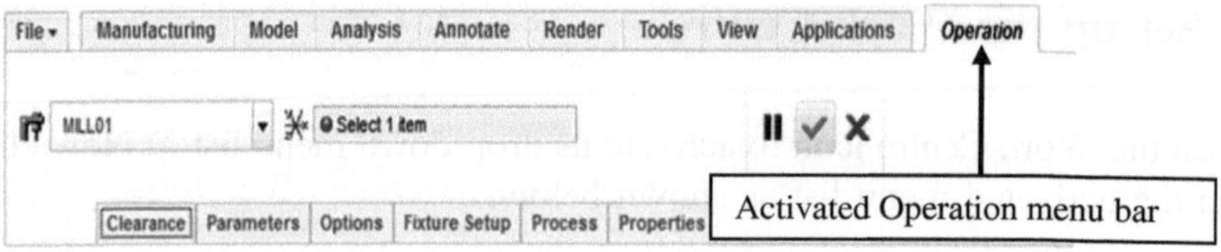

Fig. 4.19 Activated Operation dasboard

Click on the Coordinate System (Programme Zero) icon section box, and click on ACS3 on the Model Tree to add ACS3. Make sure that the created MILL01 is on the machine icon section box.

Click on the Clearance tab to activate its content ≫ On the Retract group, click on the Type section box, to activate its drop-down menu list, and on the drop-down list, click on "Plane". Click on the Workpiece top surface and the Reference section box is automatically updated by the system ≫ Type 15 mm in the Value section box as illustrated in Fig. 4.20.

Note: The Retract surface is the surface that the tool will move up to after each Milling operation.

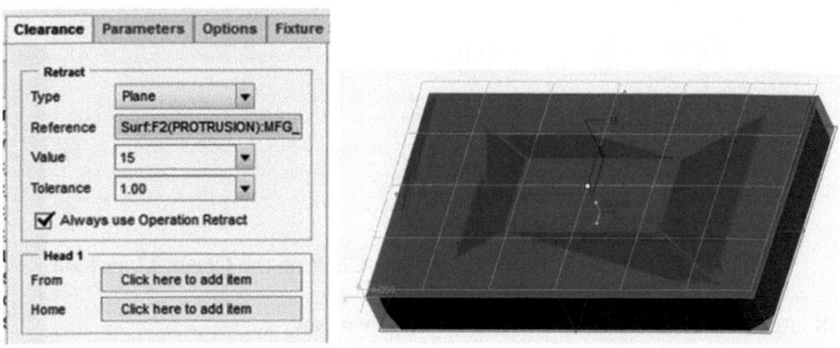

Fig. 4.20 Activated Clearance panel and the Retract plane on Worpiece surface

Now click the Check Mark icon to exit Operation setup.

Check the Model Tree as shown in Fig. 4.21 to make sure that ACS3, MILL01, and OP010 [MILL01] are in the Model Tree as created.

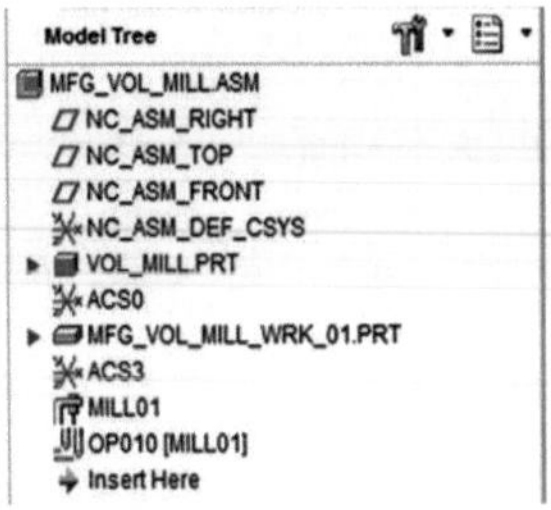

Fig. 4.21 Model Tree showing ACS3, MILL01, and OP010 [MILL01]

4.7 Set up the Cutting Tool

Click on the Cutting Tools icon as shown below.

The Tools Setup dialogue window General tab content is activated on the main graphic window.

Click on the General tab to activate its content ≫ Type the Tool name in the Name section box or leave the default Tool name ≫ Click on the Type section box to activate its menu list, now click on End Mill Tool on the activated drop-down menu list ≫ In the Material section box, type the Tool material or leave empty. Add dimensions to the Tool as illustrated in Fig. 4.22.

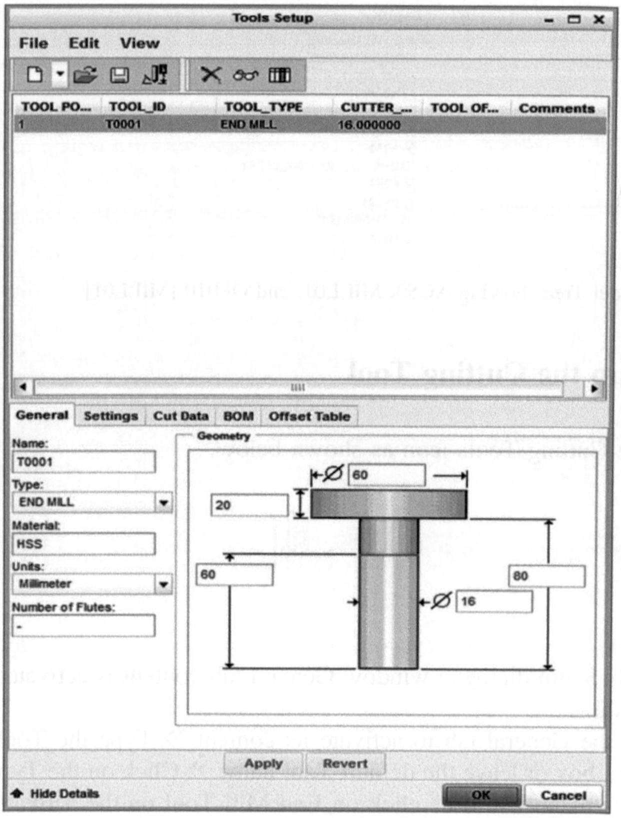

Fig. 4.22 Tools Setup dialogue window after setting up process

Now click on Apply tab.
Click on the OK tab to exit Tools Setup dialogue window.

4.8 Creating the Mill Volume

The created Mill Volume will be removed during the Volume Rough milling operation.

Click on the Mill bar to activate its ribbon menu ≫ Now, click on Mill Volume icon on the Manufacturing Geometry group as shown below.

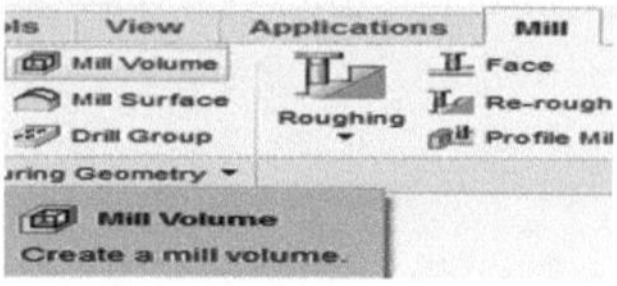

The Mill Volume dashboard is activated as shown in Fig. 4.23.

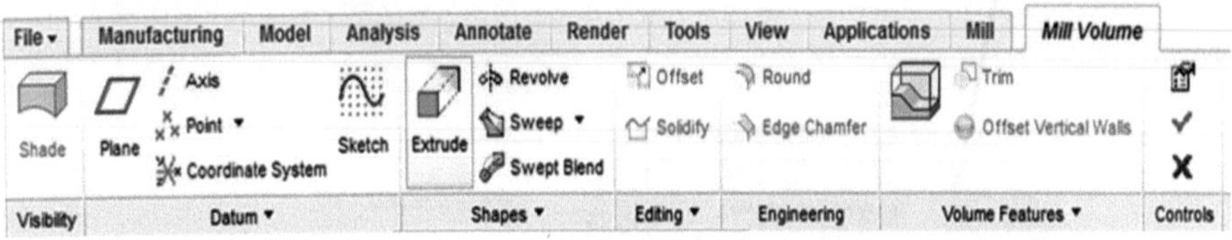

Fig. 4.23 Activated Mill Volume dashboard in concise form

Click on the Extrude icon .

The Extrude dashboard tools are activated as shown in Fig. 4.24.

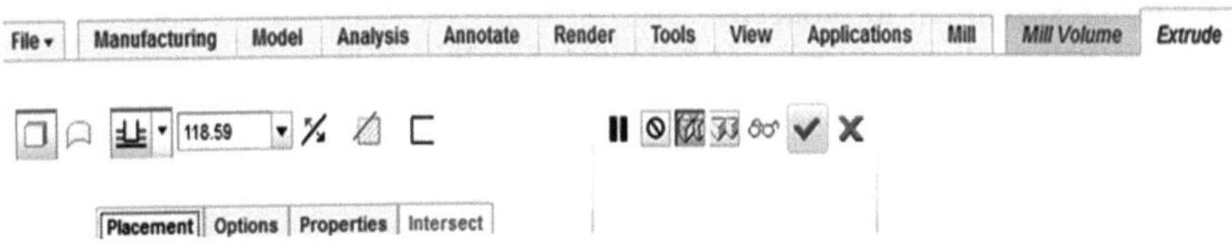

Fig. 4.24 Extrude application dashboard tools in concise form

Now, click on the Placement tab to activate its panel ≫ On the Sketch group, click on Define tab as shown in Fig. 4.25.

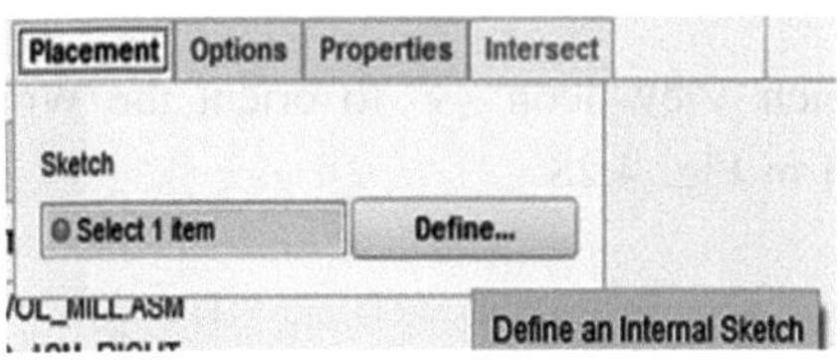

Fig. 4.25 Activated Placement panel

The Sketch dialogue box is activated on the main graphic window ≫ Now click on the Placement tab ≫ On the Sketch Plane group, click on the Plane section box and now click on the top surface of the Workpiece as indicated by the arrow below. The Reference section box is automatically updated. Make sure that the Orientation section box is set to Right as shown in Fig. 4.26.

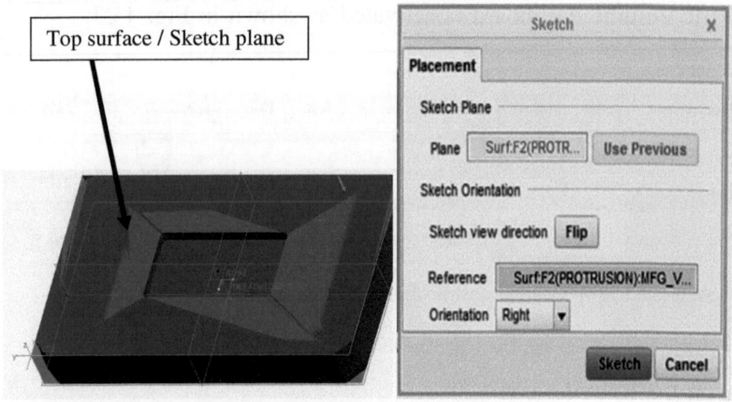

Fig. 4.26 Sketch dialogue box after adding the sketching plane

Click on Sketch tab to proceed.
The Sketch application tools are activated. See Fig. 4.27.

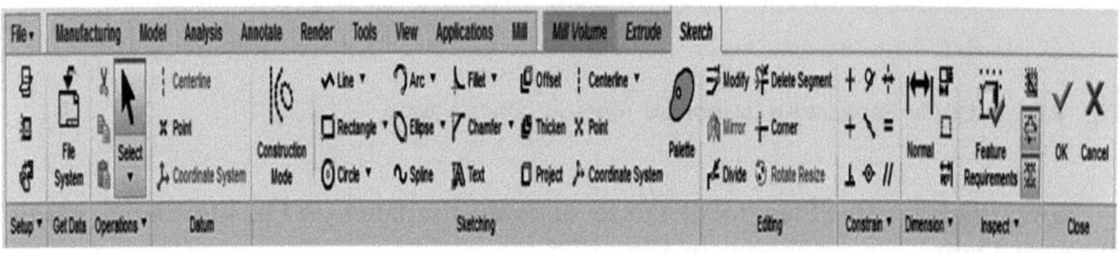

Fig. 4.27 Activated Sketch tools in concise form

Now click the Sketch View icon  to orient the Workpiece to the correct sketch plane as shown in Fig. 4.28.

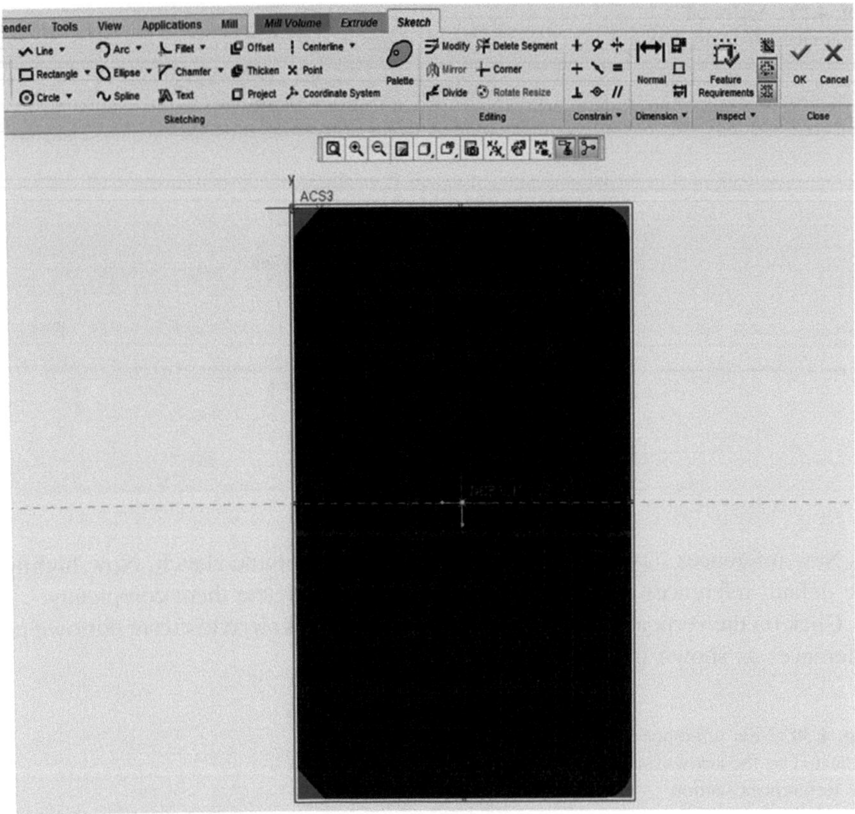

Fig. 4.28 Workpiece orientated correctly on the sketch plane

4.8.1 Create Reference

Click on the Reference icon as shown below.

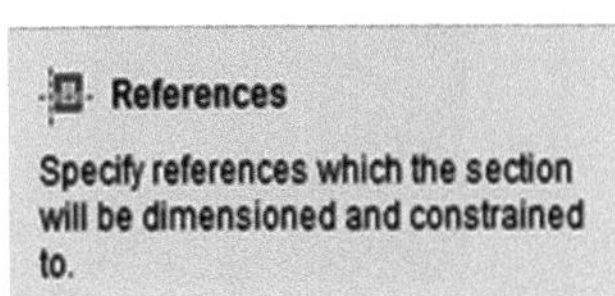

The References dialogue box is activated in its default settings on the main graphic window as shown in Fig. 4.29.

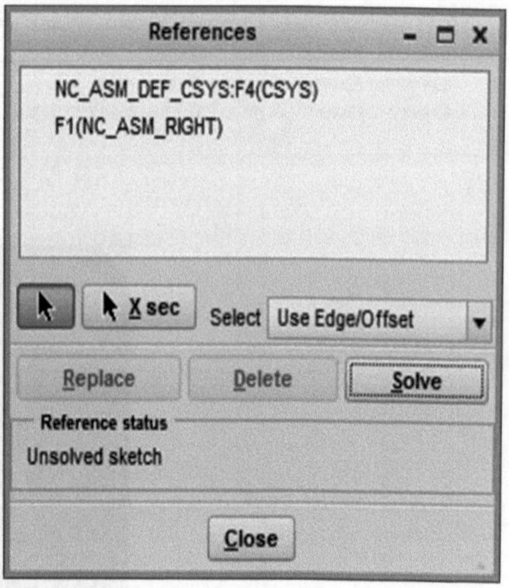

Fig. 4.29 Activated References dialogue box

New references have to be created for this Mill Volume sketch. Now highlight the default references and click on the Delete tab to delete them completely.

Click on the vertical and horizontal axis on the Workpiece to create our own new references as shown in Fig. 4.30.

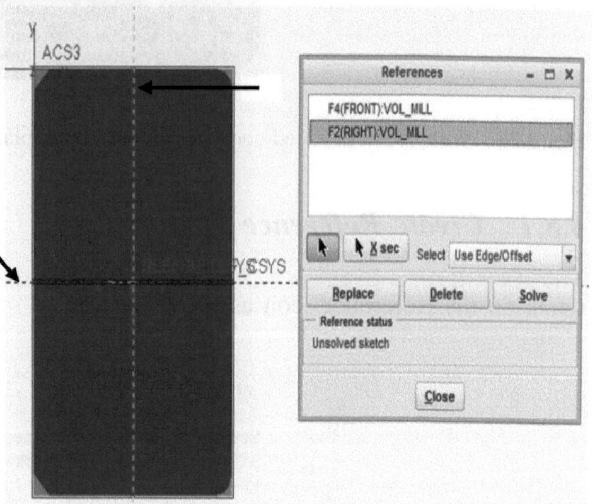

Fig. 4.30 New references as indicated by the arrows and in the References section

Click on the Close tab to exit the References dialogue box.

4.8.2 *Project Sketch*

The Project tool will enable us to project the sketched lines into the sketched plane.

Click on the Project icon on the Sketching group as shown below.

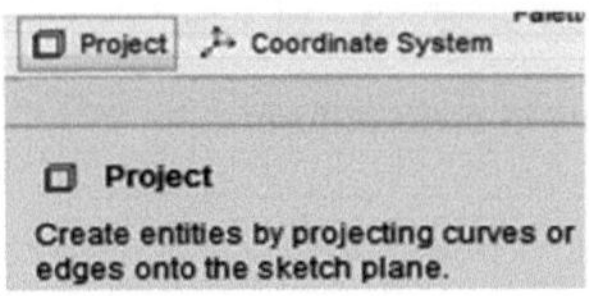

The Type dialogue box is activated as shown in Fig. 4.31.

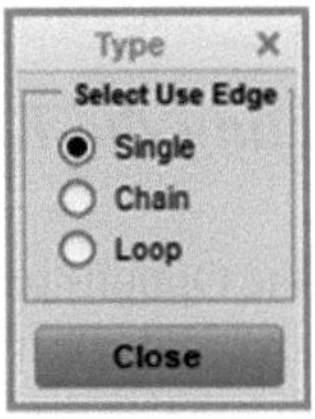

Fig. 4.31 Activated Type dialogue box

On the Type dialogue box, click on the Single radio button.

Project the internal sketched rectangle lines by clicking on the edge lines as indicated by the arrows as shown in Fig. 4.32.

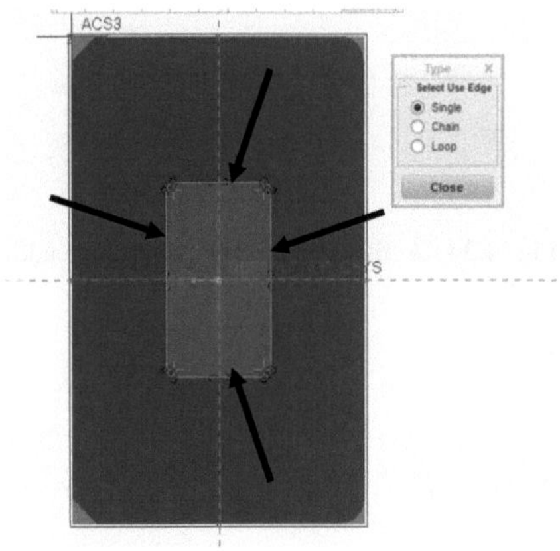

Fig. 4.32 Projecting the internal rectangle on the sketch plane

Click on the Close tab to exit the Type dialogue box.

Now click on the Check Mark/OK icon to exit the Sketch application.

The Extrude dashboard tools are activated as shown in Fig. 4.33.

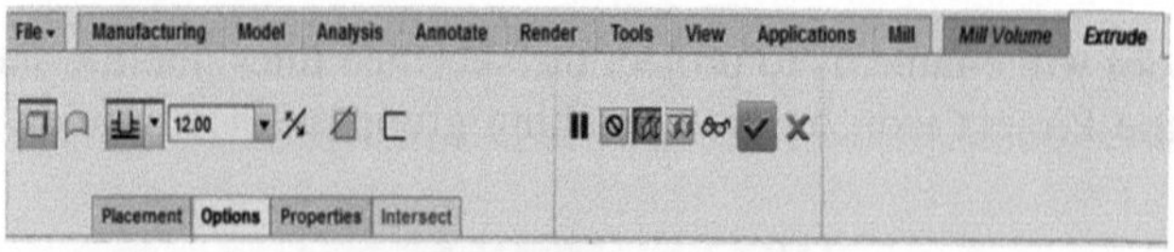

Fig. 4.33 Activated Extrude tools in concise form

Note: There are two ways to create the required Mill Volume depth as illustrated in the first and second methods as shown below.

4.8.3 *First Method of Creating the Extrude Depth*

Click on the Options tab to activate the Options panel ≫ On the Depth group, click on the Side 1 section box downward arrow, and select Blind from the activated drop-down menu list ≫ Type 12 mm as the depth value in the Side 1 value section box. Now, click on the Extrude direction arrow icon to change the extrusion direction (Fig. 4.33).

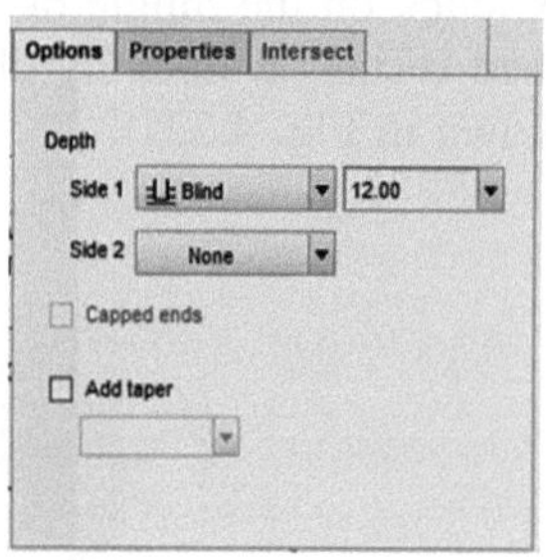

Fig. 4.34 Activated Options panel after adding the inputs

The projected and extruded internal rectangle is indicated by the arrow in Fig. 4.35.

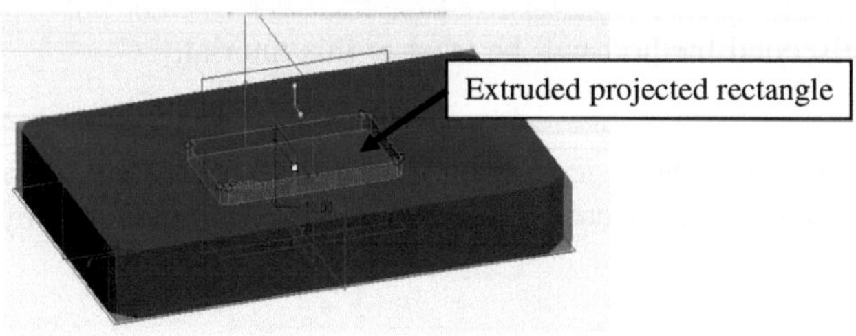

Fig. 4.35 Workpiece showing projected and extrude rectangle

Click the Check Mark icon to exit from the Extrude application.

4.8.4 Second Method of Creating the Extrude Depth

Click on the Options tab to activate the options panel ≫ On the Depth group, click on Side 1 section box, and on the activated drop-down menu list, click on "To Selected" ≫ Now, click on the base of the internal rectangle as the depth of extrusion as indicated by the arrow in Fig. 4.36.

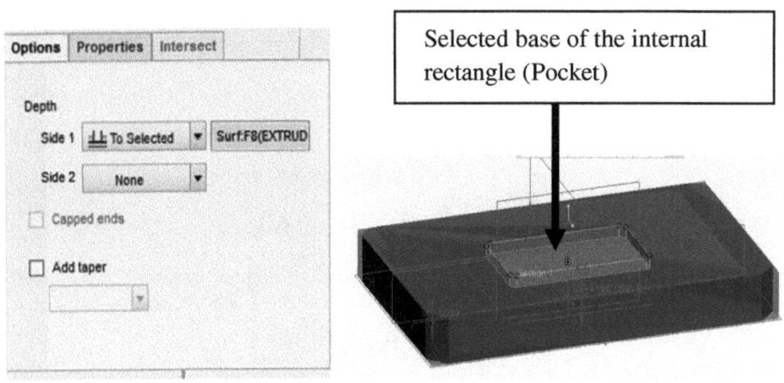

Fig. 4.36 Options panel after adding the inputs and the rectangle to be extruded

Now change the extrusion direction by clicking on the change extrude direction arrow icon , if needed to be.

Click on the Check Mark icon  to exit the Extrude application.

Note: The second method will be used in this tutorial.

Click on the Check Mark icon again to exit the Mill Volume application.

The Mill Volume is now created on the Workpiece as indicated by the arrow as shown in Fig. 4.37.

Fig. 4.37 Created Mill volume on the Workpiece

4.9 Volume Rough Milling Operation

Click on the Mill menu ≫ Click on Roughing icon tab on the Manufacturing Geometry group ≫ In the activated drop-down menu list, click on Volume Rough as shown below.

The NC SEQUENCE Menu Manager dialogue box is activated as shown in Fig. 4.38.

Fig. 4.38 Activated NC SEQUENCE and SEQ SETUP group

Check Mark Tool, Parameters, and Volume as shown in fig. 4.39.

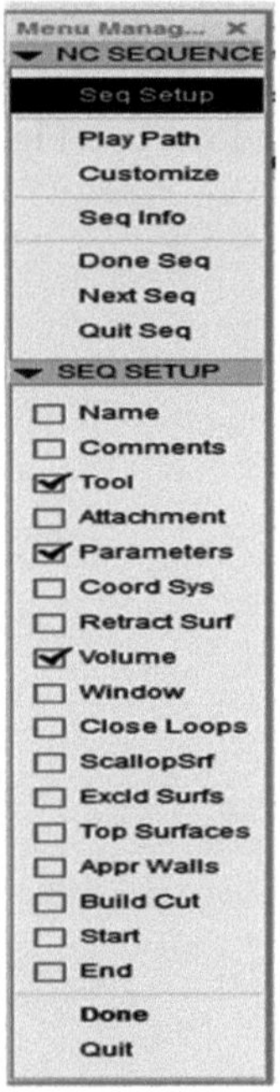

Fig. 4.39 Tools, Parameters, and Volume Check Marked

Click on Done to exit.

The Tools Setup dialogue window is activated on the main graphic window. Check to confirm that the Tool dimensions matches that of the Tool that was set up earlier in this tutorial (Figs. 4.22).

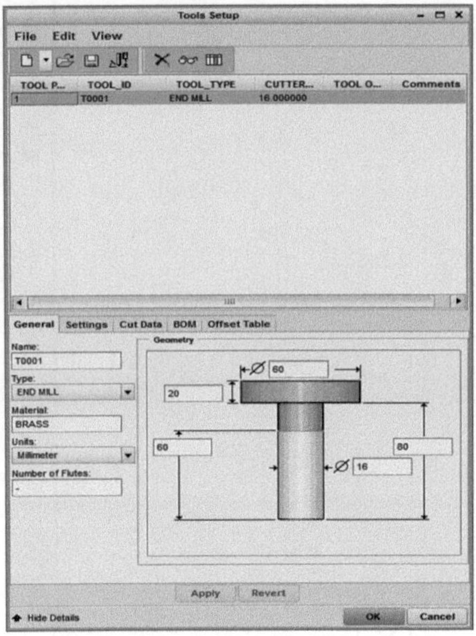

Click on OK tab to exit.

Fig. 4.40 Activated Tools Setup dialogue box

The Edit Parameters of Sequence "Volume Milling" dialogue box is activated in the main graphic window. See Fig. 4.41.

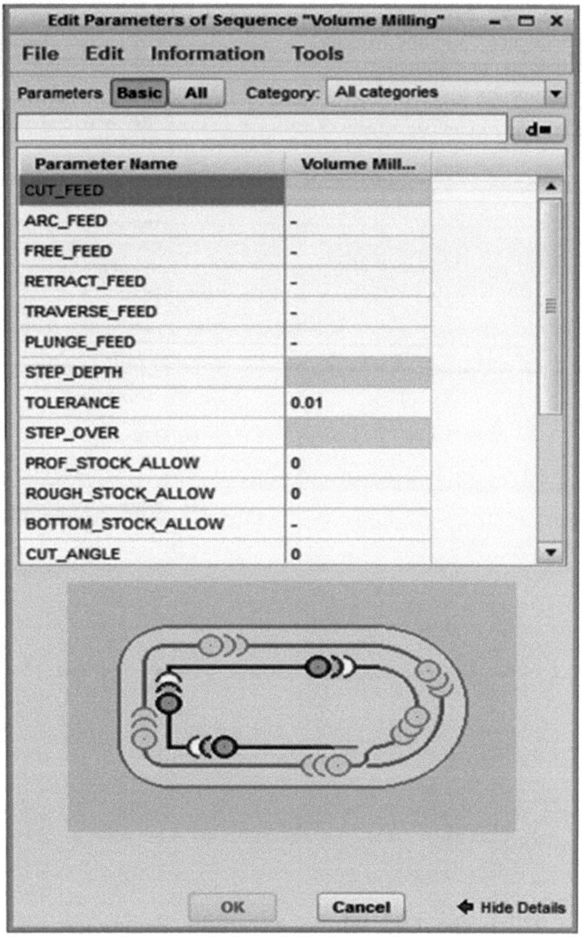

Fig. 4.41 Activated Edit Parameters of Sequence "Volume Milling" dialogue box

Input the parameter values for the Volume Rough milling operation. See Fig. 4.42.

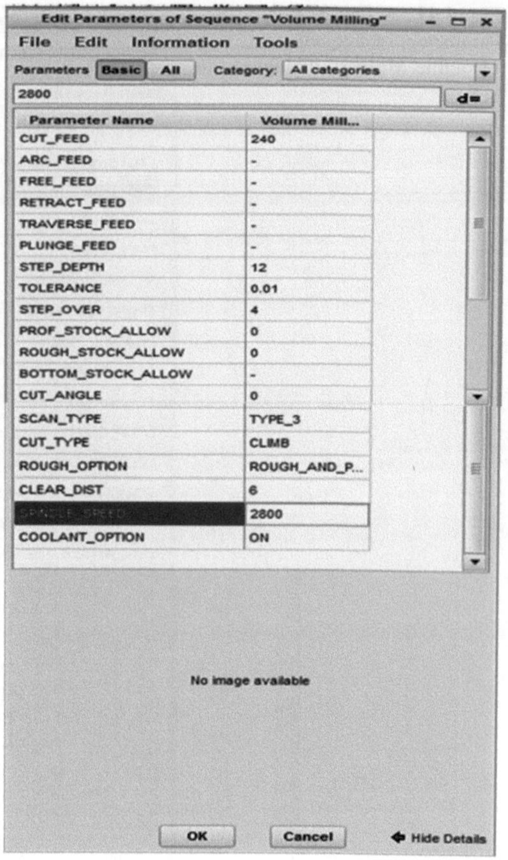

Fig. 4.42 Edit Parameters of Sequence "Volume Milling" dialogue box showing parameters given as input values

Click the OK tab to exit Edit Parameters of Sequence "Volume Milling" application.

Note: Not all the above-mentioned parameters, given as input, are accurate. They are only used for illustrative purposes in this tutorial. To obtain accurate parameter values, you have to use the appropriate formulas to calculate their values and also consult the appropriate handbook manual.

Now, click on the created Mill Volume (EXTRUDE1 [MIL_VOL_1-MILL VOLUME]) as the volume to be Volume Roughed, when prompted to do so by the system as illustrated in Fig. 4.43.

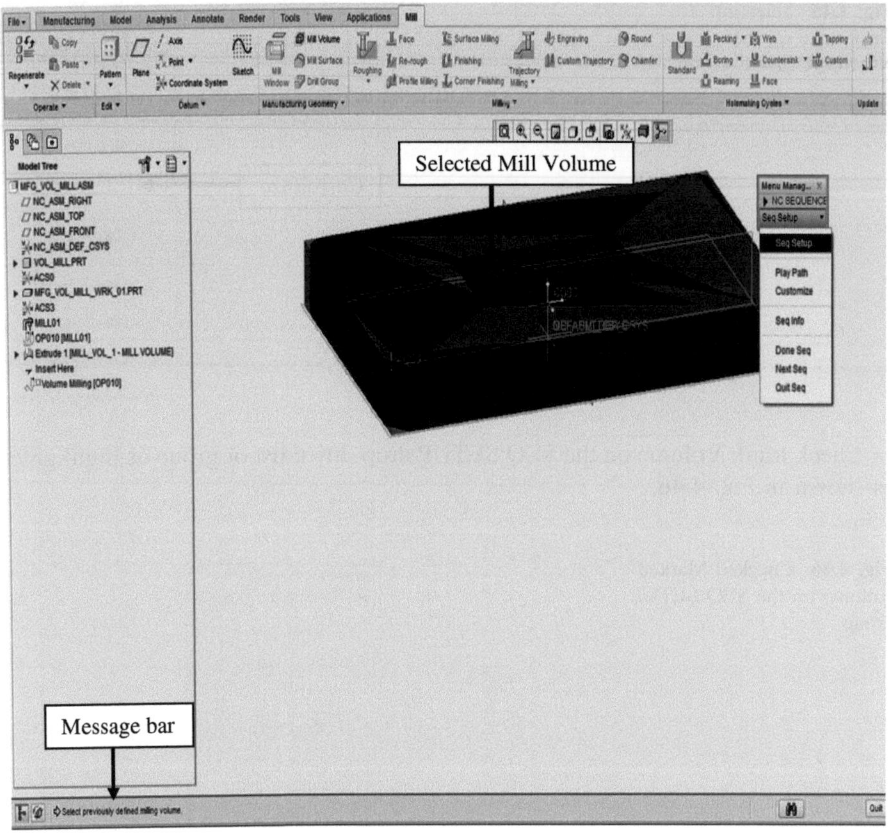

Fig. 4.43 Adding the created Mill Volume

Particular attention must be paid at this stage to the information on the Message bar on the graphic window as shown in Fig. 4.44.

Fig. 4.44 Information on the Message bar

If Mill Volume is not automatically added by the system, click on Customize in the NC SEQUENCE group as indicated by the arrow in Fig. 4.45.

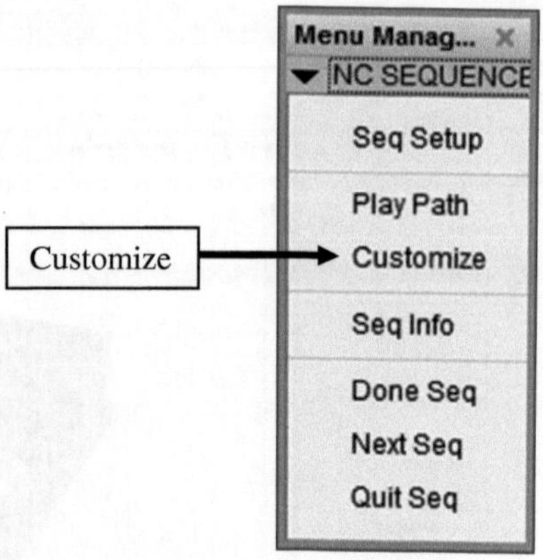

Fig. 4.45 Selecting Customize on the NC SEQUENCE group

Check Mark Volume on the SEQ SETUP drop-down list or group as highlighted as shown in Fig. 4.46.

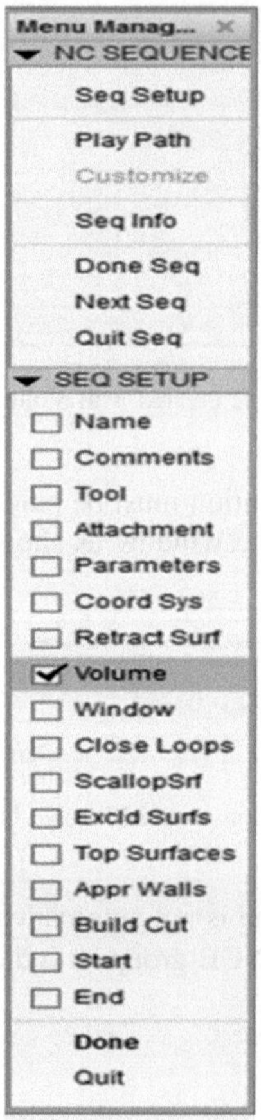

Fig. 4.46 Checked Marked Volume on the SEQ SETUP group

Click on Done.

When asked by the system to select the Volume Mill again, again click on the already created Mill Volume (EXTRUDE1 [MIL_VOL_1-MILL VOLUME]) either on the Model Tree or on the Workpiece as indicated by the arrow shown in Fig. 4.47.

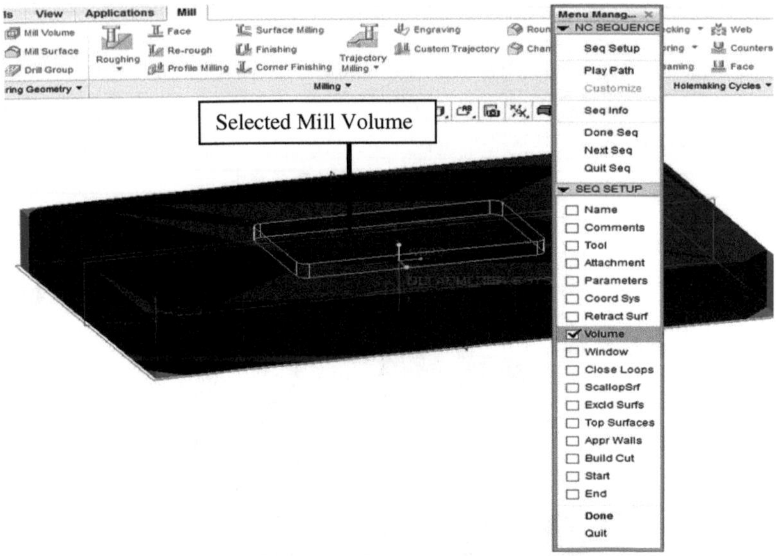

Fig. 4.47 Adding the created Mill Volume

Note: This method is used when Volume is omitted or when the system does not automatically select the created Mill Volume.

Click on Play Path in the NC SEQUENCE group and the NC SEQUENCE Menu Manager dialogue box expand as shown in Fig. 4.48.

Fig. 4.48 Arrows indicating Play Path and Screen Play

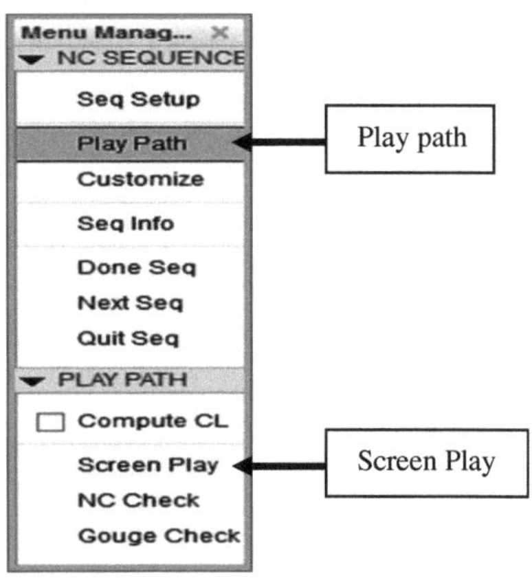

On PLAY PATH group on the Menu Manager dialogue box, click on Screen Play as indicated by the arrow in Fig. 4.48.

The cutting Tool and the PLAY PATH dialogue box are activated on the main graphic window as shown in Fig. 4.49.

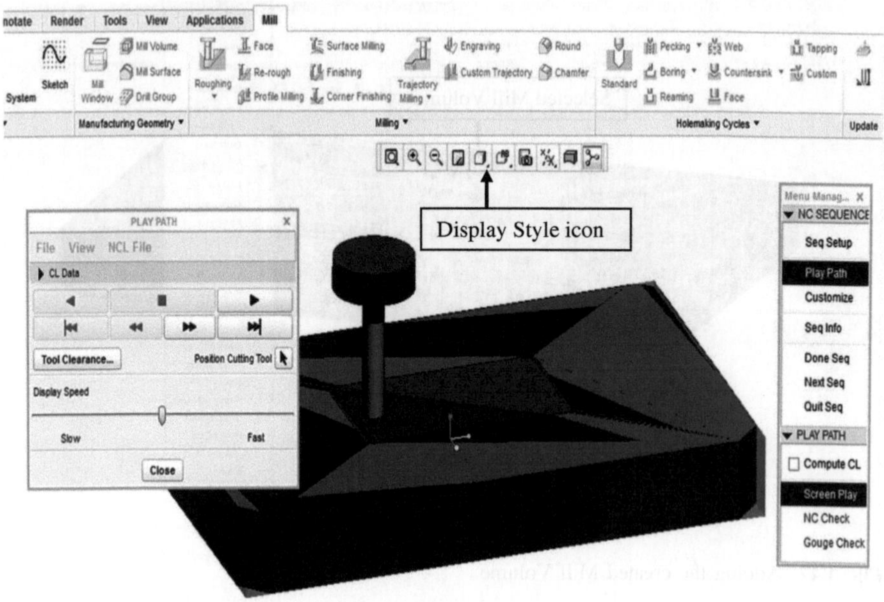

Fig. 4.49 Activated cutting Tool and Play Path dialogue box in the main graphic window

Congratulation because you are now ready to start the Volume Rough Milling Operation.

For better view of the Volume Rough Milling Operation, click on the Display Style icon and now click on Wireframe icon in the activated drop-down menu list as the display style. The Workpiece and Cutting Tool now changes to the wireframe display as shown in Fig. 4.50.

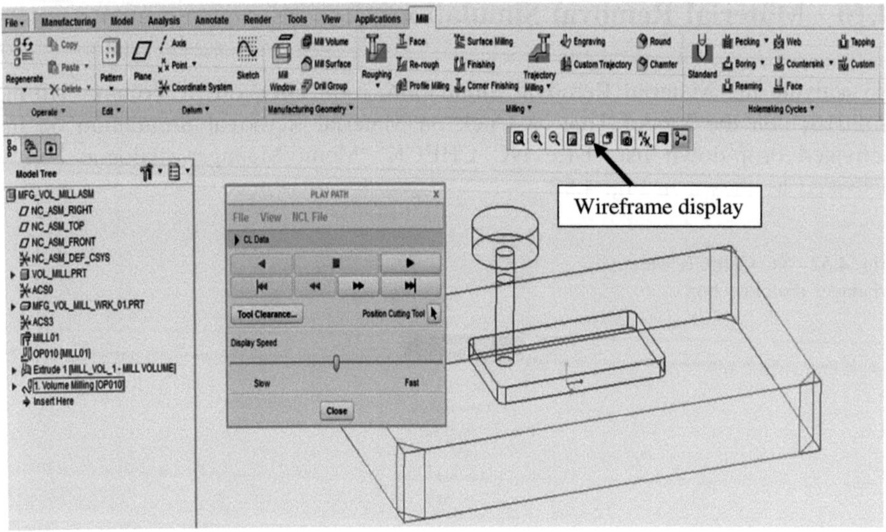

Fig. 4.50 Cutting Tool and Workpiece in wireframe display

Click the Play tab to activate the on screen simulation process.
The end of the Volume Rough milling operation is shown in Fig. 4.51.

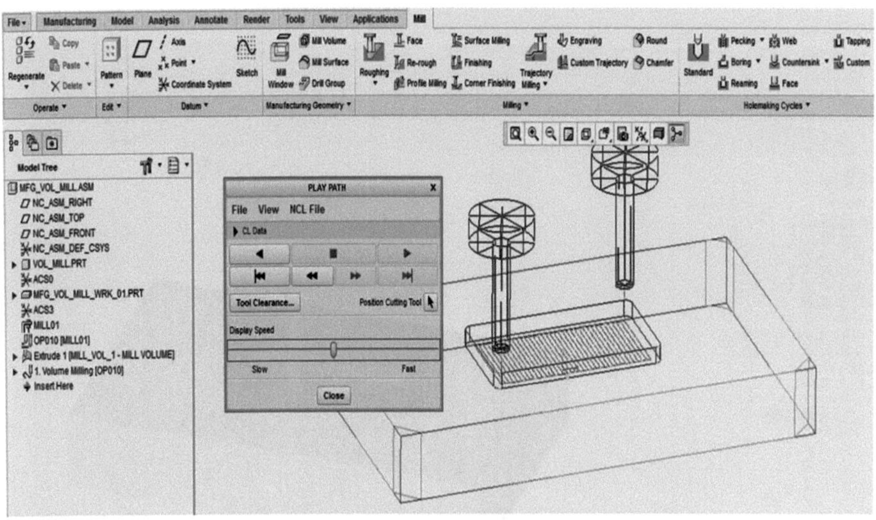

Fig. 4.51 End of the Volume Rough milling operation in wireframe display

Click on the Close tab on the PLAY PATH dialogue box to exit.
Now, click on Done Seq in the NC SEQUENCE group to exit.

4.10 Material Removal Simulation

To activate the Material Removal Simulation, right click on “1.Volume Milling [OP010]” on the Model Tree ≫ Click on Material Removal Simulation on the activated drop-down list. The “NC CHECK” Menu Manager dialogue box is activated as shown in Fig. 4.52.

Fig. 4.52 NC CHECK Menu Manager dialogue box

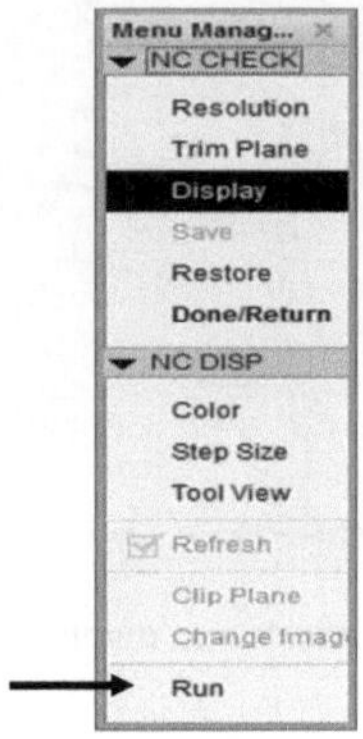

Click on Run as indicated by the arrow.

The Material Removal Simulation process is activated as shown in Fig. 4.53.

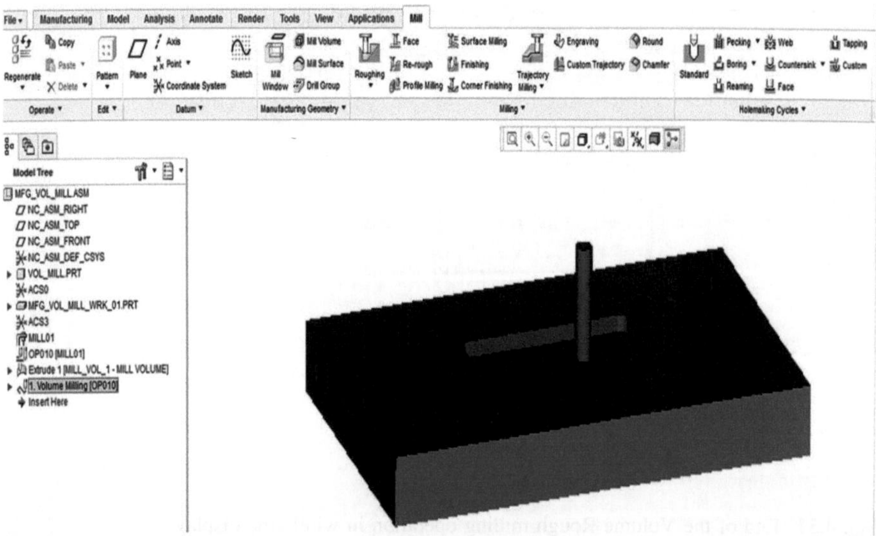

Fig. 4.53 Active Material Removal Simulation display

Click on Step-Size on the NC CHECK group in the Menu Manager dialogue box ≫ Add a lower value, as this action will slow down the simulation animation process so that you can capture the slow motion as shown above ≫ Now click on Run again.

Note: If you cannot run or activate Material Removal Simulation, then you have to create "nccheck_type" and "nccheck value". Go to previous tutorial on chapter three and follow the step-by-step guide on how to create and add nccheck.

Create Cutter Location (CL) Data

No Cutter Location data will be created for the Volume Rough milling operation in this tutorial.

Create the G-codes

No G-codes will be created for the Volume Rough Milling Operation in this tutorial.

Chapter 5
Profile Milling Operation

This chapter will highlight the importance of Profile Milling Operation after performing Volume Rough milling operation on a Workpiece.

Profile Milling is used to obtain a refined finish on vertical surfaces after Volume Rough milling operation.

The steps below will be followed when creating a Profile Milling operation.

- Start the Manufacturing Application
- Import and constrain the 3D Part
- Add Stock to the Part
- Create the Programme Zero (Machine Coordinate System)
- Add the Work centre
- Set up Operation
- Set up Profile Milling Operation Sequence (Add Tools, Parameters, etc.)
- The Cutter Location (CL) Data is generated
- The G-codes are generated

> Note: The first six steps mentioned above will be omitted in this tutorial because they have been covered in Chap. 4.

P.O. Kanife, *Computer Aided Virtual Manufacturing Using Creo Parametric*,
DOI 10.1007/978-3-319-23359-8_5

5.1 Activate Profile Milling Operation

Now click on the Profile Milling icon in the Milling group as shown below.

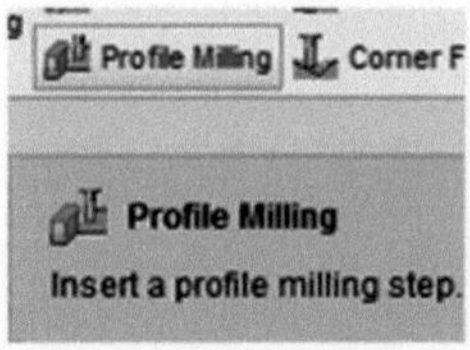

The Profile Milling dashboard tools are activated as shown in Fig. 5.1 in concise form.

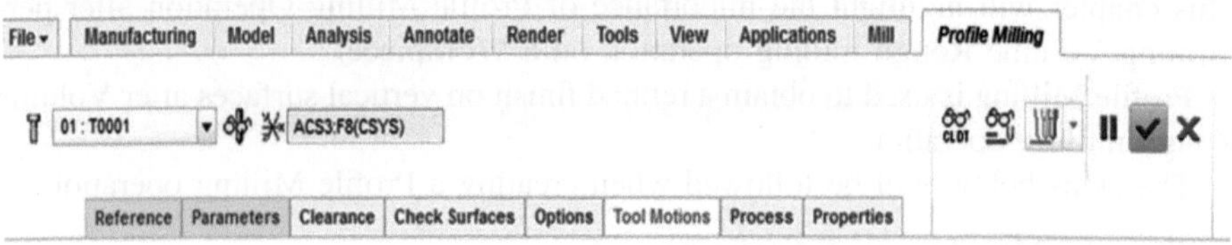

Fig. 5.1 Activated Profile Millings dashboard

Click on the Coordinate System icon section box as shown in Fig. 5.1 and now click on ACS3 in the Model Tree.

Click on Tool icon  as shown in Fig. 5.1 to define the new Tool for the Profile Milling operation.

The Tools Setup dialogue window is activated ≫ Click on the General tab to activate its content ≫ On the Name section box, type T0002 ≫ On the Type section box, click on End Mill Tool on the activated drop-down menu list ≫ On the Geometry section, add new dimensions to the Tool as shown in Fig. 5.2.

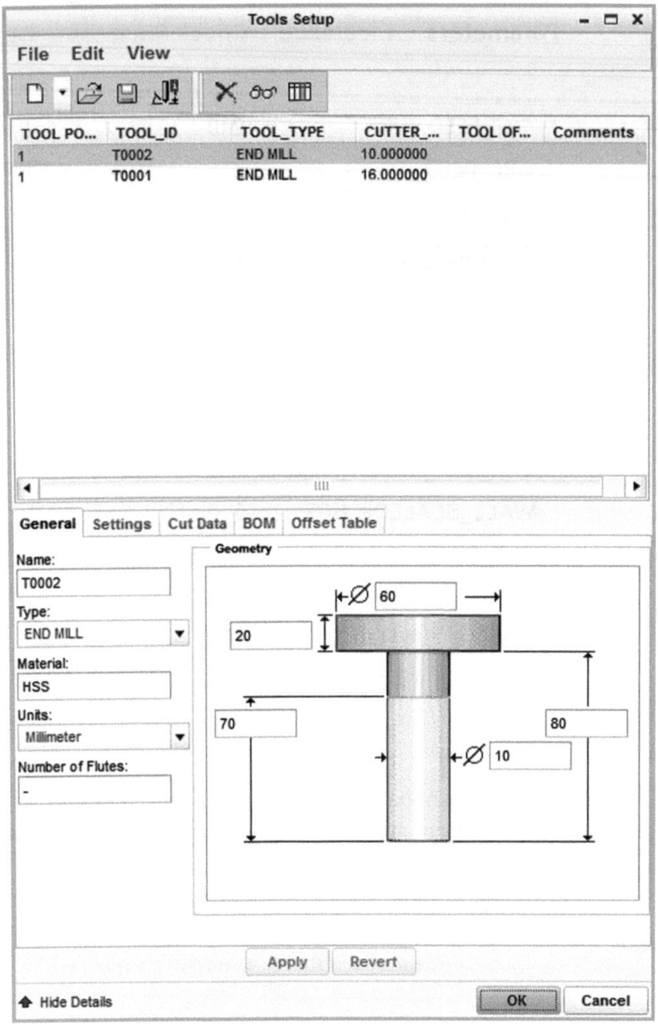

Click on Apply tab.

Fig. 5.2 Tools Setup dialogue window showing the parameters of the new Tool

Now click on OK tab to exit the Tools Setup dialogue window.

Click on the Parameters tab at the activated dashboard to activate its panel and add the parameters as shown in Fig. 5.3.

Fig. 5.3 Parameters panel with added parameters

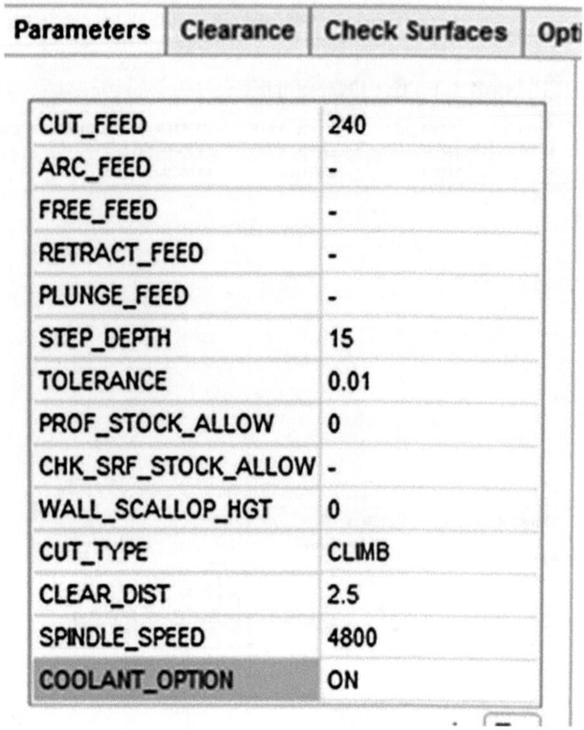

Now click on the Clearance tab on dashboard to activate its panel, and you will find that the previous parameters created are automatically generated by the system as shown (Fig. 5.4).

Fig. 5.4 Activated Clearance panel

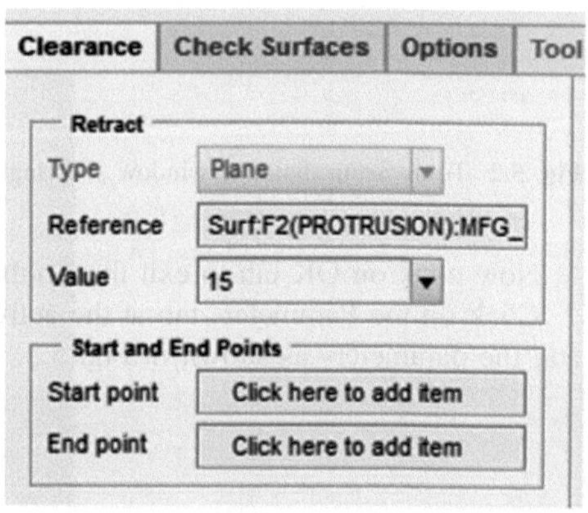

If this is not the case » On the main graphic window, click on the Active Filter downward pointing arrow to activate the drop-down list as indicated by the arrow shown on Fig. 5.5.

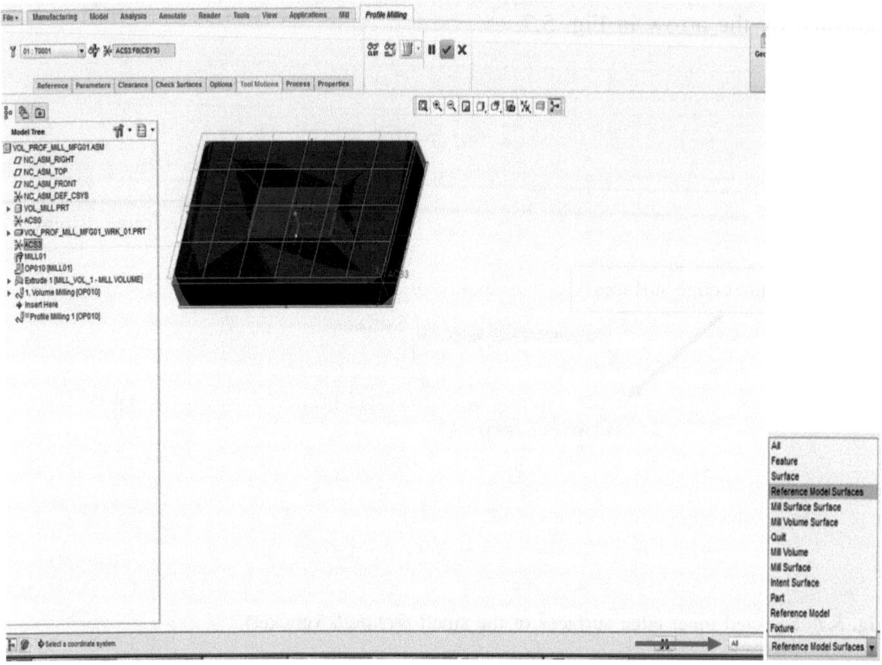

Fig. 5.5 Main graphic window indicating the Active Filter drop-down list

On the Active Filter drop-down menu list, click on the Reference Model Surfaces as highlighted in Fig. 5.6.

Fig. 5.6 Active Filter drop-down menu list

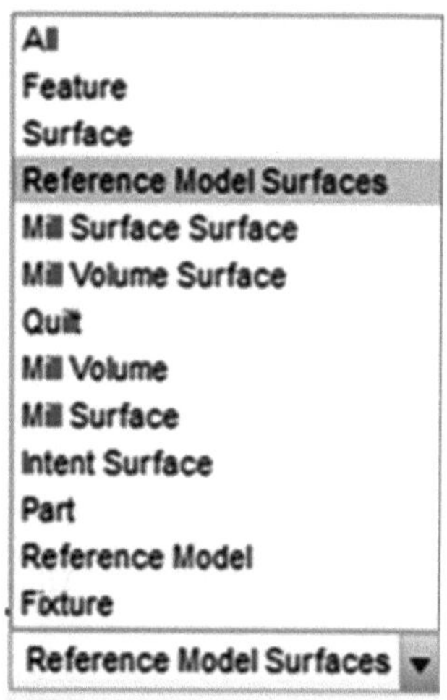

Click on the Reference tab as shown in Fig. 5.1 to activate the reference panel ≫ Click on the downward arrow in the Type section box, and now click on Surface on the activated drop-down menu list. Click on the Machining References section box, and now click on the inner edge surfaces of the internal rectangle while holding down the Ctrl key when clicking on the inner edge surfaces of the pocket as indicated by the arrow in Fig. 5.7.

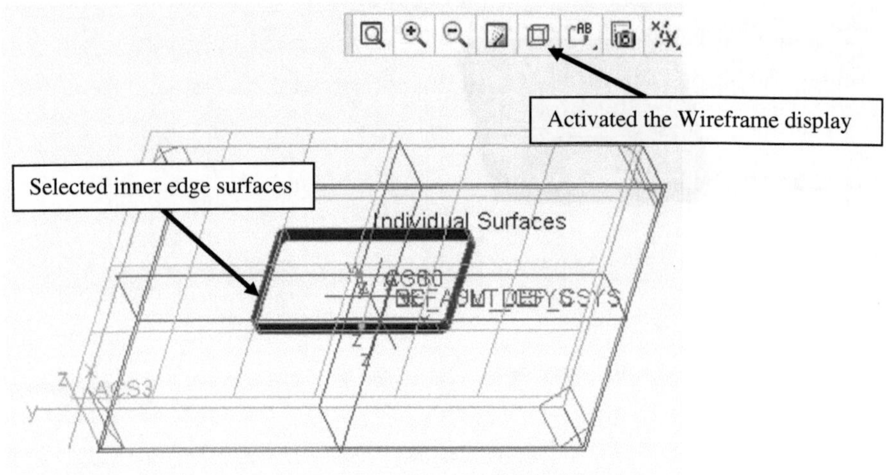

Fig. 5.7 Selected inner edge surfaces of the small *rectangle* (pocket)

Fig. 5.8 Activated Reference panel

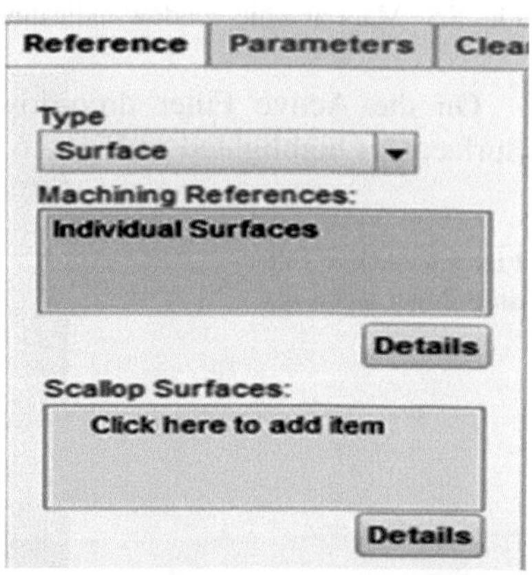

Click on the Check Mark icon to exit the Profile Milling application.

Click on Play Path icon on the Manufacturing toolbar. Alternatively, go to the Model Tree and right click on Profile Milling 1 [OP010], and then click on Play Path in the activated drop-down menu list (Fig. 5.8).

The cutting Tool and the PLAY PATH dialogue box are activated on the main graphic window as shown in Fig. 5.9.

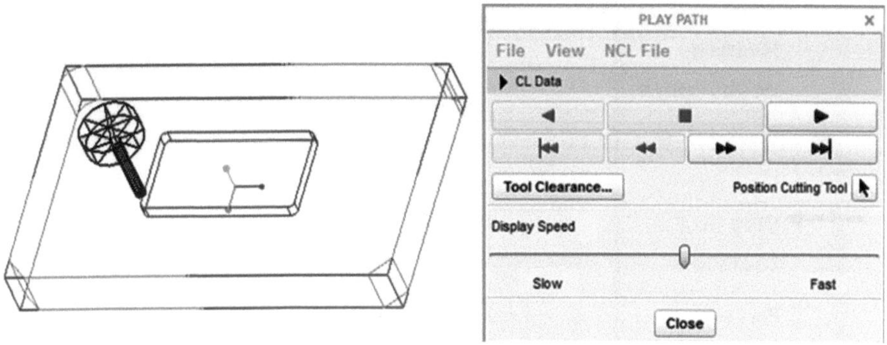

Fig. 5.9 Activated cutting Tool and workpiece in wireframe display

Click on Play tab to start Profile Milling.

5.2 Material Removal Simulation

Follow the step-by-step guide as illustrated on Chap. 4, to activate the Material Removal Simulation process.

For better view of the simulation process, click on Step Size as indicated by the arrow, and enter a step size of four (Fig. 5.10).

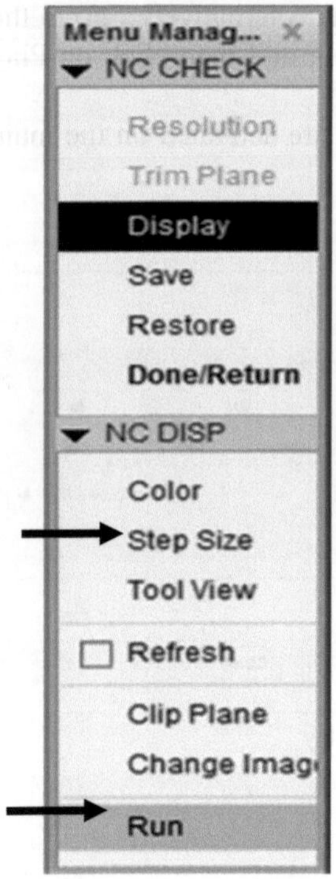

Click on Run, to start the Material Removal Simulation.

Fig. 5.10 NC CHECK Menu Manager dialogue box

The Material Removal Simulation is activated as shown in Fig. 5.11.

Fig. 5.11 Profile Milling Material Removal Simulation

Click on Done/Return in the NC CHECK group to exit.

5.3 Creating the Cuter Location (CL) Data

Click on the Save a CL File icon on the Manufacturing toolbar as shown below.

The SELECT FEAT Menu Manager dialogue box is activated on the main graphic window ≫ Click on Operation in the SELECT FEAT group and click on OP010 in the SEL MENU group (drop-down list) as highlighted in Fig. 5.12.

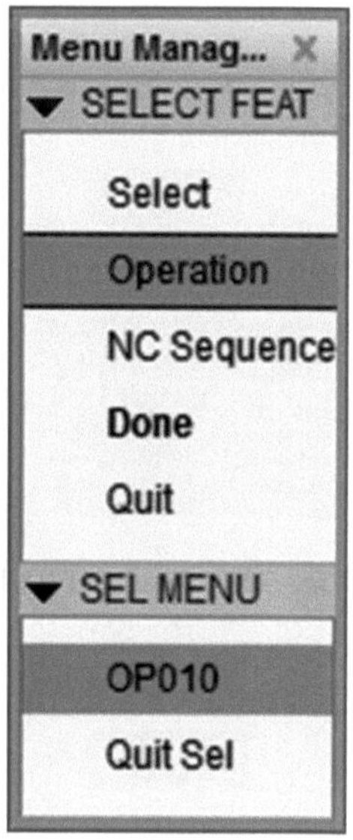

Fig. 5.12 The Menu Manager dialogue box, highlighting Operation and OP010

Once OP010 is clicked, the PATH Menu Manager dialogue box is activated on the main graphic window ≫ Click on File in PATH group and Check Mark the CL File and Interactive square boxes on the OUTPUT TYPE group as shown in Fig. 5.13

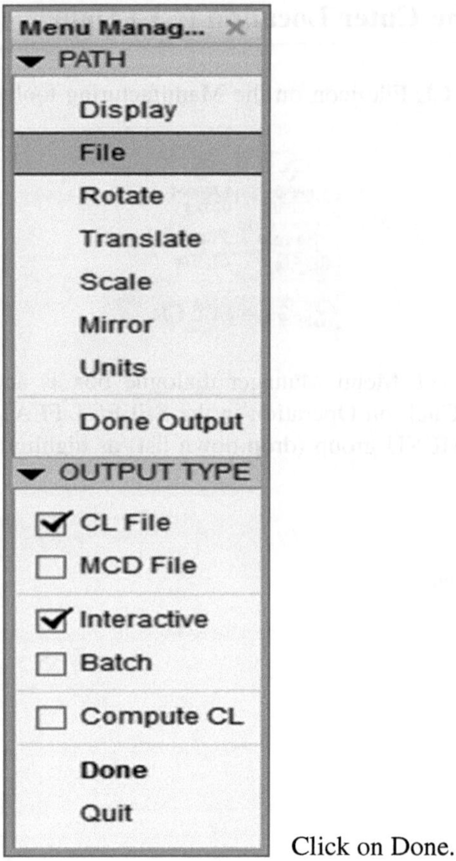

Fig. 5.13 Activated PATH Menu Manager dialogue box

The Save a Copy dialogue window is activated as shown in Fig. 5.14.

Fig. 5.14 Activated Save a Copy dialogue window in concise form

Make sure that the file name in the New Name section box is the correct one. Now save the CL File in your already chosen directory folder. You can also change the name and file location where you want to save the CL File data.

Click on the OK tab to exit the Save a Copy dialogue window.

Click on Done Output on the PATH group on the Menu Manager dialogue box as indicated by the arrow in Fig. 5.15.

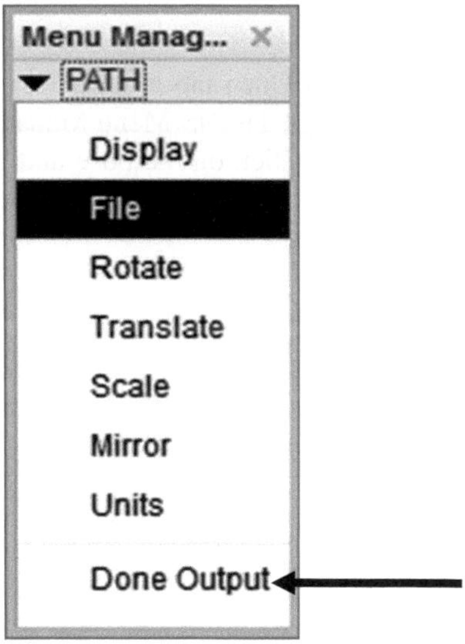

Fig. 5.15 Activating Done Output PATH group

The CL File is now created and saved in the chosen directory.

5.4 Create the G-code Data

Click on the Post a CL File icon in the Manufacturing toolbar as shown below.

The Open dialogue window is activated on the main graphic window as shown in Fig. 5.16. Make sure that the file name on the File name section box has got the extension (.ncl) at the end.

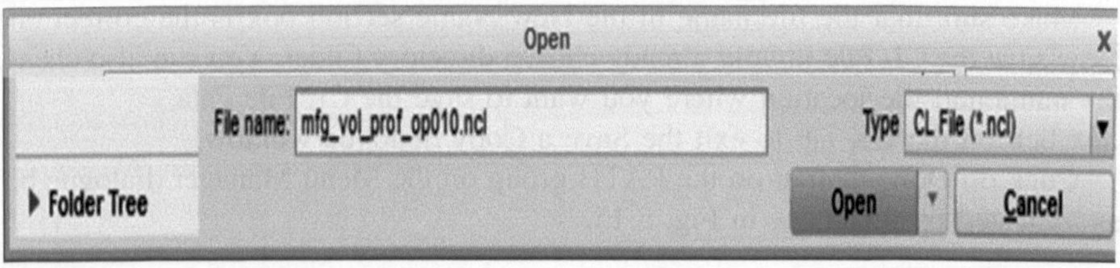

Fig. 5.16 Activated Open dialogue window in concise form

Click on Open tab to exit.

The PP OPTIONS Menu Manager dialogue box is activated on the main graphic window ≫ Click on Verbose and Trace square boxes to Check Mark them if they are not Checked Marked automatically by the system as shown in Fig. 5.17.

Fig. 5.17 Activated PP OPTIONS Menu Manager dialogue box

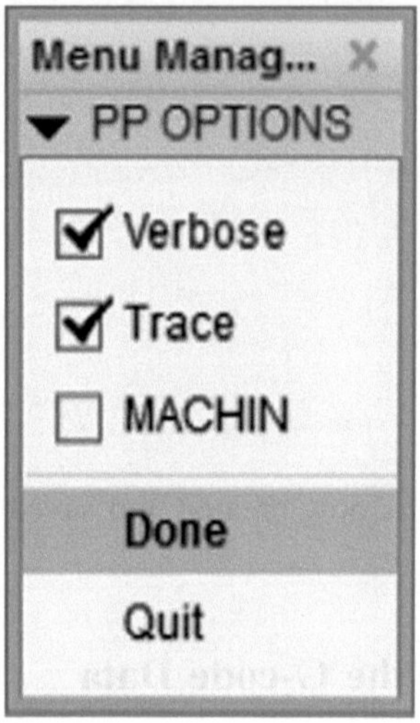

Click on Done.

The system automatically generates the Post Processor (PP) LIST as shown in Fig. 5.18.

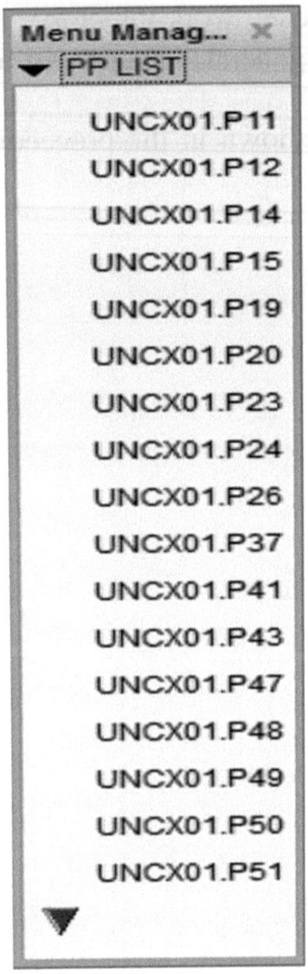

On the PP LIST group above, click on the UNCX01.P12.

Fig. 5.18 Activated PP LIST Menu Manager dialogue box

> Note: The UNCX01.P12 PP code is capable of generating G-codes for four, three and two point five (4, 3 and 2.5) axis milling operation

The INFORMATION WINDOW dialogue window is activated on the main graphic window as shown in Fig. 5.19.

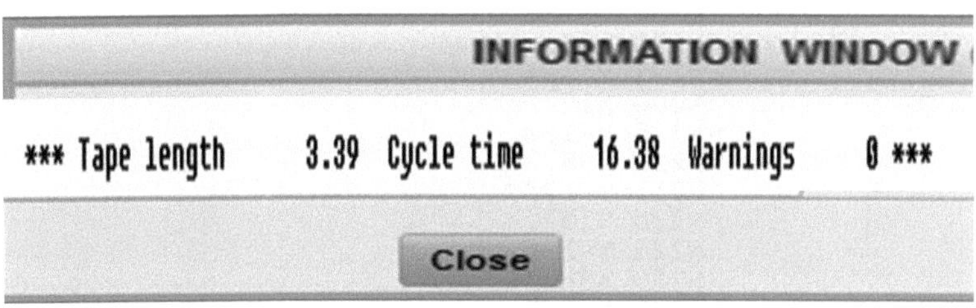

Fig. 5.19 Activated INFORMATION WINDOW in concise form

Check to see if the warning indicates programme/process error.
Click on the Close tab to exit the INFORMATION WINDOW.

The generated G-code is stored as a TAP file in the chosen work directory. The saved TAP file will be opened using Notepad to view the generated Profile Milling operation G-codes.

The G-codes command generated in this process are shown in the proceeding pages.

mfg_vol_prof_op010.tap - Notepa

File Edit Format View Help

```
N5 G71
N10 ( / MFG_VOL_MILL)
N15  G0 G17 G99
N20  G90 G94
N25  G0 G49
N30 T1 M06
N35 S2800 M03
N40 G0 G43 Z15. M08 H1
N45 X88. Y-128.
N50 Z6.
N55 G1 Z-12. F240.
N60 X162.
N65 Y-132.
N70 X88.
N75 Y-136.
N80 X162.
N85 Y-140.
N90 X88.
N95 Y-144.
N100 X162.
N105 Y-148.
N110 X88.
N115 Y-152.
N120 X162.
N125 Y-156.
N130 X88.
N135 Y-160.
N140 X162.
N145 Y-164.
N150 X88.
N155 Y-168.
N160 X162.
N165 Y-172.
N170 X88.
N175 Y-176.
N180 X162.
N185 Y-180.
N190 X88.
N195 Y-184.
N200 X162.
N205 Y-188.
N210 X88.
N215 Y-192.
N220 X162.
N225 Y-196.
N230 X88.
N235 Y-200.
N240 X162.
N245 Y-204.
N250 X88.
N255 Y-208.
N260 X162.
N265 Y-212.
```

```
N270 X88.
N275 Y-216.
N280 X162.
N285 Y-220.
N290 X88.
N295 Y-224.
N300 X162.
N305 Y-228.
N310 X88.
N315 Y-232.
N320 X162.
N325 Y-236.
N330 X88.
N335 Y-240.
N340 X162.
N345 Y-244.
N350 X88.
N355 Y-248.
N360 X162.
N365 Y-252.
N370 X88.
N375 Y-256.
N380 X162.
N385 Y-260.
N390 X88.
N395 Y-264.
N400 X162.
N405 Y-268.
N410 X88.
N415 Y-272.
N420 X162.
N425 Y-128.
N430 X88.
N435 Y-272.
N440 X162.
N445 Z15.
N450  G0 G49
N455 T2 M06
N460 S4800 M03
N465 G0 G43 Z15. M08 H2
N470 X86. Y-125.
N475 Z2.5
N480 G1 Z-12. F240.
N485 G3 X85. Y-126. I0. J-1.
N490 G1 Y-274.
N495 G3 X86. Y-275. I1. J0.
N500 G1 X164.
N505 G3 X165. Y-274. I0. J1.
N510 G1 Y-126.
N515 G3 X164. Y-125. I-1. J0.
N520 G1 X86.
N525 Z15.
N530 M30
%
```

Chapter 6
Volume Rough Milling Operation, Mill Surface and Drill Operation

The steps below will be followed when creating all the manufacturing operations in this chapter.

- Start the manufacturing application
- Import and constrain the saved part unto the manufacturing GUI environment.
- Add Stock using the automatic workpiece method
- Create programme zero (Machine coordinate system)
- Create the machine work centre
- Setup operation
- Create cutting tool
- Create mill volumes
- Volume Rough milling operation sequence is created (add tools, parameters, etc.)
- Create mill surface
- Create cut lines
- Create drill cycle (operation)
- Cutter location (CL) Data is generated for the operation
- G-codes are generated

Note: In this tutorial, the first two steps above will be omitted because by now the reader should be familiar with how to start Creo Parametric Manufacturing application, import and constrain 3D Part. Chapters 2, 3 and 4 explain these steps very well.

P.O. Kanife, *Computer Aided Virtual Manufacturing Using Creo Parametric*,
DOI 10.1007/978-3-319-23359-8_6

6.1 Volume Rough Milling Operation Using Mill Volume

6.1.1 Add Stock to the Imported Part

Click on the workpiece icon ≫ Now click on automatic workpiece on the activated drop down menu list as shown below.

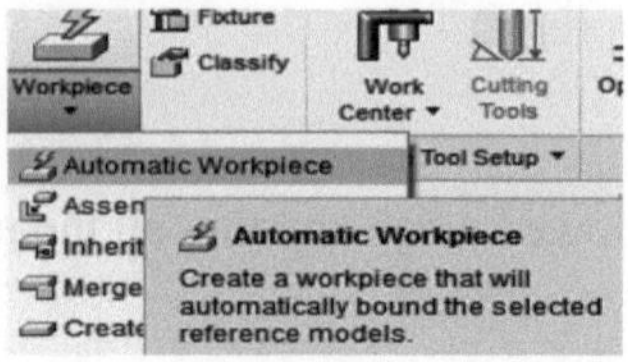

The Auto workpiece creation application tools are activated. See Fig. 6.1.

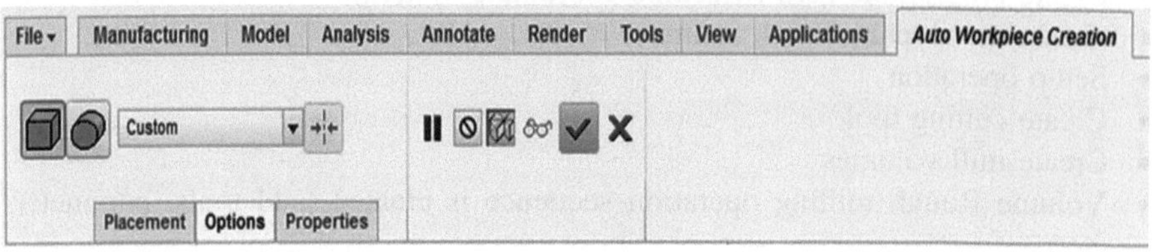

Fig. 6.1 Activated Auto workpiece Creation dashboard

Click on the Options tab and make a note of the Units, overall dimensions, linear offsets and rotation offsets section group as illustrated on Fig. 6.2.

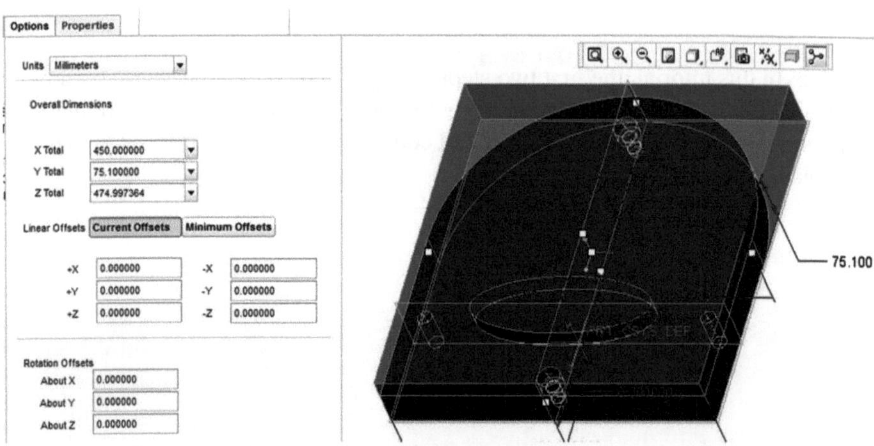

Fig. 6.2 Activated Options panel showing automatic Workpiece dimension values

Click on the check mark icon to exit.

Stock is now added to imported 3D Part as shown in Fig. 6.3.

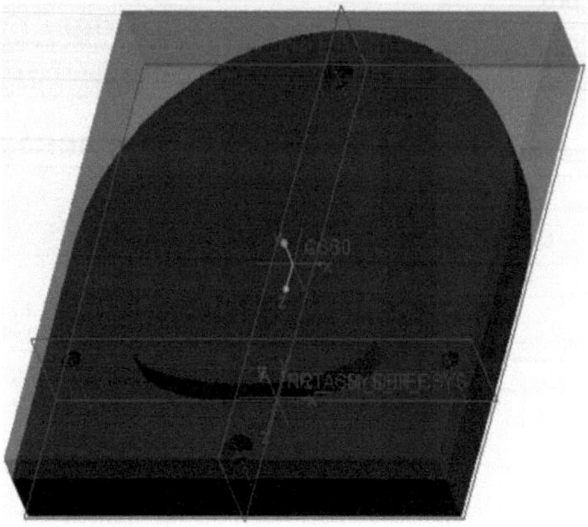

Fig. 6.3 Automatic stock added to workpiece Stock

6.1.2 *Create Coordinate System (Programme Zero)*

Click on the coordinate system icon tab as shown below.

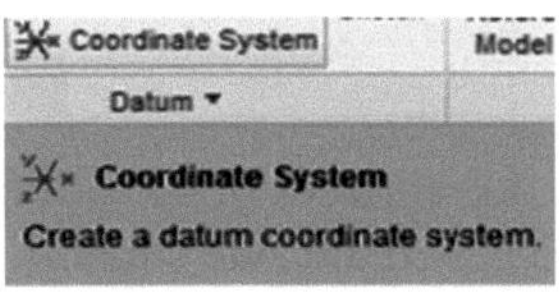

The coordinate system dialogue box is activated on the main graphic window as shown in Fig. 6.4.

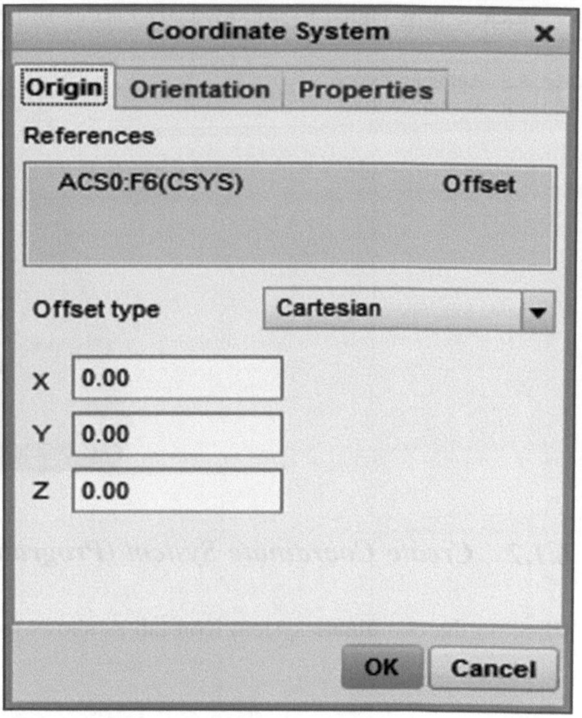

Fig. 6.4 Activated Coordinate System dialogue box

Now click on the top and side surfaces of the workpiece while holding down the Ctrl key when clicking. The clicked top side surfaces will appear on the Reference section box on the coordinate system dialogue box as illustrated on Fig. 6.5.

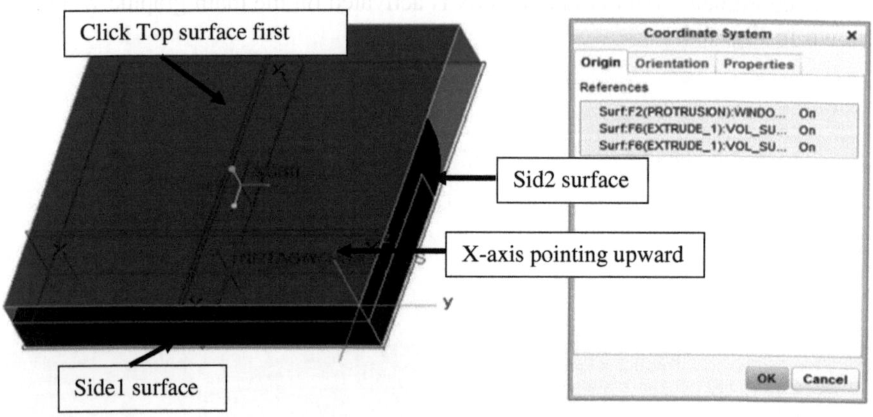

Fig. 6.5 Creating the Coordinate System on workpiece

To orient the X, Y and Z coordinate axes to the correct orientation

Click on the Orientation tab to activate its content » On the "Orient by" group, click in the radio button on the "References selection" » Click on "to determine" section box, and now click on Z-axis on the activated drop down menu list » Click on "to project" section box, and now click on X-axis on the activated drop down menu list. Now click on the Flip tab, to flip X-axis orientation. The correct orientations for the created Programme Zero axes are indicated by the arrows. See Fig. 6.6.

Fig. 6.6 Correct orientation of the created Programme Zero

Click on the OK tab on the coordinate system dialogue box to exit.

A new coordinate system (ACS1) is created. The new ACS1 created will be renamed by slow click on the ACS1 on the model tree and type ACS4. Alternatively, right click on ACS1 and click on Rename on the activated drop down menu list.

The renamed ACS4 is indicated by the arrow as shown on the model tree. See Fig. 6.7.

Fig. 6.7 Model Tree indicating the renamed coordinate system (ASC4)

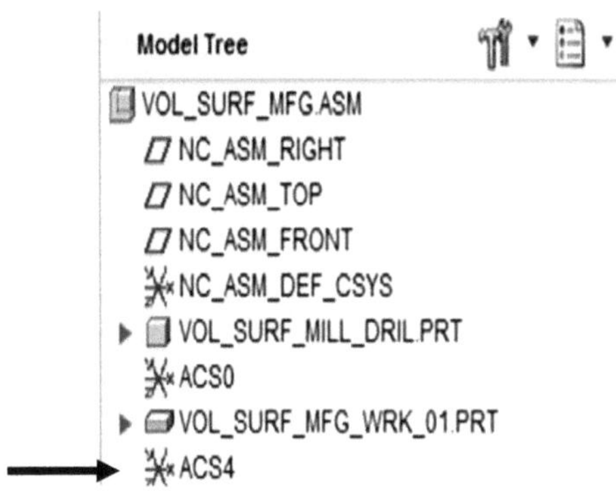

6.1.3 Setup the Work Centre

Click on the work centre icon on to activate its drop down menu list ≫ Now click on Mill on the activated drop down menu list as shown below.

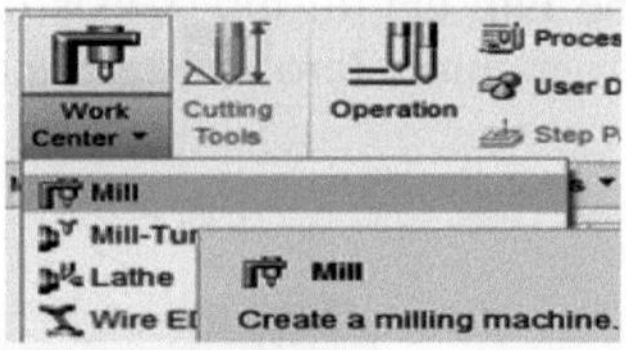

The Milling Work Centre dialogue window is activated on the main graphic window as shown in Fig. 6.8. Make note of the name, type, post processor, number of axis, etc.

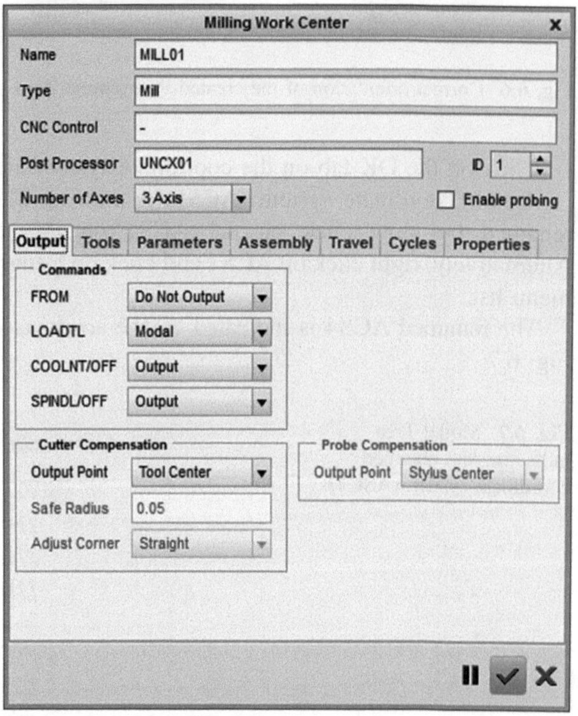

Fig. 6.8 Activated Milling Work Centre dialogue window

Click on the Check Mark icon to exit the Milling Work Centre dialogue window.

6.1.4 Setup the Operation

Click on the Operation icon as shown below.

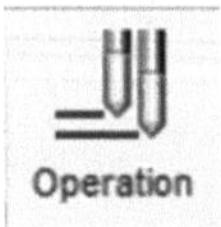

The operation dashboard tools are activated as shown. See Fig. 6.9.

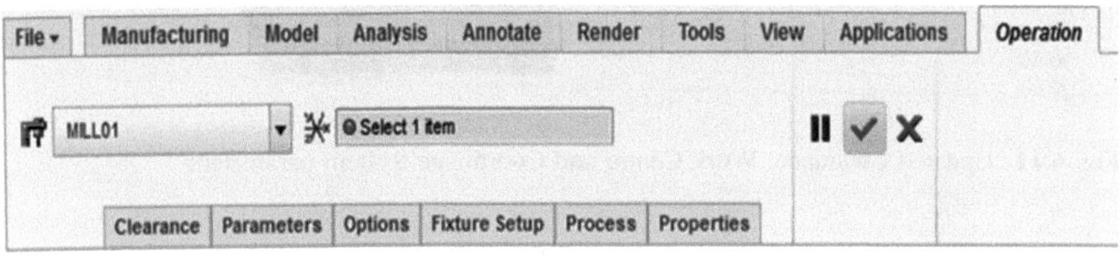

Fig. 6.9 Activated Operation dashboard in concise form

Click on the coordinate System icon section box ≫ Go to the model tree and click on the created coordinate system (ACS4), if it is not added automatically by the programme.

Click on the clearance tab to activates its panel ≫ On the Retract group, click on the Type section box to activate the drop down menu list, now click on Plane on the drop down list as shown below ≫ Click on the workpiece top surface and the Reference section box is automatically updated by the system ≫ Type 15 mm in the Value section box as the Retract distance (Figs. 6.10 and 6.11).

Fig. 6.10 Updated Clearance panel after adding parameters

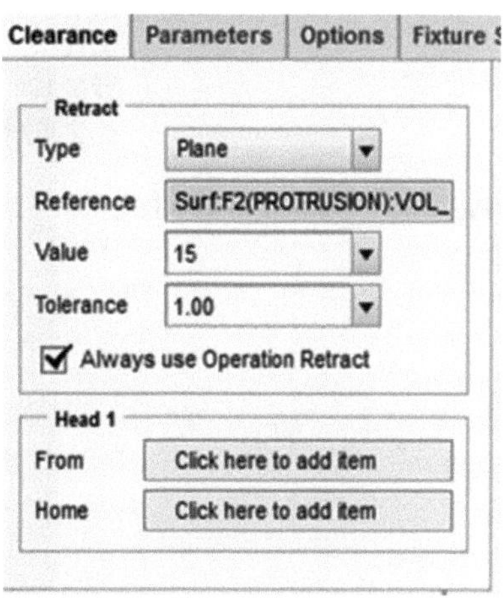

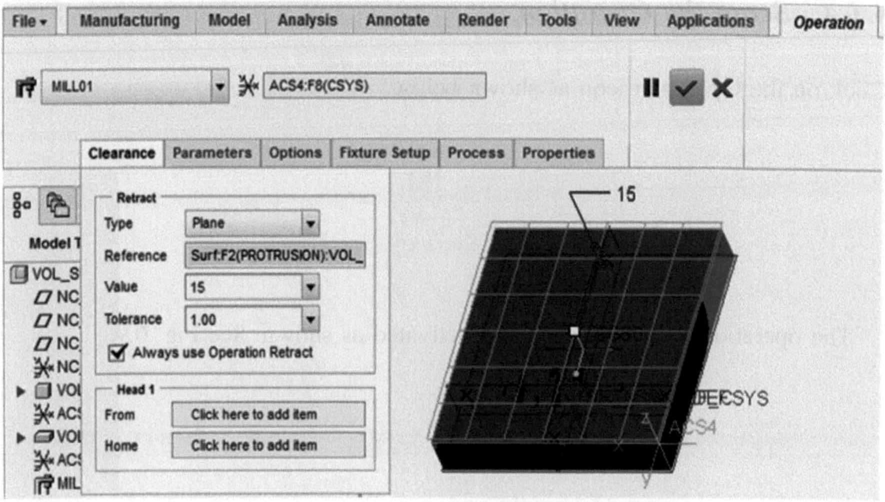

Fig. 6.11 Updated Clearance, Work Centre and Coordinate System parameters

Click on the Check Mark icon to exit Operation setup.

Check the Model Tree as shown in Fig. 6.12 to confirm that the ACS4, MILL01 and OP010 [MILL01] are on the model tree as created.

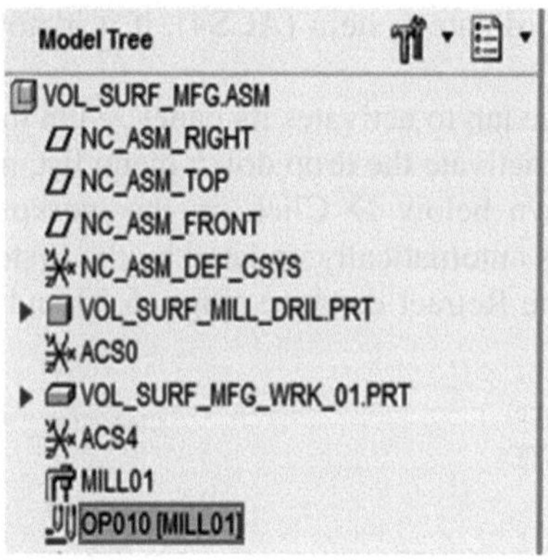

Fig. 6.12 Updated Model Tree

6.1.5 Setup the Cutting Tool

Click on the cutting tools icon 

.

The Tools Setup dialogue window is activated on the main graphic window ≫ Click on the General tab to activate its content ≫ Type the Tool name in the Name section box or leave the default Tool name ≫ Click on the Type section box, now click on End Mill Tool on the activated drop down list ≫ In the Material section box, type the Tool material or leave empty ≫ Add dimensions to the Tool. Now click on the Apply tab (Fig. 6.13).

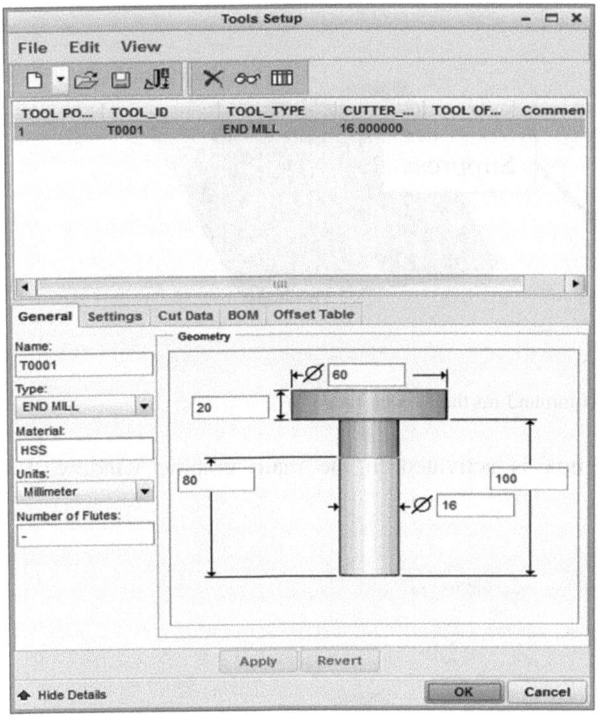

Click on OK tab to exit.

Fig. 6.13 Updated Tools Setup dialogue box

6.1.6 Hide the Holes on Part

The Holes have to be hidden before starting the Volume Rough milling operation to avoid Tool gouging.

Go to Model Tree ≫ Click on VOL_SURF_MILL_DRILL.PRT downward arrow to expand its content ≫ Select the four Holes while holding down the Ctrl key to highlight them, now right click on the mouse right button and on the drop down menu list, click on Suppress as indicated by the arrow on Fig. 6.14.

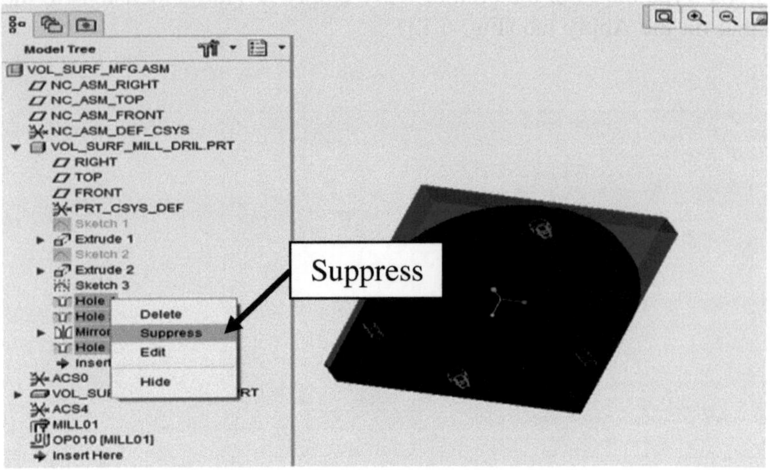

Fig. 6.14 Activating Suppress command on the Model Tree

The Suppress dialogue box is activated in the main graphic window. See Fig. 6.15.

Fig. 6.15 Activated Suppress dialogue box

Click on the OK tab on the Suppress dialogue box to exit.

All four Holes are now suppressed/hidden on both part surface and on the model tree as shown in Fig. 6.16.

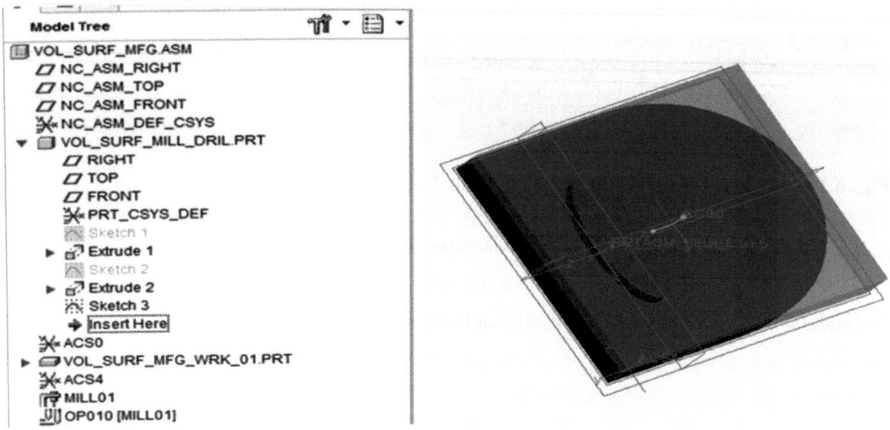

Fig. 6.16 Model Tree and Workpiece surface indicating the suppressed Holes

> Note: The suppressed Holes must be un-suppressed before carrying out the drilling operation later in this tutorial.

6.1.7 *Create Mill Volume*

Create the first mill volume

Click on the Mill menu bar ≫ Now click on the Mill Volume icon as shown below.

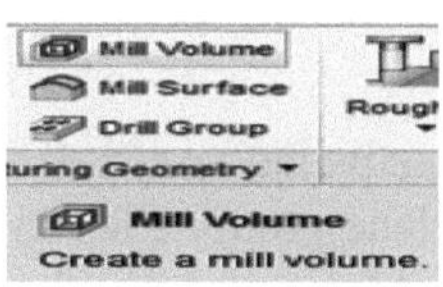

The Mill Volume application tools are activated. See Fig. 6.17.

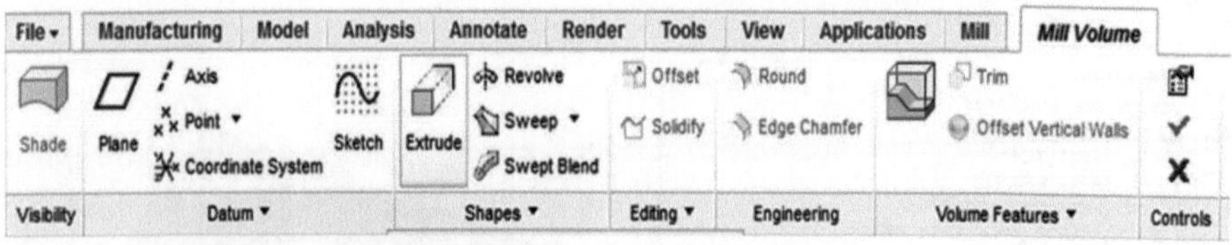

Fig. 6.17 Activated Mill Volume dashboard tools in concise form

Now click on the Extrude icon  as shown in Fig. 6.17.

The Extrude dashboard tools are activated as shown in Fig. 6.18.

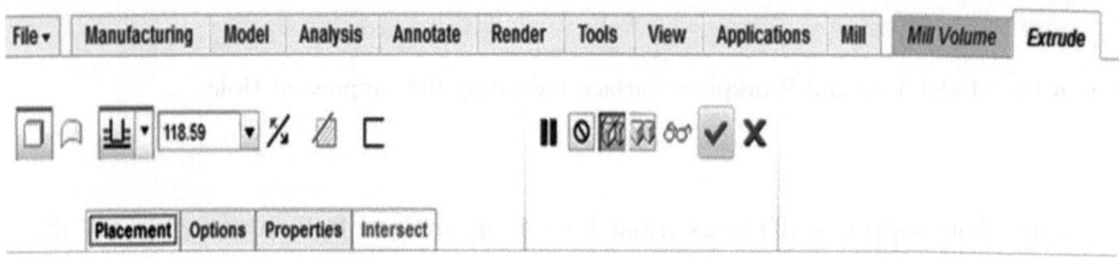

Fig. 6.18 Activated extrude dashboard in concise form

Now click on the Placement tab to activate its panel as shown in Fig. 6.19

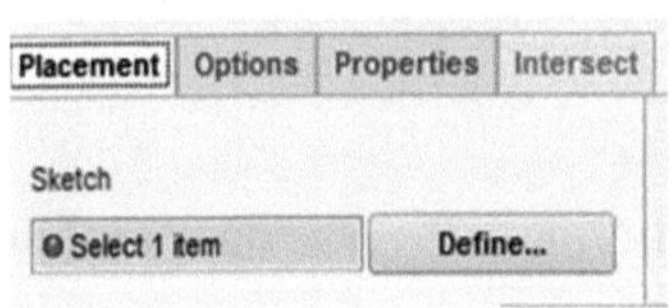

Click on the Define tab on the Sketch section box.

Fig. 6.19 Activated Placement panel

The Sketch dialogue box is activated on the main graphic window. Click on the Placement tab, on the Sketch Plane group, click in the plane section box, and now click on the Workpiece surface. Make sure that the orientation is set to Right in the Orientation section box as shown in Fig. 6.20.

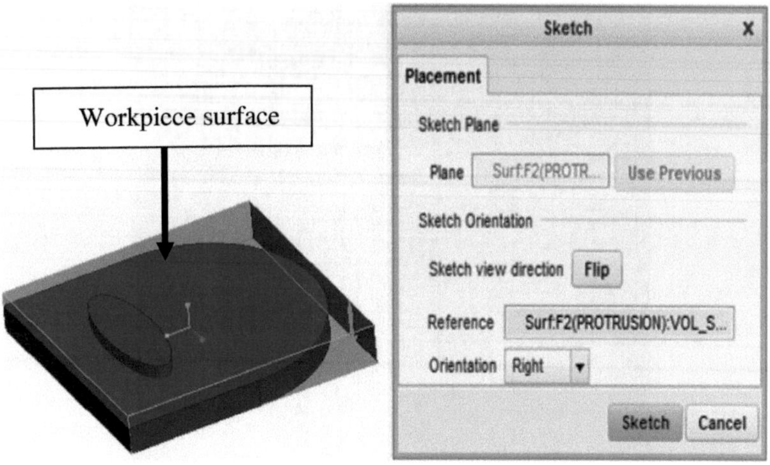

Fig. 6.20 Updated sketch dialogue box after selecting sketch plane

Click on the Sketch tab on the Sketch dialogue box to exit.
The Sketch tools are activated as shown in Fig. 6.21.

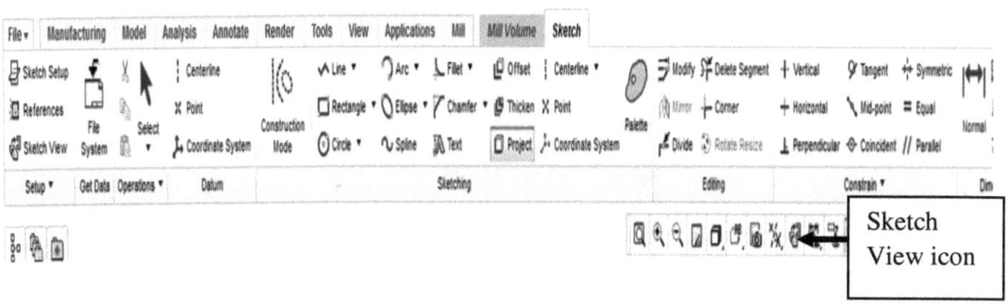

Fig. 6.21 Sketch tools

Now click on the Sketch View icon  to orient the Workpiece on the correct sketch plane as shown in Fig. 6.22.

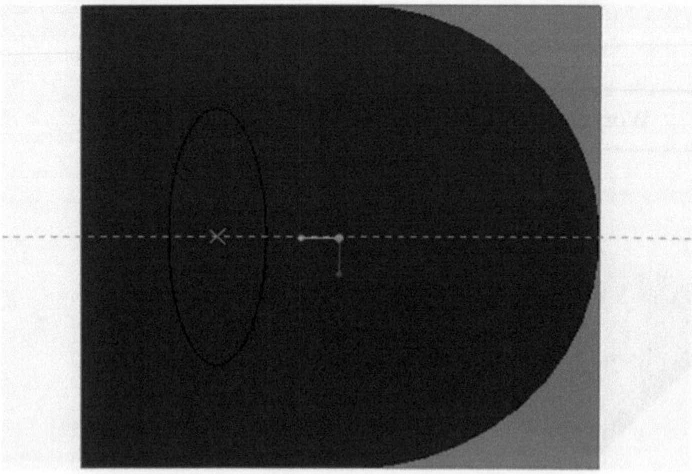

Fig. 6.22 Correct orientation of Workpiece

Create Reference

Click on the References icon as shown below.

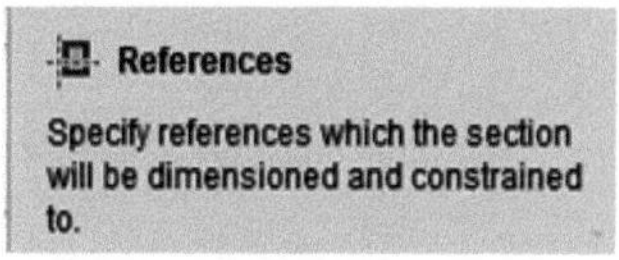

The References dialogue box is activated on the main graphic window ≫ Click on the edges of the workpiece to create the reference axes as indicated by the arrows as shown in Fig. 6.23.

Fig. 6.23 Adding references

Click on the Close tab to exit the Reference dialogue box.

Project Sketch

Click on the Project icon as indicated below.

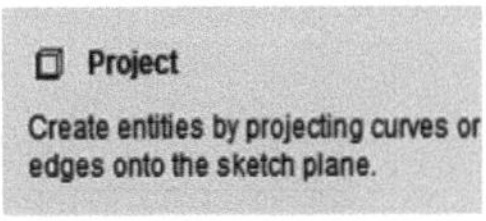

The Type dialogue box is activated as shown in Fig. 6.24.

Fig. 6.24 Type dialogue box with active Single radio button

On the Select Use Edge group, click in the Single radio button if it is not automatically activated by the system as shown in the Type dialogue box on Fig. 6.24.

Click on the edges of the Ellipse as indicated by the arrow on Fig. 6.25.

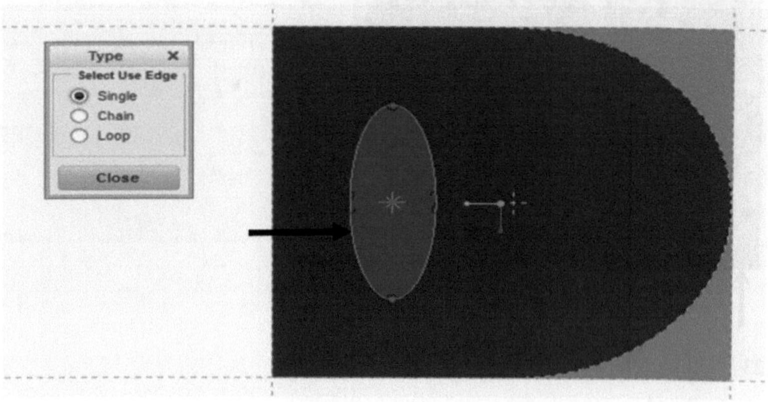

Fig. 6.25 Highlighted Ellipse after selection

Now, click on the lines and curves on the outer part edges as indicated by the arrows shown in Fig. 6.26. Delete non-intersecting lines if necessary.

Fig. 6.26 Projecting outer edges and curves on Part

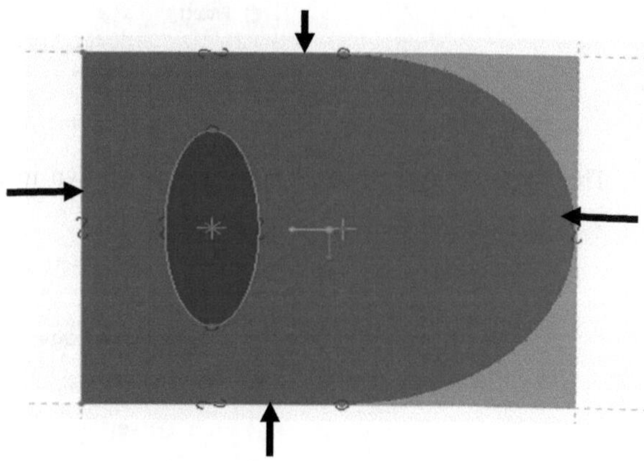

Click on the Close tab to exit Type dialogue box.

Click on the OK/Check Mark icon to exit the Sketch application.

The Extrude dashboard tools are activated as shown in Fig. 6.27.

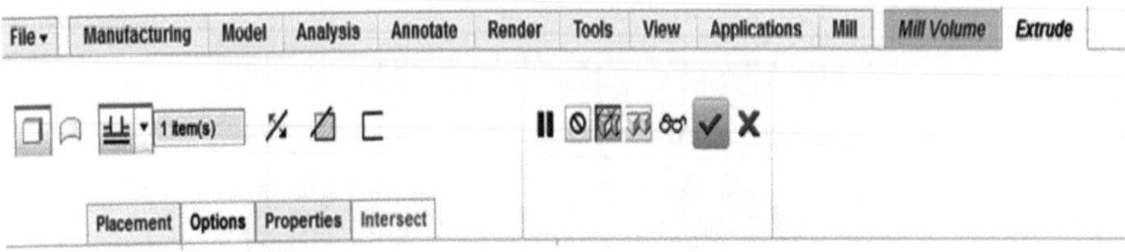

Fig. 6.27 Activated Extrude tools in concise form

Click on the Options tab to activate the options panel ≫ On the Depth group, click on the Side 1 section box to activate its drop down menu list, now click on "To Selected" on the drop down menu list ≫ Click on the Part surface as the depth of extrusion as indicated by the arrow on Fig. 6.28.

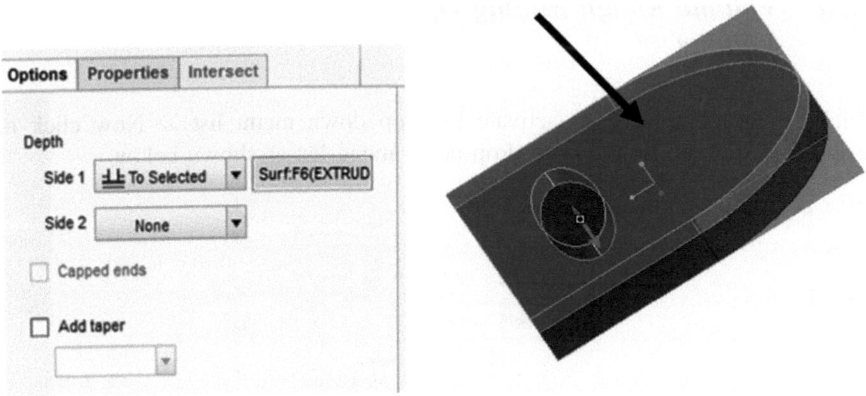

Fig. 6.28 Creating the depth of extrusion on the Workpiece

> Click on the change extrude direction arrow icon  if need be, so that the extrusion direction arrow now faces downward.

Click on the Check Mark icon  to exit the Extrude application.

Click on the Check Mark icon again to exit the Mill Volume application.

The workpiece changes on the main graphic window as shown in Fig. 6.29.

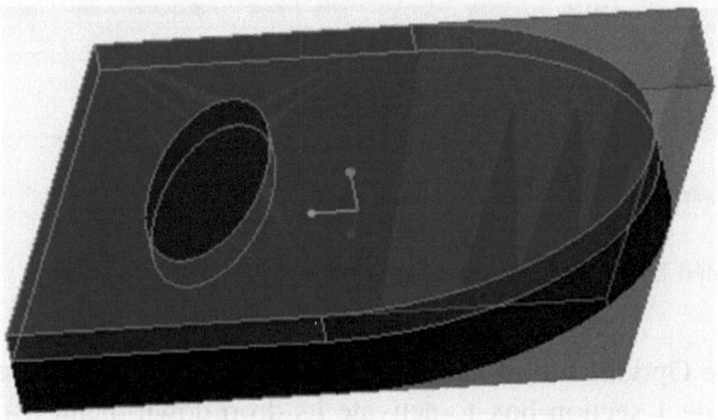

Fig. 6.29 Mill Volume now created

6.1.8 Volume Rough Milling Operation for the First Mill Volume

Click on Roughing icon to activate its drop down menu list ≫ Now click on Volume Rough on the activated drop down menu list as shown below.

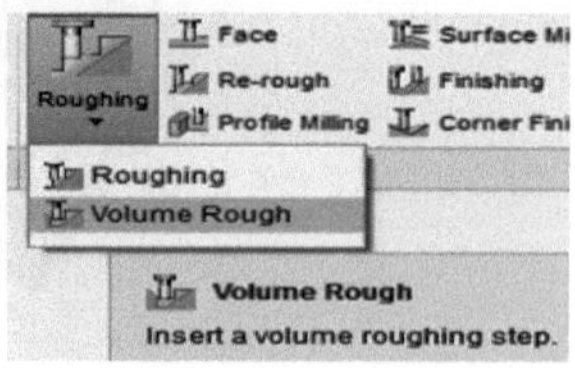

The NC SEQUENCE Menu Manager dialogue box is activated on the main window.

Check mark tool, parameters and volume on the SEQ SETUP group as shown in Fig. 6.30.

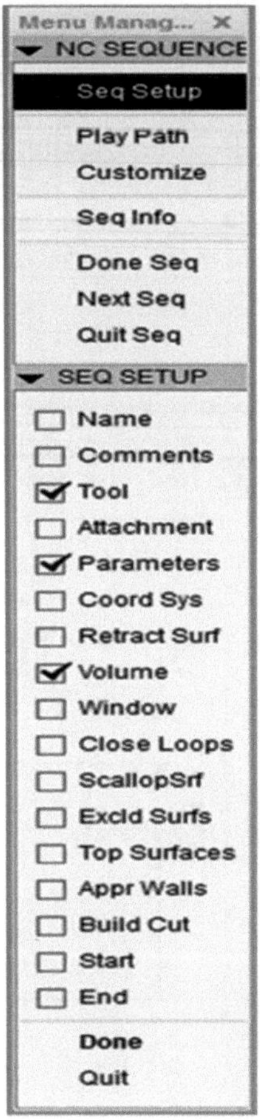

Fig. 6.30 NC SEQUENCE Menu Manager dialogue box with Tool, Parameters and Volume Check Marked

Click on Done to exit.

The Tools Setup dialogue window is activated again on the main graphic window as shown in Fig. 6.31. Check to confirm that the Tool dimensions matches that of the Tool that was setup earlier on in this tutorial. See Fig. 6.13.

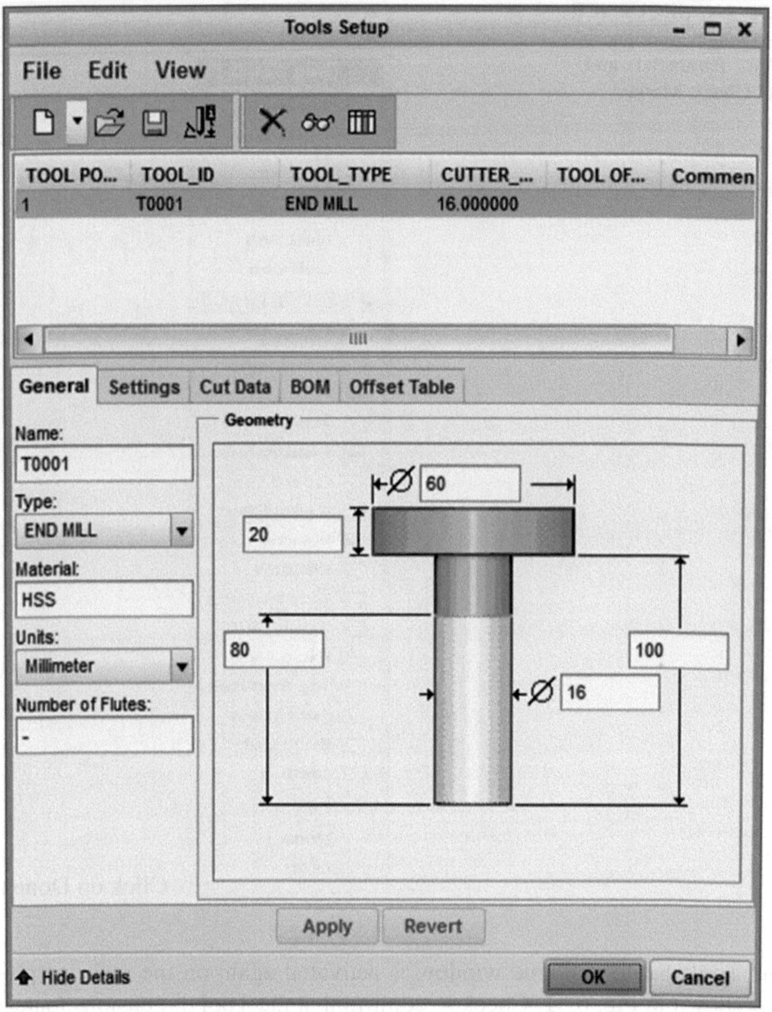

Fig. 6.31 Activated Tools Setup dialogue window showing Tool dimensions

Click on the OK tab to exit.

The Edit parameters of sequence "Volume Milling" dialogue box is activated on the main graphic window as shown in Fig. 6.32.

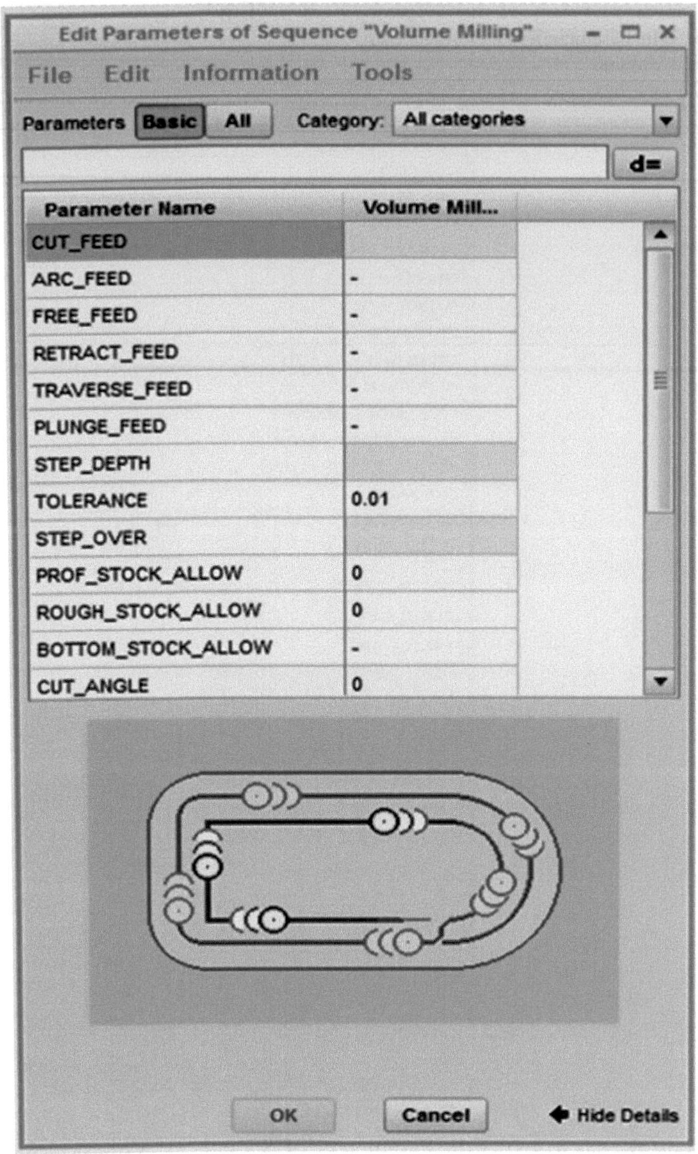

Fig. 6.32 Activated Edit Parameters of Sequence "Volume Milling" dialogue box

Input the parameter values for the Volume Rough Milling operation as shown on the Edit Parameters of Sequence "Volume Milling" dialogue box. See Fig. 6.33.

Fig. 6.33 Parameters values added to the Edit Parameters of Sequence "Volume Milling" dialogue box

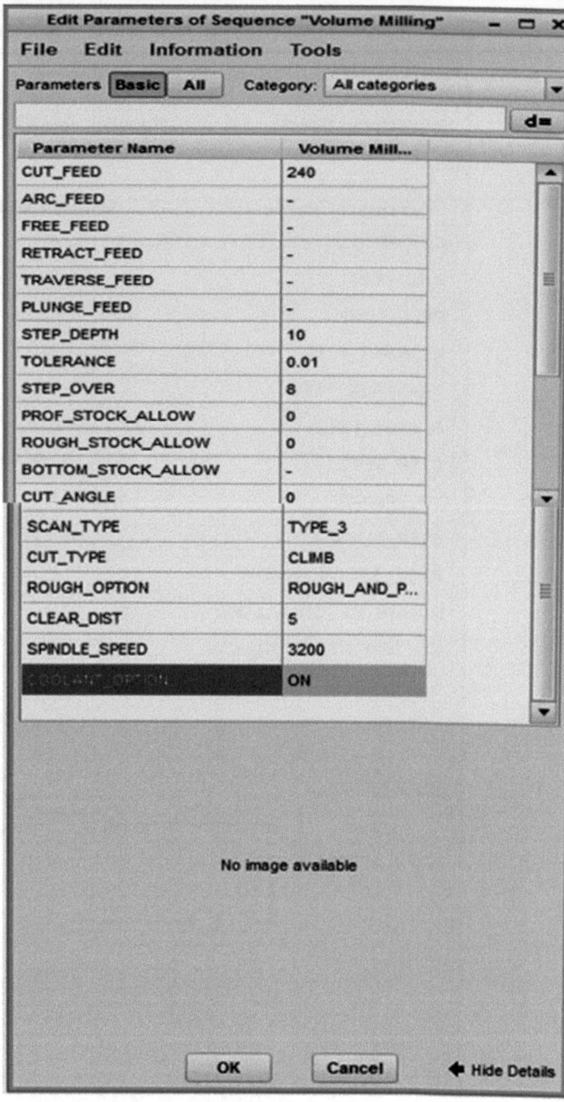

Click the OK tab to exit Edit Parameters of Sequence "Volume Milling" dialogue box.

Note: Not all the inputted parameters values on the Edit Parameters of Sequence "Volume Milling" dialogue box above are accurate. They are only used for illustrative purposes in this tutorial. To obtain accurate parameter values, you have to use the appropriate formulas to calculate their values and also consult the appropriate handbook manual.

Now click on the created Mill Volume as the volume to be Volume Roughed when prompted to do so by the system as illustrated on Fig. 6.34.

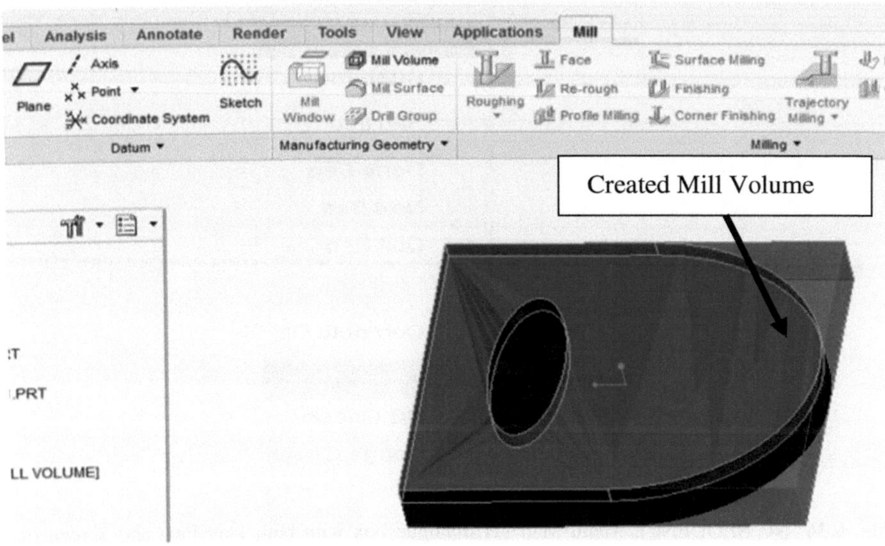

Fig. 6.34 Selected created Mill Volume

Particular attention should be paid at this stage on the message bar as shown in Fig. 6.35.

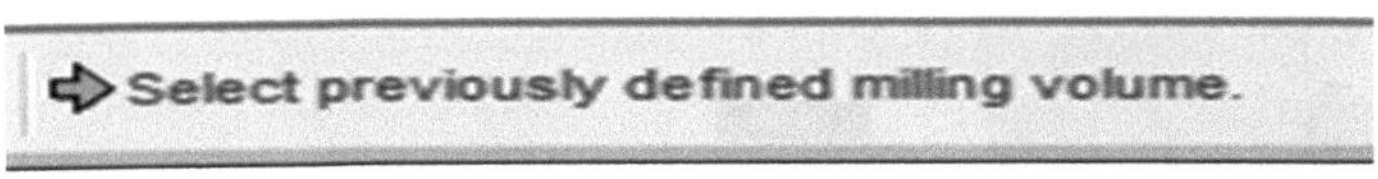

Fig. 6.35 Information on the Message bar

Click on Play Path on the NC SEQUENCE group >> Now click on Screen Play on the PLAY PATH group as indicated by the arrows on Fig. 6.36.

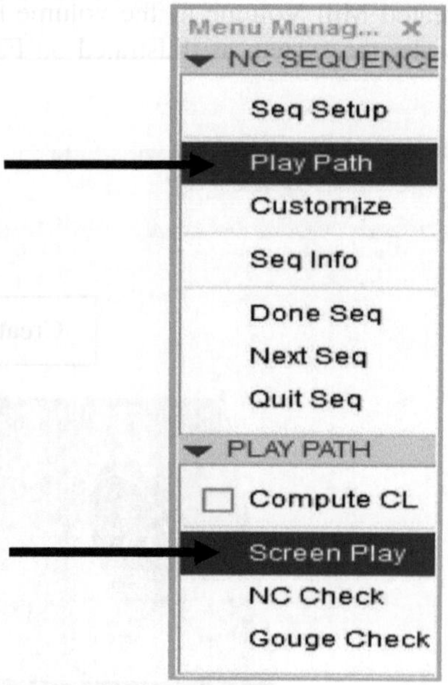

Fig. 6.36 NC SEQUENCE Menu Manager dialogue box with both Play Path and Screen Play highlighted

The PLAY PATH dialogue box and cutting tool are activated on the main graphic window as shown in Fig. 6.37.

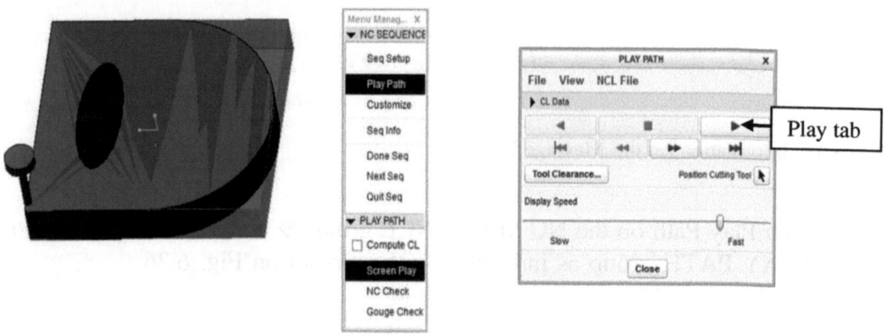

Fig. 6.37 Activated cutting Tool and PLAY PATH dialogue box

For better view of the volume Rough milling Operation, click on Display View icon, and click on Wireframe on the drop down list as the display style.

Now clik on the Play tab to start the on screen Volume Rough milling operation.

End of the on screen Volume Rough milling machining simulation is as shown in Fig. 6.38 in wireframe display.

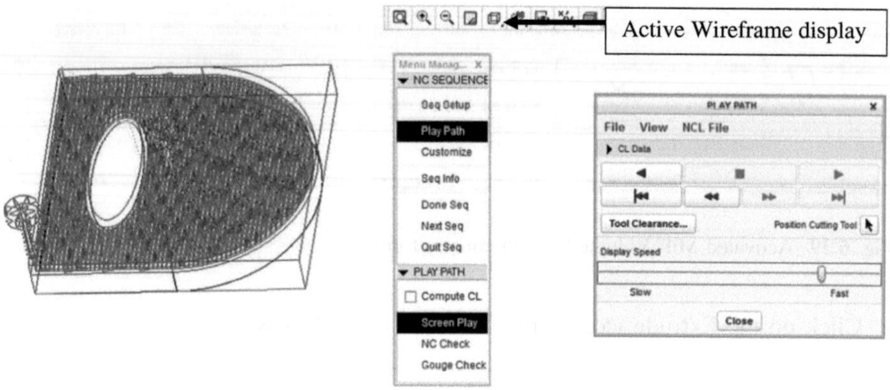

Fig. 6.38 End of the Volume Rough milling operation in wireframe display

Click on the Close tab to exit the PLAY PATH dialogue box.

Now click on the Done Seq on the NC SEQUENCE group to exit Volume Rough milling operation.

6.1.9 Create the Second Mill Volume

Hide the first created Mill Volume

The first Mill Volume has to be hidden to avoid unnecessary confusion.

Now go to Model Tree and right click on Extrude 1 [Mill_Vol_1–MILL VOLUME] ≫ Click on Hide on the drop down list. The first created Mill Volume is now hidden.

Click on the Mill menu bar to activate the Mill toolbar ≫ Now click on the Mill Volume icon as shown below.

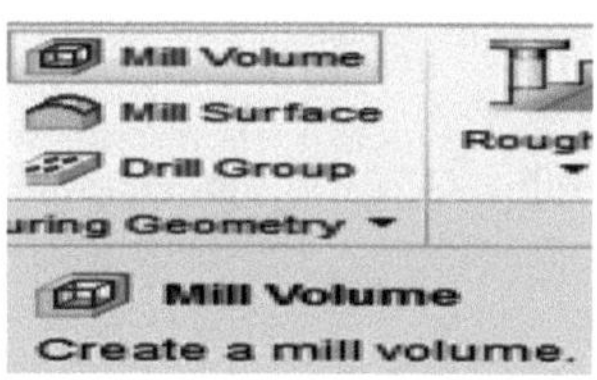

The Mill Volume application dashboard is activated as shown in Fig. 6.39.

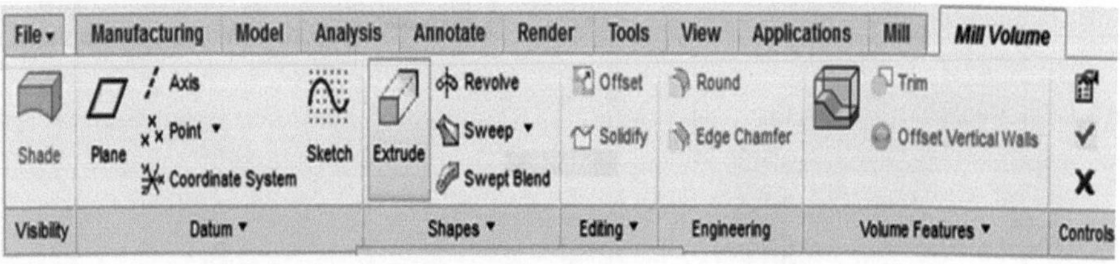

Fig. 6.39 Activated Mill Volume tools in concise form

Click on the Extrude icon in Fig. 6.39 as shown below.

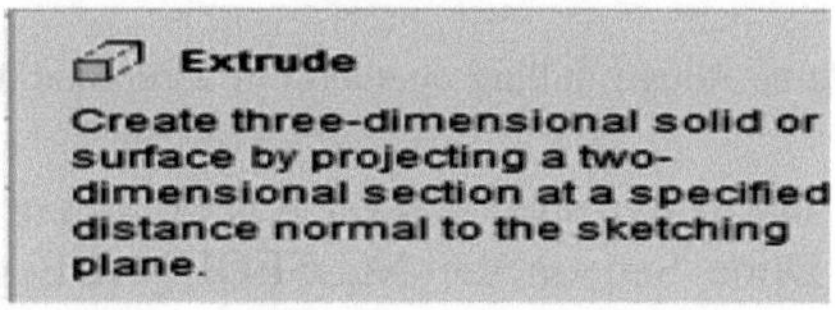

The Extrude application tools are activated as shown in Fig. 6.40.

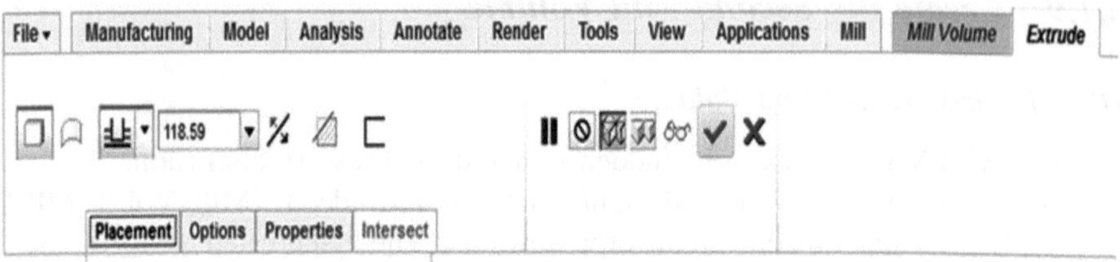

Fig. 6.40 Activated extrude dashboard

Now click on the Placement tab to activate its panel as shown in Fig. 6.41.

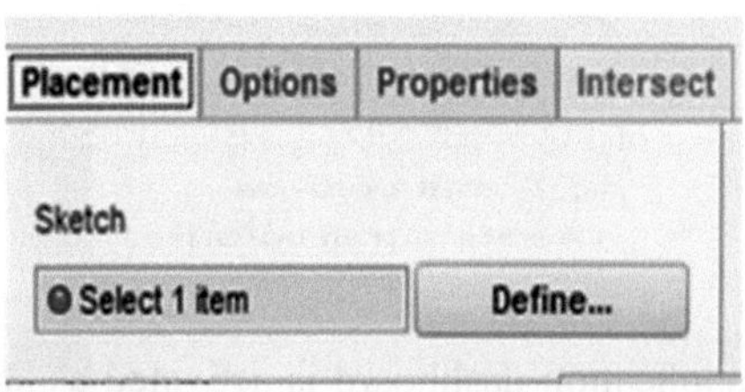

Fig. 6.41 Activated placement panel

Click on the Define tab on the Sketch section.

The sketch dialogue box is activated on the main graphic window. Click on the Placement tab, on the Sketch Plane group, click in the Plane section box, and click on the Workpiece surface. Make sure that the Orientation is set to Right in the Orientation section box as shown in Fig. 6.42. Alternatively, click on use previous tab.

Fig. 6.42 Sketch dialogue box after selecting workpiece surface as the sketch plane

Click on the Sketch tab on the Sketch dialogue box to exit.

The sketch application tools are activated as shown in Fig. 6.43.

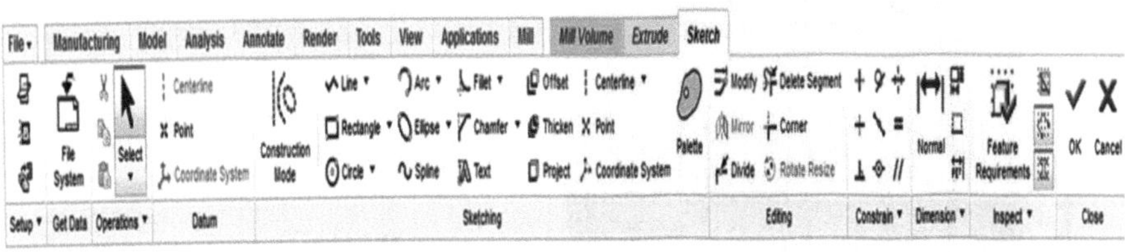

Fig. 6.43 Sketch tools in concise form

Now click the Sketch View icon  to orient the Workpiece to the correct sketch plane as shown in Fig. 6.44.

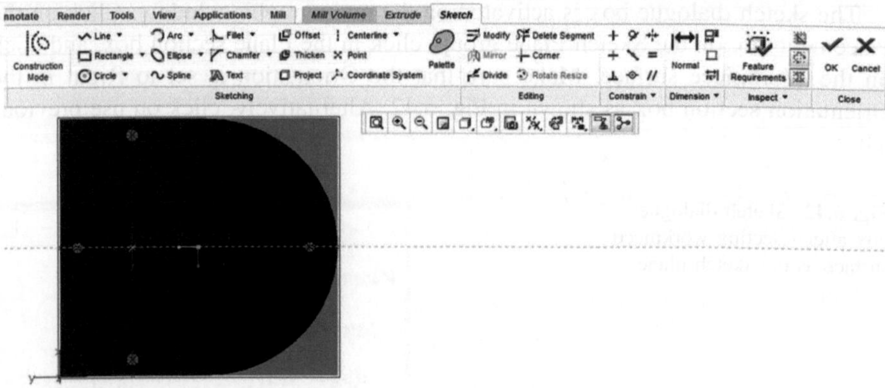

Fig. 6.44 Correct orientation of workpiece

Create Reference

Click on the Reference icon  in Fig. 6.44.

The References dialogue box is activated on the main graphic window ≫ Click on the edges of the workpiece and the horizontal line to create the reference axes as indicated by the arrows as shown in Fig. 6.45.

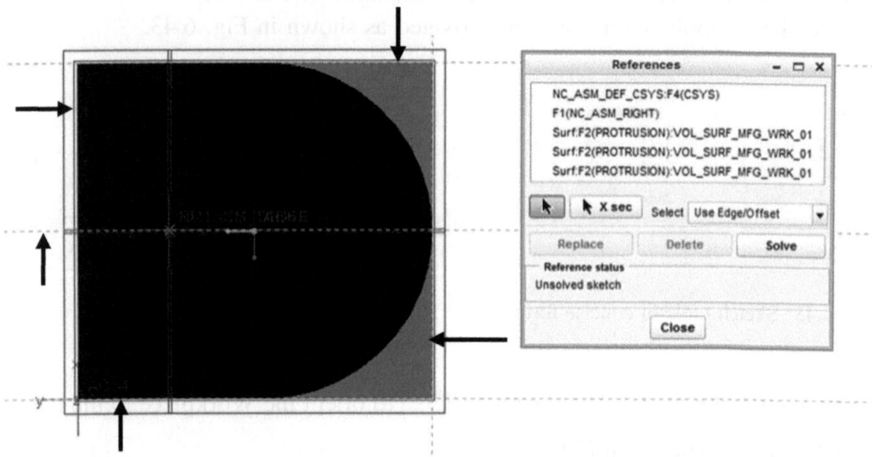

Fig. 6.45 Adding references

Click on the Close tab to exit the References dialogue box.

Project the Sketch

Click on the Project icon 

in Fig. 6.44.

The Type dialogue box is activated on the main graphic window ≫ On the Select Use Edge group, click in the Single radio button as shown in Fig. 6.46, if it is not selected automatically by the system.

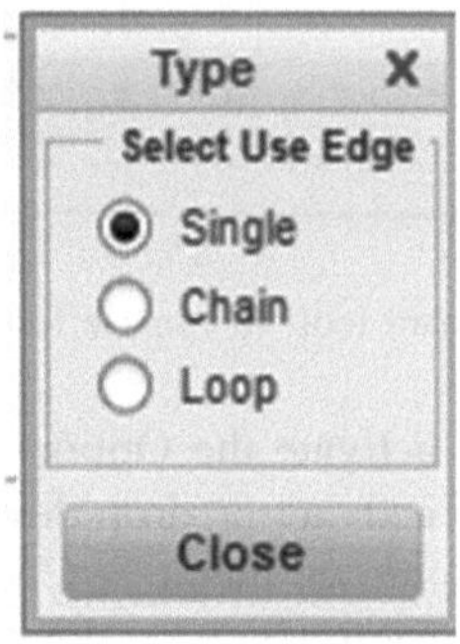

Fig. 6.46 Type dialogue box with active Single radio button

Now click on the curve edge to project it to the Workpiece surface ≫ Use the Line tool to complete the sketching of the Mill Volume as highlighted on Fig. 6.47.

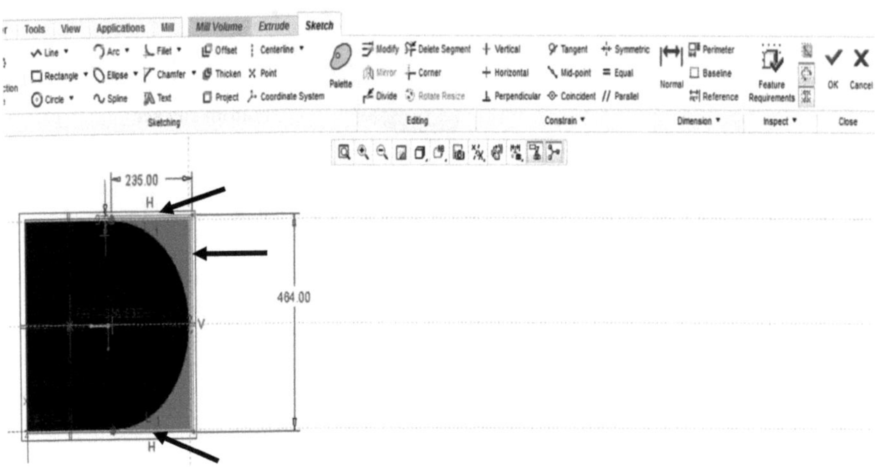

Fig. 6.47 Projected outer edge lines and curve on Workpiece

Click on OK/Check Mark icon to exit the Sketch application.

Note: The Mill Volume sketched is outside the Workpiece dimension, so that Tool can remove enough material without damaging the edge of the Part.

The Extrude tools are activated as shown in Fig. 6.48.

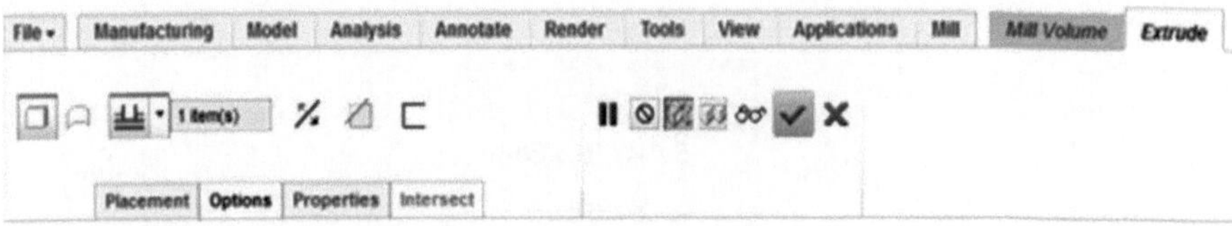

Fig. 6.48 Activated extrude dashboard tools in concise form

Click on the options tab to activate the Options panel ≫ On the Depth group, click on Side 1 section box, to activate its drop down menu list, now click on "To Selected" on the activated drop down menu ≫ Click on the bottom surface of the workpiece as the depth of extrusion. Click on the extrude direction icon to change the direction of extrusion as illustrated on Fig. 6.49.

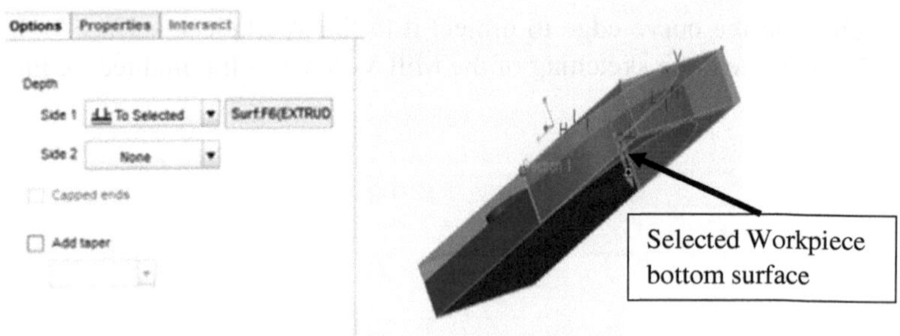

Fig. 6.49 Activated options panel and the active created Mill Volume

Click on the Check Mark icon to exit the Extrude application.
Click on the Check Mark icon again to exit the Mill Volume application.
The arrow as shown in Fig. 6.50 indicates the created Mill Volume.

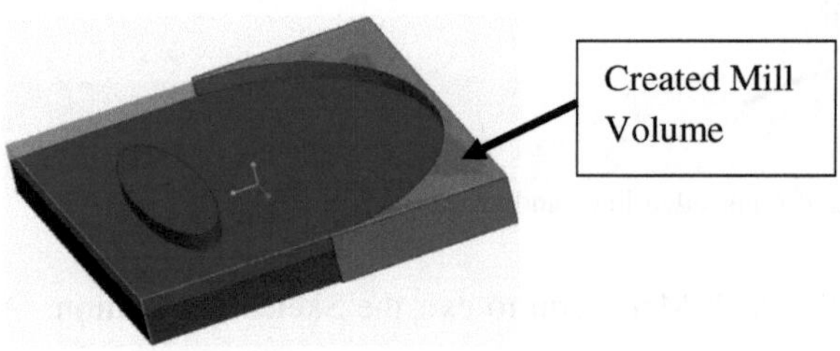

Fig. 6.50 Created second Mill Volume

6.1.10 Volume Rough Milling Operation

Click on Volume Rough on the drop down menu list as illustrated below.

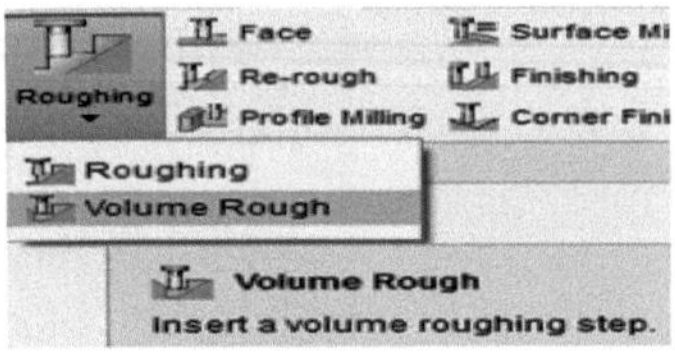

The NC SEQUENCE Menu Manager dialogue box is activated on the main graphic window ≫ Check Mark Parameters and Volume on the SEQ SETUP group as indicated by the arrows on Fig. 6.51.

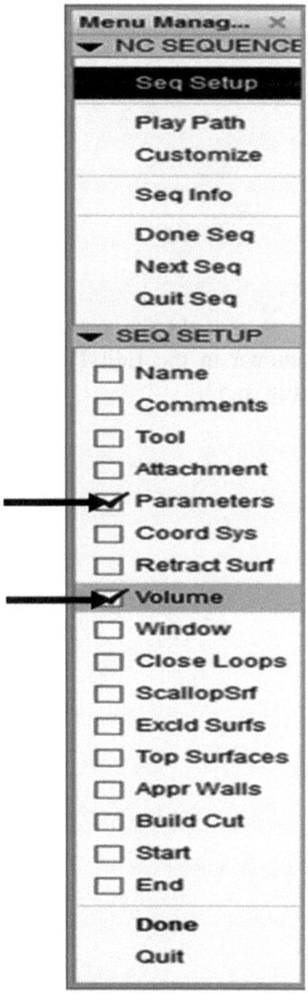

Fig. 6.51 NC SEQUENCE Menu Manager dialogue box with Parameters and Volume Check Marked

Click on Done on the SEQ SETUP group on NC SEQUENCE Menu Manager dialogue box.

The Edit Parameters of Sequence “Volume Milling” dialogue box is activated on the main graphic window as shown in Fig. 6.52.

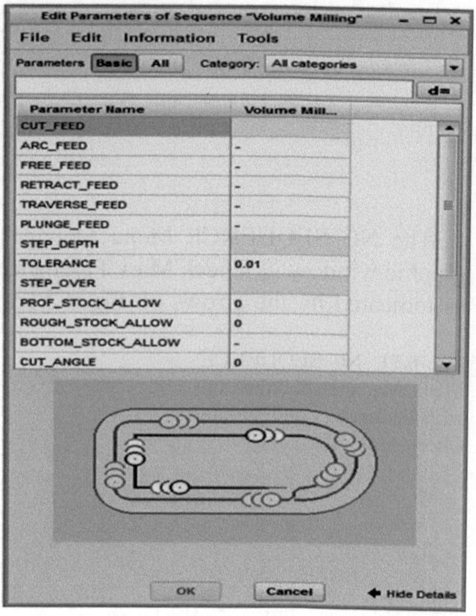

Fig. 6.52 Activated Edit Parameters of Sequence “Volume Milling” dialogue box

Input/add the parameter values for the Volume Rough Milling operation as shown in the Edit Parameters of Sequence “Volume Milling” dialogue box on Fig. 6.53.

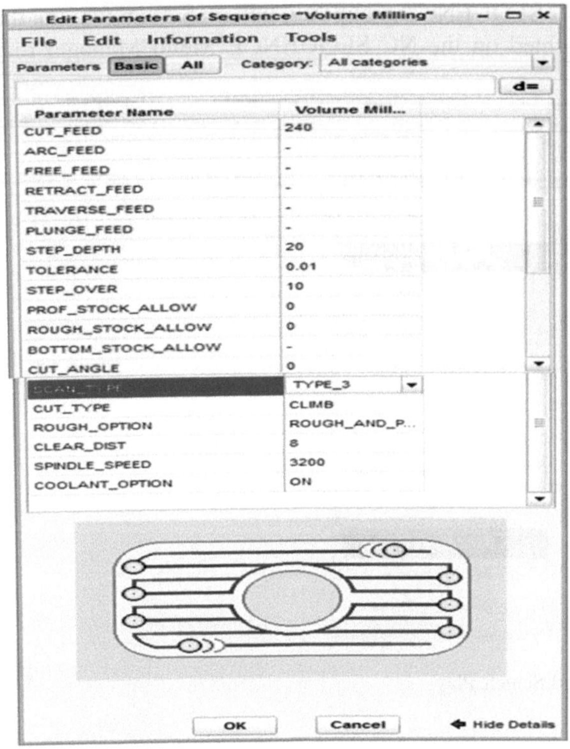

Now click on OK tab to exit.

Fig. 6.53 Volume Rough milling parameters values

Now click on the created Mill Volume as the volume to be Volume Roughed when prompted to do so by the system.

Note: Particular attention should be paid at this stage to the information on the message bar.

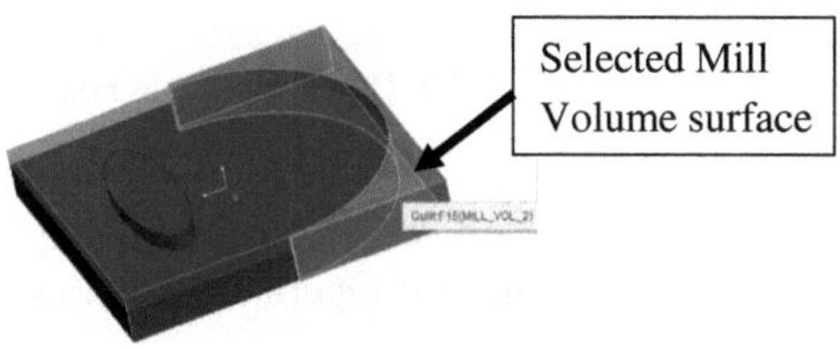

Fig. 6.54 Created Mill Volume selected when requested by the system

Click on Play Path on the NC SEQUENCE group ≫ Click on Screen Play on the PLAY PATH group as highlighted on the NC SEQUENCE Menu Manager dialogue box on Fig. 6.55.

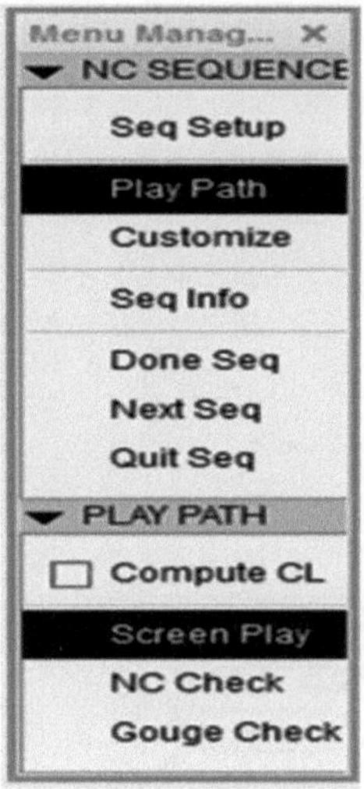

Fig. 6.55 Highlighted Play Path and Screen Play

The cutting Tool and the PLAY PATH dialogue box are activated. See Fig. 6.56.

Fig. 6.56 Activated cutting Tool and the PLAY PATH dialogue box

Click on the Play tab to start the on screen Volume Rough milling operation. The end of the Volume Rough Milling Operation is shown in Fig. 6.57.

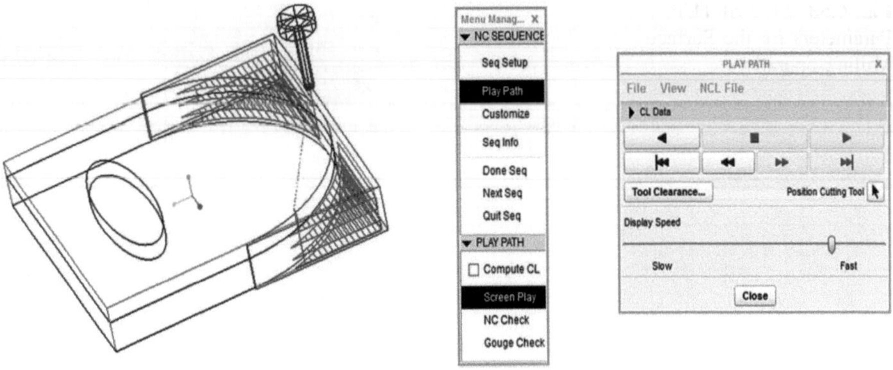

Fig. 6.57 End of Volume Rough Milling operation in wireframe display

Now click on the Close tab.

Now click on the Done Seq on the NC SEQUENCE group to exit the Volume Rough milling operation.

6.2 Surface Milling Operation

Surface Milling is used for smoothing out surfaces after Volume Rough milling operation.

Click on Mill menu bar to activate the Mill toolbars.

Click on Surface Milling icon as shown below.

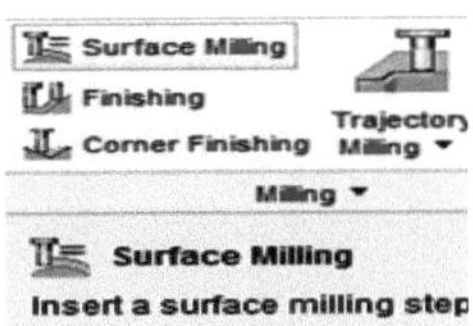

The NC SEQUENCE Menu Manager dialogue box appears immediately on the main graphic window as shown in Fig. 6.58.

Check mark tool, parameters, surfaces and define cut on the SEQ SETUP group as illustrated on Fig. 6.58.

Fig. 6.58 SEQ SETUP Parameters for the Surface Milling operation

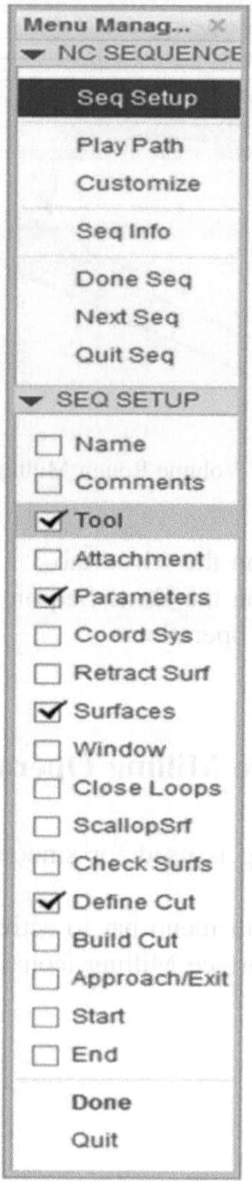

Click on Done.

The Tools Setup dialogue window is activated on the main graphic window ≫ Now define the new Tool for Surface Milling ≫ Click on New icon downward arrow and click on the End Mill Tool on the drop down list ≫ Add new dimensions to the Tool. See Fig. 6.59.

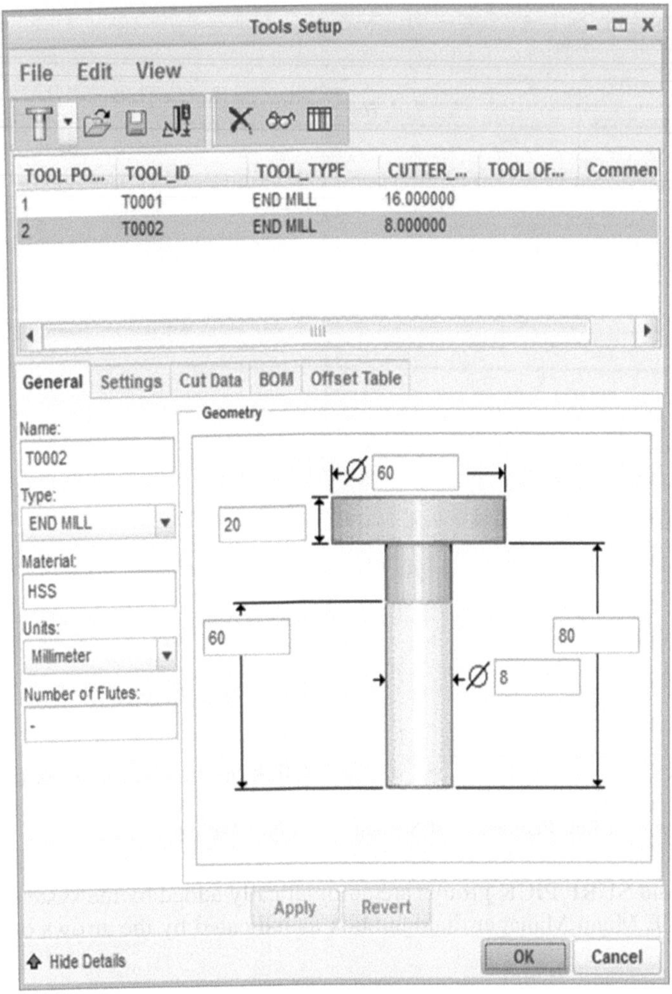

Click on Apply tab.

Fig. 6.59 New End Mill for Surface Milling operation

Click on OK tab to exit the Tools Setup dialogue window.

The Edith Parameters of Sequence "Surface Milling" dialogue window is activated ≫ Input parameters values for the Surface Milling process as shown in Fig. 6.60.

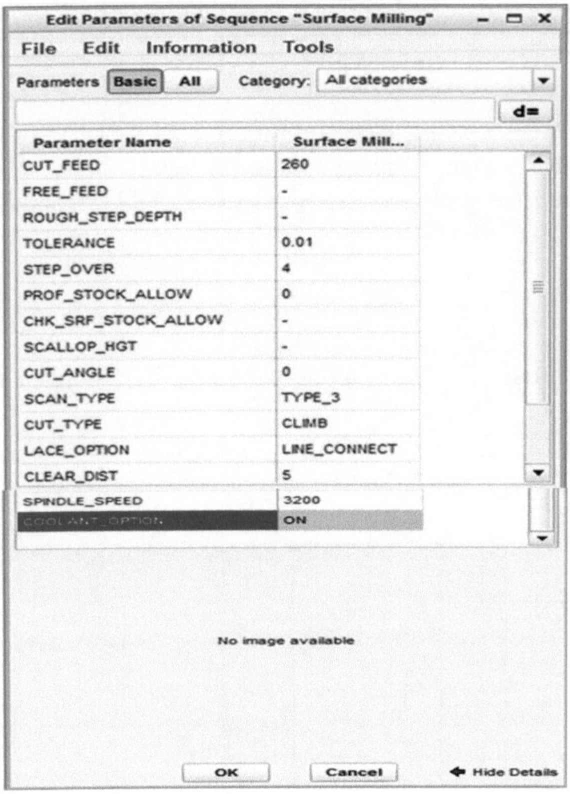

Click on the OK tab to exit.

Fig. 6.60 Parameter values for Edit Parameters of Sequence "Surface Milling"

NC SEQ SURFS and SURF PICK groups are automatically added by the system to the NC SEQUENCE Menu Manager dialogue box as indicated by the arrows on Fig. 6.61.

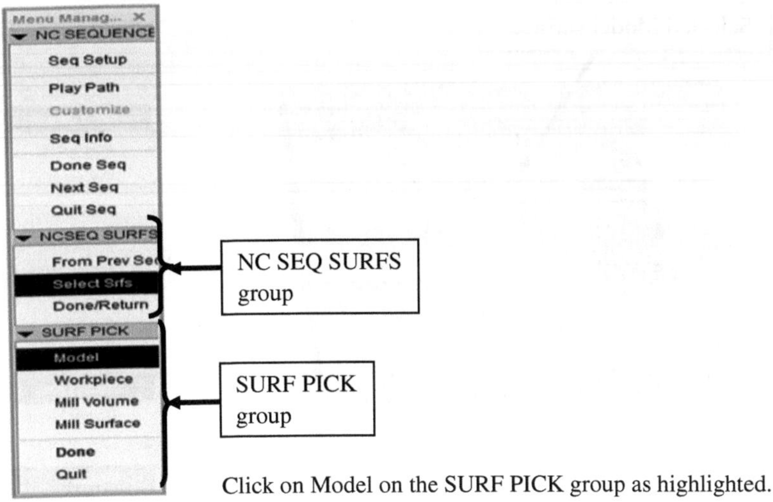

Fig. 6.61 NC SEQ SURFS and SURF PICK groups

Note: To hide EXTRUDE 2 [MILL_VOL_2—MILL VOLUME], select and highlight it and then right click and select hide. In this tutorial, the Mill volume will not be hidden.

Now click at the surface to be the selected Model as indicated by the arrow on Fig. 6.62.

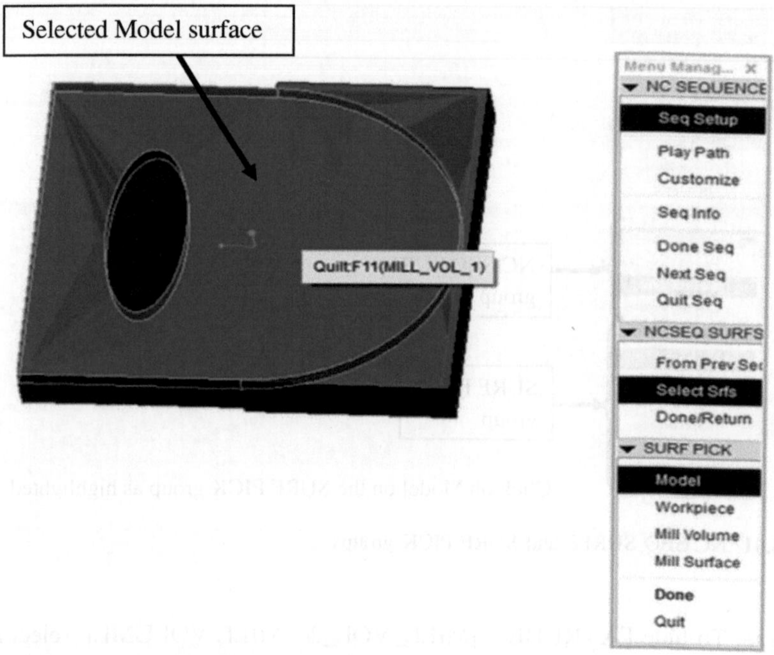

Fig. 6.62 Surface picked Model

Click on Done on the SURF PICK group on the NC SEQUENCE Menu Manager dialogue box.

The Select dialogue box is activated on the main graphic window as shown in Fig. 6.63.

Fig. 6.63 Activated Select dialogue box

Now click on the Workpiece surface to add as the surfaces to be Surface Milled as indicated by the arrow on Fig. 6.64.

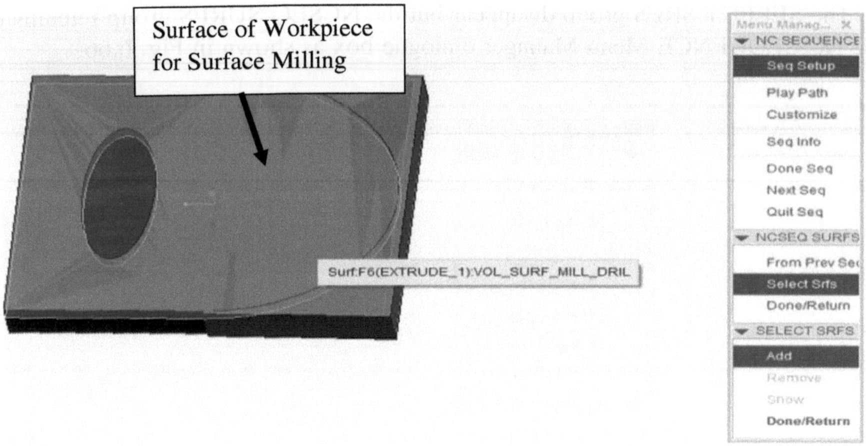

Fig. 6.64 Added Workpiece surface for Surface Milling

Hold down the Ctrl key and now select the Ellipse surface to add as the surfaces to be Surface Milled as indicated by the arrow on Fig. 6.65.

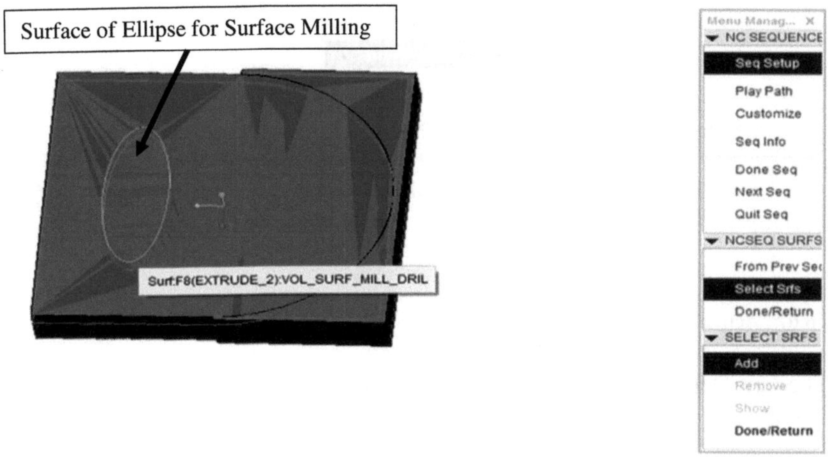

Fig. 6.65 Added Ellipse surface for the Surface Milling operation

Click on the OK tab on the Select dialogue box ≫ Now click on Done/Return on the SELECT SRFS group.

The SELECT SRFS group disappear but the NCSEQ SURFS group remains on the NC SEQUENCE Menu Manager dialogue box as shown in Fig. 6.66.

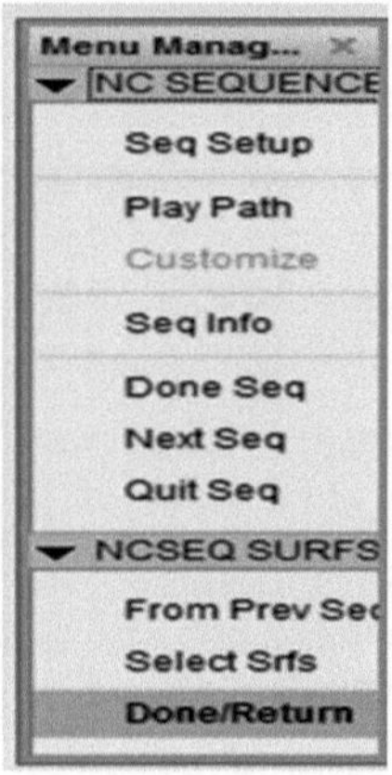

Click on Done/Return.

Fig. 6.66 Exiting NCSEQ SURFS group on Menu Manager dialogue box

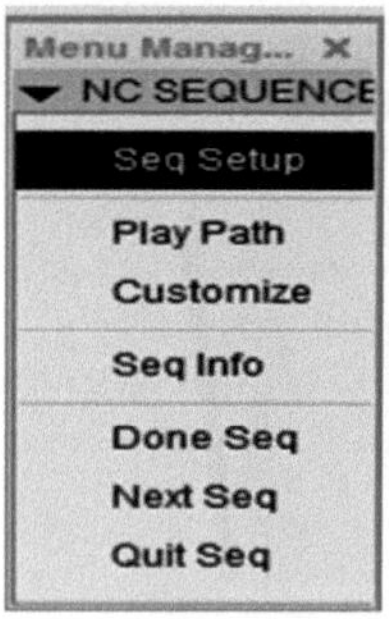

Fig. 6.67 Active NC SEQUENCE group

The Cut Definition dialogue box is activated on the main graphic window. See Fig. 6.68.

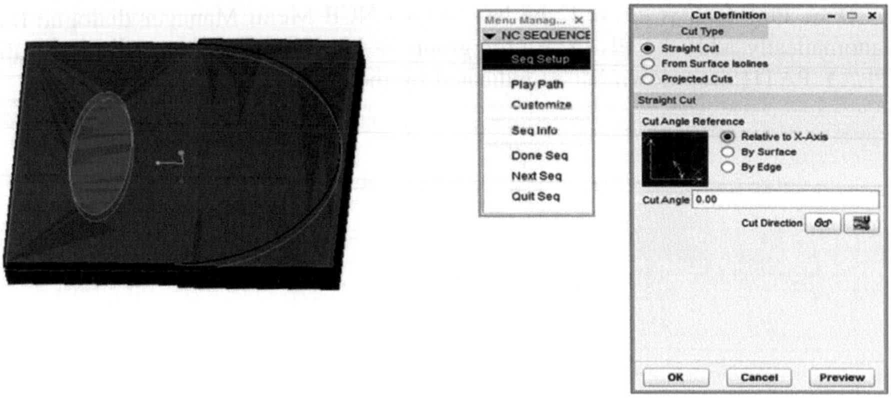

Fig. 6.68 Activated cut definition dialogue box on main window

In this tutorial, click in the Reference to X-axis radio button to make it active on the Cut Angle Reference section, on the Straight Cut group.

Click OK tab to exit the Cut Definition dialogue box.

Now click on Play Path on the NC SEQUENCE drop down list as highlighted on Fig. 6.69.

Fig. 6.69 Highlighted Play Path on the NC SEQUENCE group

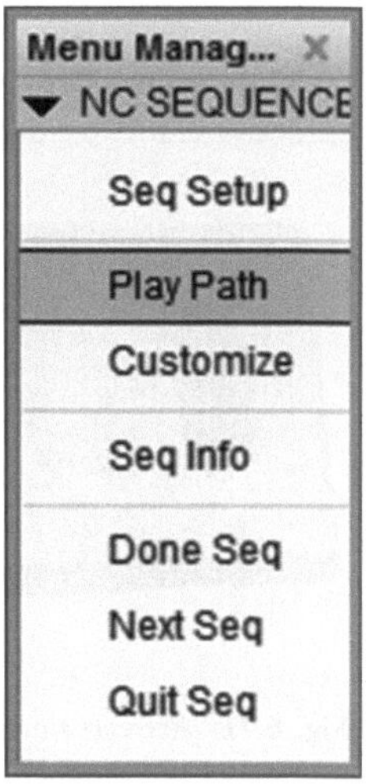

Once Play Path is clicked, the NC SEQUENCE Menu Manager dialogue box automatically adds the PLAY PATH group ≫ Now click on Screen Play on the PLAY PATH drop down list as indicated by the arrow on Fig. 6.70.

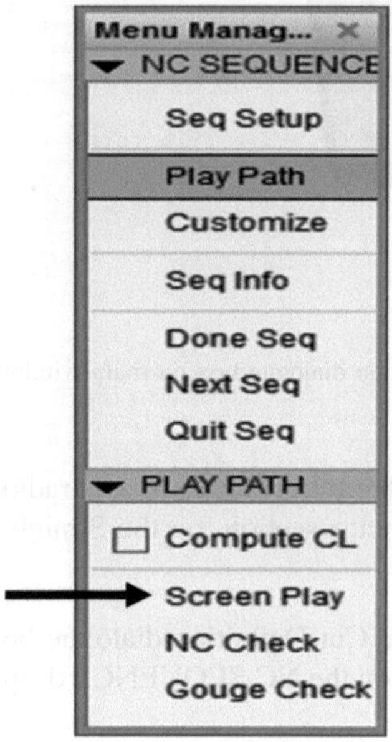

Fig. 6.70 Activated PLAY PATH group on the NC SEQUENCE dialogue box

The cutting Tool and the PLAY PATH dialogue box are activated. See Fig. 6.71.

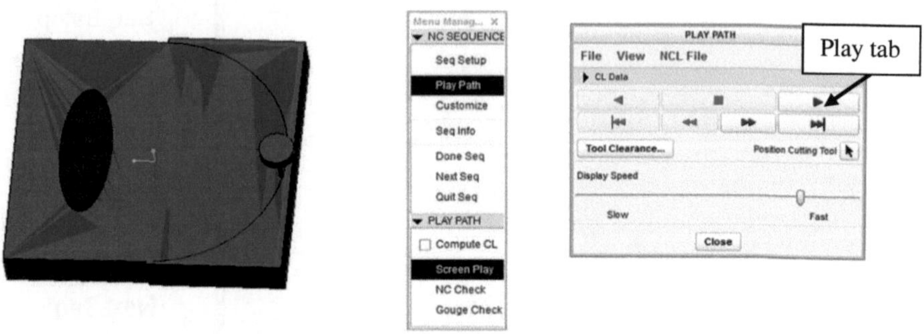

Fig. 6.71 Activated cutting Tool and the PLAY PATH dialogue box

Now click on the Play tab to start the Surface Milling operation.

The end of the Surface Milling operation is shown in Fig. 6.72 in wireframe display.

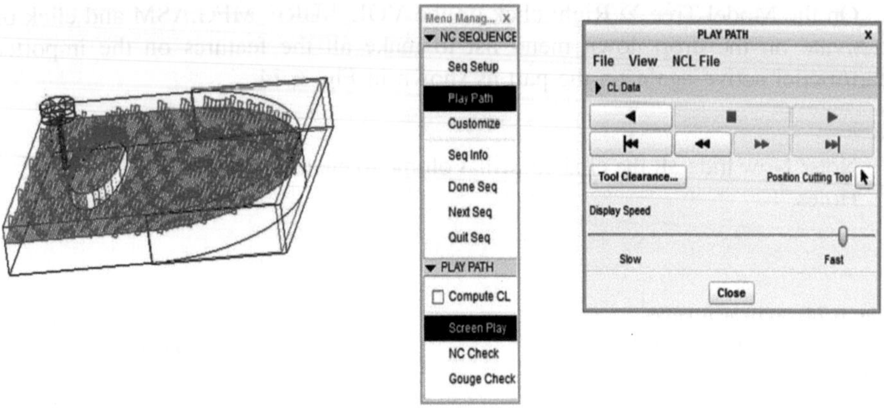

Fig. 6.72 End of on screen Surface Milling operation

6.3 Drilling of the Holes

6.3.1 Activate/Un-suppress the Holes

Before carrying out the drilling of the Holes, the suppressed Holes have to be un-suppressed.

Go to the Model Tree ≫ Right click on VOL_SURF_MILL_DRILL. PRT ≫ Click on Activate on the activated drop down options list to make the Part active.

Now go to Operations and click on the downward pointing arrow ≫ Click on Resume All on the activated drop down menu list as shown in Fig. 6.73.

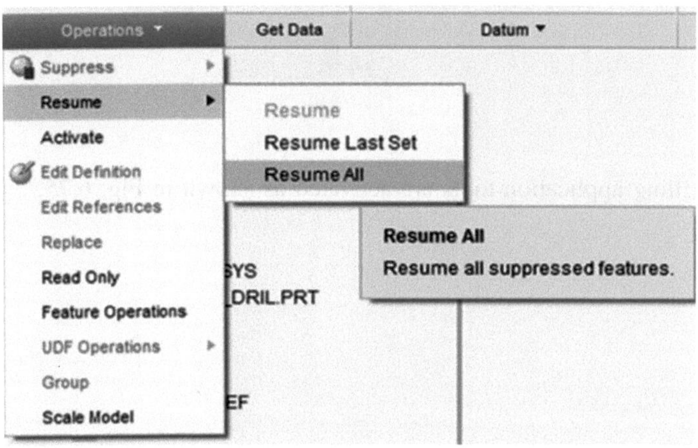

Fig. 6.73 Operations drop down list

On the Model Tree ≫ Right click on the VOL_SURF_MFG.ASM and click on Activate on the drop down menu list to make all the features on the imported part/model active again on the part as shown in Fig. 6.74.

Note: Now hide all the created Mill Volume to enable proper view of the four Holes.

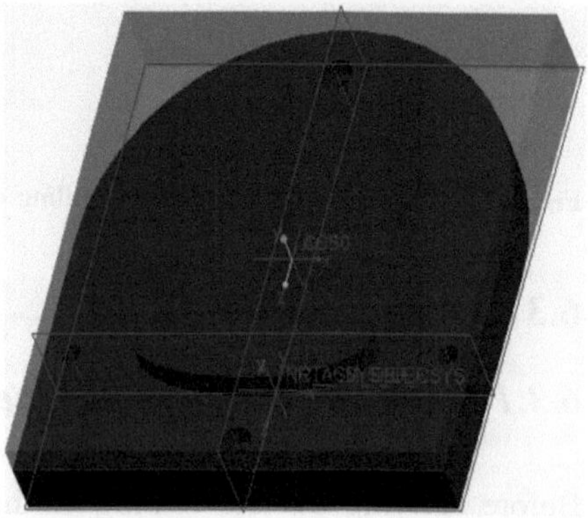

Fig. 6.74 Activated Holes are now active again on the Part

6.3.2 Create Drilling Operation

Click on Mill menu bar ≫ Now click on Standard drill tool icon as shown below.

The Drilling application tools are activated as shown in Fig. 6.75.

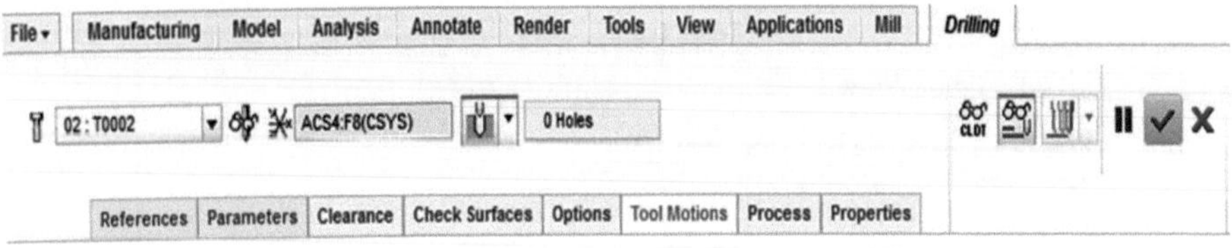

Fig. 6.75 Activated Drilling dashboard

Click on the Tool icon  in Fig. 6.75 to create a new tool for the Standard drilling operation.

The Tools setup dialogue window is activated on the main window ≫ To define the new Tool click on New icon downward pointing arrow, and now click on the basic drill tool on the drop down list. Add new dimensions to the tool, name and material (Fig. 6.76).

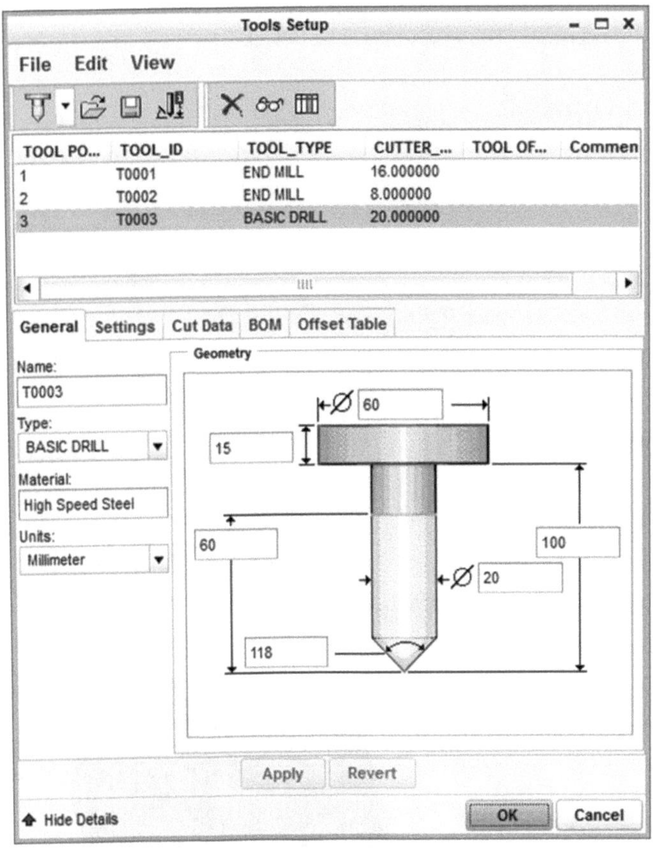

Click on Apply tab.

Fig. 6.76 Drilling Tool parameters

Click on OK tab to exit the Tools Setup operation as shown in Figs. 6.76.

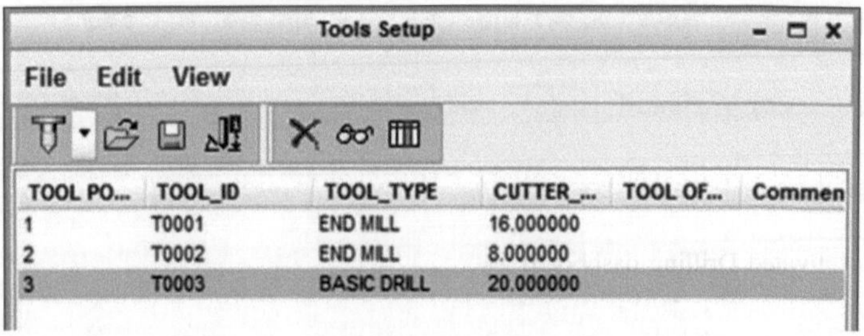

Fig. 6.77 Created BASIC DRILL Tool T0003 highlighted

Click on References tab to activate its panel ≫ Click on the downward pointing arrow on the Type section box and now click on Axes on the drop down menu list ≫ Click in the Hole section box, and now click on the axes of the highlighted Holes on Fig. 6.79. hold down the Ctrl key while selecting the four axes. The start and end section box is set at Automatic by default as shown in Fig. 6.78.

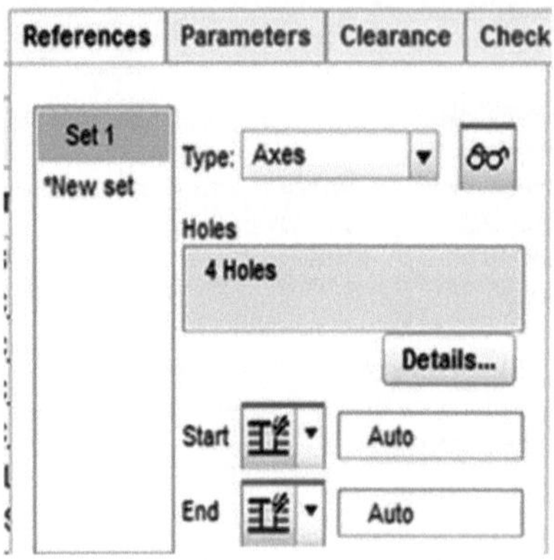

Fig. 6.78 Updated References panel

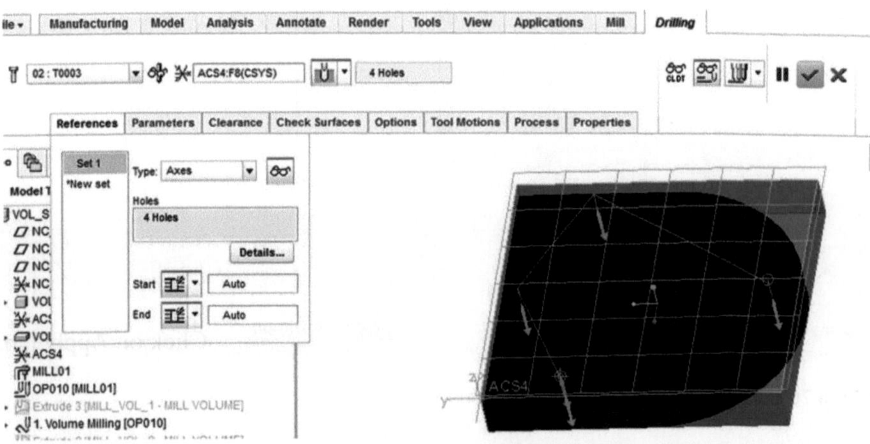

Fig. 6.79 Drilling main graphic window after adding the basic drill parameters

Click on the Check Mark icon ✓ to accept the inputs and exit the Drilling application.

Click on Parameters tab to add parameter values as shown in Fig. 6.80

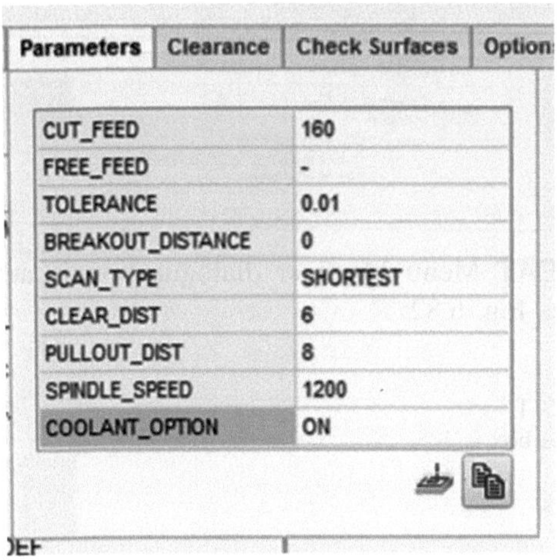

Fig. 6.80 Added parameters values for basic drilling

Click on the Check Mark to exit.

Click on the Manufacturing menu bar to activate its ribbon ≫ Click on Play Path. Alternatively, go to the model tree, and right click on Drilling 1(OP010), and now click on Play Path on the activated drop down menu list.

The Drilling Tool and PLAY PATH dialogue box are activated as shown in Fig. 6.81.

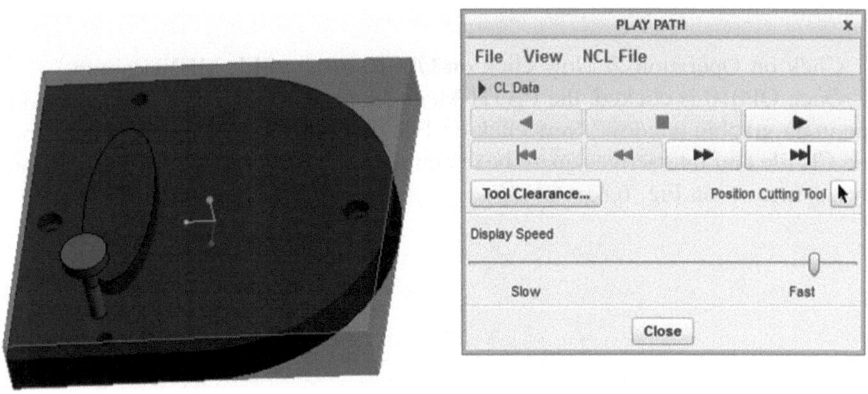

Fig. 6.81 Activated Drilling Tool and PLAY PATH dialogue box

Click on Play tab arrow on the Play Path dialogue box to start the on screen drilling operation.

Click on Close tab to exit.

6.4 Create the Cutter Location (CL) Data

Click on the Save a CL File icon illustrated below.

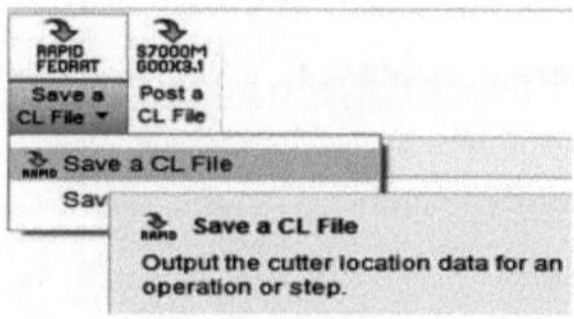

The SELECT FEAT Menu Manager dialogue box is activated on the main graphic window. See Fig. 6.82.

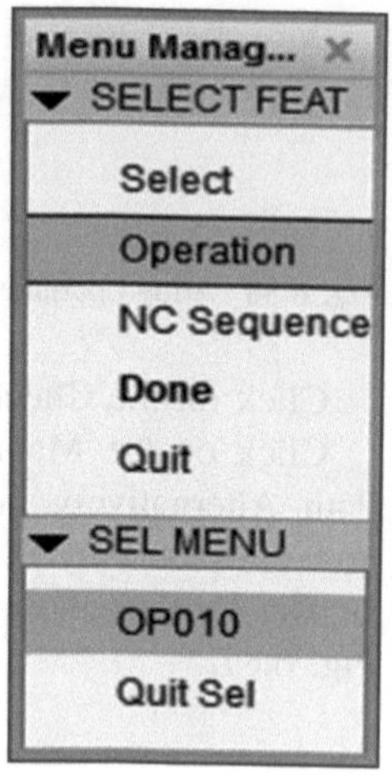

Fig. 6.82 SELECT FEAT Menu Manager dialogue box

Click on Operation ≫ Now click on OP010 on the SEL MENU group.

Once OP010 is clicked, the PATH Menu Manager dialogue box is activated on the main graphic window. Now click on File on the PATH group and Check mark the CL file and interactive square boxes on the OUTPUT TYPE group as indicated by the arrows on Fig. 6.83.

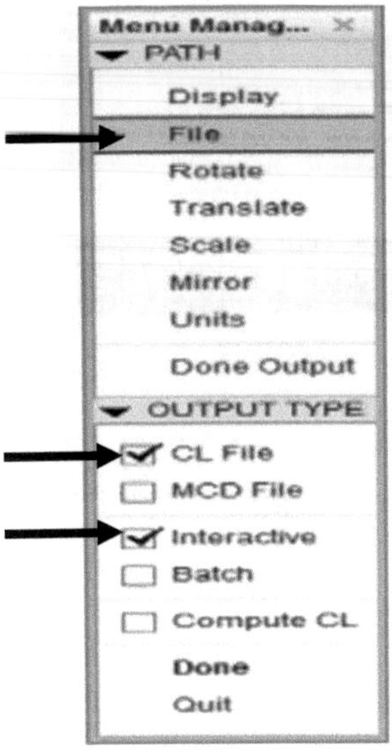

Click on Done.

Fig. 6.83 PATH dialogue box with CL File and Interactive Check Marked

The Save a Copy dialogue window is activated on the main window. See Fig. 6.84.

Note: You can also change the name and file directory where you want to save the CL file data.

Fig. 6.84 Activated Save a Copy dialogue window in concise form

Click on the OK tab to save the CL file in your already chosen directory folder.

CL data file is now created and saved in the chosen directory.

Click on Done Output on the PATH group on the Menu Manager dialogue box as indicated by the arrow shown in Fig. 6.85.

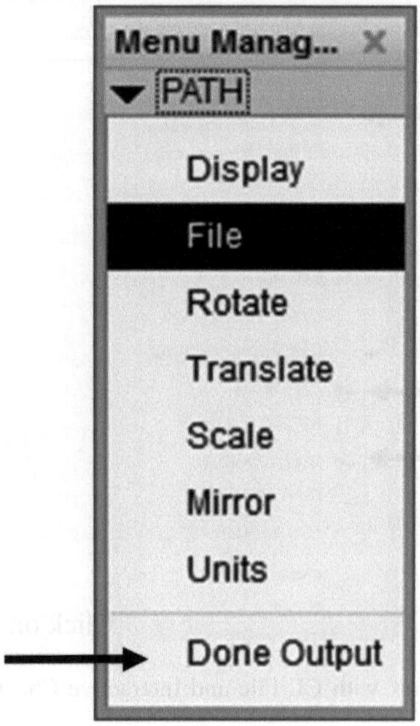

Fig. 6.85 PATH group on the Menu Manager dialogue box

6.5 Create the G-Code Data

Click on the Post a CL File icon as shown below.

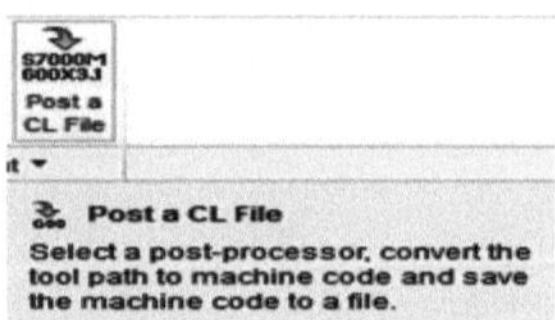

The Open dialogue window is activated on the main graphic window. See Fig. 6.86.

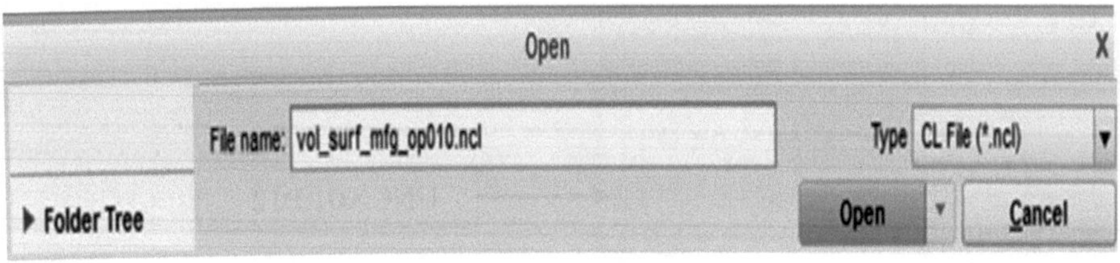

Fig. 6.86 Activated open dialogue window in concise form

Click on the Open tab to exit the Open dialogue window.

The PP OPTIONS Menu Manager dialogue box is activated on the main graphic window ≫ Check Mark Verbose and Trace square boxes only if they are now automatically Check Marked by the system as shown in Fig. 6.87.

Fig. 6.87 Activated PP OPTIONS dialogue box

Click on Done

The PP LIST Menu Manager dialogue box is activated on the main graphic window as shown in Fig. 6.88.

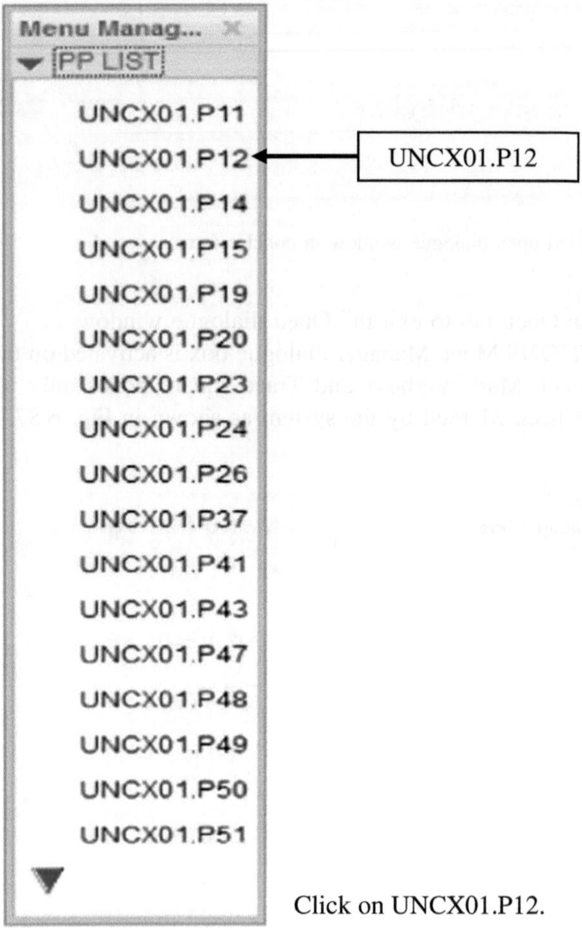

Fig. 6.88 Activated PP LIST Menu Manager dialogue box

The INFORMATION WINDOW dialogue window is activated in concise form as shown in Fig. 6.89. Check to see if there are any generated error warnings.

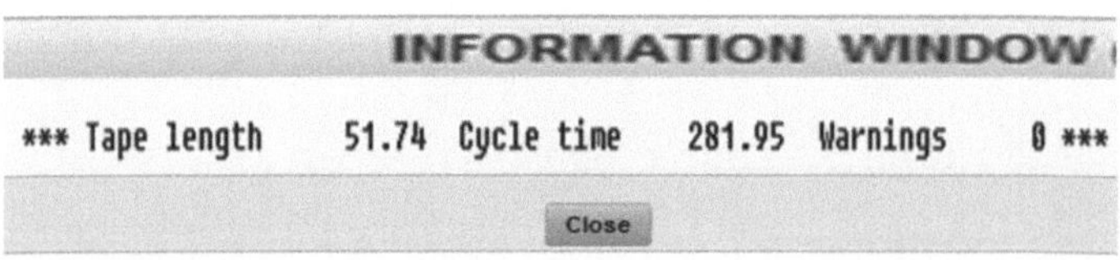

Fig. 6.89 Activated Save a Copy dialogue window in concise form

Click on the Close tab to exit.

The generated G-codes are stored as a TAP file in the chosen work directory. Open the TAP file using Notepad to view the generated G-codes.

Some of the G-Codes generated for this process are shown in the proceeding pages.

op010.tap - Notepad

File Edit Format View Help

```
N5 G71
N10 ( / VOL_SURF_MFG)
N15  G0 G17 G99
N20  G90 G94
N25  G0 G49
N30 T1 M06
N35 S3200 M03
N40 G0 G43 Z15. M08 H1
N45 X8. Y-8.
N50 Z5.
N55 G1 Z-10. F240.
N60 X442.
N65 Y-15.914
N70 X8.
N75 Y-23.828
N80 X442.
N85 Y-31.743
N90 X8.
N95 Y-39.657
N100 X442.
N105 Y-47.571
N110 X8.
N115 Y-55.485
N120 X442.
N125 Y-63.399
N130 X8.
N135 Y-71.314
N140 X442.
N145 X8.
N150 Y-79.228
N155 X156.136
N160 G3 X147.867 Y-81.29 I49.559 J-216.4
N165 G1 X147.867 Y-81.29 Z-10.
N170 G3 X139.663 Y-83.7 I49.186 J-182.538
N175 G1 X139.663 Y-83.7 Z-10.
N180 G3 X129.941 Y-87.142 I46.866 J-147.828
N185 G1 X129.941 Y-87.142 Z-10.
N190 X8.
N195 Y-95.056
N200 X113.582
N205 G3 X109.829 Y-97.466 I39.503 J-65.66
N210 G1 X109.829 Y-97.466 Z-10.
N215 G3 X104.702 Y-101.367 I34.064 J-50.08
N220 G1 X104.702 Y-101.367 Z-10.
N225 G3 X102.902 Y-102.971 I31.879 J-37.611
N230 G1 X102.902 Y-102.971 Z-10.
N235 X8.
N240 Y-110.885
N245 X96.263
N250 X95.998 Y-111.317
N255 G3 X93.94 Y-115.389 I25.114 J-15.247
N260 G1 X93.94 Y-115.389 Z-10.
N265 G3 X92.8 Y-118.799 I24.09 J-9.949
N270 G1 X92.8 Y-118.799 Z-10.
N275 X8.
N280 Y-126.713
N285 X92.061
N290 X92.156 Y-127.749
N295 G3 X92.607 Y-130.405 I24.66 J2.816
N300 G1 X92.607 Y-130.405 Z-10.
N305 G3 X93.359 Y-133.065 I25.277 J5.716
N310 G1 X93.359 Y-133.065 Z-10.
N315 X93.853 Y-134.396
N320 X93.947 Y-134.627
N325 X8.
N330 Y-142.542
N335 X98.713
N340 X99.698 Y-143.723
```

op010.tap - Notepad

File Edit Format View Help

```
N345 G3 X102.229 Y-146.389 I33.141 J28.937
N350 G1 X102.229 Y-146.389 Z-10.
N355 X103.657 Y-147.72
N360 X105.199 Y-149.05
N365 X106.858 Y-150.378
N370 X106.963 Y-150.456
N375 X8.
N380 Y-158.37
N385 X119.832
N390 X122.109 Y-159.484
N395 X124.846 Y-160.735
N400 X127.728 Y-161.967
N405 X130.756 Y-163.175
N410 G3 X137.253 Y-165.511 I55.855 J145.114
N415 G1 X137.253 Y-165.511 Z-10.
N420 X139.615 Y-166.284
N425 X8.
N430 Y-174.198
N435 X174.093
N440 X8.
N445 Y-182.113
N450 X442.
N455 Y-190.027
N460 X8.
N465 Y-197.941
N470 X442.
N475 Y-205.855
N480 X8.
N485 Y-213.769
N490 X442.
N495 Y-221.684
N500 X8.
N505 Y-229.598
N510 X442.
N515 Y-237.512
N520 X8.
N525 Y-245.426
N530 X442.
N535 Y-250.
N540 X441.974 Y-253.341
N545 X8.026
N550 X8.026 Y-253.341 Z-10.
N555 G3 X8.292 Y-261.255 I216.974 J3.341
N560 G1 X441.708
N565 G2 X441.152 Y-269.169 I-216.708 J11.255
N570 G1 X441.152 Y-269.169 Z-10.
N575 X8.848
N580 X8.848 Y-269.169 Z-10.
N585 G3 X9.697 Y-277.083 I216.152 J19.169
N590 G1 X440.303
N595 G2 X439.159 Y-284.997 I-215.303 J27.083
N600 G1 X439.159 Y-284.997 Z-10.
N605 X10.841
N610 X10.841 Y-284.997 Z-10.
N615 G3 X12.285 Y-292.912 I214.159 J34.997
N620 G1 X437.715
N625 G2 X435.964 Y-300.826 I-212.715 J42.912
N630 G1 X435.964 Y-300.826 Z-10.
N635 X14.036
N640 X14.036 Y-300.826 Z-10.
N645 G3 X16.101 Y-308.74 I210.964 J50.826
N650 G1 X433.899
N655 G2 X431.51 Y-316.654 I-208.899 J58.74
N660 G1 X431.51 Y-316.654 Z-10.
N665 X18.49
N670 X18.49 Y-316.654 Z-10.
N675 G3 X21.214 Y-324.568 I206.51 J66.654
N680 G1 X428.786
```

op010.tap - Notepad

File Edit Format View Help

```
N680 G1 X428.786
N685 G2 X425.713 Y-332.483 I-203.786 J74.568
N690 G1 X425.713 Y-332.483 Z-10.
N695 X24.287
N700 X24.287 Y-332.483 Z-10.
N705 G3 X27.725 Y-340.397 I200.713 J82.483
N710 G1 X422.275
N715 G2 X418.453 Y-348.311 I-197.275 J90.397
N720 G1 X418.453 Y-348.311 Z-10.
N725 X31.547
N730 X31.547 Y-348.311 Z-10.
N735 G3 X35.777 Y-356.225 I193.453 J98.311
N740 G1 X414.223
N745 G2 X409.557 Y-364.139 I-189.223 J106.225
N750 G1 X409.557 Y-364.139 Z-10.
N755 X40.443
N760 X40.443 Y-364.139 Z-10.
N765 G3 X45.579 Y-372.054 I184.557 J114.139
N770 G1 X404.421
N775 G2 X398.774 Y-379.968 I-179.421 J122.054
N780 G1 X398.774 Y-379.968 Z-10.
N785 X51.226
N790 X51.226 Y-379.968 Z-10.
N795 G3 X57.437 Y-387.882 I173.774 J129.968
N800 G1 X392.563
N805 G2 X385.725 Y-395.796 I-167.563 J137.882
N810 G1 X385.725 Y-395.796 Z-10.
N815 X64.275
N820 X64.275 Y-395.796 Z-10.
N825 G3 X71.827 Y-403.71 I160.725 J145.796
N830 G1 X378.173
N835 G2 X369.798 Y-411.625 I-153.173 J153.71
N840 G1 X369.798 Y-411.625 Z-10.
N845 X80.202
N850 X80.202 Y-411.625 Z-10.
N855 G3 X89.554 Y-419.539 I144.798 J161.625
N860 G1 X360.446
N865 G2 X349.897 Y-427.453 I-135.446 J169.539
N870 G1 X349.897 Y-427.453 Z-10.
N875 X100.103
N880 X100.103 Y-427.453 Z-10.
N885 G3 X112.182 Y-435.367 I124.897 J177.453
N890 G1 X337.818
N895 G2 X323.647 Y-443.282 I-112.818 J185.367
N900 G1 X323.647 Y-443.282 Z-10.
N905 X126.353
N910 X126.353 Y-443.282 Z-10.
N915 G3 X143.703 Y-451.196 I98.647 J193.282
N920 G1 X306.297
N925 G2 X282.983 Y-459.11 I-81.297 J201.196
N930 G1 X282.983 Y-459.11 Z-10.
N935 X167.017
N940 X275.909 Y-174.198
N945 X442.
N950 Y-166.284
N955 X310.386
N960 G3 X320.083 Y-162.849 I-46.915 J147.812
N965 G1 X320.083 Y-162.849 Z-10.
N970 G3 X328.486 Y-159.201 I-44.493 J113.999
N975 G1 X328.486 Y-159.201 Z-10.
N980 X330.183 Y-158.37
N985 X442.
N990 Y-150.456
N995 X343.043
N1000 G3 X345.298 Y-148.633 I-36.936 J48.002
N1005 G1 X345.298 Y-148.633 Z-10.
N1010 G3 X348.488 Y-145.678 I-31.879 J37.611
N1015 G1 X348.488 Y-145.678 Z-10.
```

op010.tap - Notepad

File Edit Format View Help

```
N1120 X353.737 Y-110.885 Z-10.
N1125 X442.
N1130 Y-102.971
N1135 X347.084
N1140 X346.343 Y-102.28
N1145 X344.801 Y-100.95
N1150 X343.142 Y-99.622
N1155 X341.36 Y-98.298
N1160 X339.452 Y-96.979
N1165 X337.413 Y-95.666
N1170 X336.398 Y-95.056
N1175 X442.
N1180 Y-87.142
N1185 X320.038
N1190 X319.244 Y-86.825
N1195 G3 X312.747 Y-84.489 I-55.855 J-145.114
N1200 G1 X312.747 Y-84.489 Z-10.
N1205 G3 X305.671 Y-82.28 I-57.187 J-170.802
N1210 G1 X305.671 Y-82.28 Z-10.
N1215 G3 X298.035 Y-80.226 I-57.352 J-197.97
N1220 G1 X298.035 Y-80.226 Z-10.
N1225 X294.016 Y-79.265
N1230 X293.855 Y-79.228
N1235 X442.
N1240 X358. Y-125.
N1245 G2 X357.485 Y-129.981 I-24.401 J.005
N1250 G1 X357.485 Y-129.981 Z-10.
N1255 G2 X356.06 Y-134.611 I-25.515 J5.319
N1260 X354.002 Y-138.683 I-27.172 J11.175
N1265 G1 X354.002 Y-138.683 Z-10.
N1270 G2 X351.951 Y-141.699 I-29.11 J17.593
N1275 G1 X351.951 Y-141.699 Z-10.
N1280 G2 X348.488 Y-145.678 I-31.913 J24.282
N1285 G1 X348.488 Y-145.678 Z-10.
N1290 G2 X345.298 Y-148.633 I-35.069 J34.656
N1295 G1 X345.298 Y-148.633 Z-10.
N1300 G2 X340.171 Y-152.534 I-39.191 J46.179
N1305 G1 X340.171 Y-152.534 Z-10.
N1310 G2 X335.61 Y-155.423 I-43.256 J63.25
N1315 G1 X335.61 Y-155.423 Z-10.
N1320 G2 X328.486 Y-159.201 I-48.074 J82.06
N1325 G1 X328.486 Y-159.201 Z-10.
N1330 G2 X320.083 Y-162.849 I-52.896 J110.351
N1335 G1 X320.083 Y-162.849 Z-10.
N1340 G2 X310.337 Y-166.3 I-56.612 J144.377
N1345 G1 X310.337 Y-166.3 Z-10.
N1350 G2 X302.133 Y-168.71 I-57.39 J180.128
N1355 X290.027 Y-171.615 I-57.828 J214.338
N1360 G1 X290.027 Y-171.615 Z-10.
N1365 G2 X280.119 Y-173.509 I-53.514 J253.061
N1370 G1 X280.119 Y-173.509 Z-10.
N1375 G2 X265.922 Y-175.581 I-48.956 J285.864
N1380 G1 X265.922 Y-175.581 Z-10.
N1385 G2 X253.754 Y-176.821 I-38.587 J318.115
N1390 G1 X253.754 Y-176.821 Z-10.
N1395 G2 X236.16 Y-177.824 I-28.273 J341.162
N1400 G1 X236.16 Y-177.824 Z-10.
N1405 G2 X217.072 Y-177.911 I-11.162 J354.565
N1410 G1 X217.072 Y-177.911 Z-10.
N1415 G2 X206.998 Y-177.542 I7.819 J350.361
N1420 G1 X206.998 Y-177.542 Z-10.
N1425 G2 X197.048 Y-176.887 I17.378 J340.118
N1430 G1 X197.048 Y-176.887 Z-10.
N1435 G2 X182.545 Y-175.39 I25.543 J318.56
N1440 G1 X182.545 Y-175.39 Z-10.
N1445 G2 X173.271 Y-174.069 I37.047 J293.239
N1450 G1 X173.271 Y-174.069 Z-10.
N1455 G2 X164.398 Y-172.51 I42.567 J268.323
```

```
N1455 G2 X164.398 Y-172.51 I42.567 J268.323
N1460 G1 X164.398 Y-172.51 Z-10.
N1465 G2 X155.984 Y-170.735 I46.555 J241.488
N1470 G1 X151.965 Y-169.774
N1475 X151.965 Y-169.774 Z-10.
N1480 G2 X144.329 Y-167.72 I49.716 J200.024
N1485 G1 X144.329 Y-167.72 Z-10.
N1490 G2 X137.253 Y-165.511 I50.111 J173.011
N1495 G1 X137.253 Y-165.511 Z-10.
N1500 G2 X130.756 Y-163.175 I49.358 J147.45
N1505 G1 X127.728 Y-161.967
N1510 X124.846 Y-160.735
N1515 X122.109 Y-159.484
N1520 X119.518 Y-158.216
N1525 X117.068 Y-156.934
N1530 X114.759 Y-155.639
N1535 X112.587 Y-154.334
N1540 X110.548 Y-153.021
N1545 X108.64 Y-151.702
N1550 X106.858 Y-150.378
N1555 X105.199 Y-149.05
N1560 X103.657 Y-147.72
N1565 X102.229 Y-146.389
N1570 X102.229 Y-146.389 Z-10.
N1575 G2 X99.698 Y-143.723 I30.61 J31.603
N1580 G1 X98.587 Y-142.39
N1585 X98.587 Y-142.39 Z-10.
N1590 G2 X96.653 Y-139.724 I27.877 J22.253
N1595 G1 X96.653 Y-139.724 Z-10.
N1600 G2 X95.083 Y-137.06 I26.513 J17.417
N1605 X93.853 Y-134.396 I25.49 J13.388
N1610 G1 X93.359 Y-133.065
N1615 X93.359 Y-133.065 Z-10.
N1620 G2 X92.607 Y-130.405 I24.525 J8.376
N1625 G1 X92.607 Y-130.405 Z-10.
N1630 G2 X92.156 Y-127.749 I24.209 J5.472
N1635 G1 X92.156 Y-127.749 Z-10.
N1640 G2 X92. Y-125. I24.091 J2.749
N1645 G1 X92. Y-125. Z-10.
N1650 G2 X92.515 Y-120.019 I24.401 J-.005
N1655 G1 X92.515 Y-120.019 Z-10.
N1660 G2 X93.94 Y-115.389 I25.515 J-5.319
N1665 G1 X93.94 Y-115.389 Z-10.
N1670 G2 X95.998 Y-111.317 I27.172 J-11.175
N1675 X98.049 Y-108.301 I29.11 J-17.593
N1680 G1 X98.049 Y-108.301 Z-10.
N1685 G2 X101.512 Y-104.322 I31.913 J-24.282
N1690 G1 X101.512 Y-104.322 Z-10.
N1695 G2 X104.702 Y-101.367 I35.069 J-34.656
N1700 G1 X104.702 Y-101.367 Z-10.
N1705 G2 X109.829 Y-97.466 I39.191 J-46.179
N1710 G1 X109.829 Y-97.466 Z-10.
N1715 G2 X114.39 Y-94.577 I43.256 J-63.25
N1720 G1 X114.39 Y-94.577 Z-10.
N1725 G2 X121.514 Y-90.799 I48.074 J-82.06
N1730 G1 X121.514 Y-90.799 Z-10.
N1735 G2 X129.917 Y-87.151 I52.896 J-110.351
N1740 G1 X129.917 Y-87.151 Z-10.
N1745 G2 X139.663 Y-83.7 I56.612 J-144.377
N1750 G1 X139.663 Y-83.7 Z-10.
N1755 G2 X147.867 Y-81.29 I57.39 J-180.128
N1760 G1 X147.867 Y-81.29 Z-10.
N1765 G2 X159.973 Y-78.385 I57.828 J-214.338
N1770 G1 X159.973 Y-78.385 Z-10.
N1775 G2 X169.881 Y-76.491 I53.514 J-253.06
N1780 G1 X169.881 Y-76.491 Z-10.
N1785 G2 X184.078 Y-74.419 I48.956 J-285.864
N1790 G1 X184.078 Y-74.419 Z-10.
```

Chapter 7
Volume Rough Milling Using Mill Window and Surface Milling Operation

The Volume Rough Milling using Mill Window and the Surface Milling operation will be covered in this chapter.

7.1 Volume Rough Milling Using Mill Window Operation

The steps below will be covered in this tutorial.

- Import and constrain the 3D Part
- Suppress all the four Holes
- Add Stock
- Create Programme Zero (Machine Coordinate System)
- Create the Work centre
- Setup Operation
- Create Cutting Tool
- Create Mill Window
- Volume Rough Milling Operation Sequence is created
- Create Surface Milling Operation Sequence

Note: In this tutorial, the first step above will be omitted because by now the reader should be familiar with how to import and constrain 3D Part. Chapters 2, 3 and 4 explain these steps very well.

To suppress all the four holes, go to the Model Tree ≫ Click on the arrow on VOL_SURF_MILL_DRIL.PRT (The imported and constrained 3D Part) to activate its menu list ≫ To highlight all the four Holes, click on each holes while holding down the Ctrl key and then right click ≫ On the activated drop down menu list, select Suppress and all the holes are now suppressed.

P.O. Kanife, *Computer Aided Virtual Manufacturing Using Creo Parametric*,
DOI 10.1007/978-3-319-23359-8_7

7.1.1 *Add Stock to the Imported Part Using Assemble Workpiece Method*

Click on Workpiece icon >> Now click on Assemble Workpiece on the activated drop down menu list as shown below.

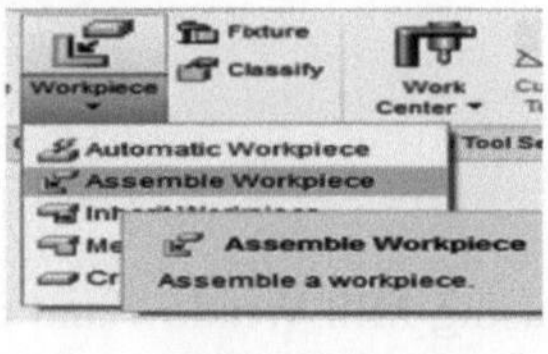

The Open dialogue window is activated >> Select the already created and saved Workpiece >> Click on Open tab to import unto the GUI window.

The Component Placement dashboad tools are activated as shown in Fig. 7.1.

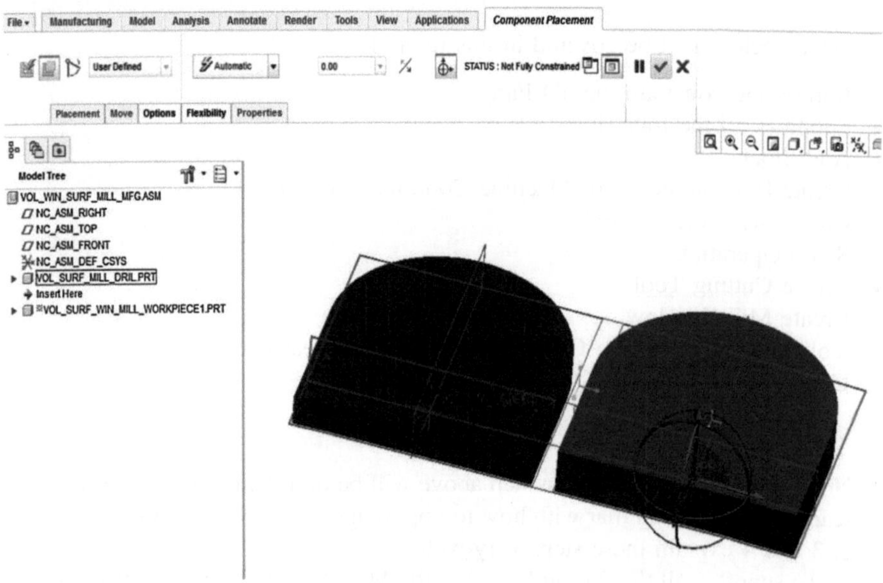

Fig. 7.1 Activated Component Placement showing the Part and Workpiece

To merge and constrain the Part and Workpiece together >> Right click on the graphic window and select Default Constraint from the drop down menu list >> Now click on the Check Mark icon to accept and exit >> The Part and Workpiece are automatically assembled together as shown in Fig. 7.2.

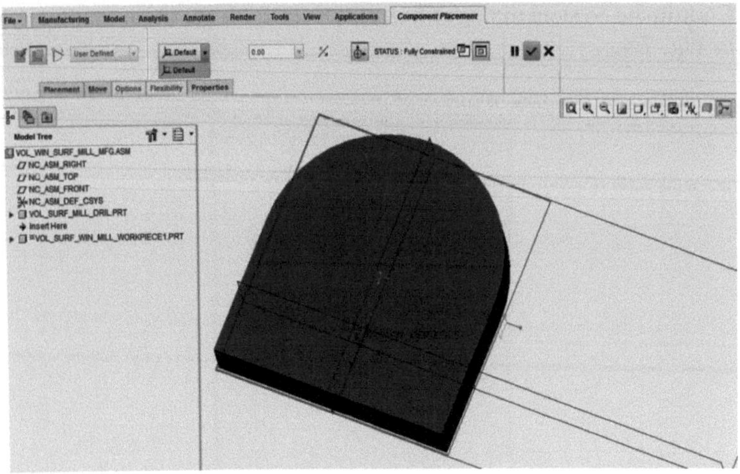

Fig. 7.2 Part and Workpiece are fully assembled and constrained

Stock is now added to Part as shown in Fig. 7.3.

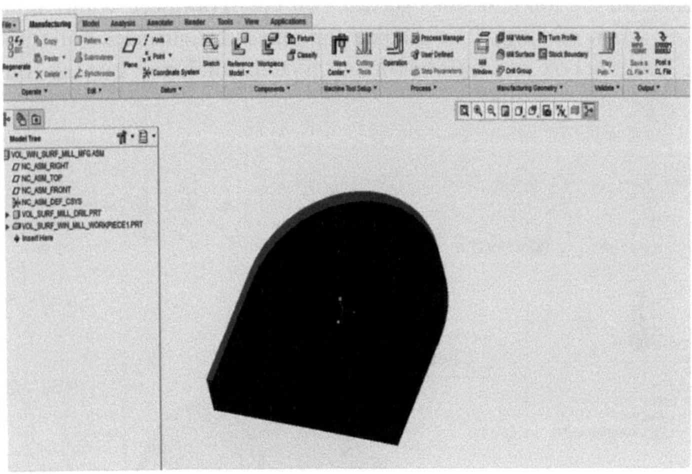

Fig. 7.3 Part and Workpiece assemble together

7.1.2 Create Coordinate System for the Workpiece

Click on the Coordinate System icon as shown below.

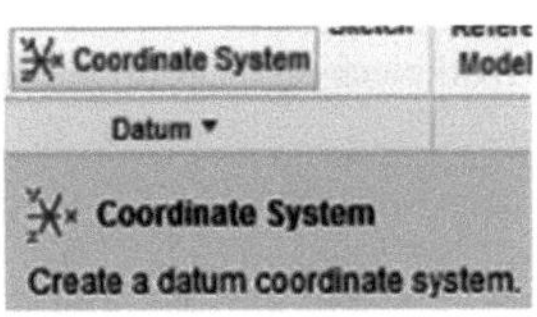

The Coordinate System dialogue box is activated on the main graphic window as shown in Fig. 7.4.

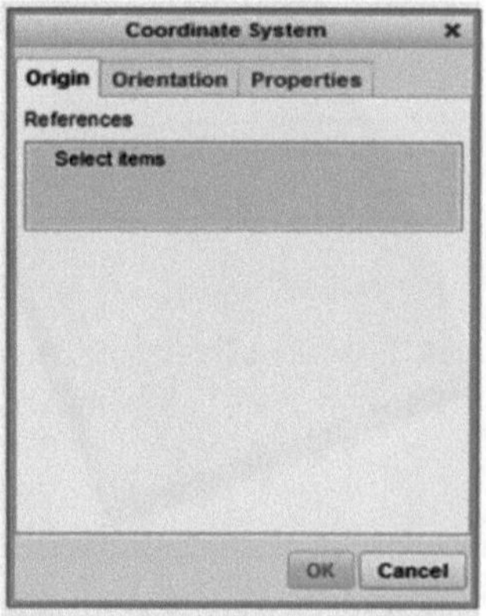

Fig. 7.4 Activated Coordinate System dialogue box

Now click on the top and sides surfaces on the Workpiece while holding down the Ctrl key when clicking. The clicked top and sides surfaces will appear on the Reference section box on the Coordinate System dialogue box as illustrated on Fig. 7.5.

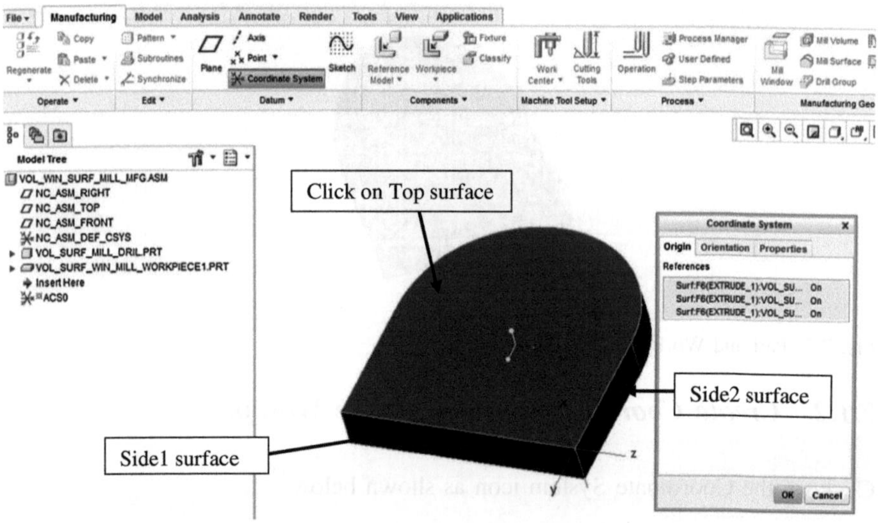

Fig. 7.5 Selected References on Workpiece and Coordinate System dialogue box

The created *X*, *Y* and *Z*-axes of the Coordinate System are not in the correct orientation as shown in Fig. 7.6.

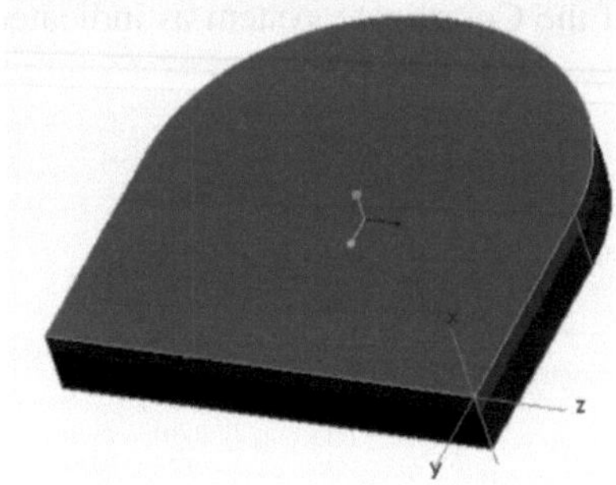

Fig. 7.6 Wrong orientations of *X*, *Y* and *Z*-axes of the Coordinate System

To orient the X, Y and Z coordinate axes to the correct orientation

Click on the Orientation tab on the Coordinate System dialogue box ≫ On the "Orient by" group, click in the radio button on the "References selection" ≫ Click on "to determine" section box, and now click on *Z*-axis on the activated drop down menu list ≫ Click on "to project" section box to activate its drop down menu list, and now click on *X*-axis on the drop down list. Now click on the Flip tab, to flip *X*-axis orientation. The arrows as shown below indicate the correct orientations for the created Programme Zero axes. See Fig. 7.7.

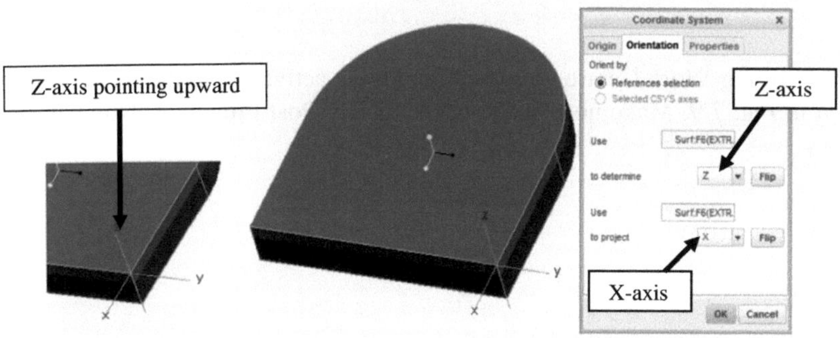

Fig. 7.7 Correct orientations of *X*, *Y* and *Z*-axes of the Coordinate System

Click on the OK tab to exit Coordinate System dialogue box.

A new ASC1 Coordinate System (Programme Zero) is created.

The new ACS1 created will be re-named to PZ1 in this tutorial. To rename ACS1, slow click on ACS1 on the Model Tree and type PZ1. Alternatively, right click on ACS1 and click on Rename on the activated drop down menu list and type PZ1 as the new name of the Coordinate system as indicated by the arrow shown in Fig. 7.8.

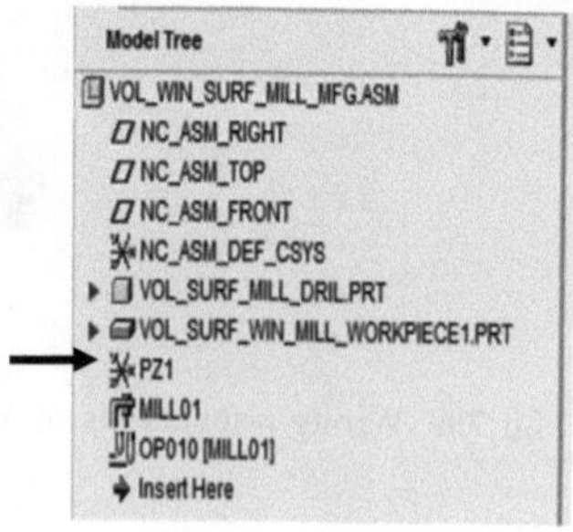

Fig. 7.8 Model Tree indicating the new Coordinate System PZ1

7.1.3 *Setup the Work Centre*

Click on Work Centre icon ≫ Now click on Mill on the activated drop down menu list as illustrated below.

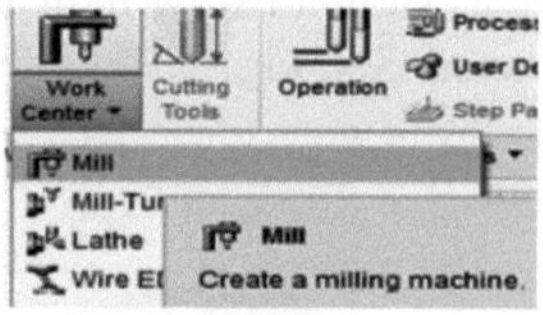

The Milling Work Centre dialogue window is activated on the main window as shown in Fig. 7.9. Make note of the Name, Type, Post Processor, Number of Axis etc.

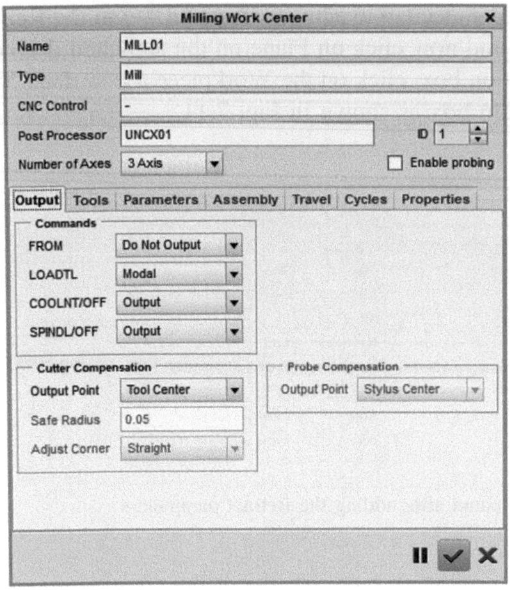

Fig. 7.9 Activated Milling Work Centre dialogue window

Click on the Check Mark icon to exit.

7.1.4 Setup the Operation

Click on the Operation icon as shown below

The Operation tools are activated ≫ Now click on the created Coordinate System that is PZ1 on the Model Tree to add it unto the Coordinate System icon section box (Fig. 7.10).

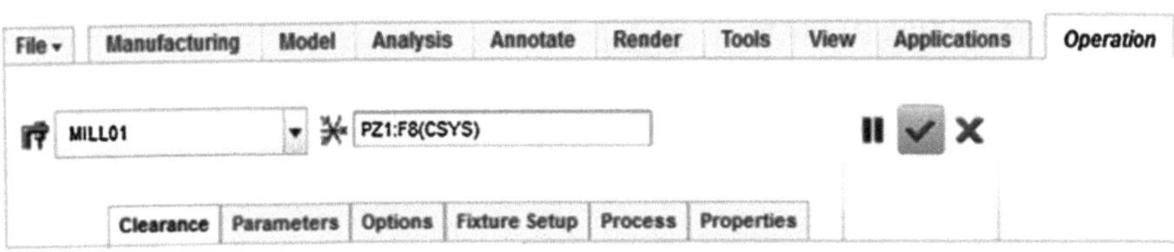

Fig. 7.10 Updated Operation ribbon toolbar after adding MILL01 and PZ1

Click on the Clearance tab to activate its panel ≫ On the Retract group, click on Type section box and now click on Plane on the activated drop down menu list. In the Reference section box, click on the Workpiece top surface ≫ Now type 10 mm in the Value section box as shown in Fig. 7.11.

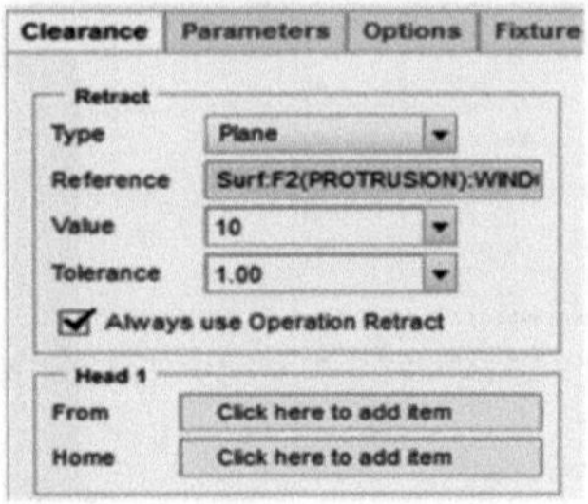

Fig. 7.11 Clearance panel after adding the Retract parameters

Click on the Check Mark icon to exit the Operation application.

Check the Model Tree as shown below to see if the PZ1, MILL01 and OP010 [MILL01] are on the Model Tree as created (Fig. 7.12).

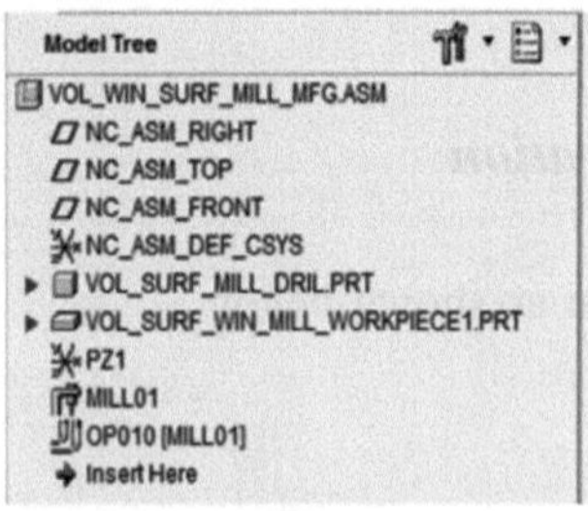

Fig. 7.12 Model Tree indicating the created PZ1, MILL01 and OP010

7.1.5 Setup the Cutting Tool

Click on the Cutting Tools icon

The Tools Setup dialogue window is activated on the main graphic window ≫ Click on the General tab to activate its contents ≫ On the Name section box type

T0001 ≫ Click on the Type section box to activate its drop down menu list ≫ Now click on Milling Tool on the drop down list. Add dimensions to the Tool. See Fig. 7.13.

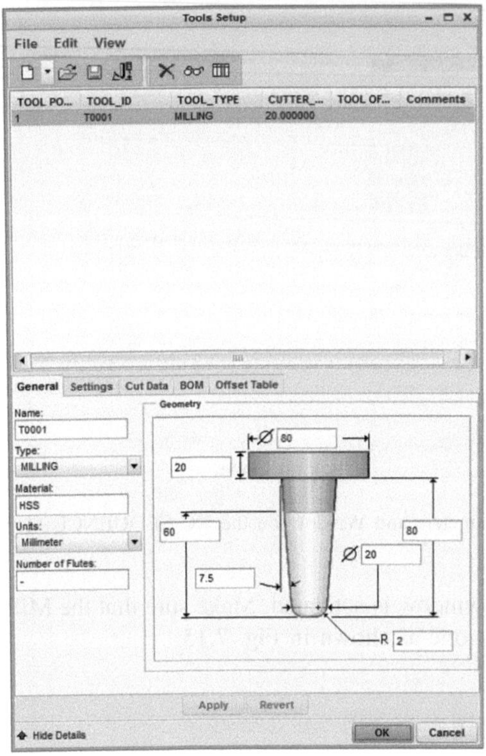

Click on Apply tab.

Fig. 7.13 Tools Setup dialogue window after adding the Tool parameters

Click on the OK tab to exit Tools Setup dialogue window

7.1.6 Volume Rough Milling Using Mill Window

Click on the Mill menu bar ≫ Now click on the Roughing icon on the Milling group ≫ Click on Volume Rough on the activated drop down menu list as illustrated below.

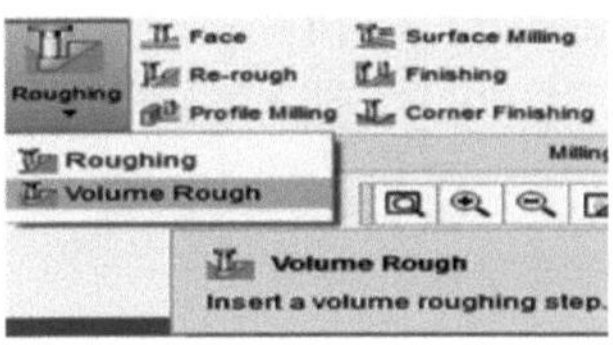

The NC SEQUENCE Menu Manager dialogue box is activated on the main graphic window ≫ Check Mark Tool, Parameters and Window on the SEQ SETUP group as shown in Fig. 7.14.

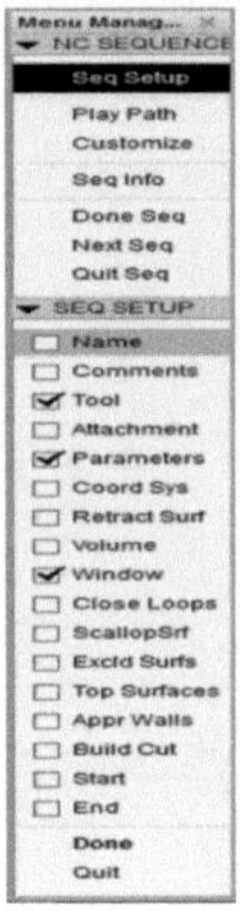

Click on Done.

Fig. 7.14 Check Mark Tool, Parameters and Window on the NC SEQUENCE Menu Manager dialogue box

The Tools Setup dialogue window is activated. Make sure that the Milling Tools that is activated is the correct one as shown in Fig. 7.15.

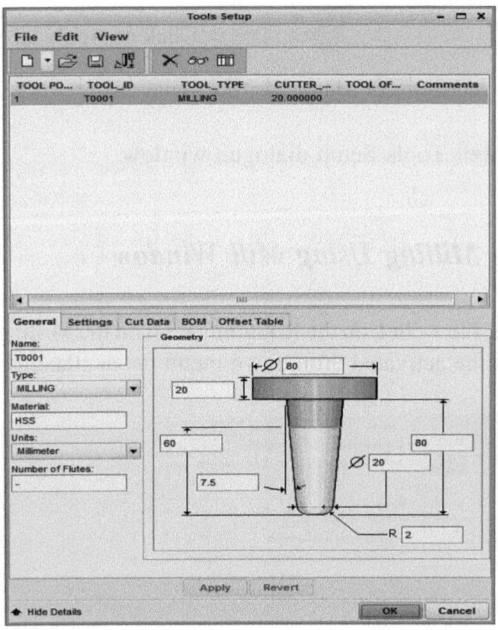

Click on OK tab.

Fig. 7.15 Activated Milling Tool parameters

The Edit Parameters of Sequence "Volume Milling" is activated ≫ Input parameters value as shown in Fig. 7.16.

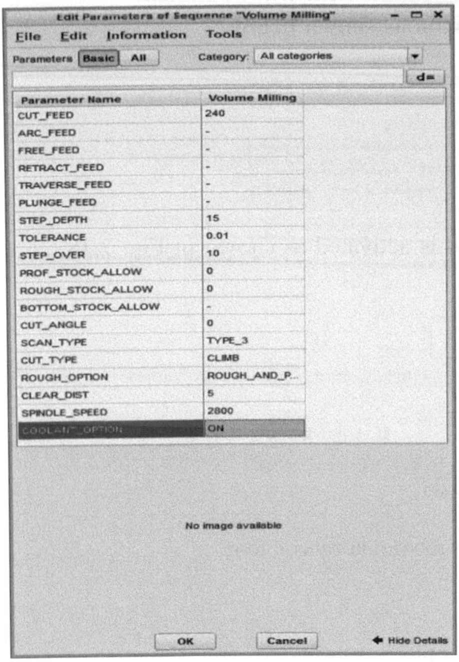

Click on OK tab to exit.

Fig. 7.16 Edit Parameters of Sequence "Volume Milling" parameters

The DEFINE WIND group is automatically added unto NC SEQUENCE Menu Manager dialogue box by the system. The Select Wind is highlighted on DEFINE WIND group, and programme is asking for the Mill Window to be defined or selected (Fig. 7.17).

Fig. 7.17 Select Wind highlighted on DEFINE WIND group

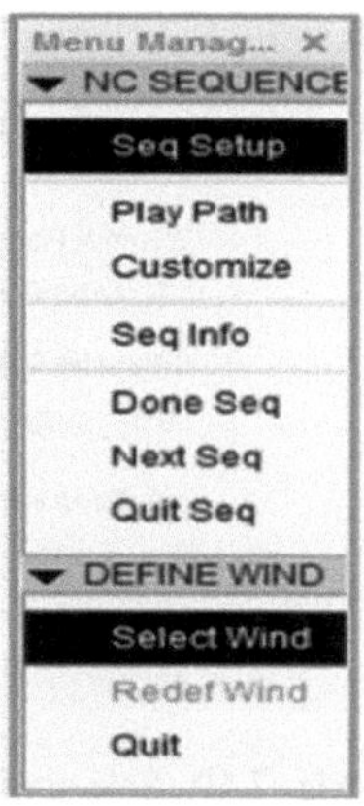

7.1.7 Create Window Mill

All material below the created Window Mill will be milled.

Click on the Mill Window icon as shown below.

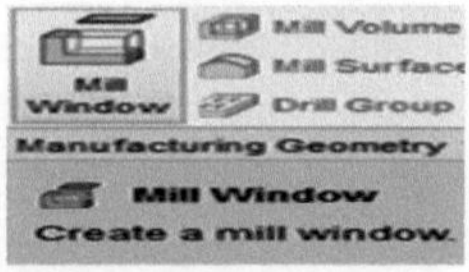

The Mill Window dashboard is activated as shown in Fig. 7.18.

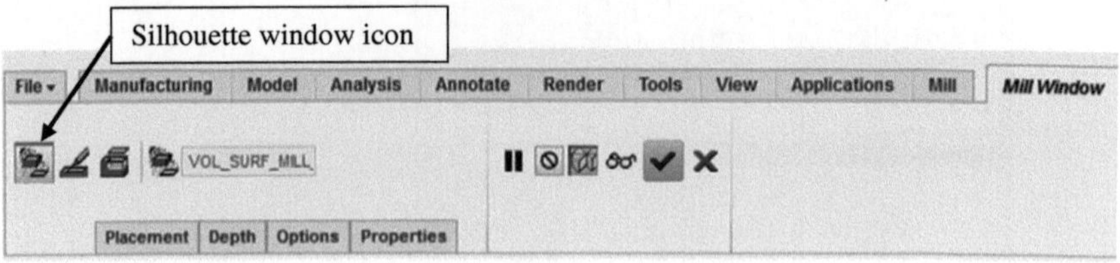

Fig. 7.18 Activated Mill Window dashboard in concise form

Make sure the Silhouette Window type icon  is activated. See Fig. 7.18.

Click on the Placement tab to make its panel active ≫ Click on the Window Plane section box ≫ Now click on the surface of the Workpiece as indicated by the arrow shown below. Do this if the Workpiece surface is not selected automatically by the system. See Fig. 7.19.

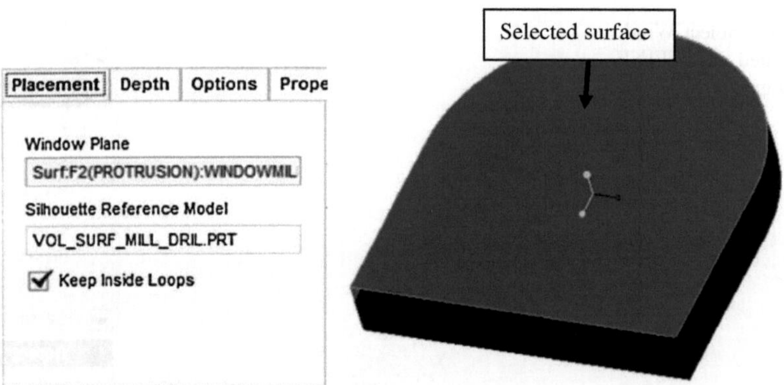

Fig. 7.19 Selecting the Placement parameters and as indicated by the arrow

Click on the Depth tab to make its panel active ≫ Check Mark "Specify the depth" square box ≫ Click on the Depth Options section box, now click on 'To Selected' on the activated drop down menu list ≫ Select the Part surface as indicated by the arrow. This is the height of the created Window Mill to the Part surface as shown in Fig. 7.20.

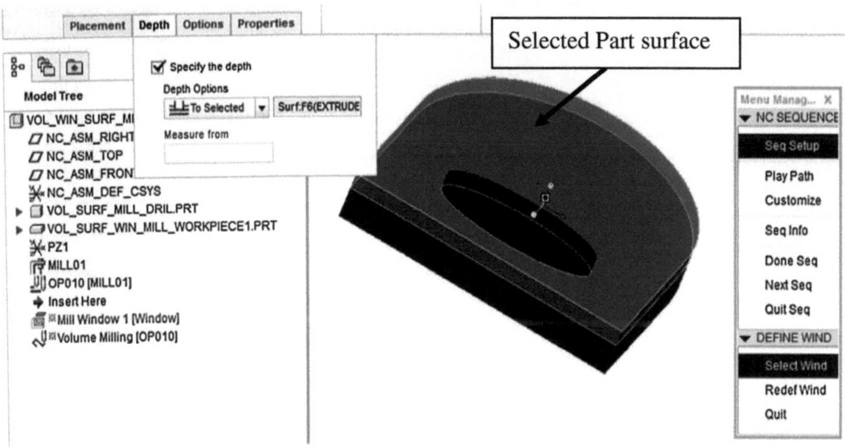

Fig. 7.20 Activated depth panel indicating added parameters

Click on the Options tab to activate its panel ≫ On the Machine with tool group, click in "On window contour" radio button to make it active. Make sure that the adjust geometry collection with window square box is Checked Marked as shown in Fig. 7.21.

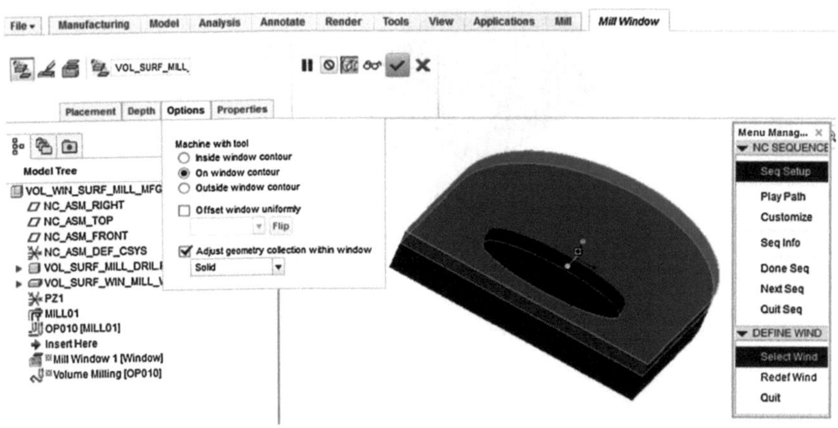

Fig. 7.21 Activated Options panel indicating added parameters

Click on the Check Mark icon to accept and proceed.

On the NC SEQUENCE Menu Manager dialogue box, click on Play Path and click on Screen Play on the PLAY PATH group as highlighted on Fig. 7.22.

Fig. 7.22 Play Path and Screen Play activated

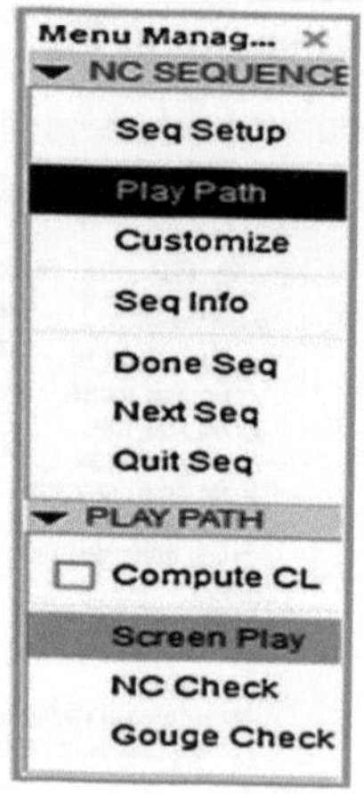

The milling Tool and the PLAY PATH dialogue box are activated on the main graphic window as shown in Fig. 7.23.

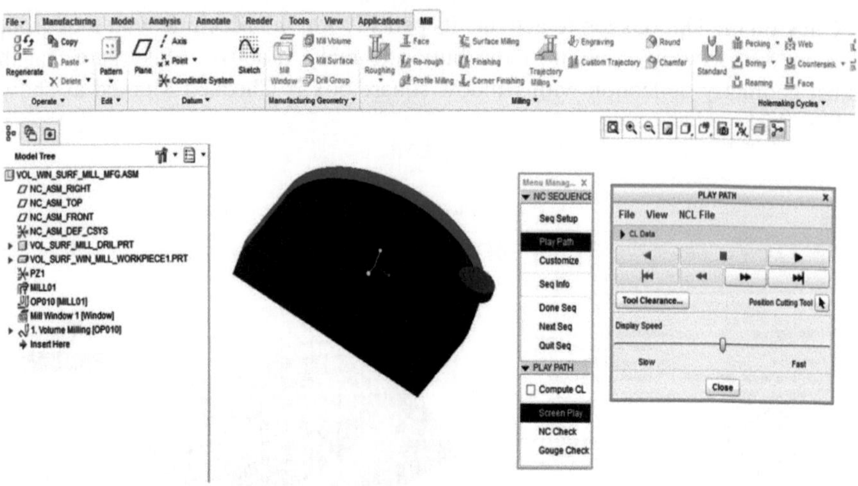

Fig. 7.23 Activated milling Tool and PLAY PATH dialogue box

Click on the Play tab on the PLAY PATH dialogue box to start the on screen milling operation.

The end of the Volume Rough milling operation using Window Milling method is shown in Fig. 7.24 in Wireframe display.

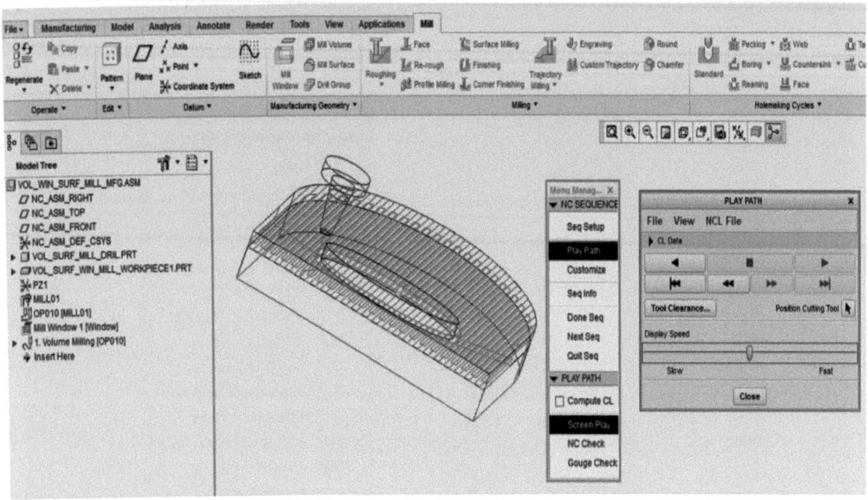

Fig. 7.24 End of Volume Rough milling using Window Milling process

Click on the close tab to exit.
Now click on Done Seq on the NC SEQUENCE group to exit the operation.

7.2 Create Surface Milling

Surface Milling process is use to smoothen surfaces after Roughing operation.
Click on the Surface Milling icon as shown below.

The NC SEQUENCE Menu Manager dialogue box is activated again on the main window ≫ Now Check Mark Tool, Parameters, Surfaces and Define Cut as shown in Fig. 7.25.

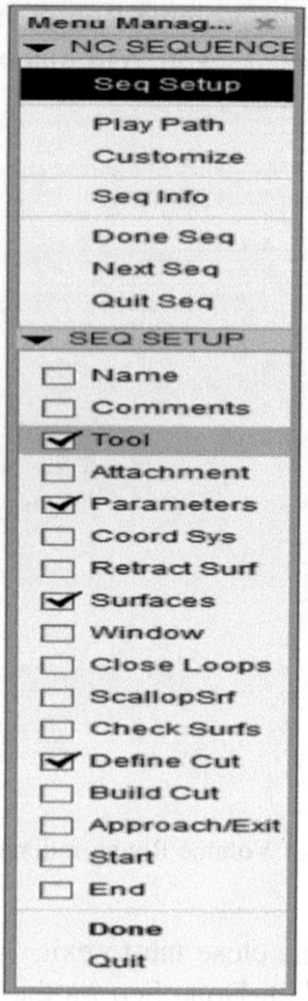

Fig. 7.25 Tool, Parameters, Surfaces and Define Cut parameters check marked

Click on Done.

The Tools Setup dialogue window is activated again on the main graphic window ≫ Seclect the previous Milling cutting Tool type ≫ Add dimensions only when its missing. Alternatively you can create a new End Mill Tool. See Fig. 7.26.

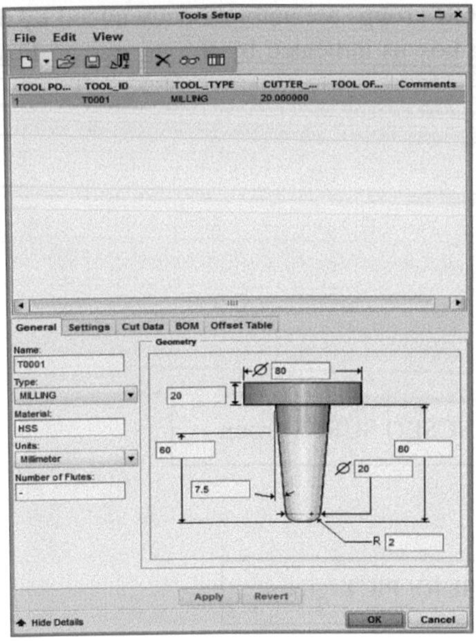

Click on the OK tab to exit.

Fig. 7.26 End Mill Tool dimensions

The Edit Parameters of Sequence "Surface Milling" dialogue box is activated on the main graphic window ≫ Input parameters values as shown in Fig. 7.27.

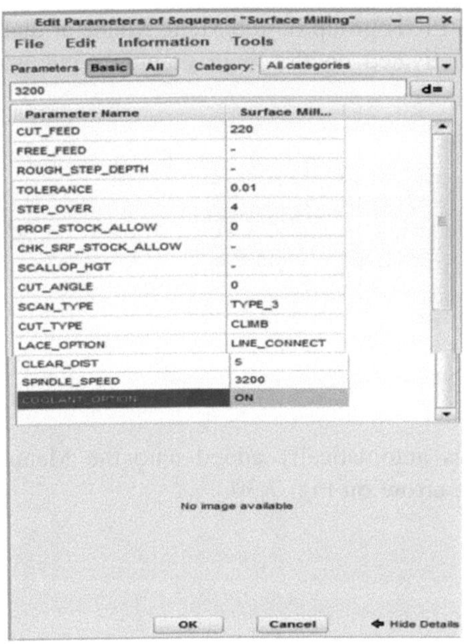

Click on OK tab to exit.

Fig. 7.27 Parameters values for the Surface Milling operation

NCSEQ SURFS and SURF PICK groups are automatically added by the system unto the Menu Manager dialogue box as indicated by the arrows on Fig. 7.28.

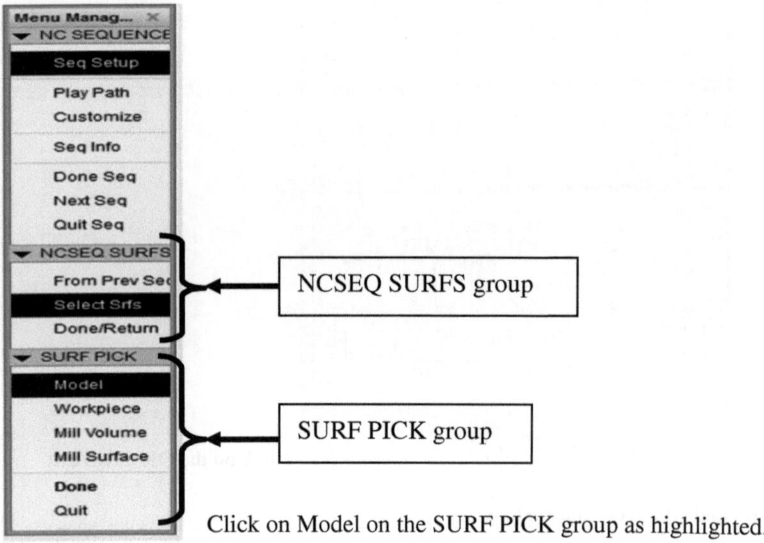

Fig. 7.28 NCSEQ SURFS and SURF PICK groups on the Menu Manager

Now click on the Part to be the selected Model.

Click on Done on the SURFS PICK group on the Menu Manager dialogue box.

The Select dilaogue box is activated on the main graphic window shown in Fig. 7.29.

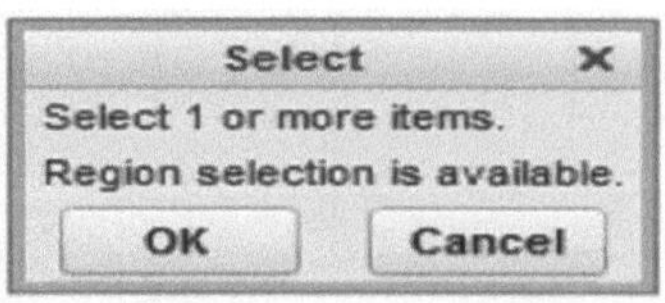

Fig. 7.29 Activated Select dialogue box

The SELECT SRFS group is automatically added unto the Menu Manager dialogue box as indicated by the arrow on Fig. 7.30.

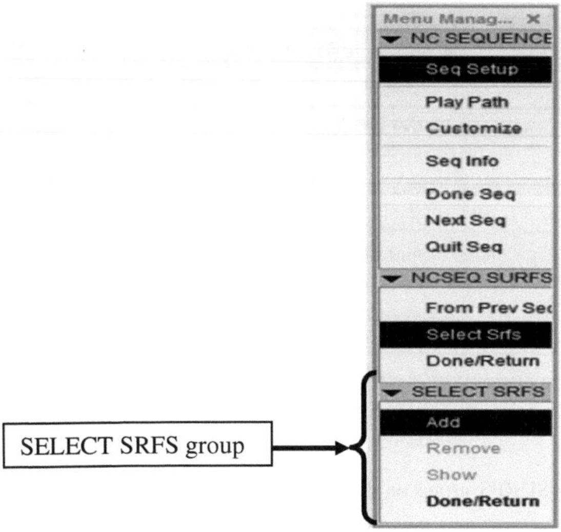

Fig. 7.30 SELECT SRFS group automatically added unto the Menu Manager dialogue box

Now click on the Ellipse surface to add as the surfaces to be Surface Milled ≫ Hold down the Ctrl key and select the Part surface to add as the surfaces to be Surface Milled as indicated by the arrows on Fig. 7.31.

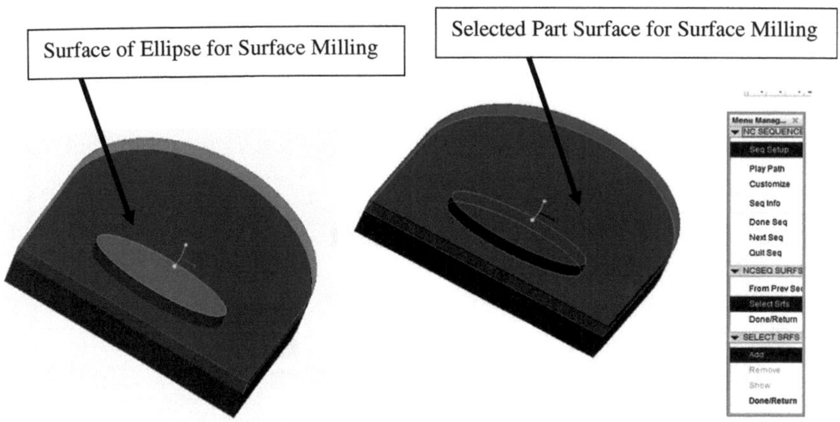

Fig. 7.31 Selected Ellipse and Part surface for Surface Milling operation

Click on Done/Return on the SELECT SRFS group

The SELECT SRFS group disappear but the NCSEQ SURFS group remains on the Menu Manager dialogue box as shown in Fig. 7.32.

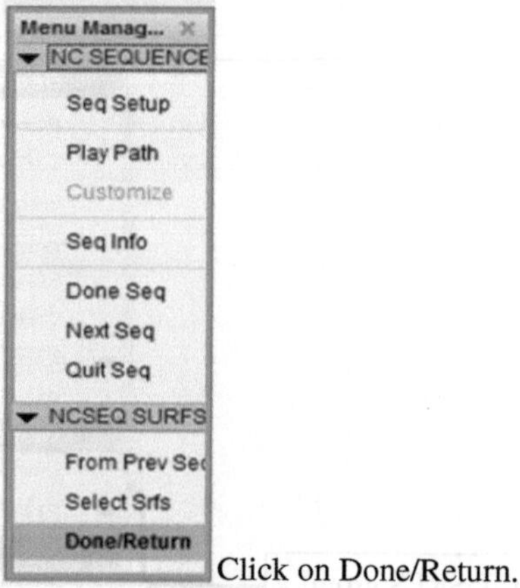

Fig. 7.32 NCSEQ SURFS group on the Menu Manager dialogue box

The Cut Definition dialogue box is activated ≫ Activate "By Surface" ribbon icon on the Cut Angle Reference group and now select a Planar surface or plane when prompted to do so by the system. See Fig. 7.33.

Note: The Cut Definition dialogue box is activated on the main graphic window ≫ In this tutorial, click in the By Surface radio button to make it active on the Cut Angle Reference section, on the Straight Cut group ≫ Now select a surface that is perpendicular to the Retract Plane as requested on the message bar.

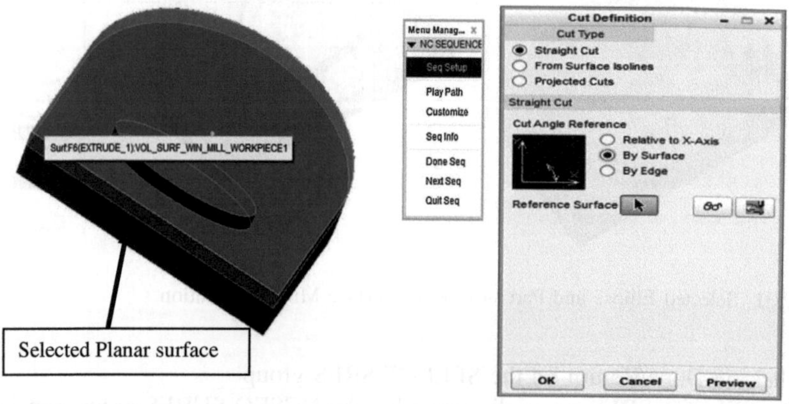

Fig. 7.33 Selected planar surface and activated Cut Definition dialogue box

Click on the OK tab on the Cut Definition dialogue box to proceed.

The NCSEQ SURFS group disappear and the NC SEQUENCE Menu Manager dialogue box automatically changes as shown in Fig. 7.34.

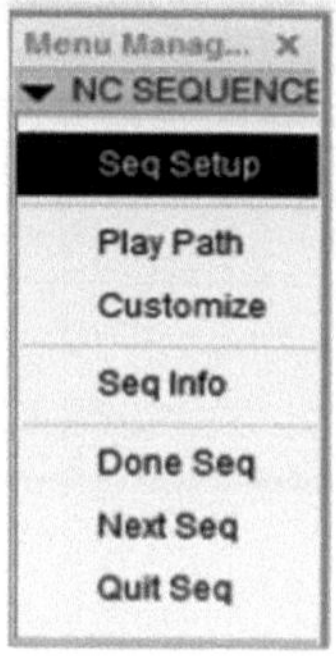

Fig. 7.34 NC SEQUENCE group on Menu Manager dialogue box

7.2.1 Activate the Surface Milling Machining Process

Click on Play Path on the NC SEQUENCE group as indicated by the arrow on Fig. 7.35.

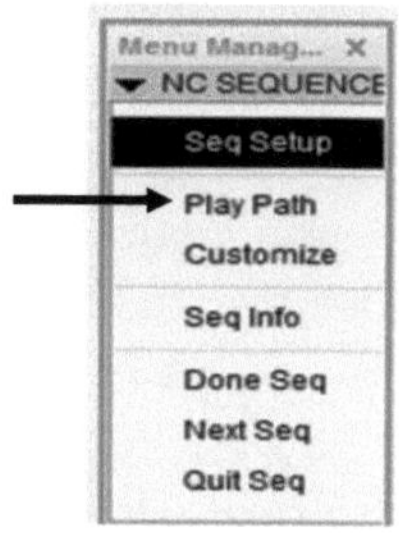

Fig. 7.35 Clicked Play Path on the NC SEQUENCE group

Once Play Path is clicked, the NC SEQUENCE Menu Manager dialogue box automatically adds the PLAY PATH group ≫ Now click on Screen Play on the PLAY PATH group as highlighted on Fig. 7.36.

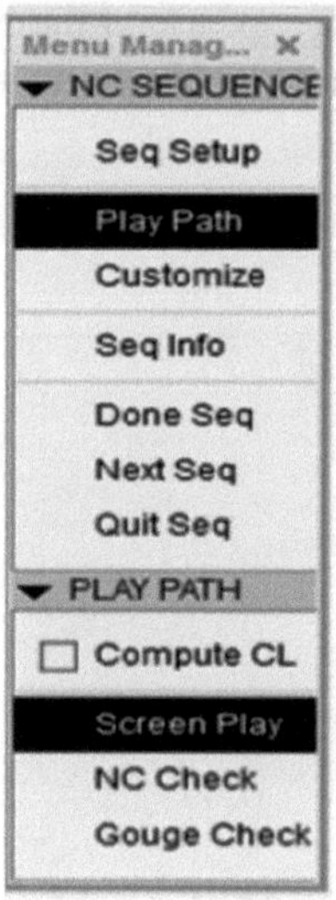

Fig. 7.36 Activating Screen Play on the Menu Manager dialogue box

The cutting Tool, the PLAY PATH and Menu Manager dialogue boxes are shown the main graphic window in Fig. 7.37.

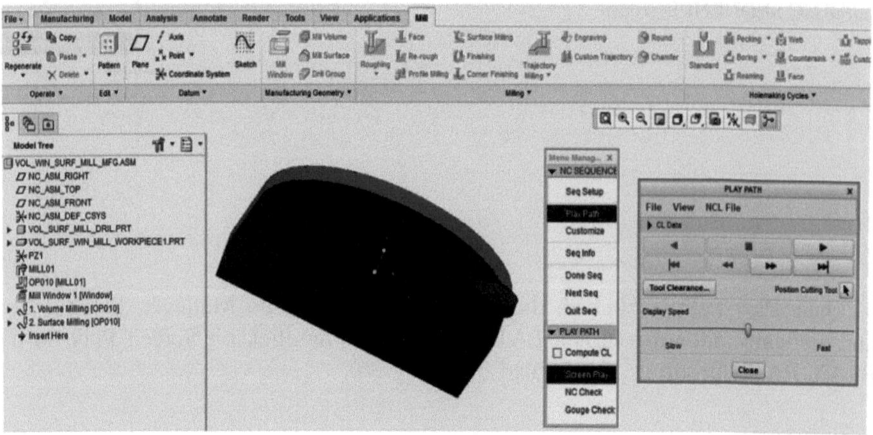

Fig. 7.37 Activated Tool and PLAY PATH dialogue box

Click on the Play tab to start the Surface Milling operation.

The end of the Surface Milling process in Wireframe display is as shown in Fig. 7.38.

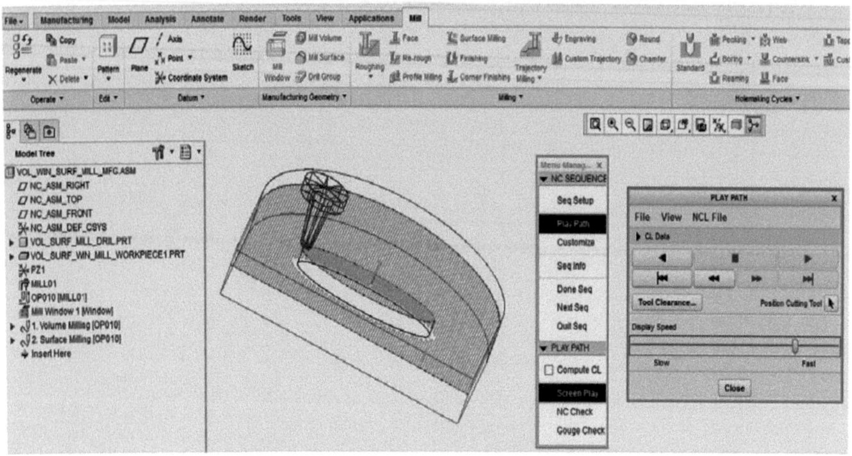

Fig. 7.38 End of the Surface Milling Operation in Wireframe display

Click on Close tab to exit the PLAY PATH dialogue box.

Now click on Done Seq on the NC SEQUENCE group on the Menu Manager dialogue box to exit the Surface Milling operation.

Chapter 8
Expert Machinist

This machining operation is a two-and-half axis machining process.

The following steps will be used to achieve the goals in this tutorial:

- Start Expert Machining
- Import the 3D Part into the Graphic Window
- Add Stock
- Create Programme Zero (Machine Coordinate System)
- Create the Machining Application Work Centre
- Set up Operation
- Create Cutting Tools
- Create Tool Path
- Create Material Removal Sequences for the operations
- Generate Cutter Location (CL) Data
- Generate the G-codes

8.1 Start Expert Machinist

Click on New icon .

The new dialogue box is activated on the main window as shown in Fig. 8.1.

P.O. Kanife, *Computer Aided Virtual Manufacturing Using Creo Parametric*,
DOI 10.1007/978-3-319-23359-8_8

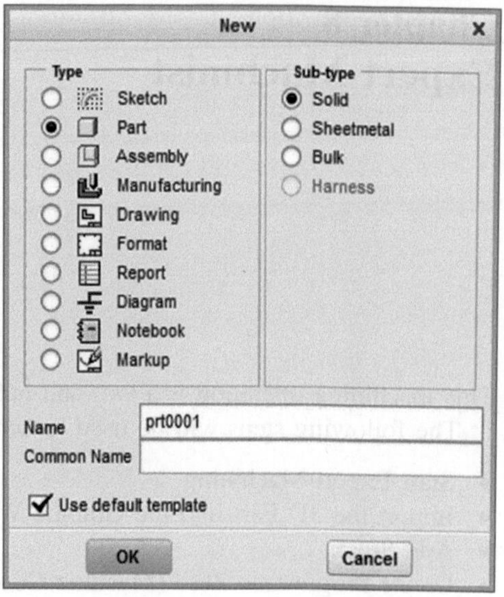

Fig. 8.1 Activated New dialogue box

On the Type group, click on the Manufacturing radio button to make it active ≫ On the Sub Type group, click on the Expert Machinist radio button to make it active ≫ Type the name of the Part in the Name section box ≫ Click on the square box on the Use default template to clear the square box as shown on the New dialogue box in Fig. 8.2.

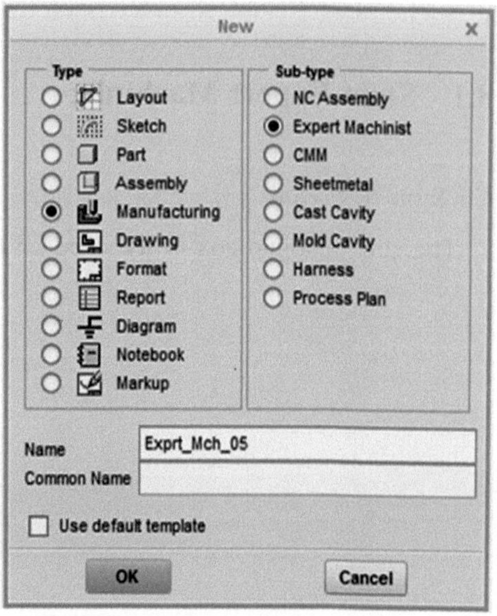

Fig. 8.2 New dialogue box with Manufacturing and Expert Machinist active

Click on the OK tab to proceed.

The New File Options dialogue window is activated on the main window ≫ Click on the "mmns_mfg_nc" on the Template group ≫ On the Parameters group, type your name in the "MODELLED_BY" section box and type Expert Machinist in the DESCRIPTION section box as shown in Fig. 8.3.

Note: You can type whatever you want in this section.

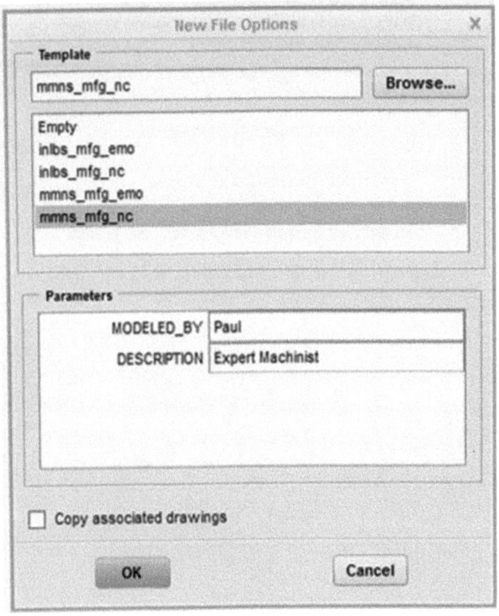

Click on the OK tab to proceed.

Fig. 8.3 New File Options dialogue box parameters

The NC-WIZARD dialogue window is activated on the main window. See Fig. 8.4.

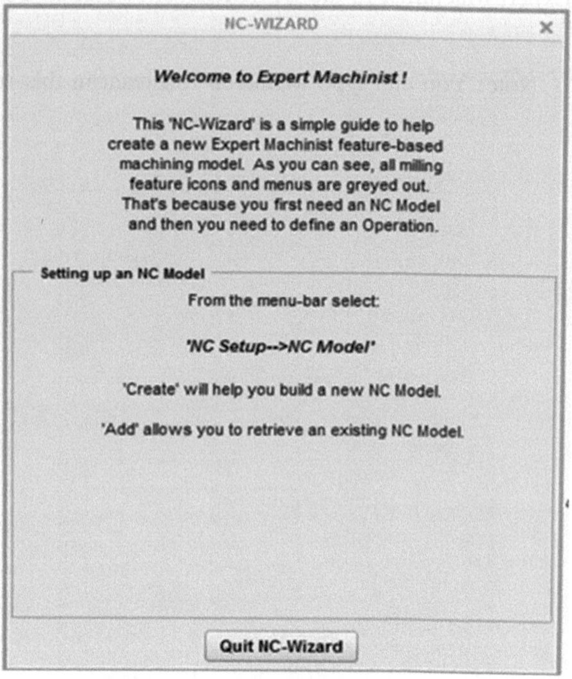

Fig. 8.4 Activated NC-WIZARD dialogue window

Click on Quit NC-WIZARD tab to exit the NC-WIZARD dialogue window.

Expert Machinist ribbon is activated on the main graphic window as shown in Fig. 8.5.

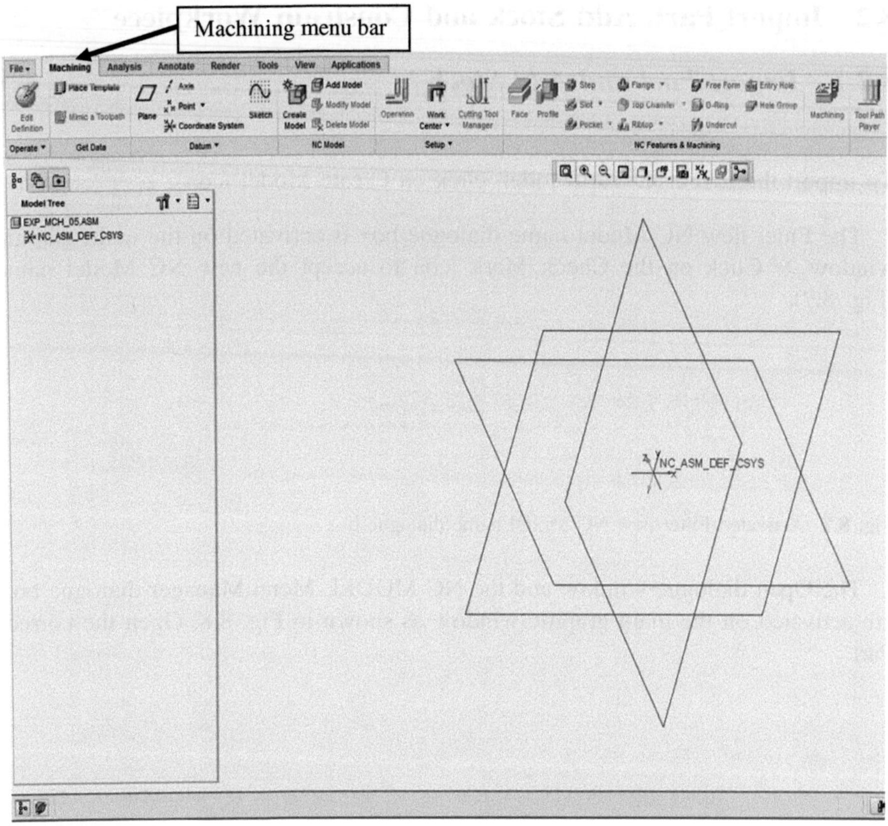

Fig. 8.5 Activated Expert Machinist main graphic user interface window

If Machining is not shown on the Menu bar, click on Applications Menu bar to activate its content ≫ Click on Expert Machinist icon as indicated by the arrow to activate its application as shown in Fig. 8.6.

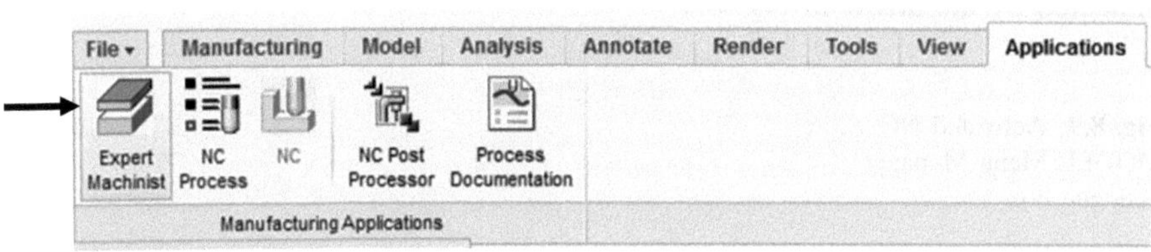

Fig. 8.6 Activated Applications ribbon

8.2 Import Part, Add Stock and Constrain Workpiece

8.2.1 Import Part and Add Stock

To import the saved 3D Part/Model, click on Create Model icon 

.

The Enter new NC Model name dialogue box is activated on the main graphic window ≫ Click on the Check Mark icon to accept the new NC Model name (Fig. 8.7).

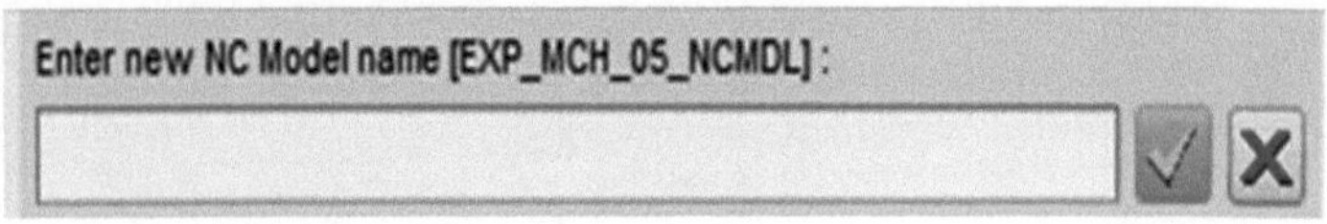

Fig. 8.7 Activated Enter new NC Model name dialogue box

The Open dialogue window and the NC MODEL Menu Manager dialogue box are activated on the main graphic window as shown in Fig. 8.8. Open the correct Part.

Fig. 8.8 Activated Open dialogue window in concise form

Click on the Open tab to exit.

The Part is imported into the graphic window ≫ On the active NC MODEL Menu Manager dialogue box ≫ Click on Create Stock on the NC MODEL group as indicated by the arrow in Fig. 8.9.

Fig. 8.9 Activated NC MODEL Menu Manager dialogue box

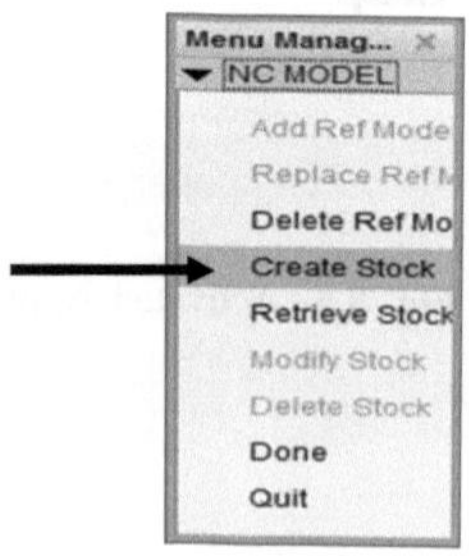

The Auto Workpiece Creation application is activated on the main graphic window ≫ Click on the Options tab to activate its panel. Take note of the Overall Dimensions section. On the Linear Offsets group, click on the Current Offsets tab ≫ Add values in the X, Y and Z section boxes as shown in Fig. 8.10.

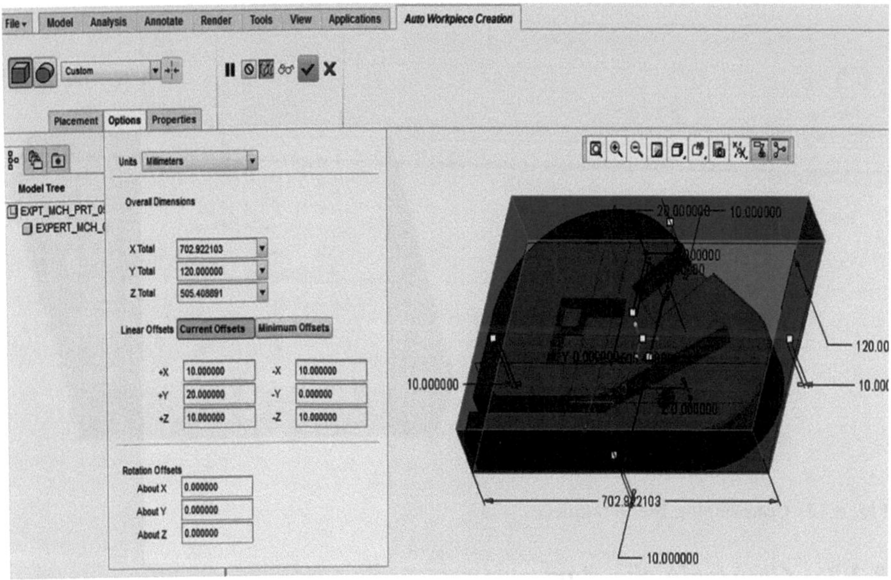

Fig. 8.10 Activated Options panel and rectangular Stock dimensions on Part

Click on the Check Mark icon to create Stock and exit the Auto Workpiece Creation application.

Fig. 8.11 Active NC MODEL Menu Manager dialogue box

Now click on Done on the NC MODEL group as highlighted in Fig. 8.11.

Stock is now added into the Workpiece and the Component Placement dashboard is now activated as shown in Fig. 8.12.

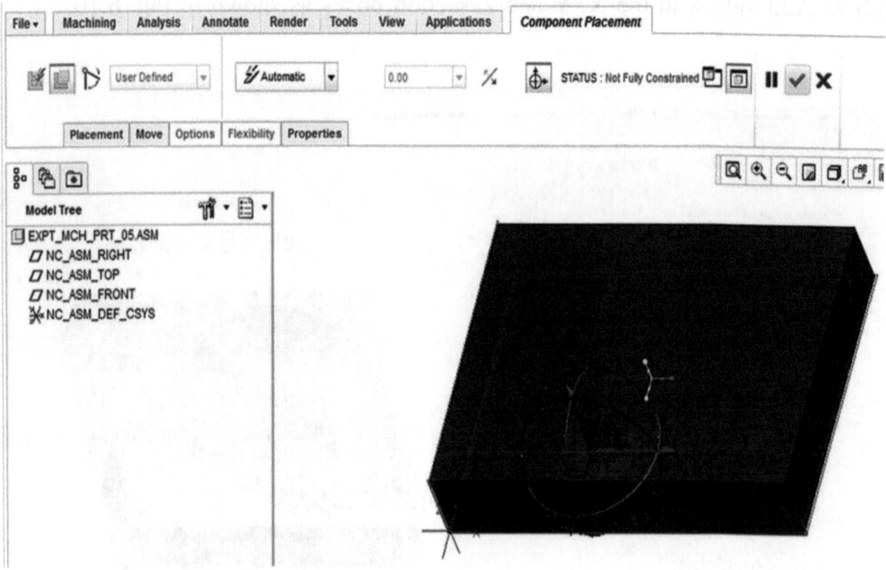

Fig. 8.12 Constraining the Workpiece

8.2.2 *Constrain the Part*

To constrain the Workpiece ≫ Right click on the mouse button on the main graphic window and click on Default Constraint on the drop-down list as illustrated in Fig. 8.13.

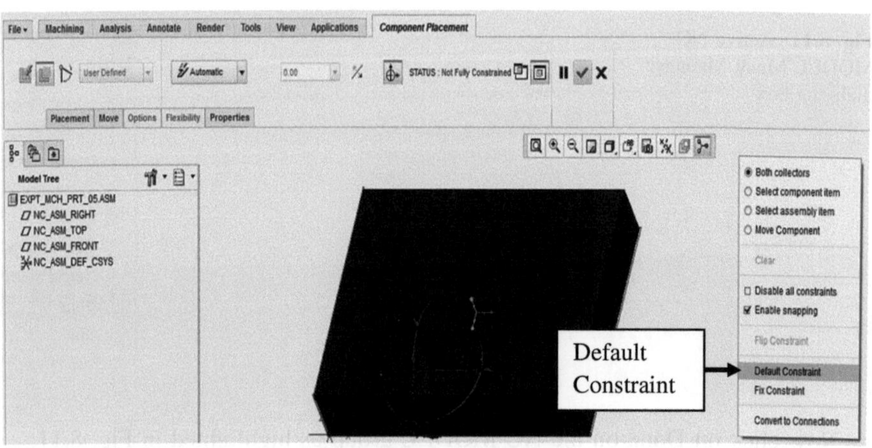

Fig. 8.13 Activating the Default Constraint

The Workpiece is now fully constrained as indicated on the STATUS information bar as indicated by the arrow as shown in Fig. 8.14.

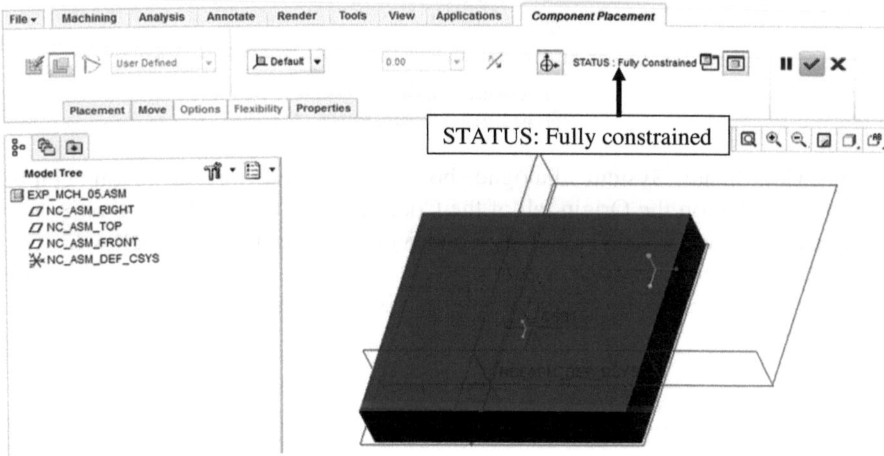

Fig. 8.14 Workpiece is now fully constrained

Click on the Check Mark icon.

Workpiece changes on the Machining main graphic window as shown in Fig. 8.15.

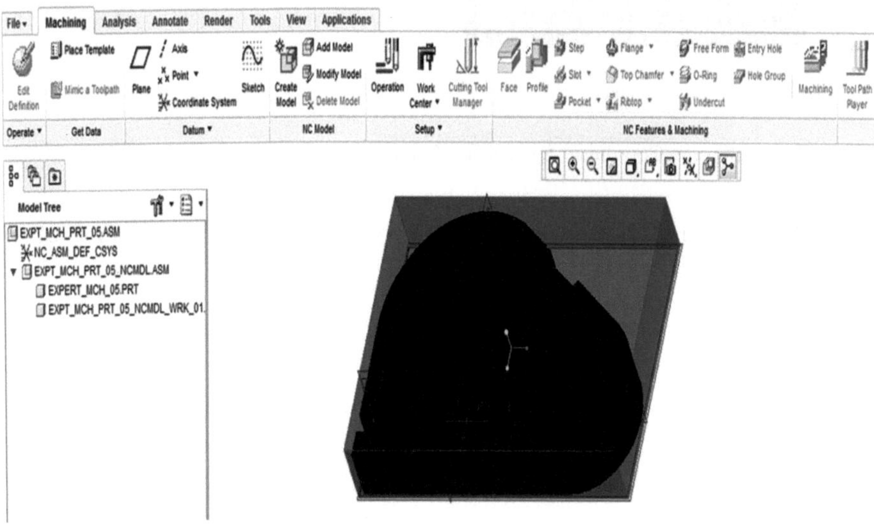

Fig. 8.15 Stock created and fully constrained on the Machining graphic window

8.3 Create Coordinate System or Programme Zero

Click on Coordinate System icon Fig. 8.15 as shown below.

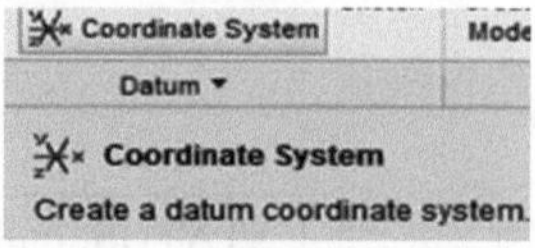

The Coordinate System dialogue box is activated on the main graphic window ≫ Click on the Origin tab of the Coordinate System dialogue box ≫ Now click on the top surface and two sides of the Workpiece as illustrated below to select the references. The selected references are automatically added into the References section on the Coordinate System dialogue box. See Fig. 8.16.

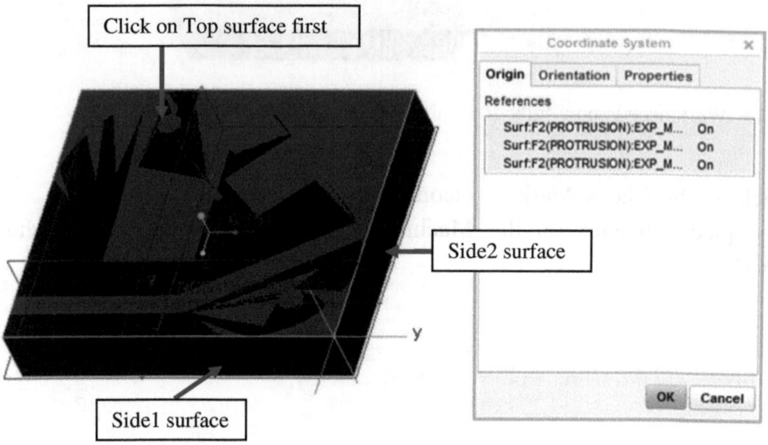

Fig. 8.16 Creating the Coordinate System X, Y and Z axes

The Orientation of the above Coordinate System axes as shown in Fig. 8.17 is not correct because X-axis is pointing upward.

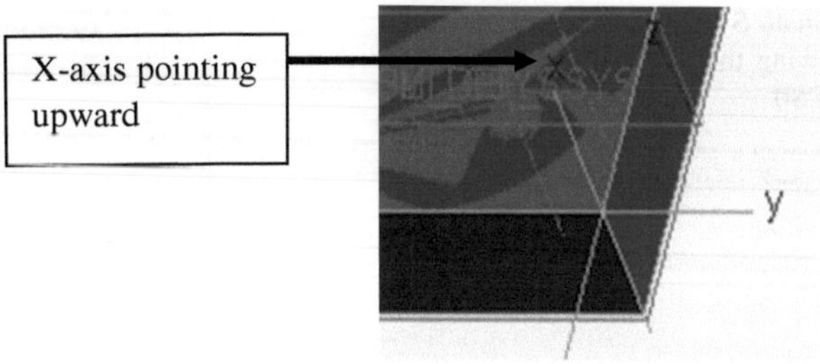

Fig. 8.17 Wrong orientations of X, Y and Z axes of the Coordinate System

To orient the X, Y and Z coordinate axes to the correct orientation

Click on the Orientation tab on the Coordinate System dialogue box. On the "Orient by" group, click on the radio button on the "References selection" ≫ Click on the downward pointing arrow on the "to determine" section box and then click on Z axis on the activated drop-down menu list ≫ Click on the downward pointing arrow on the "to project" section box and then click on X axis on the activated drop-down menu list. Now click on the Flip tab, to flip X axis orientation. The arrows as shown in Fig. 8.18 indicate the correct orientations for the created Programme Zero axes.

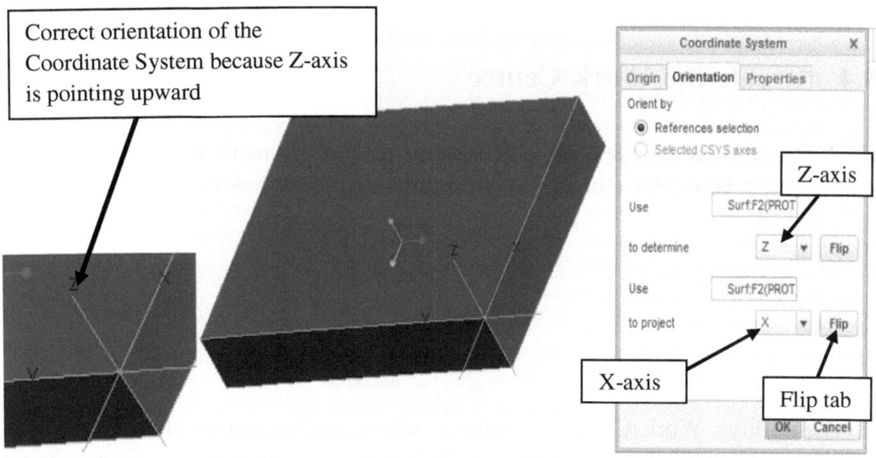

Fig. 8.18 Correct orientations of X, Y and Z axes of the Coordinate System

Click on the Properties tab on the Coordinate System dialogue box, to check if the name of the created Coordinate System (ASC0) is in the Name section box. See Fig. 8.19.

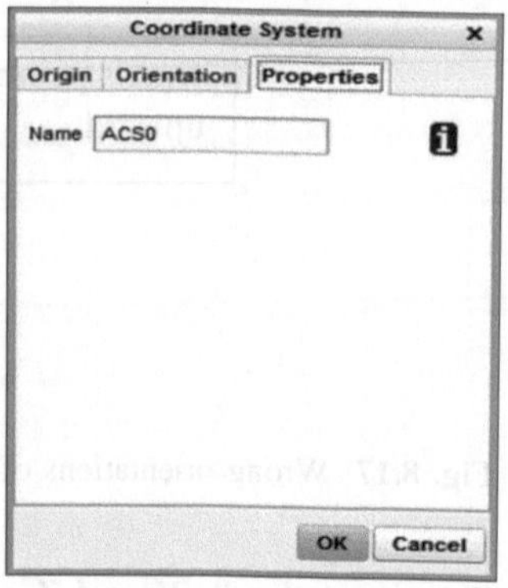

Fig. 8.19 Coordinate System dialogue box showing the newly created ACS0

To change the name of the new Coordinate System (ACS0), just click on the Name section box and type a new name.

In this tutorial, the new Coordinate System name (ACS0) will not be changed.

Click on the OK tab to exit the Coordinate System dialogue box.

8.4 Create the Work Centre

Click on the Work Centre icon to activate its drop-down menu list ≫ Now click on the Machine Tool Manager on the drop-down list as shown below.

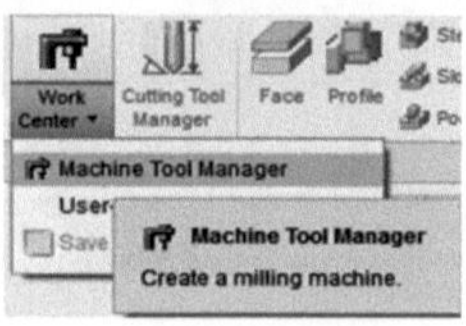

The Milling Work Centre dialogue window is activated on the main graphic window as shown below. Take note of the Name, Type, Post Processor, Number of Axes and section boxes. See Fig. 8.20.

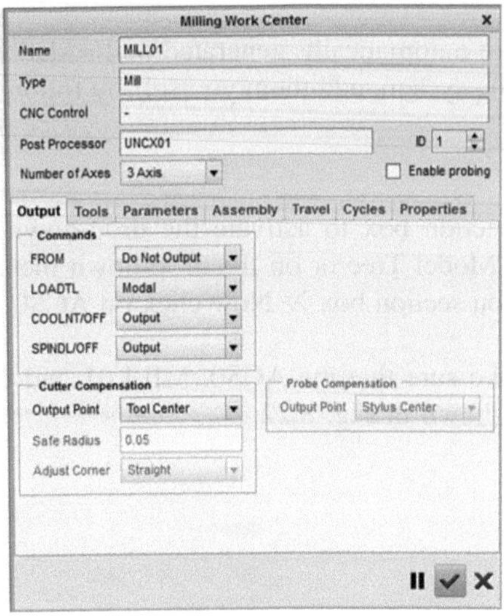

Click on Check Mark icon to exit

Fig. 8.20 Activated Milling Work Centre dialogue window

Note: The Cutting Tools Setup dialogue window can also be activated on the Milling Work Centre dialogue window shown in Fig. 8.20. To activate, just click on the Tools tab.

8.5 Set up Operation

Click on the Operation icon 

.

The Operation application tools are activated. Check to see if the MILL01 and ACS0 are automatically generated in their respective section boxes as indicated by the arrows shown in Fig. 8.21.

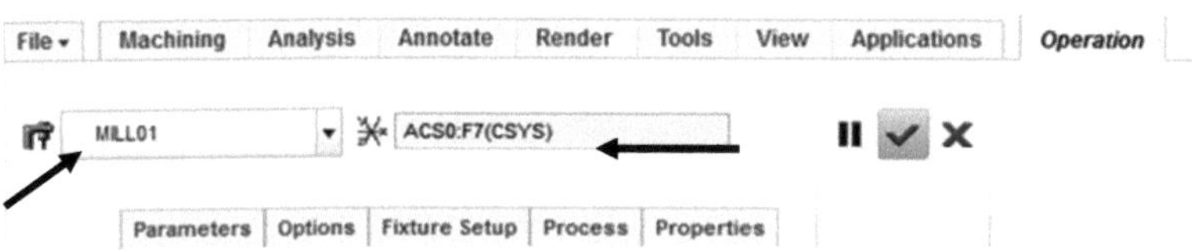

Fig. 8.21 Activated Operation tools in concise form

Now click on the Check Mark icon to exit Operation setup

Note: If MILL01 and ACS0 are not automatically generated and added into three respective section boxes by the system, add them yourself by following the step-by-step guide below.

Click on the Work Centre icon section box to activate the drop-down menu list ≫ Now click on MILL01 on the Model Tree or on the drop-down menu list.

Click on the Coordinate System icon section box ≫ Now click on ACS0 on the Model Tree.

Now check the Model Tree to make sure that the ACS0, MILL01 and OP010 [MILL01] are on the Model Tree as shown in Fig. 8.22.

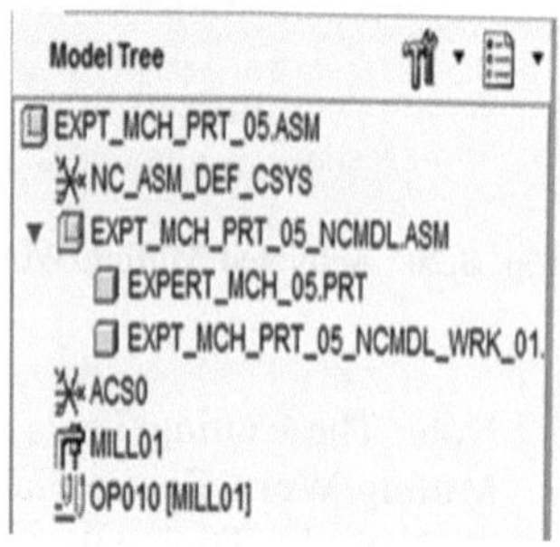

Fig. 8.22 Model Tree showing the created ACS0, MILL01 and OP010 [MILL01]

8.6 Create Cutting Tools

Click on Cutting Tool Manager icon as shown below.

The Tools Setup dialogue window is activated on the main graphic window as shown in Fig. 8.23.

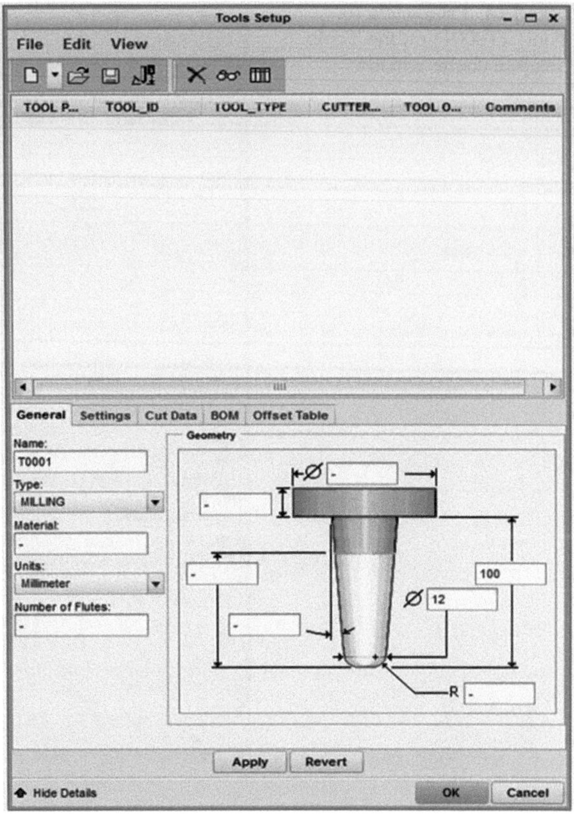

Fig. 8.23 Activated Tools Setup window

Add Tool1 (T0001)

Click on the General tab to activate its panel ≫ Click on the Name section box and type T0001 ≫ Click on the Type section box to activate the drop-down menu list, now click on DRILLING ≫ Click on the Material section box and type TUNGSTEIN CARBIDE as the material (type the name of the material you are using). Make sure that the Unit is set to whatever unit you are working with. Add dimensions to the Tool on the Geometry section, and now click on Apply tab to add T0001. See Fig. 8.24.

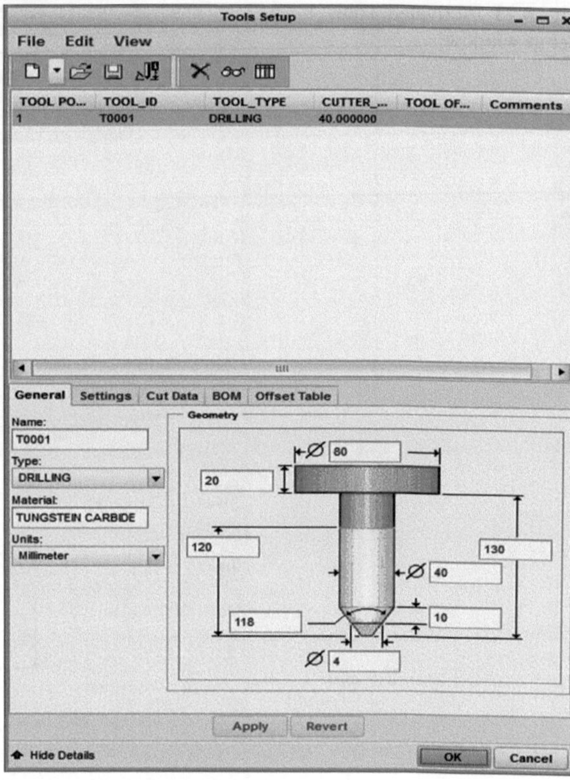

Fig. 8.24 Parameters for Tool1 (T0001) on the Tools Setup dialogue window

Add Tool2 (T0002)

Click on the General tab to activate its panel if it is not active already ≫ Click on the Name section box and type T0002 ≫ Click on the Type section box to activate the drop-down list, now click on END MILL ≫ Click on the Material section box and type TUNGSTEIN CARBIDE as the material (type the name of the material you are using). Make sure that the Unit is set to whatever unit you are working with. Add dimensions to the Tool on the Geometry section, and now click on Apply tab to add T0002. See Fig. 8.25.

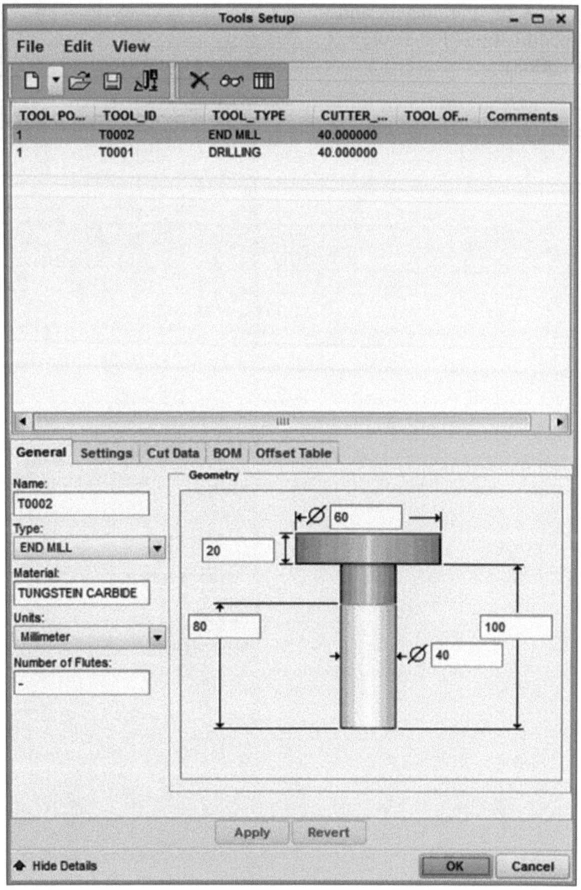

Fig. 8.25 Parameters for Tool2 (T0002) on the Tools Setup dialogue window

Add Tool3 (T0003)

Click on the General tab to activate its panel if it is not active ≫ Click on the Name section box and type T0003 ≫ Click on the Type section box to activate the drop-down list, now click on MILLING ≫ Click on the Material section box and type TUNGSTEIN CARBIDE as the material (type the name of the material you are using). Make sure that the Unit is set to whatever unit you are working with. Add dimensions to the Tool on the Geometry section, and now click on Apply tab to add T0003. See Fig. 8.26.

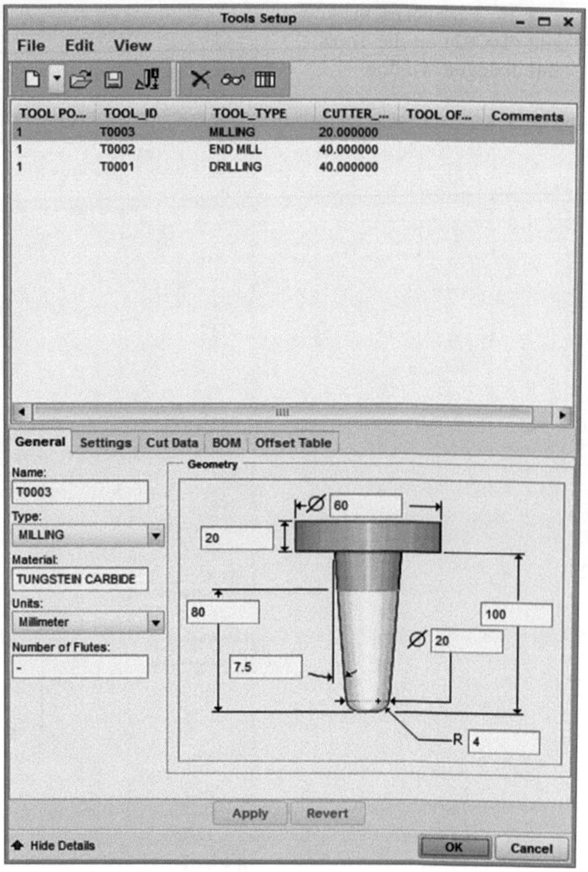

Fig. 8.26 Parameters and specifications for Tool3 (T0003)

Adding Tool4 (T0004)

Click on the Name section box and type T0004 ≫ Click on the Type section box to activate the drop-down list, now click on END MILL ≫ Click on the Material section box and type TUNGSTEIN CARBIDE as the material (type the name of the material you are using). Make sure the that Unit is set to whatever unit you are working with. Add dimensions to the Tool on the Geometry section, and now click on Apply tab to add T0004. See Fig. 8.27.

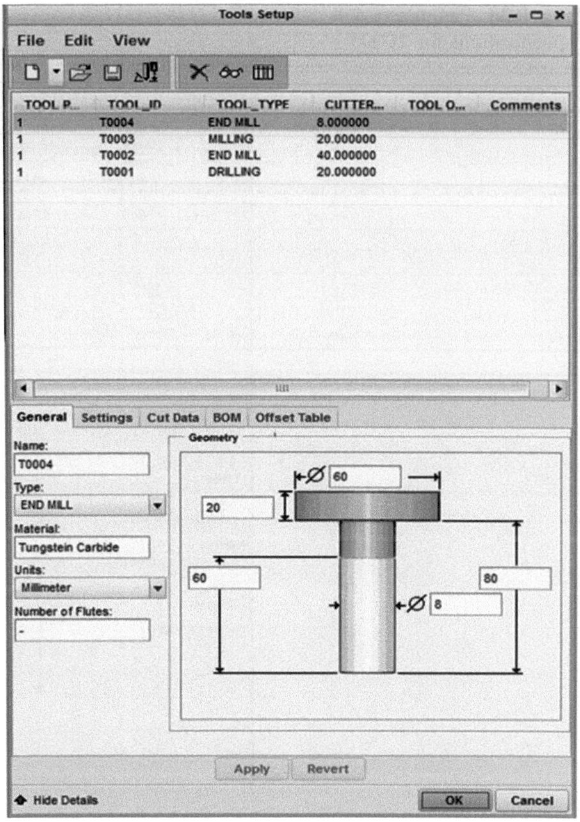

Fig. 8.27 Parameters and specifications for T0004

Add Tool5 (T0005)

Click on the Name section box and type T0005 ≫ Click on the Type section box to activate the drop-down list, now click on CORNER ROUNDING ≫ Click on the Material section box and type TUNGSTEIN CARBIDE as the material (type the material name you are using). Make sure that the Unit is set to whatever unit you are working with. Add dimensions to the Tool on the Geometry section, and now click on Apply tab to add T0005. See Fig. 8.28.

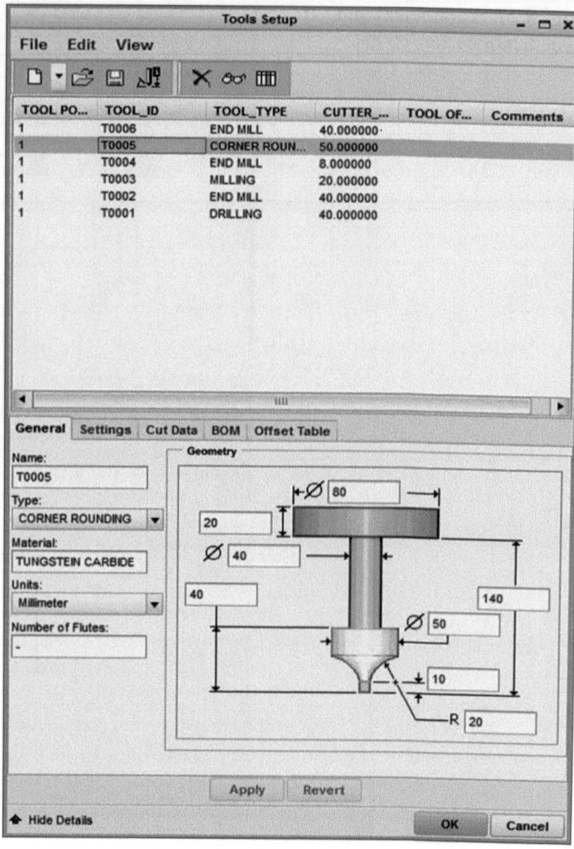

Fig. 8.28 Parameters and specifications for T0005

Add Tool6 (T0006)

Click on the Name section box and type T0006 ≫ Click on the Type section box to activate the drop-down list, now click on END MILL ≫ Click on the Material section box and type the material name you using. Make sure that the Unit is set to whatever unit you are working with. Add dimensions to the Tool on the Geometry section, and now click on Apply tab to add T0006. See Fig. 8.29.

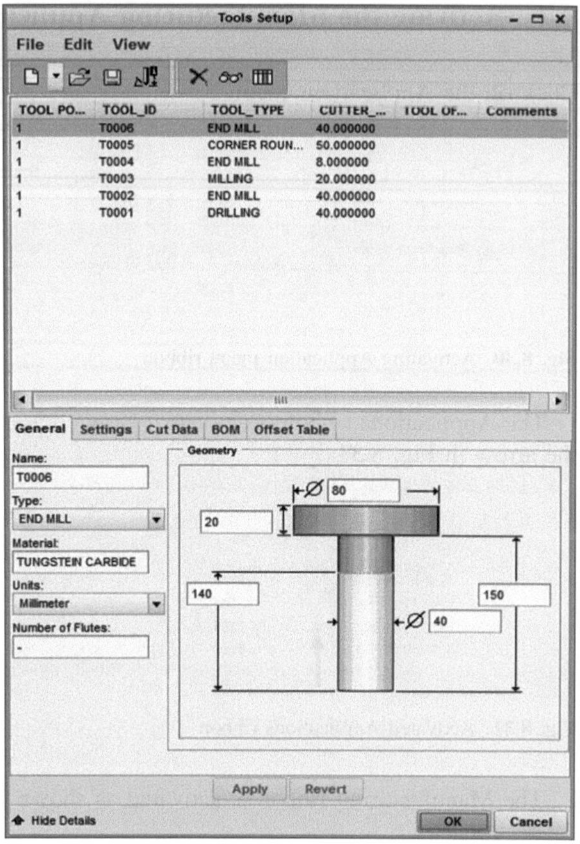

Fig. 8.29 Parameters and specifications for T0006

Now click on OK tab to exit.

Note: Not all the Tools created in the above Tools Setup process will be used in this tutorial.

Before creating the material removal sequences in Expert Machinist, the created Pocket, Steps and Slot on Part will be hidden. Mill Volume will be created and Volume Rough milling will be used to machine the exert shape as will be illustrated on the proceeding pages.

8.7 Activate the Manufacturing Application

Click on the Applications menu bar as indicated by the arrow in Fig. 8.30.

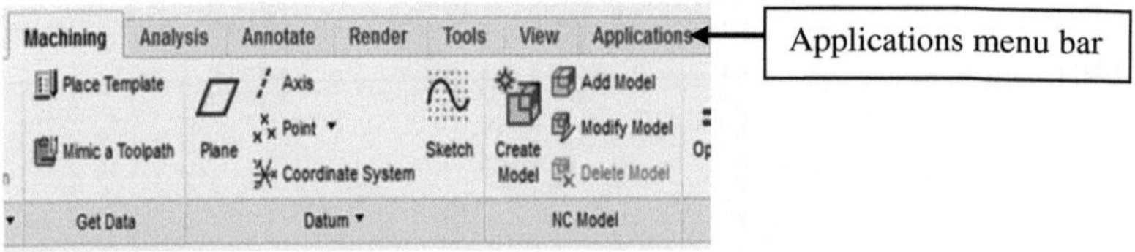

Fig. 8.30 Activating Application menu ribbon

The Applications ribbon is activated. Now click on the NC icon as indicated by the arrow in Fig. 8.31.

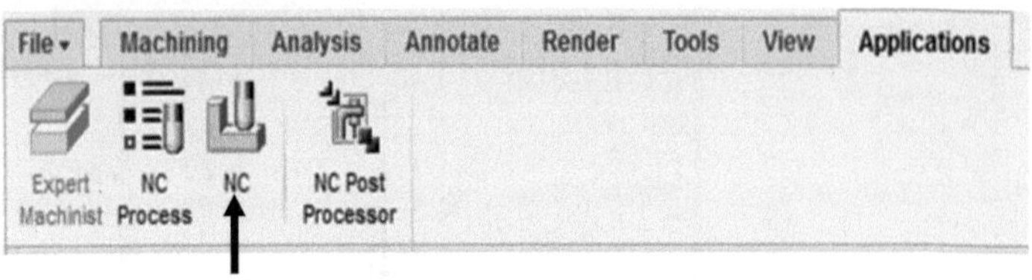

Fig. 8.31 Activated Applications ribbon

The Manufacturing ribbon is activated as shown in Fig. 8.32.

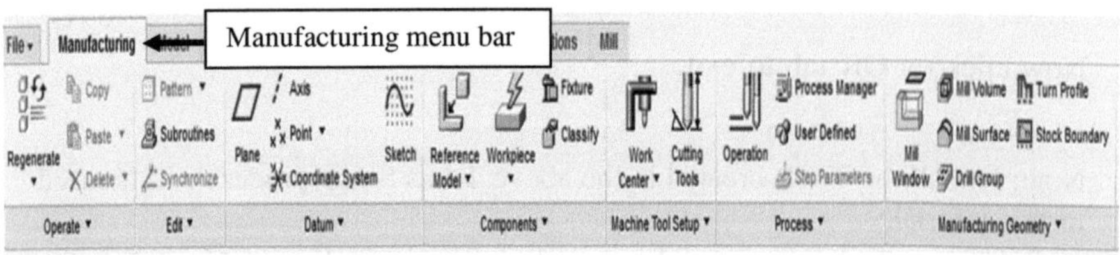

Fig. 8.32 Activated Manufacturing ribbon

8.7.1 Hide the Holes, Pocket, Step and Slots on Part

The Holes, Pocket, Step and Slots on Part have to be hidden before starting the Volume Rough milling operation to avoid Tool gouging.

Go to Model Tree ≫ Click on EXPT_MCH_PRT_05_NCMDL.ASM downward pointing arrow to expand its content as indicated by the arrow as shown in Fig. 8.33.

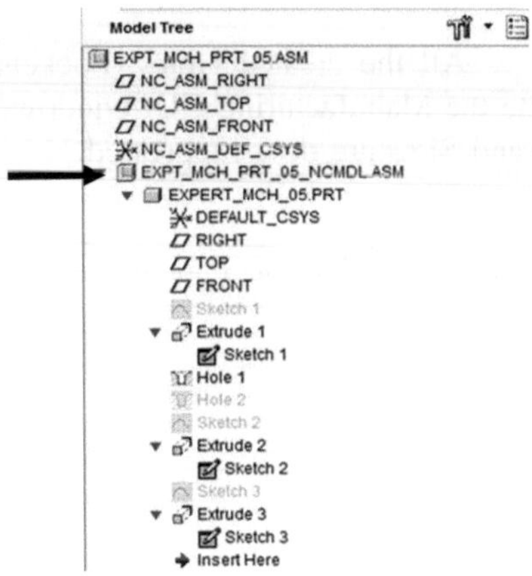

Fig. 8.33 Expanded contents of EXPT_MCH_PRT_05_NCMDL.ASM

Click on EXPERT_MCH_05.PRT downward arrow sign to expand its content ≫ Click on the Holes, Pocket, Step and Slots while holding down the Ctrl key ≫ Now right click on the mouse button and on the drop-down menu list click on Suppress.

The Suppress dialogue box is activated on the main window as shown in Fig. 8.34.

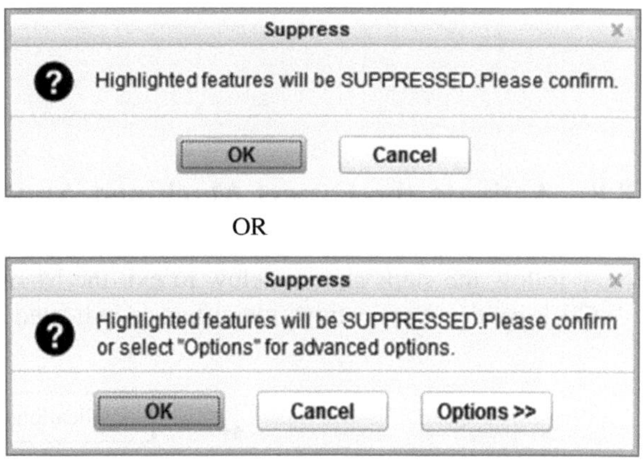

Fig. 8.34 Activated Suppress dialogue box

Note: Alternatively, to hide all the Holes, Pocket, Step and Slots ≫ Right click on EXPERT_MCH_05.PRT and then select Activate on the menu list ≫ Right click on EXPERT_MCH_05.PRT again and now select Open from the menu list. The EXPERT_MCH_05.PRT is open/activated in a new GUI window ≫ Highlight the 2 Holes and the Extruded shapes ≫ Right click and select suppress » All the created Holes, Pockets and Slots are suppressed. Now activate the Manufacturing GUI window. Observe that all the Holes, Pocket, Step and Slots are now suppressed.

The Holes, Pocket, Step and Slots on Part are now suppressed as shown in Fig. 8.35.

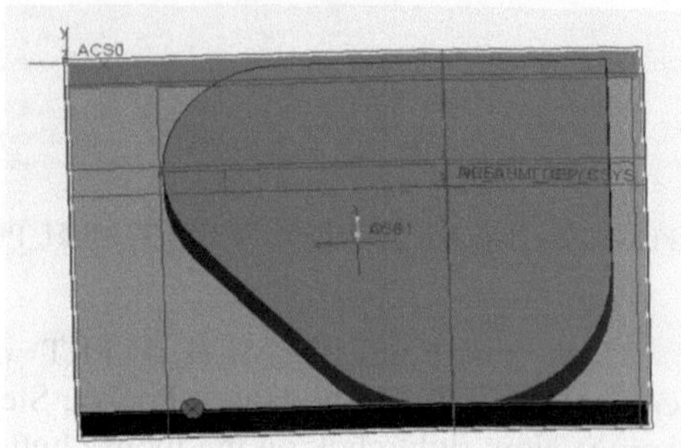

Fig. 8.35 Part with suppressed Holes, Pocket, Step and Slots

Note: The suppressed holes will be un-suppressed before carrying out the Pocket, Step, Slots and Drilling operation later.

8.8 Activate the Expert Machinist Application

Now follow the steps shown below to exit the Manufacturing application.

Click on the Applications menu bar as indicated by the arrow in Fig. 8.36.

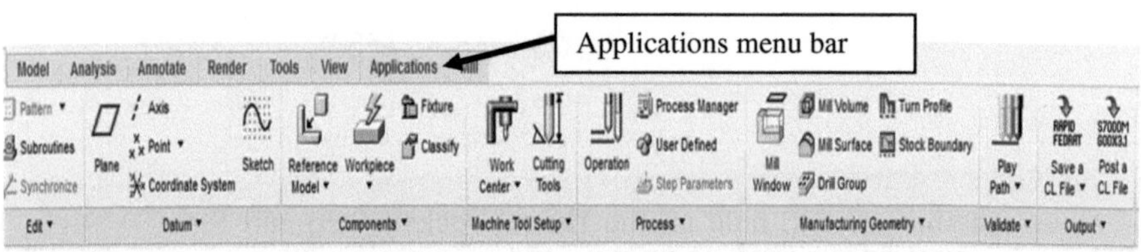

Fig. 8.36 Active Manufacturing ribbon

The Applications ribbon is activated ≫ Now click on Expert Machinist icon as indicated by the arrow as shown in Fig. 8.37.

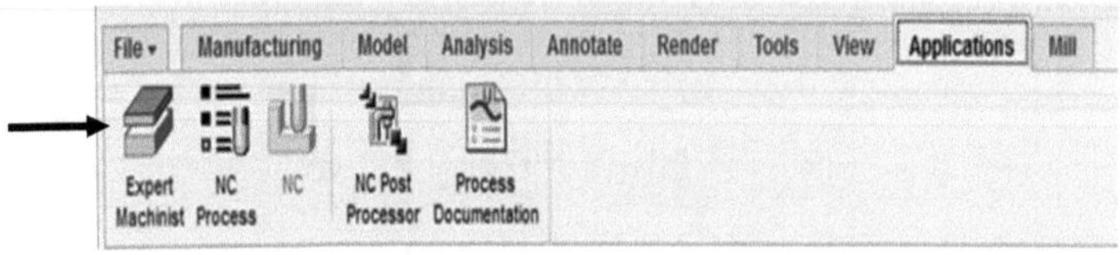

Fig. 8.37 Activated Applications ribbon

The Machining tools are activated as shown in Fig. 8.38.

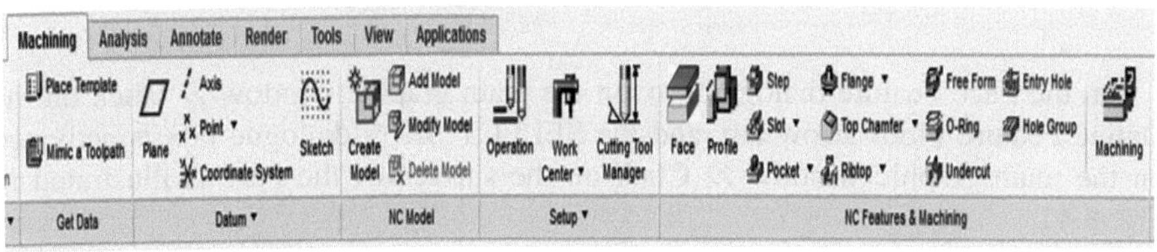

Fig. 8.38 Activated Machining ribbon in concise form

8.8.1 *Create Face Milling Material Removal Sequence*

Material removal sequences of operations are created as described in the following steps below.

On the NC Features and Machining group on the Machining toolbar, click on the Face icon as indicated by the arrow as shown in Fig. 8.39.

Fig. 8.39 NC Features and Machining group

The Face Feature dialogue box is activated on the main graphic window as shown in Fig. 8.40.

Fig. 8.40 Activated Face Feature dialogue box

On the Face Feature dialogue box on the main graphic window ≫ Click on the Define Feature Floor arrow tab, and the SELECT SRFS dialogue box is activated on the main graphic window ≫ Click on the surface of the Part as illustrated in Fig. 8.41.

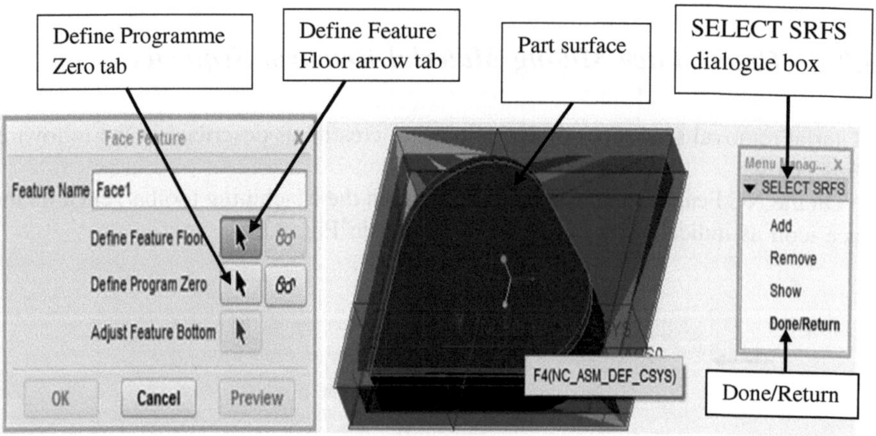

Fig. 8.41 Creating the Face Milling operation

Now click on Define Programme Zero arrow tab as indicated by the arrow in Fig. 8.41 and then click on ACS0 on the Model Tree.

Click on Done/Return on the SELECT SRFS group on the Menu Manager dialogue box, only if it did not disappear automatically by itself.

Now click on OK tab on the Face Feature dialogue box.

Note: The Material Removal Display icon as shown below enables user to show or hide material that will be removed.

8.8.2 *Create Tool Path for Face1 Milling Operation*

Go to the Model Tree and right click on FACE1 [OP010] ≫ Now click on Create Toolpath on the activated drop-down menu list as shown in Fig. 8.42.

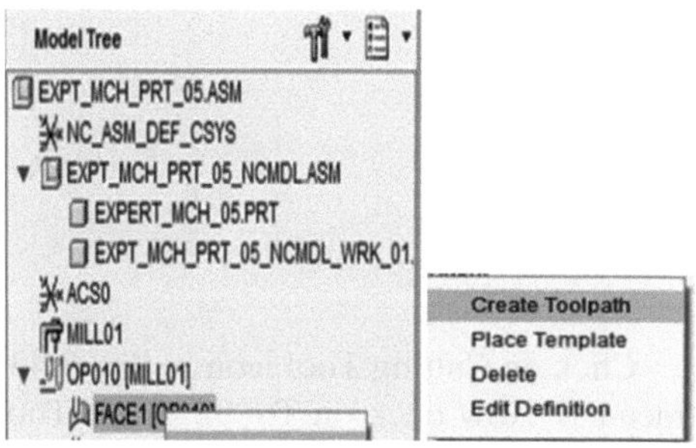

Fig. 8.42 Model Tree and FACE1 [OP010] drop-down menu list

The Face Milling dialogue window is activated on the main graphic window as shown in Fig. 8.43.

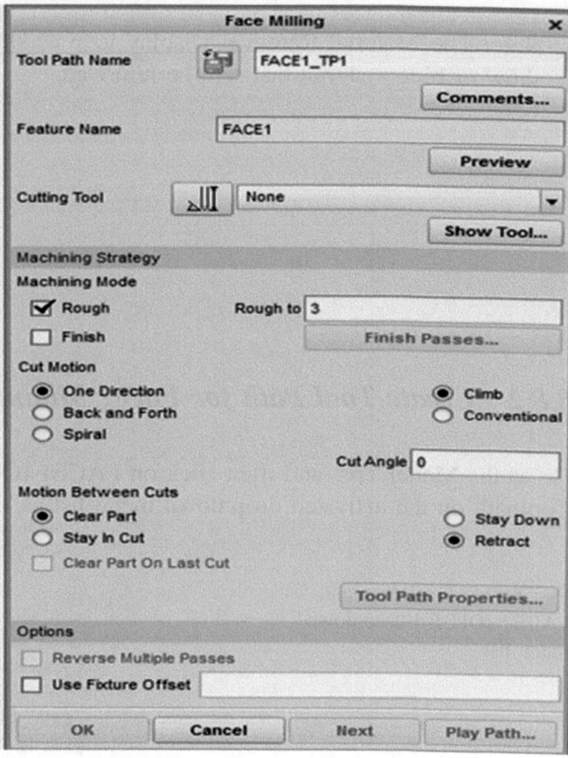

Fig. 8.43 Activated Face Milling dialogue window

Click on Cutting Tool icon in Fig. 8.43 section box to activate its drop-down list menu ≫ Now click on T0002 as the Tool for the Face Milling operation as illustrated in Fig. 8.44.

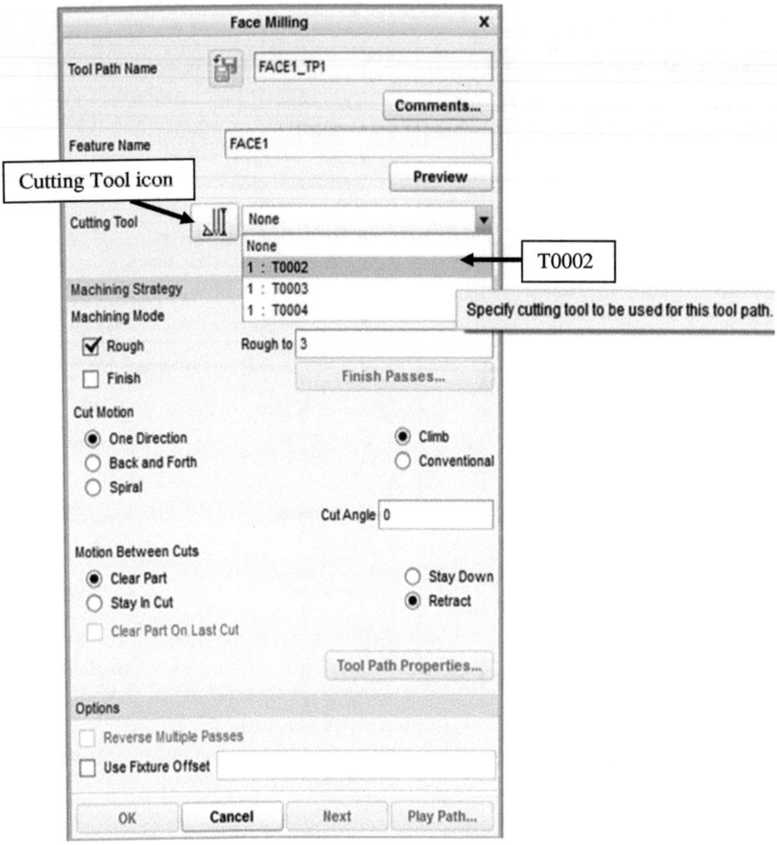

Fig. 8.44 Activating and adding the Cutting Tool for the Face Milling operation

Activate Tool Path Properties

Click on the Tool Path Properties tab on the Face Milling dialogue window as indicated by the arrow in Fig. 8.45 to activate the Tool Path Properties dialogue window on the main graphic window as shown in Fig. 8.45.

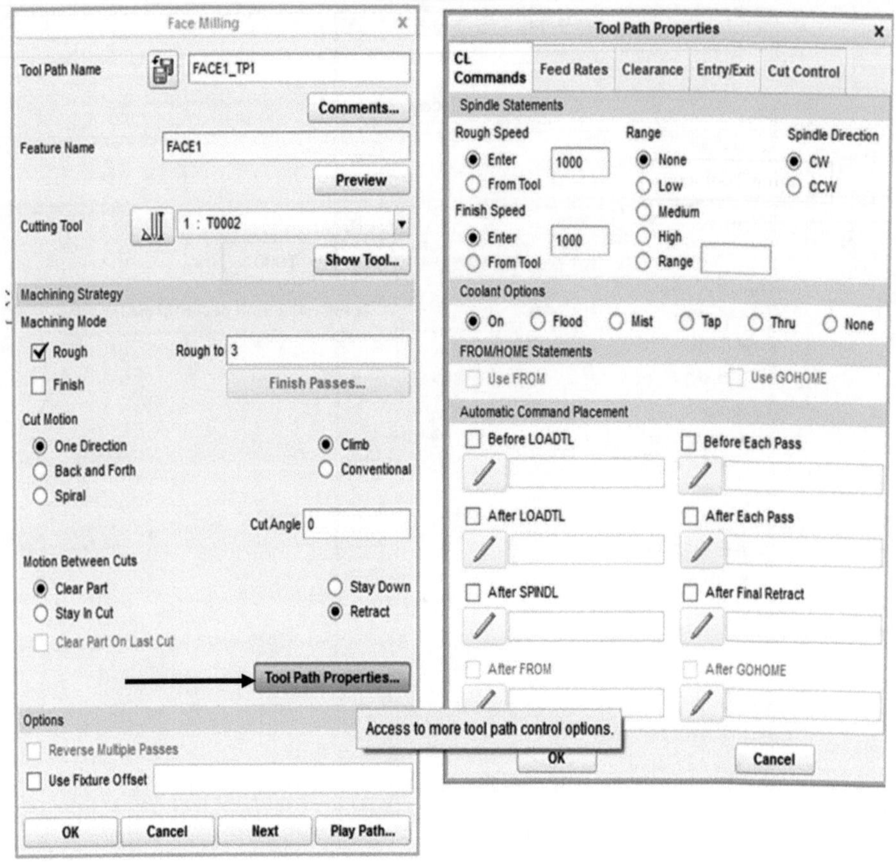

Fig. 8.45 Activated Tool Path Properties and Face Milling dialogue windows

Add Cutting Speed

Click on the CL Commands tab on the activated Tool Path Properties dialogue window.

On the Rough Speed section of the Spindle Statements group, click on the Enter radio button to make it active ≫ Enter value in the Rough Speed section box.

Click on the Enter radio button to make it active, and enter values in the Finish Speed section box. The arrows indicate the illustration as shown in Fig. 8.46.

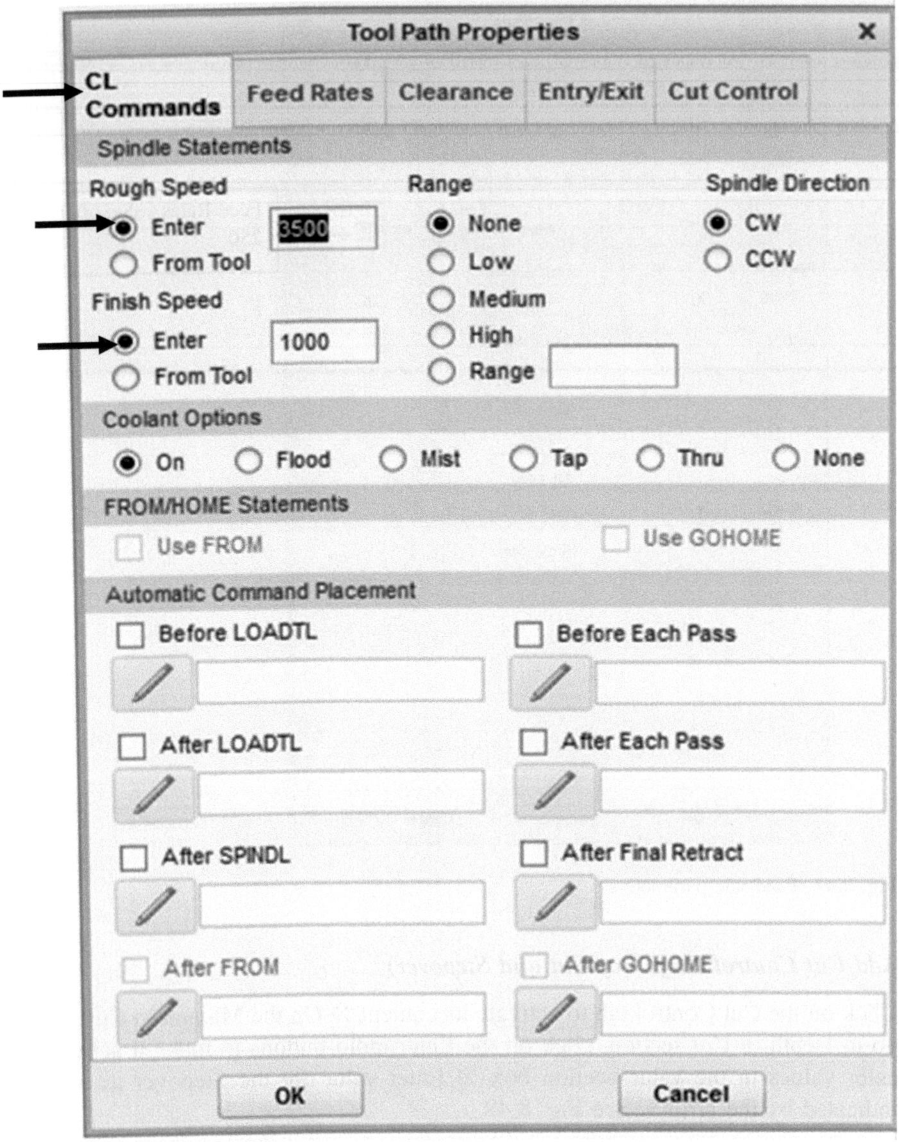

Fig. 8.46 Spindle speed parameters for the Face Milling operation

Add Feed Rates

Click on the Feed Rates tab to activate its content ≫ On the Feed Rates group, go to Rough Cutting section box and enter the Rough Cutting Feed Rate value as indicated by the arrow in Fig. 8.47.

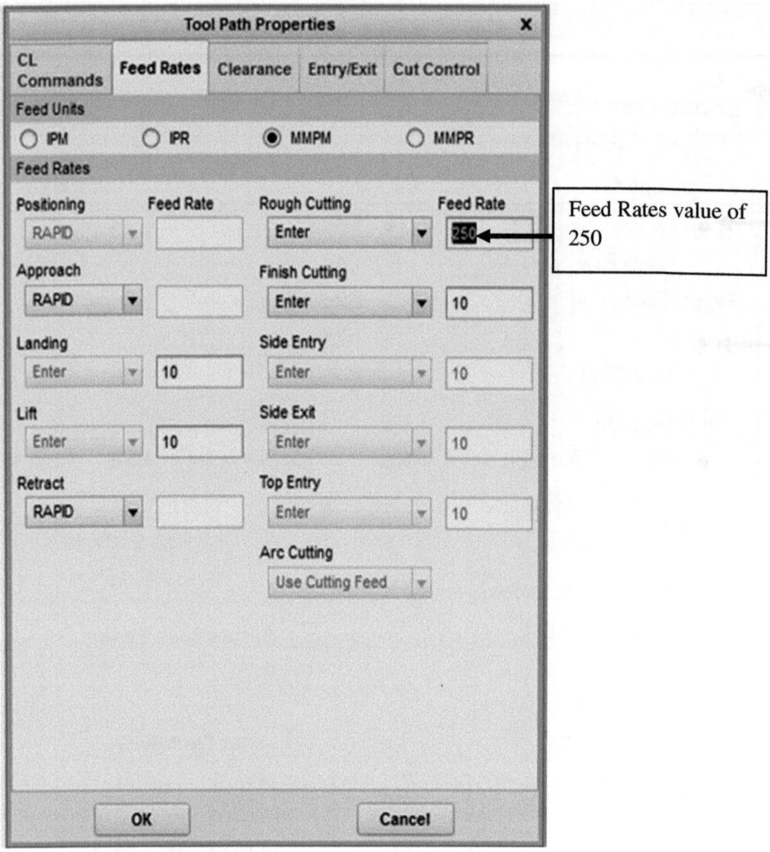

Fig. 8.47 Feed rate parameters

Add Cut Control (Depth of Cut and Stepover)

Click on the Cut Control tab to activate its content ≫ On the Maximum Cut group, go to Depth of Cut section, click on the Enter radio buttons to make it active and enter values in the value section box ≫ Enter value for the Stepover as well as indicated by the arrows. See Fig. 8.48.

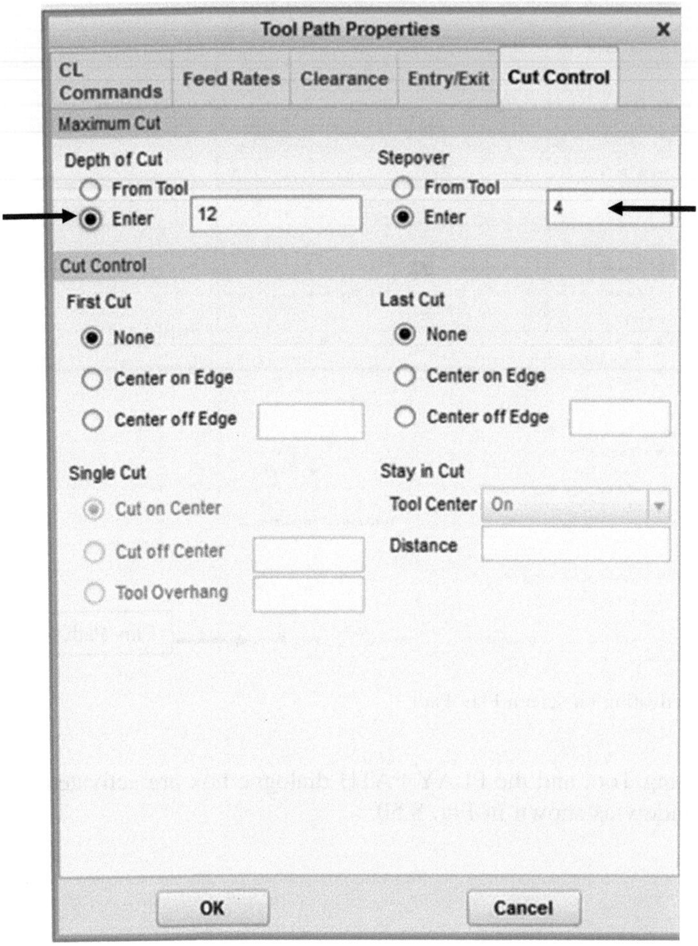

Fig. 8.48 Cut Control parameters

Click on the OK tab to exit.

8.8.3 Activate the Face Milling Play Path Simualtion

Click on Play Path tab on the Face Milling dialogue as indicated by the arrow shown in Fig. 8.49.

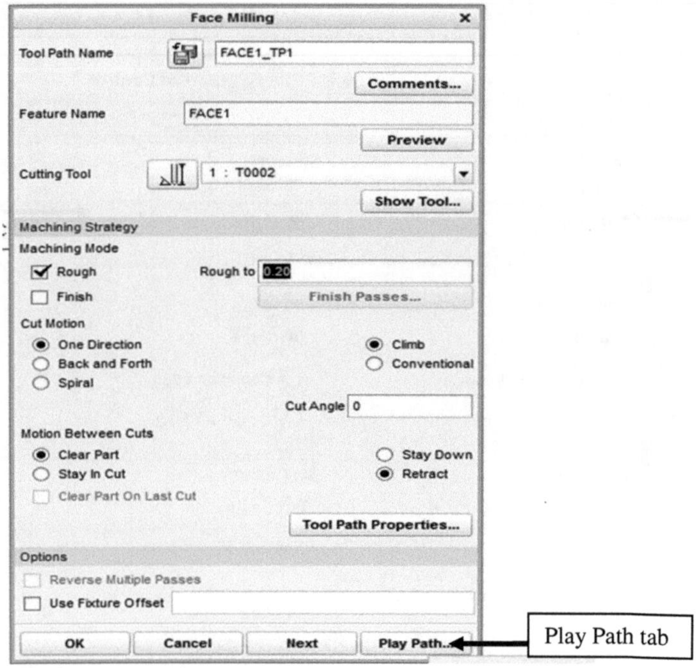

Fig. 8.49 Activating on-screen Play Path

The cutting Tool and the PLAY PATH dialogue box are activated on the main graphic window as shown in Fig. 8.50.

Fig. 8.50 Activated cutting Tool and PLAY PATH dialogue box

Click on the Play tab to start the Face Milling operation.
The end of the Face Milling operation is as shown in Fig. 8.51.

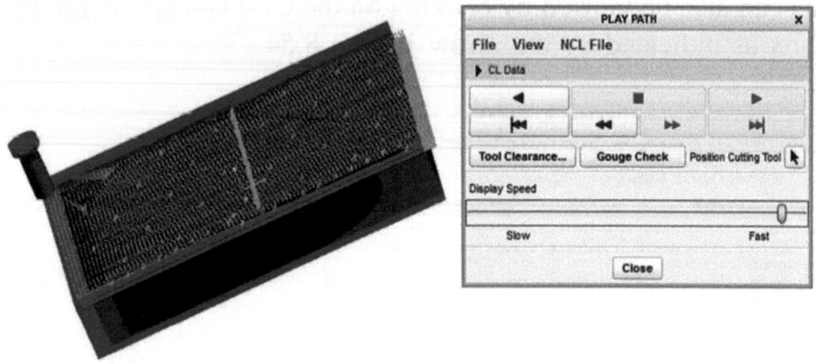

Fig. 8.51 End of Face Milling operation

Note: Do not exit the Face Milling Operation yet.

8.8.4 View the Material Removal Simulation Process

Click on the View tab on the PLAY PATH dialogue box to activate its content. Click on NC Check on the activated drop-down list. Now click on Play tab on the PLAY PATH dialogue box to start the material removal simulation of the Face Milling operation (Figs. 8.52 and 8.53).

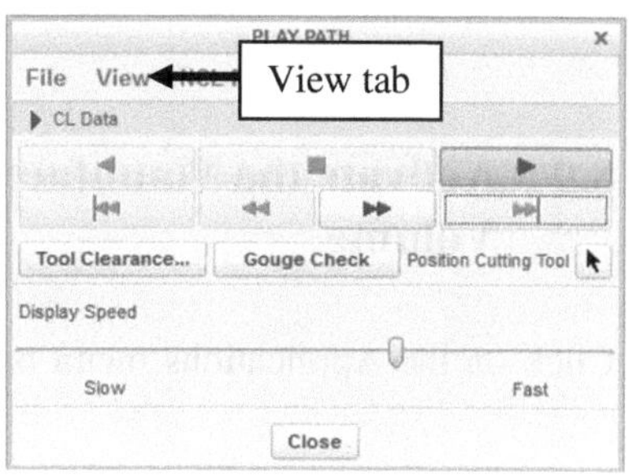

Fig. 8.52 Activating NC Check through the View tab drop-down menu list

Fig. 8.53 End of Material Removal Simulations on Workpiece

CL Data can also be viewed by clicking on the CL Data tab on the PLAY PATH dialogue box as indicated by the arrow in Fig. 8.54.

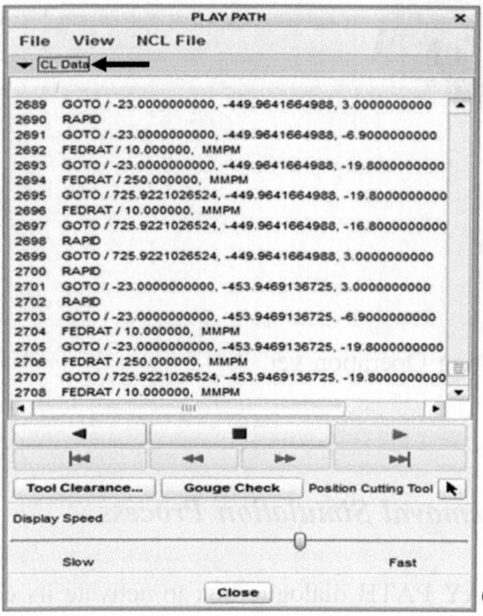

Click on the Close tab to exit.

Fig. 8.54 CL data content

Now click on the OK tab on the Face Milling dialogue box to exit.

8.9 Activate the Manufacturing Application to Create Mill Volume

Click on the Applications menu bar as indicated by the arrow in Fig. 8.55.

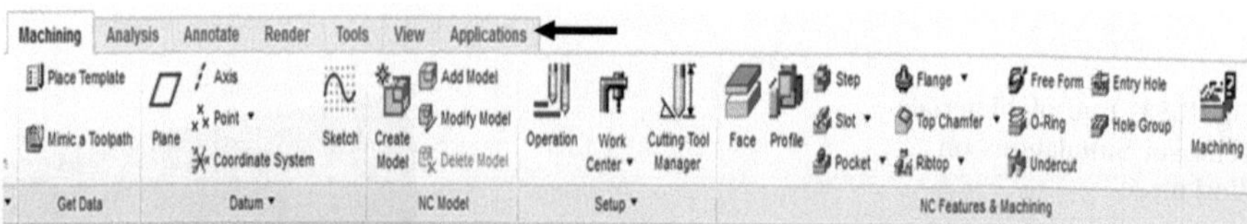

Fig. 8.55 Machining ribbon

Now click on the NC icon as indicated by the arrow in Fig. 8.56.

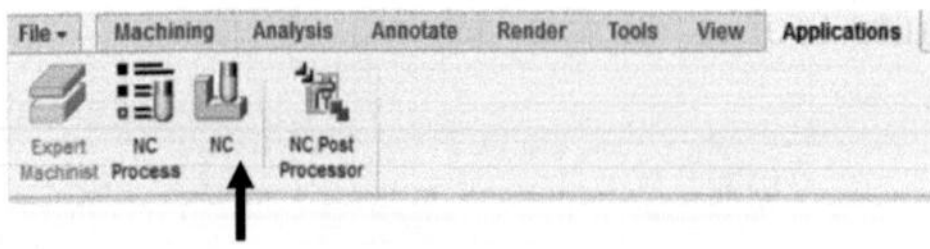

Fig. 8.56 Activated Applications ribbon

The Manufacturing tools are activated as shown in Fig. 8.57.

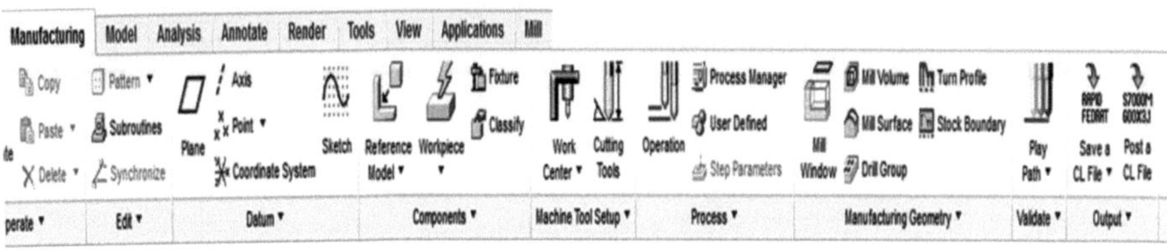

Fig. 8.57 Activated Manufacturing ribbon

8.9.1 *Create the Mill Volume*

The created Mill Volume will be removed during the Volume Rough milling process.

Click on the Mill menu bar ≫ Now select Mill Volume icon on the Manufacturing Geometry group as shown below.

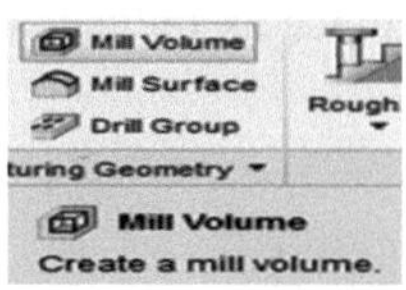

The Mill Volume ribbon is activated as shown in Fig. 8.58.

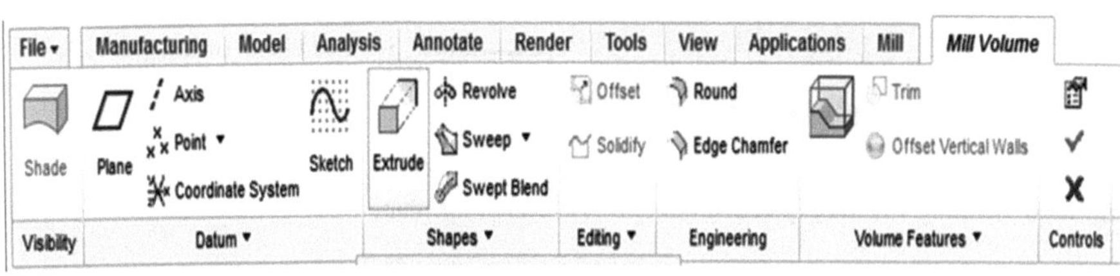

Fig. 8.58 Mill Volume ribbon

Click on the Extrude tool as shown below.

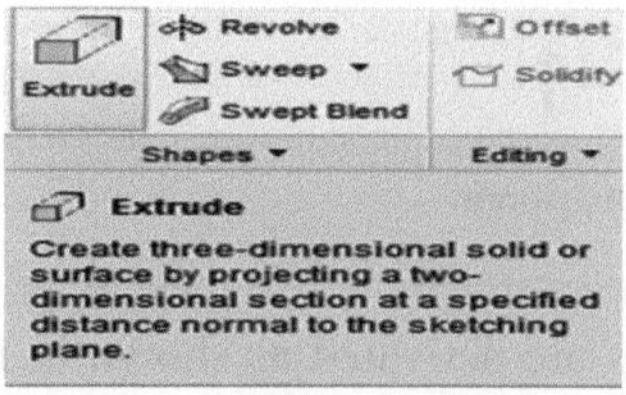

The Extrude dashboard tools are activated as shown in Fig. 8.59.

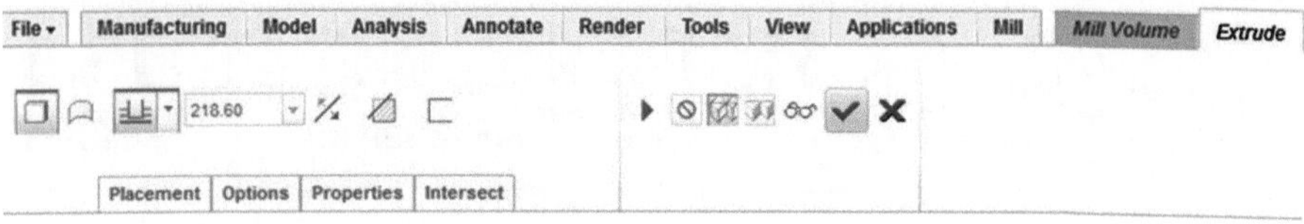

Fig. 8.59 Activated Extrude dashboard

Click on the Placement tab to activate its content ≫ Now click on the Define tab. The Sketch dialogue box is activated on the main graphic window.

Now click on the Placement tab on the Sketch dialogue box ≫ On the Sketch Plane group, click on the Plane section box and click on the Surface of the Workpiece as the sketching plane as illustrated in Fig. 8.60.

Note: If the Reference box is not activated automatically by the system, select a planar surface on the Workpiece as indicated by the arrow in Fig. 8.60.

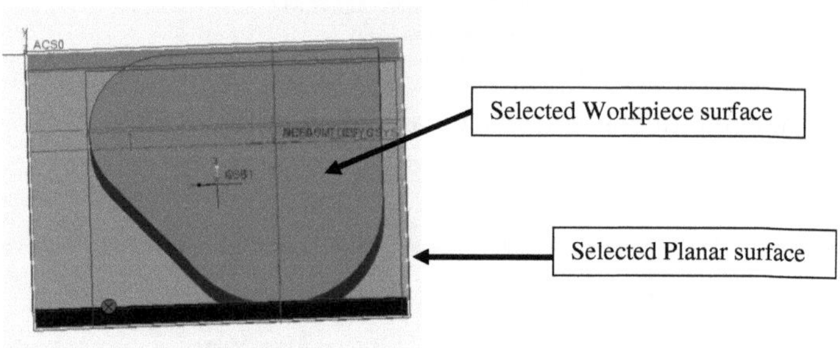

Fig. 8.60 Adding the Workpiece surface as the sketching plane

Click on the Sketch tab on the Sketch dialogue box to proceed.
The Sketch application tools are activated as shown in Fig. 8.61.

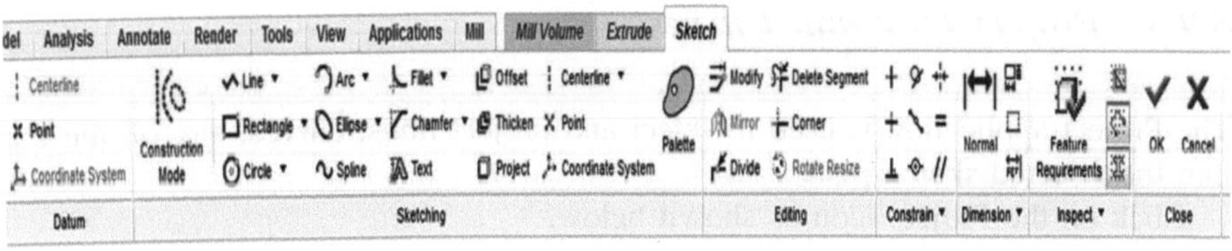

Fig. 8.61 Sketching tools

Click on the Sketch view icon  to orient/position the Workpiece on the correct sketching plane/orientation.

8.9.2 *Create References*

Click on the References icon .

The References dialogue box is activated on the main graphic window ≫ Click on the edges of the Workpiece, the vertical and horizontal axes to create the references as illustrated by the arrows in Fig. 8.62.

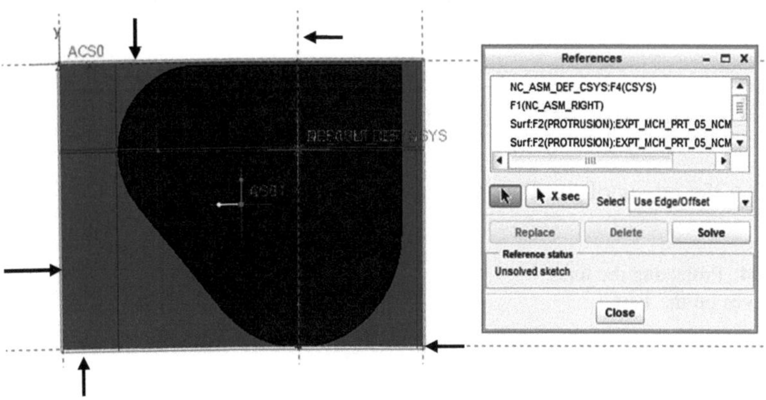

Fig. 8.62 Creating references on the Workpiece

Now click on the Close tab to exit References dialogue box.

8.9.3 Project Line and Curves

The Project application is used to select and project lines, curves, etc. on the Part into the selected sketch plane.

Click on the Project icon as shown below.

The Type dialogue box is activated on the main graphic window as shown in Fig. 8.63.

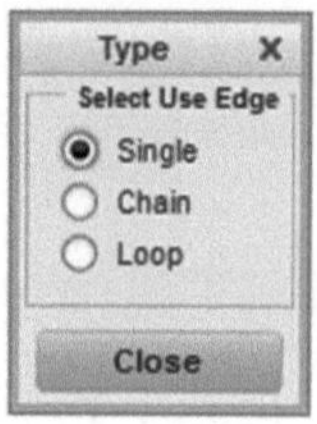

Fig. 8.63 Activated Type dialogue box

Click on the Single ribbon on the Select Use Edge group on the Type dialogue box to make it active, if it is not automatically activated by the system.

8.9.4 Create the Mill Volume Sketch

Click on the edges of the Part as illustrated by the arrows shown in Fig. 8.64.

Fig. 8.64 Projecting the lines and curves on the Part

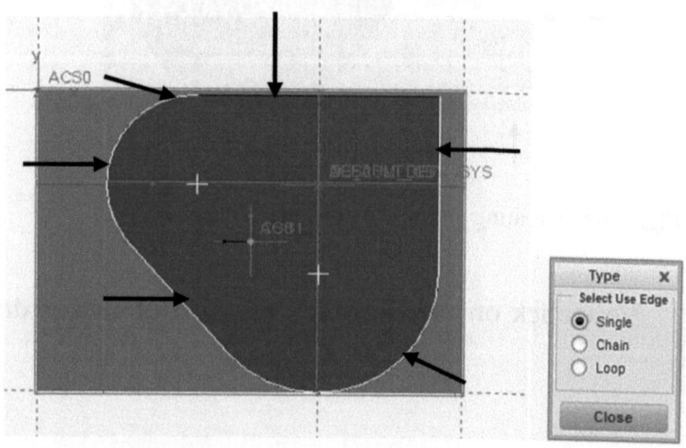

Click on the Close tab on the Type dialogue box.

To sketch the rectangle, use the Corner Rectangle tool on the Sketch tools. Alternatively, use Line and Delete Segment tools to sketch the rectangle as illustrated in Fig. 8.65.

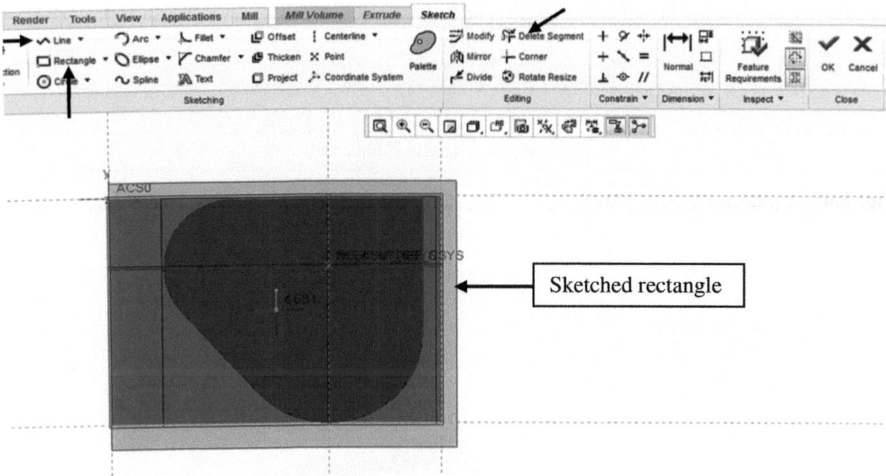

Fig. 8.65 Sketched external rectangle

Now click on the OK/Check Mark icon to exit the Sketch application. The Extrude dashboard tools are activated as shown in Fig. 8.66.

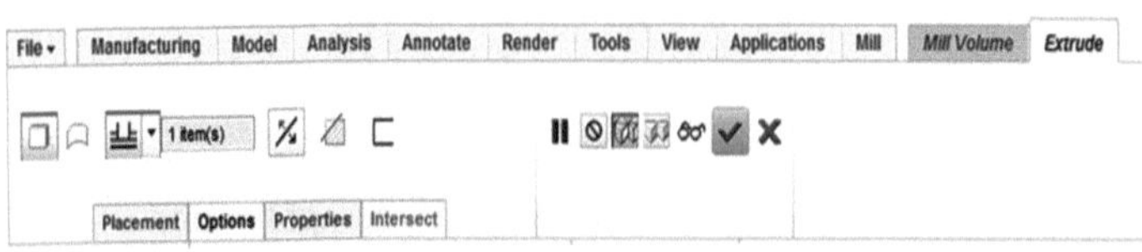

Fig. 8.66 Activated Extrude dashboard

Click on the Options tab to activate its panel >> On the Depth group, click on the downward pointing arrow on Side 1 section box, click on "To Selected" on the activated drop-down menu list >> Now click on the bottom of the Workpiece as the depth of extrusion. See Figs. 8.67 and 8.68.

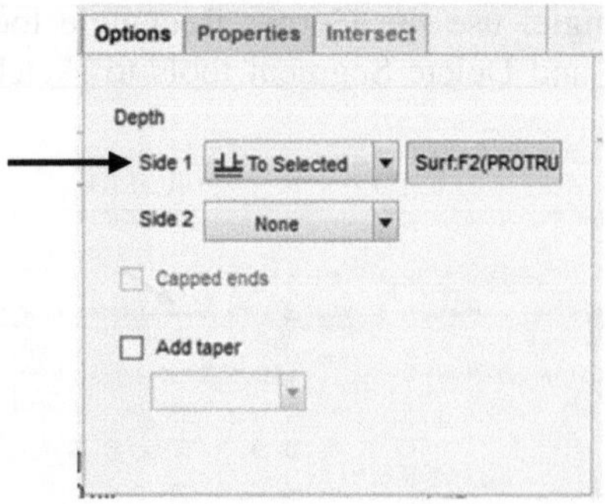

Fig. 8.67 Updated Options panel

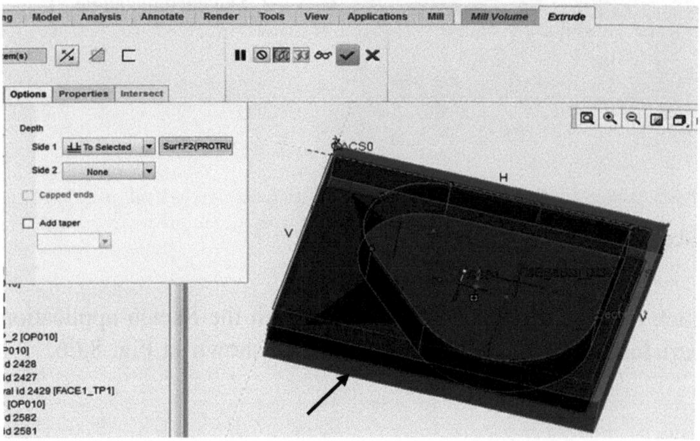

Fig. 8.68 Extruding the sketched Mill Volume

Click on the Check Mark icon to exit the Extrude application.
Click on the Check Mark icon to exit the Mill Volume application.
The created Mill Volume is shown in Fig. 8.69.

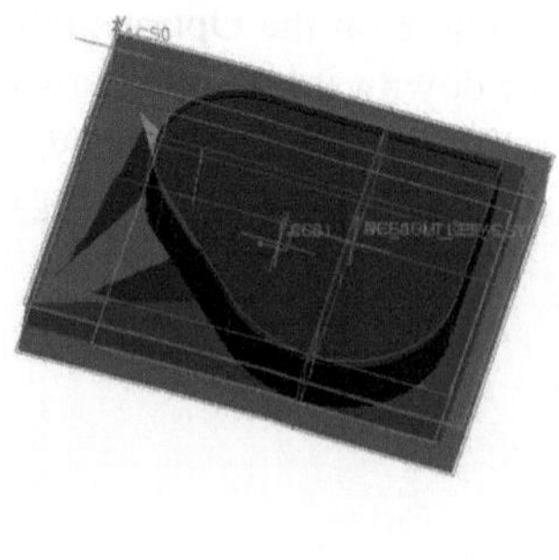

Fig. 8.69 Created Mill Volume

8.10 Create Volume Rough Milling Operation

Click on Roughing icon to activate its drop-down menu ≫ Now click on Volume Rough as illustrated below.

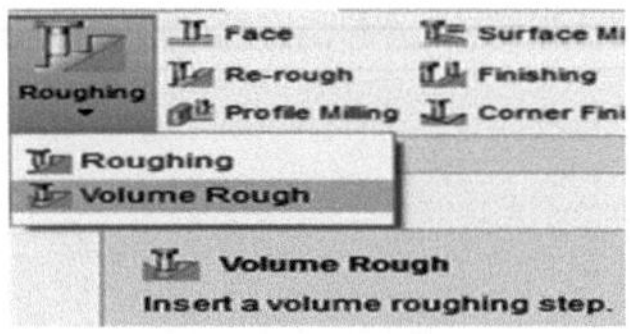

The NC SEQUENCE Menu Manager dialogue box is activated on the main graphic window ≫ Check Mark Tool, Parameters, Retract Surf and Volume on the SEQ SETUP group by clicking on them as shown in Fig. 8.70.

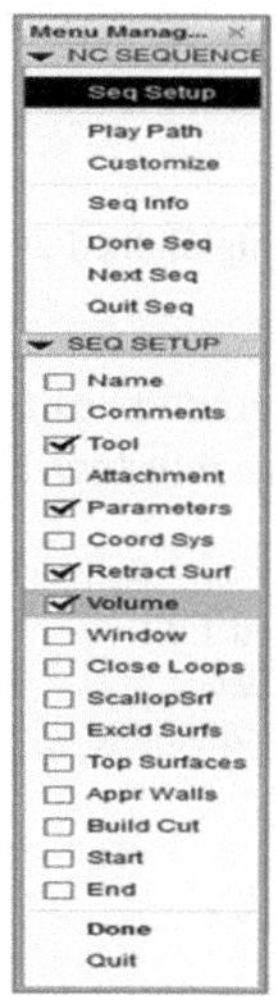

Fig. 8.70 Tool, Parameters, Retract Surf and Volume checked marked

Click on Done.

The Tools Setup dialogue window is activated on the main graphic window ≫ Click on T0006 already created as the cutting Tool as highlighted in Fig. 8.71.

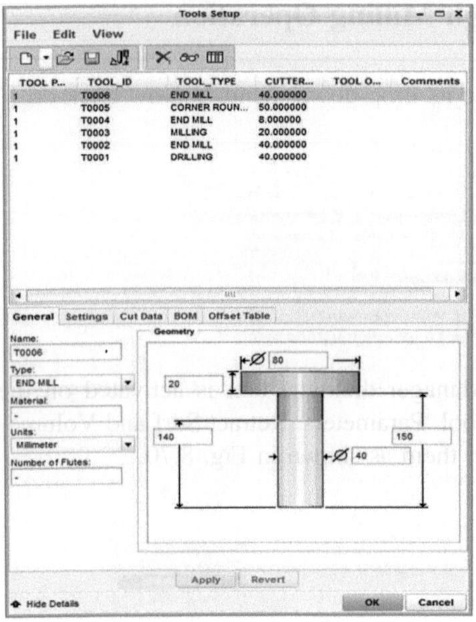

Click on the OK tab to proceed.

Fig. 8.71 T0006 highlighted as the cutting Tool

The Edit Parameters of Sequence 'Volume Milling' dialogue box is activated on the main graphic window as shown in Fig. 8.72.

Fig. 8.72 Activated Edit Parameters of Sequence "Volume Milling" dialogue box

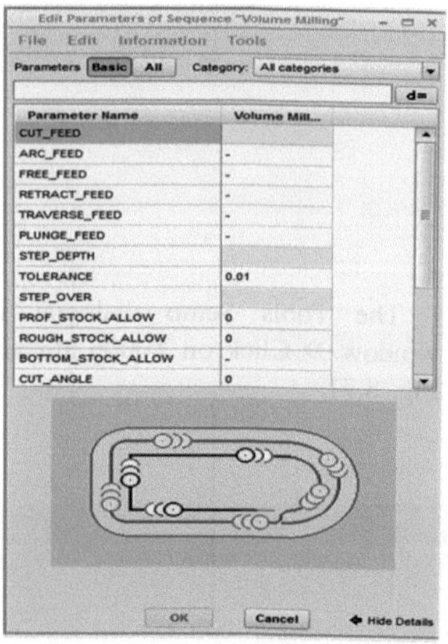

Input parameter values for the Volume Rough Milling are shown in Fig. 8.73.

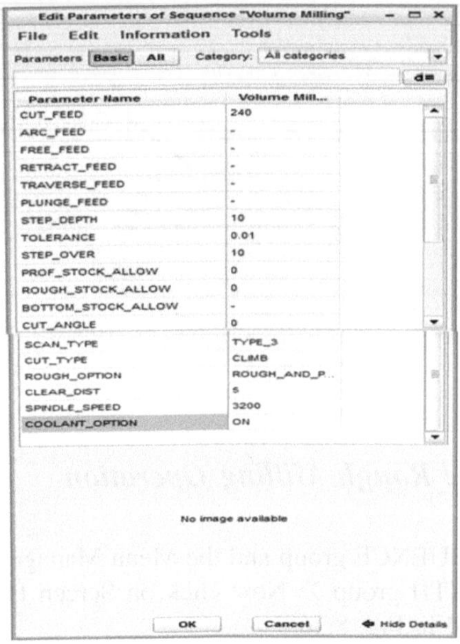

Click on the OK tab to exit.

Fig. 8.73 Edit Parameters of Sequence "Volume Milling" parameters

The Retract Setup dialogue box is activated on the main graphic window ≫ Click on the Retract tab to activate its content ≫ Click on the downward pointing arrow on the Type section box, and click on Plane on the activated drop-down menu list. Click on the Orientation section box to activate its content, now click on Z to set orientation to Z direction. Enter 15 mm in the Value section box as illustrated in Fig. 8.74.

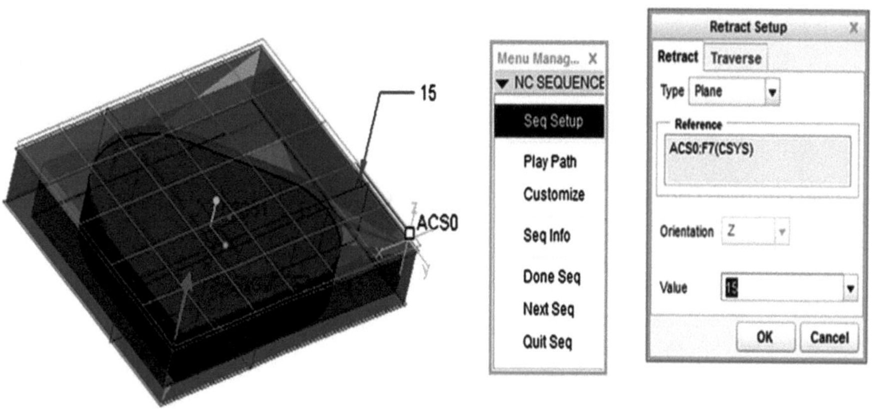

Fig. 8.74 Creating the Retract Setup parameters

Click on the OK tab to exit the Retract Setup dialogue box.

On the main graphic Message bar, system is asking for the new created Mill Volume to be selected ≫ Now click on the created Mill Volume as indicated by the arrow in Fig. 8.75.

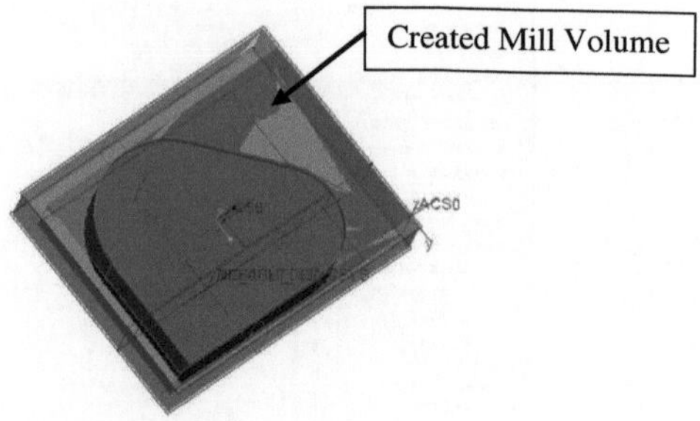

Fig. 8.75 Adding the created Mill Volume

8.10.1 Activate the Volume Rough Milling Operation

Click on Play Path on the NC SEQUENCE group and the Menu Manager dialogue box expands to add the PLAY PATH group ≫ Now click on Screen Play on the PLAY PATH group as highlighted in Fig. 8.76.

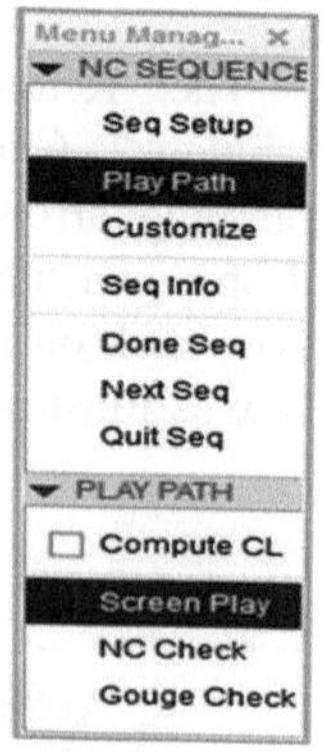

Fig. 8.76 Play Path and Screen Play highlighted

The cutting Tool and PLAY PATH dialogue box are activated as shown in Fig. 8.77.

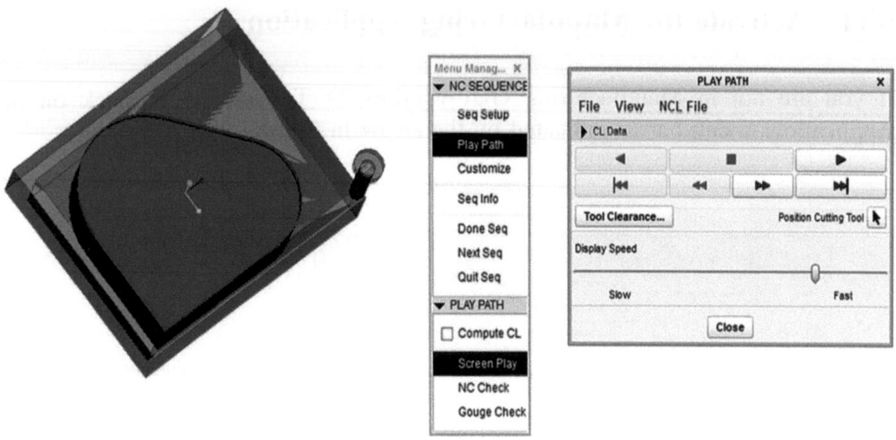

Fig. 8.77 Activated cutting Tool and PLAY PATH dialogue box on main window

For better view of the Volume Rough Milling operation ≫ Click on the Display Style icon, then click on Wireframe on the activated drop-down list as the display style.

Click on the Play tab to start the Volume Milling operation.

The end of Volume Rough milling operation is shown in Fig. 8.78 in wireframe display.

Fig. 8.78 End of Volume Rough milling operation

Click on the Close tab to exit the PLAY PATH dialogue box.

Now click on the Done Seq on the NC SEQUENCE group to exit the Menu Manager dialogue box.

8.11 Activate the Manufacturing Application

If you are not in Manufacturing GUI window ≫ To activate it, click on the Applications menu bar as indicated by the arrow in Fig. 8.79.

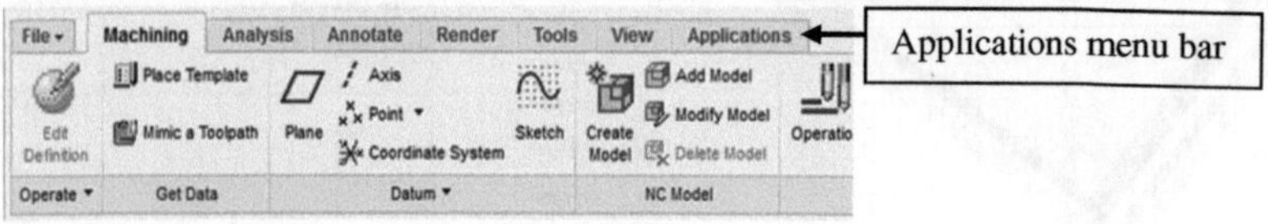

Fig. 8.79 Activating Applications tools

The Applications ribbon is activated. Now click on the NC icon as indicated by the arrow in Fig. 8.80.

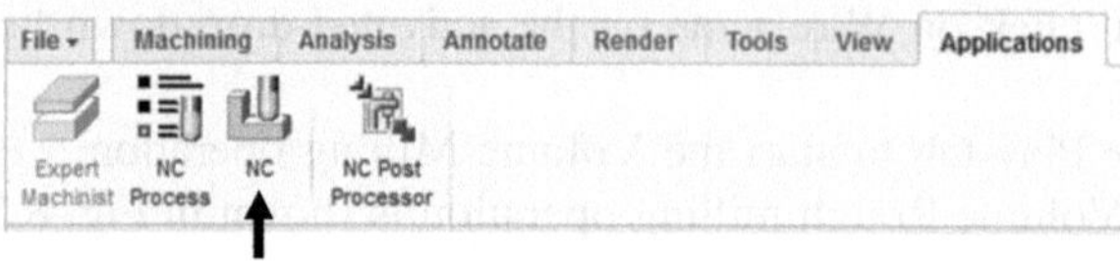

Fig. 8.80 Activated Applications ribbon

The Manufacturing ribbons are activated as shown in Fig. 8.81.

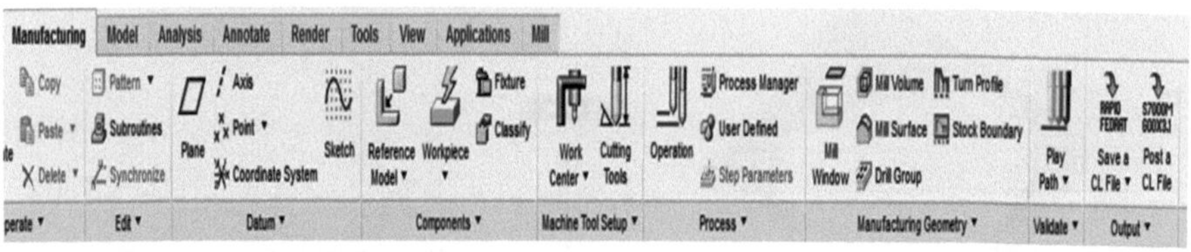

Fig. 8.81 Activated Manufacturing ribbons

8.11.1 Un-suppress the Hidden Holes, Pocket, Step and Slots

Go to Model Tree and click on EXPT_MCH_PRT_05_NCMDL.ASM to activate its content ≫ Click on EXPERT_MCH_05.PRT on the activated drop-down menu list to expand its content, now right click on the mouse and on the activated menu options, and click on Activate to make the Part active again.

Go to Operations group and click on it to activate its menu list ≫ Click on Resume on the drop-down menu list ≫ Now click on Resume All on the activated drop-down list as illustrated in Fig. 8.82.

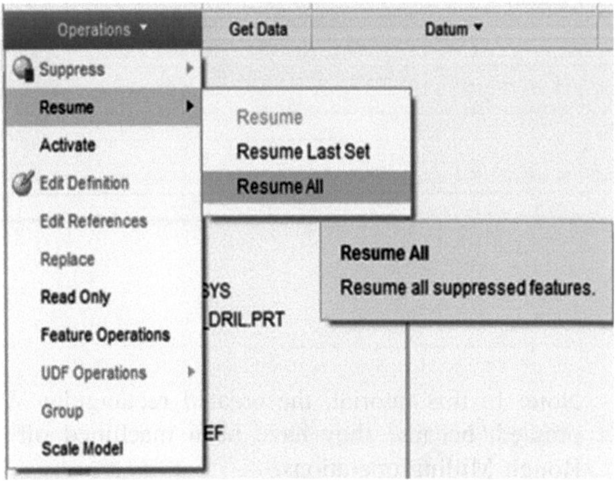

Fig. 8.82 Resuming all suppressed features on Part

All the hidden holes, pocket, step and slots on Part are active again as shown in Fig. 8.83.

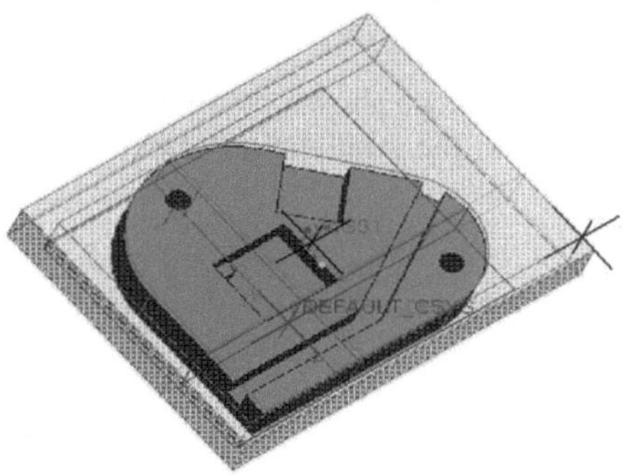

Fig. 8.83 All features on Part resumed

8.11.2 *Make Workpiece Active on the Manufacturing Window*

Go to the Model Tree and click on the EXPT_MCH_PRT_05.ASM ≫ Now click on the right mouse and then click on Activate on the drop-down list to make the assemble the active Workpiece on the Manufacturing main graphic window.

The Part automatically updates itself in the Manufacturing main graphic window as shown in Fig. 8.84.

Fig. 8.84 Activated hidden Part features

Note: In this tutorial, the created rectangular Workpiece will not be suppressed, because they have been machined off via the Face and Volume Rough Milling operations.

8.12 Activate the Expert Machining Application

Follow the steps shown below to exit the Manufacturing application.

Click on the Applications menu bar as indicated by the arrow in Fig. 8.85.

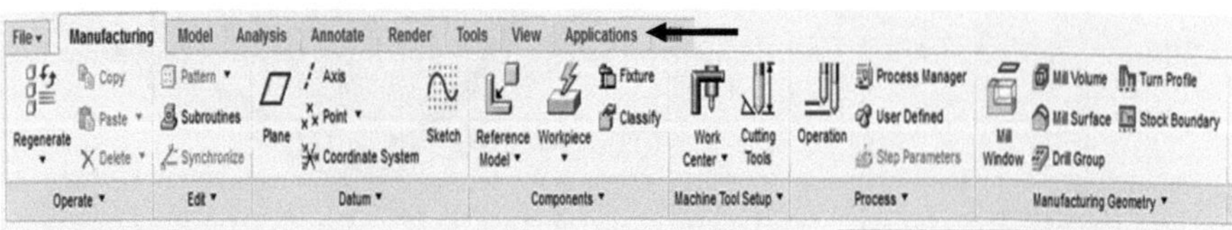

Fig. 8.85 Activating Applications

The Applications ribbon is activated ≫ Now click on the Expert Machinist icon as indicated by the arrow in Fig. 8.86.

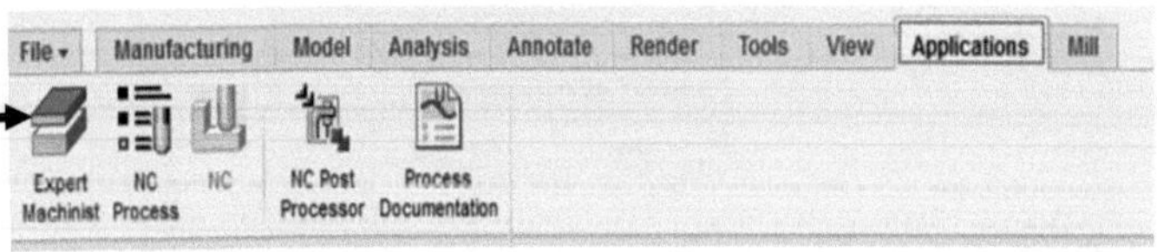

Fig. 8.86 Activated Applications ribbon

The Machining ribbon tools are activated as shown in Fig. 8.87 in concise form.

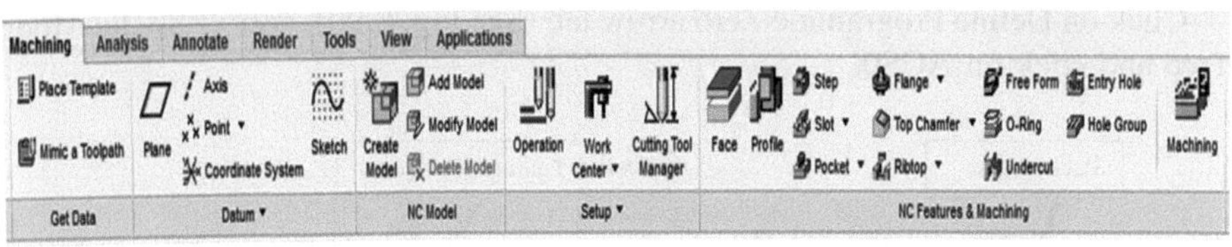

Fig. 8.87 Activated Machining ribbon

8.12.1 *Define the Pocket Milling Feature*

Click on the Pocket icon 

in Fig. 8.87 to activate the drop-down menu list ≫ Now click on Pocket on the drop-down menu list.

The Pocket Feature dialogue box is activated on the main graphic window ≫ Type Pocket1 in the Feature Name section box as shown in Fig. 8.88.

Fig. 8.88 Activated Pocket Feature dialogue box

On the Pocket Feature dialogue box, click on the Define Feature Floor arrow tab and the Select dialogue box is activated on the main window as shown in Fig. 8.89.

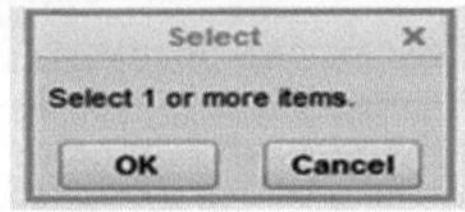

Fig. 8.89 Activated Select dialogue box

Now click on the base of the Pocket1 feature as indicated by the arrow in Fig. 8.90.

Click on the OK on the select dialogue box if it does not disappear automatically.

Click on Define Programme Zero arrow tab (See Fig. 8.90), now go to the Model Tree and click on ACS0.

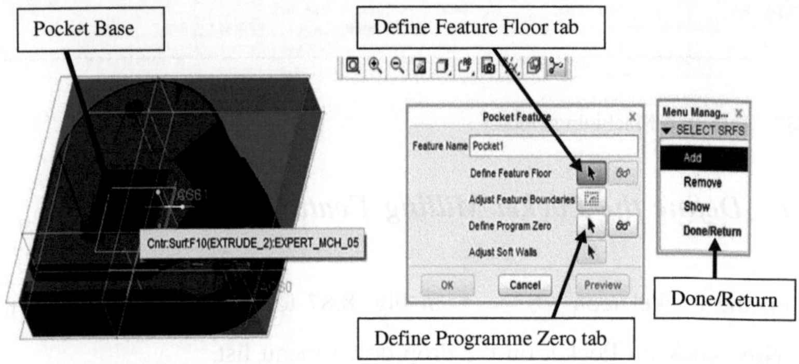

Fig. 8.90 Defining the Pocket Feature parameters

Click on Done/Return on the SELECT SRFS group on Menu Manager dialogue box.

Click on OK tab on the Pocket Feature dialogue box.

8.12.2 *Define the Step Milling Feature*

Click on the Step icon as shown below.

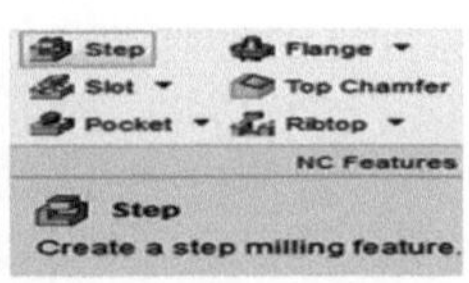

The Step Feature dialogue box is activated on the main graphic window ≫ Type Step1 in the Feature Name section box as shown in Fig. 8.91.

Fig. 8.91 Activated Step Feature dialogue box

On the Step Feature dialogue box, click on the Define Feature Floor arrow tab. The Select dialogue box is activated on the main window as shown in Fig. 8.92.

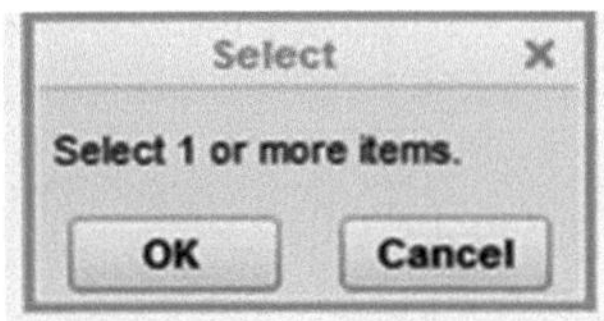

Fig. 8.92 Activated Select dialogue box

Now click on the base of Step1 as indicated by the arrow in Fig. 8.93.

Click on the OK tab on the select dialogue box if it does not disappear automatically after selecting Step1.

Click on Define Programme Zero arrow tab and click on ACS0 on the Model Tree.

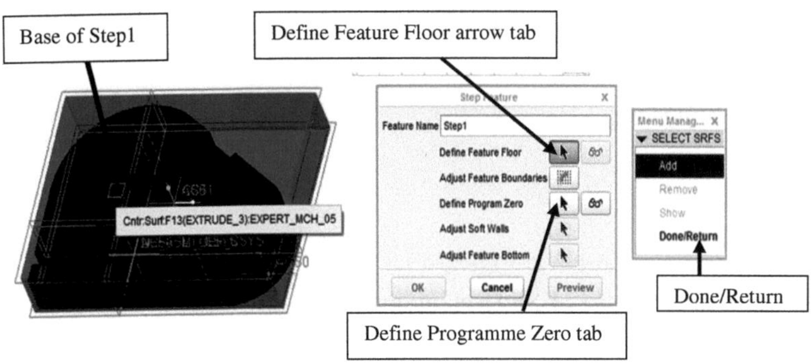

Fig. 8.93 Defining the Step1 Feature parameters

Click on Done/Return on the SELECT SRFS group Menu Manager dialogue box.

Now click on the OK tab to exit the Step Feature dialogue box.

8.12.3 *Define the Slots Milling Features*

Slot1 creation

Click on the Slot icon to activate its drop-down menu list ≫ Now click on Slot on the drop-down menu list as shown below.

The Slot Feature dialogue box is activated on the main graphic window ≫ On the Slot Feature dialogue box, type Slot1 in the Feature Name section box as shown in Fig. 8.94.

Fig. 8.94 Activated Slot Feature dialogue box

On the Slot Feature dialogue box, click on the Define Feature Floor arrow tab. The Select dialogue box is activated on the main window as shown (Fig. 8.95).

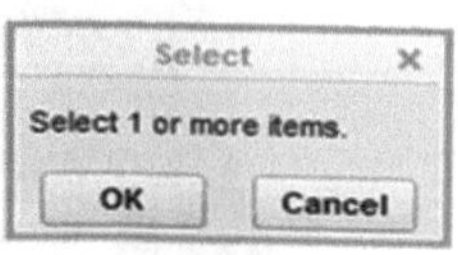

Fig. 8.95 Activated Select dialogue box

Now click on the base of Slot1 as indicated by the arrow in Fig. 8.96.

Click on OK tab on the select dialogue box if it does not disappear automatically.

Click on Define Programme Zero arrow tab and click on ACS0 on the Model Tree.

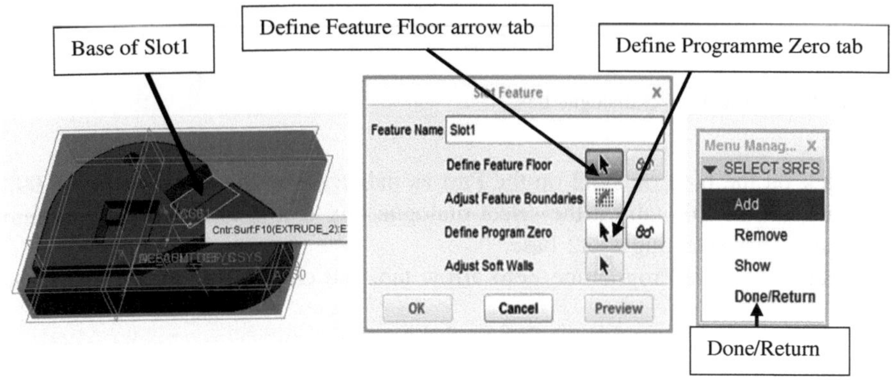

Fig. 8.96 Adding parameters for Slot1 features

Now click on Done/Return on the SELECT SRFS group on the Menu Manager dialogue box.

Now click on the OK tab to exit the Slot Feature dialogue box.

Slot2 creation

Click on the Slot icon to activate its drop-down menu list ≫ Now click on Slot on the drop-down list as shown below.

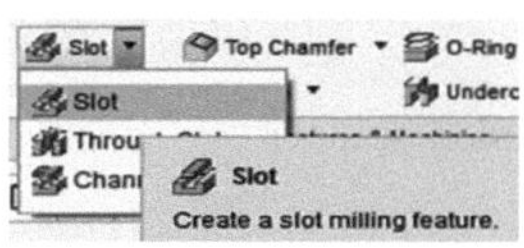

The Slot Feature dialogue box is activated on the main graphic window ≫ On the Slot Feature dialogue box, type Slot2 in the Feature Name section box as shown in Fig. 8.97.

Fig. 8.97 Activated Slot dialogue box indicating the typed Slot2

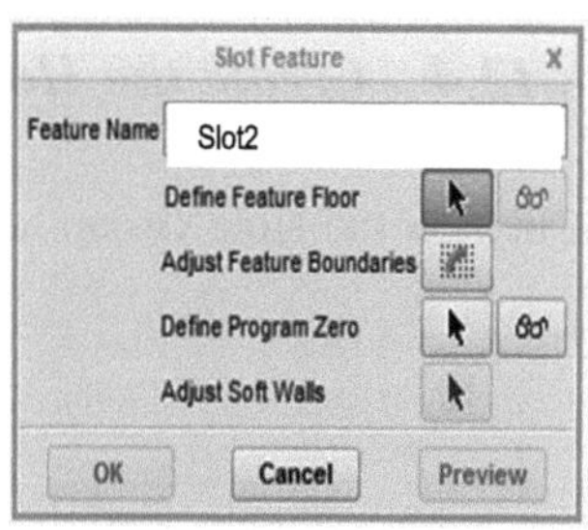

Now click on the Define Feature Floor arrow tab in Fig. 8.97 and the Select dialogue box is activated on the main window as shown in Fig. 8.98.

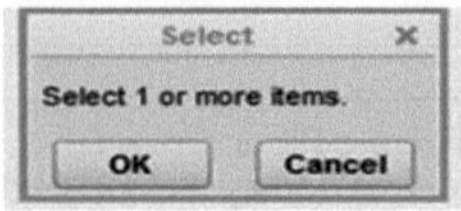

Fig. 8.98 Activated select dialogue box

Click on the base of Slot2 on the Part as indicated by the arrow in Fig. 8.99.

Click on the OK tab on the select dialogue box if it does not disappear automatically after selecting Slot2 base.

Click on Define Programme Zero arrow tab, and click on ACS0 on the Model Tree.

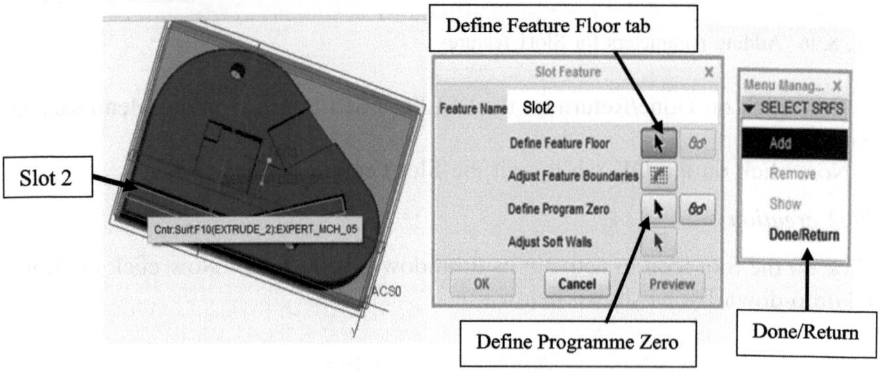

Fig. 8.99 Adding parameters for Slot2 features

Click on Done/Return on the SELECT SRFS group on the Menu Manager dialogue box.

Now click on the OK tab to exit the Slot Feature dialogue box.

8.12.4 *Create the Hole Group*

Click on the Hole Group icon as shown below.

The Drill Group dialogue box is activated on the main graphic window. See Fig. 8.100.

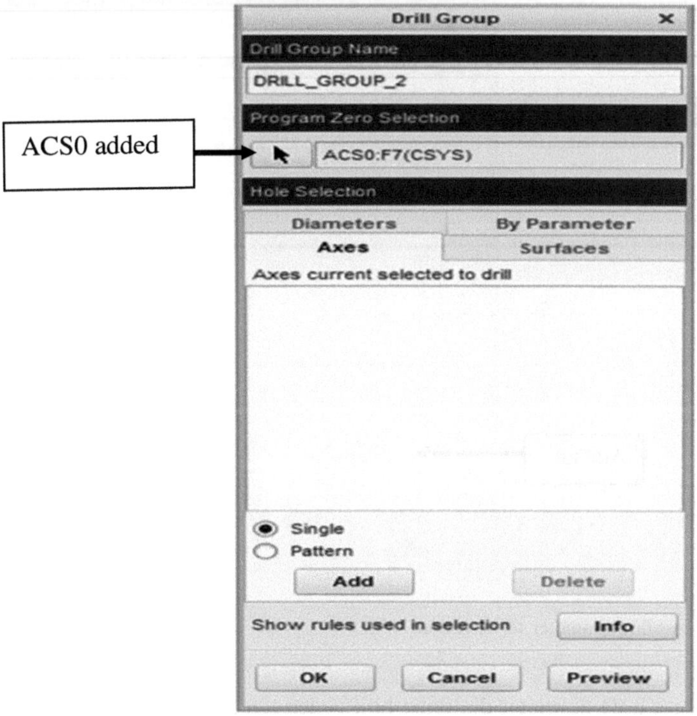

Fig. 8.100 Drill Group dialogue box

Click on the Drill Group Name section box and type the "DRILL_GROUP_2" as shown in Fig. 8.100.

On Programme Zero Selection section, click on the arrow tab, and now click on ACS0 on the Model Tree.

On the Hole Selection group, click on the Diameters tab and also on the Add tab to add diameter as indicated by the arrows in Fig. 8.101.

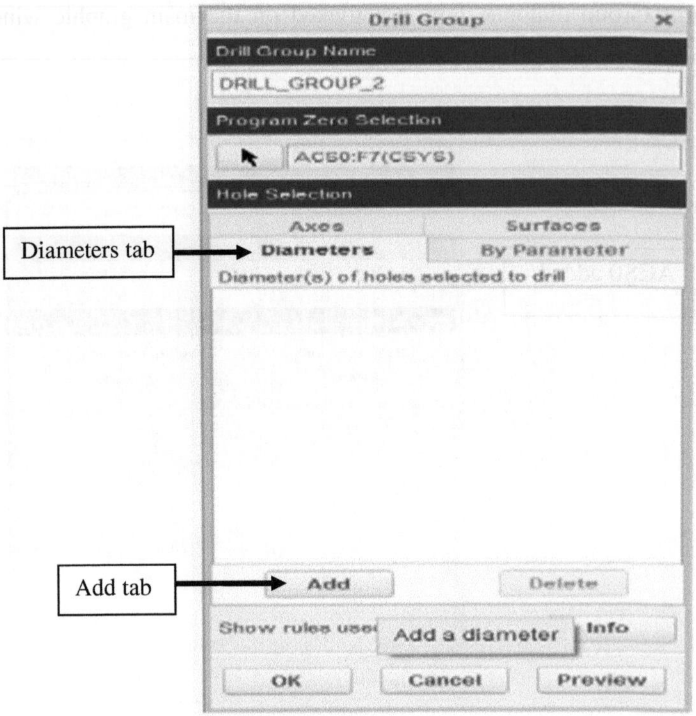

Fig. 8.101 Adding diameter to be drilled

The Select hole diameter dialogue box is activated on the main window with the exact diameter of the hole as shown in Fig. 8.102.

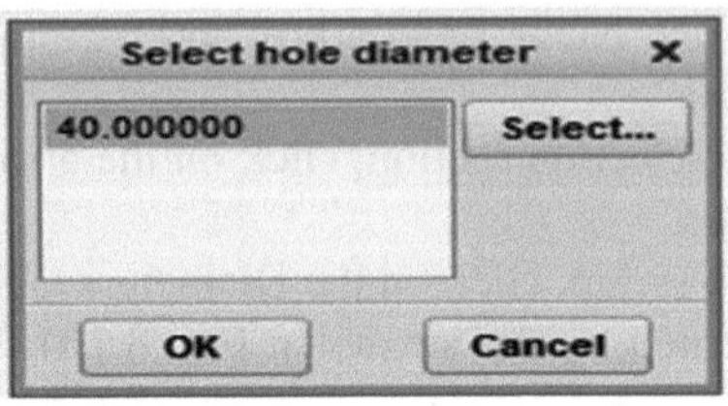

Fig. 8.102 Activated Select hole diameter dialogue box

Click on the Select tab, and then click on the diameter of the hole/holes on the Part, if the hole diameter is not generated automatically by the system.

Click on the OK tab to exit Select hole diameter dialogue box.

All holes with 40 mm diameter are now automatically added into the "Diameters of holes selected to drill" section box as illustrated in Fig. 8.103.

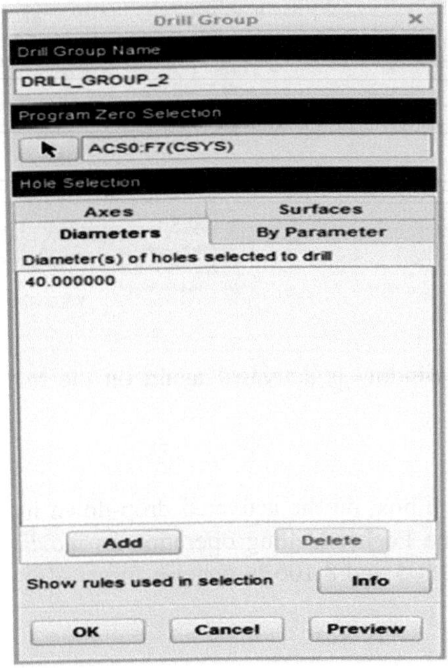

Click on the OK tab to exit.

Fig. 8.103 Diameter automatically added into the Diameters section box

8.13 Create Tool Paths for Pocket, Step, Slots and Holes

At this stage, parameters will be added to each Toolpath Sequence.

Click on Material Removal Display icon  to deactivate if activated.

8.13.1 Create Tool Path Sequence for Pocket Milling

Go to the Model Tree and right click on POCKET1 [OP010] as shown below ≫ Now click on Create Toolpath on the drop-down list as indicated by the arrow in Fig. 8.104.

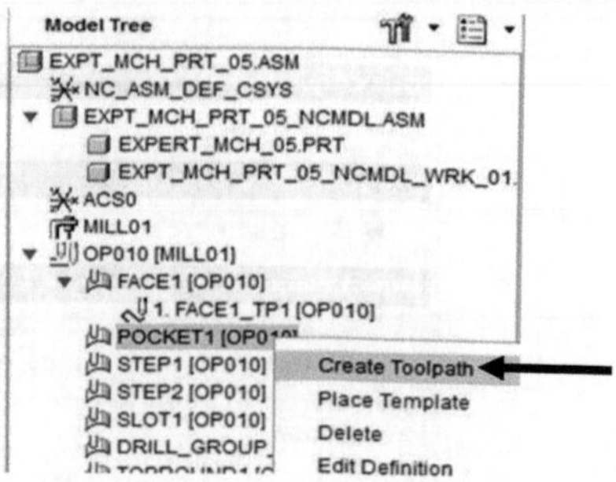

Fig. 8.104 Activating Create Toolpath on POCKET1 [OP010] drop-down menu list

The Pocket Milling dialogue window is activated again on the main graphic window.

Add Cutting Tool

Click on Cutting Tool icon section box, on the activated drop-down list, click on T0004 as the Cutting Tool for the Pocket Milling operation as indicated by the arrows on Fig. 8.105. See Figs. 8.105 and 8.106 in concise form.

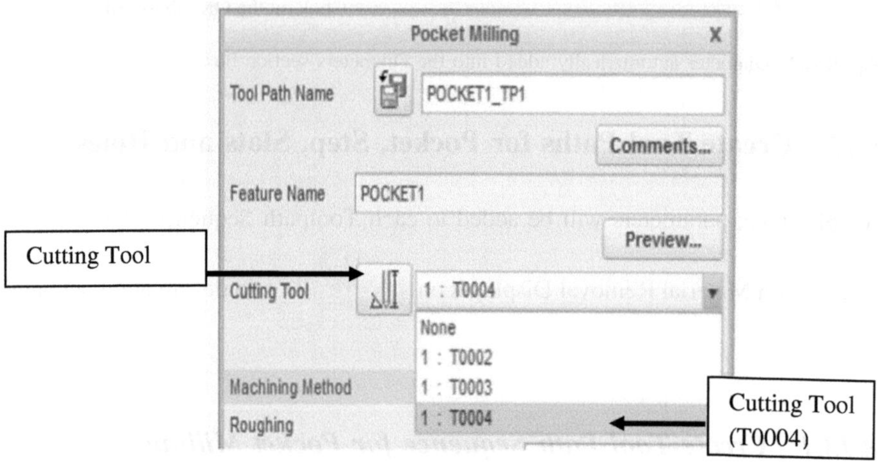

Fig. 8.105 T0004 is added as the Cutting Tool

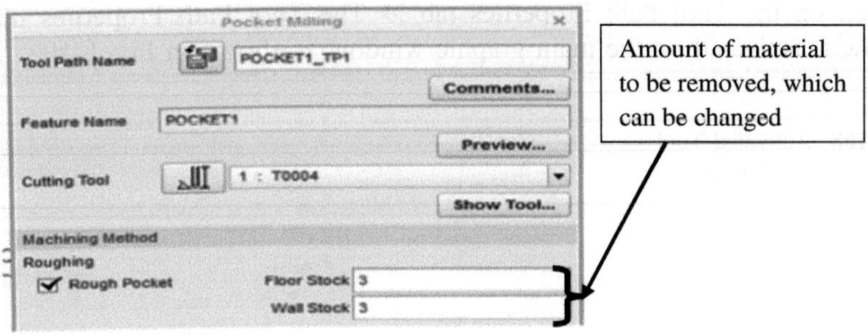

Fig. 8.106 Amount of material removal section

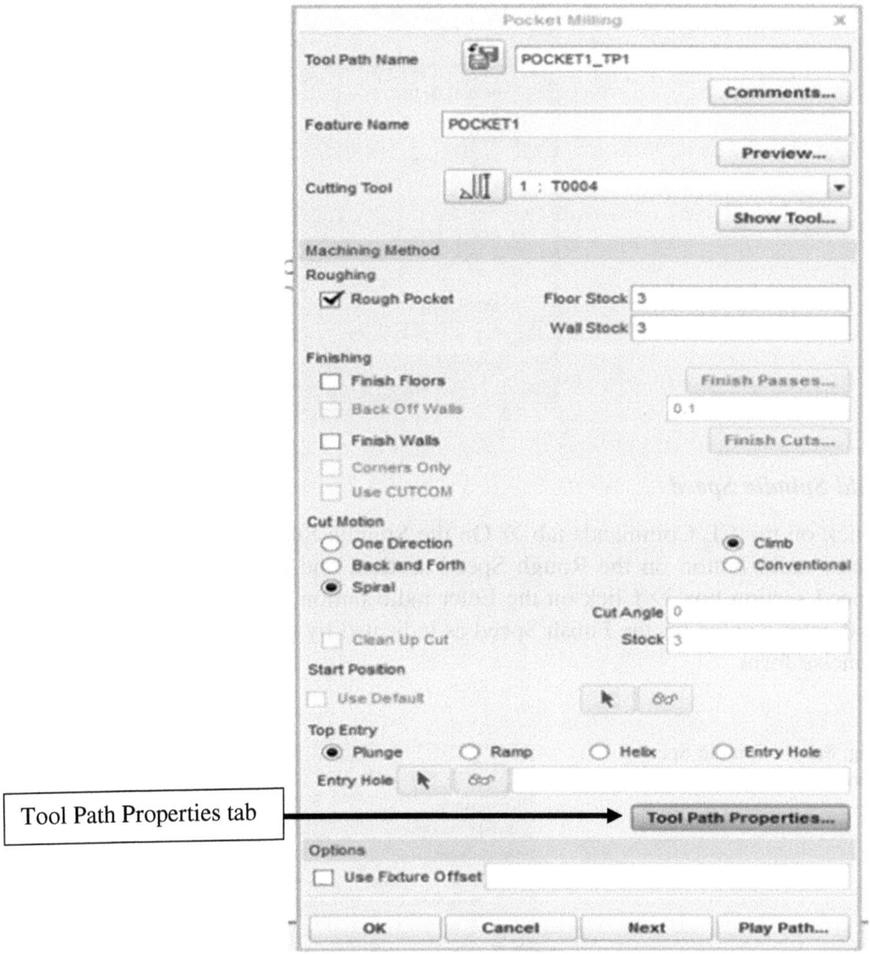

Fig. 8.107 Activating Tool Path Properties from Pocket Milling dialogue window

Click on the Tool Path Properties tab ≫ The Tool Path Properties dialogue window is activated on the main graphic window as shown in Fig. 8.108.

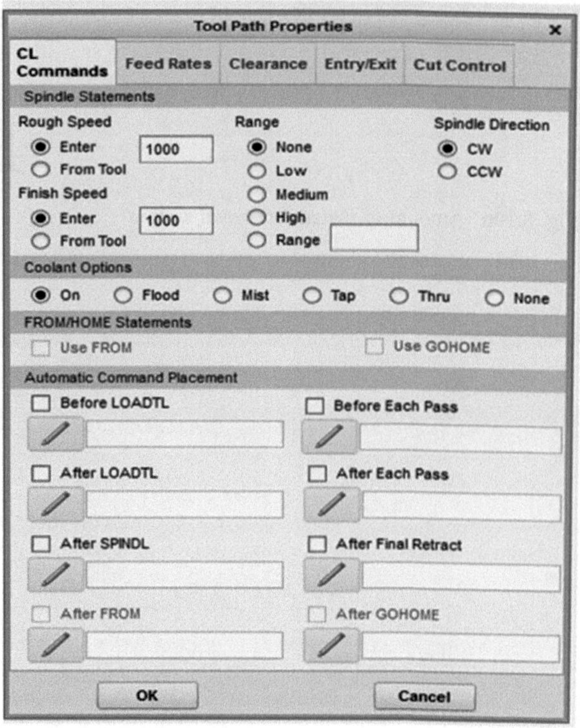

Fig. 8.108 Activated Tool Path Properties dialogue window

Add Spindle Speed

Click on the CL Commands tab ≫ On the Spindle Statements group, click on the Enter radio button on the Rough Speed section, and enter a value in the Rough Speed section box ≫ Click on the Enter radio button on the Finish Speed section and enter a value for the Finish Speed as indicated by the arrows. See Fig. 8.109 in concise form.

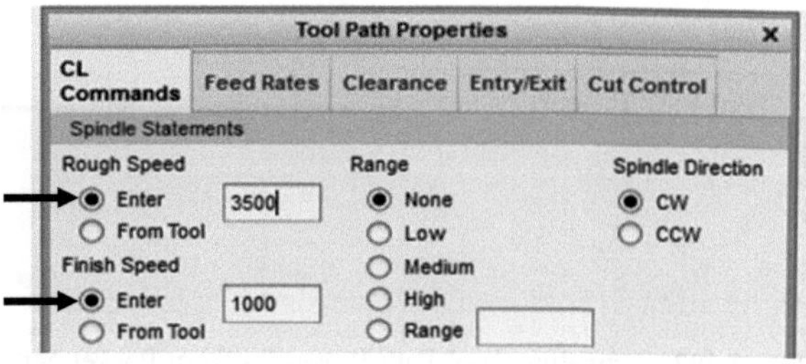

Fig. 8.109 Spindle Speed parameters

Adding Feed Rates

Click on the Feed Rates tab ≫ On the Feed Rates group, go to Rough Cutting section box and enter the Rough Cutting Feed Rate value ≫ Go to the Finish Cutting section box and enter a value for the Finish Cutting Feed Rate, as illustrated by the arrows in Fig. 8.110.

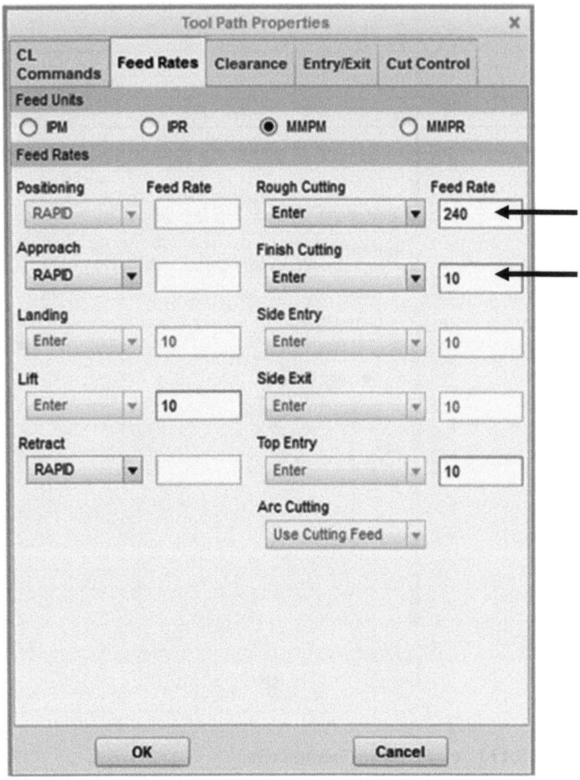

Fig. 8.110 Feed Rates parameters

Adding Cut Control

Click on the Cut Control tab to activate its content ≫ On the Maximum Cut group, go to Depth of Cut section, click on the Enter radio button to make it active and enter values in the value section box ≫ Click on the Enter radio button on the Stepover section and enter a value for the Stepover, as illustrated by the arrows in Fig. 8.111.

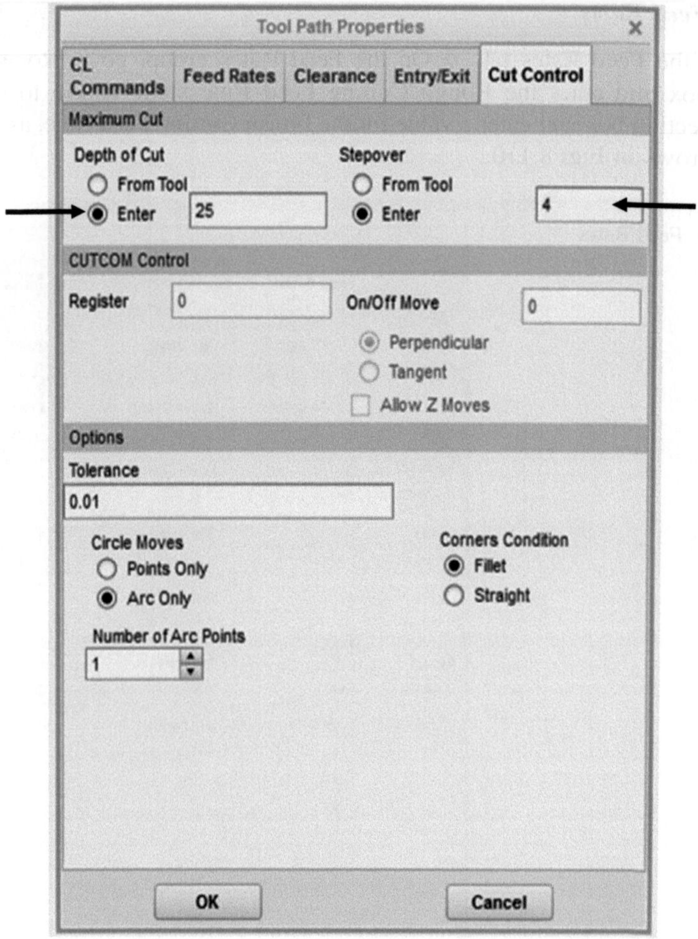

Fig. 8.111 Cut Control parameters

Click on the OK Tab to exit the Tool Path Properties dialogue window.

Activate Cutting Tool and PLAY PATH dialogue box for Pocket Milling

Click on Play Path tab on the Pocket Milling dailogue window to activate the on-screen Pocket Milling operation as indicated by the arrow in Fig. 8.112.

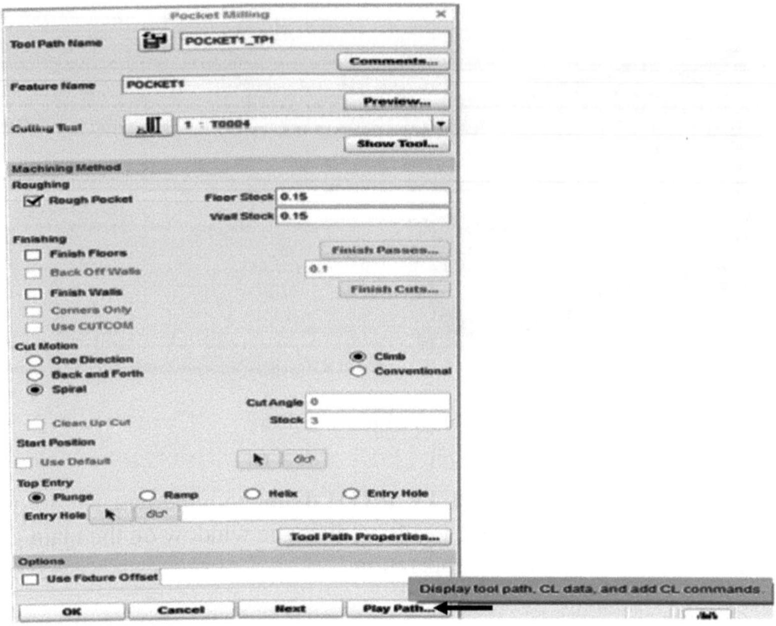

Fig. 8.112 Activating Play Path on Pocket Milling dialogue window

The cutting Tool and the PLAY PATH dialogue box are activated on the main graphic window as shown in Fig. 8.113.

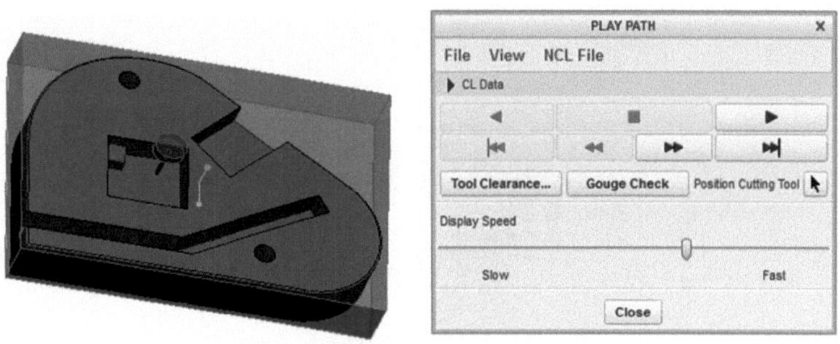

Fig. 8.113 Activated cutting Tool and PLAY PATH dialogue box

Click on the Play tab to start the Pocket Milling operation.
End of the Pocket Milling operation is as shown in Fig. 8.114.

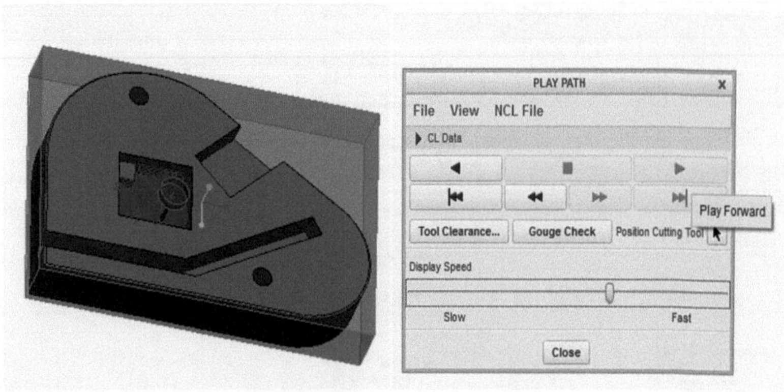

Fig. 8.114 Pocket Milling operation highlighted in *red* colours

Click on the Close tab on the PLAY PATH dialogue window.

Click on the OK tab on the Pocket Milling dialogue window on the main graphic window to exit the Pocket Milling operation.

8.13.2 *Create Tool Path for Step Milling*

Go to the Model Tree and right click on STEP1 [OP010] ≫ Now click on Create Toolpath on the drop-down menu list as indicated by the arrow as shown in Fig. 8.115.

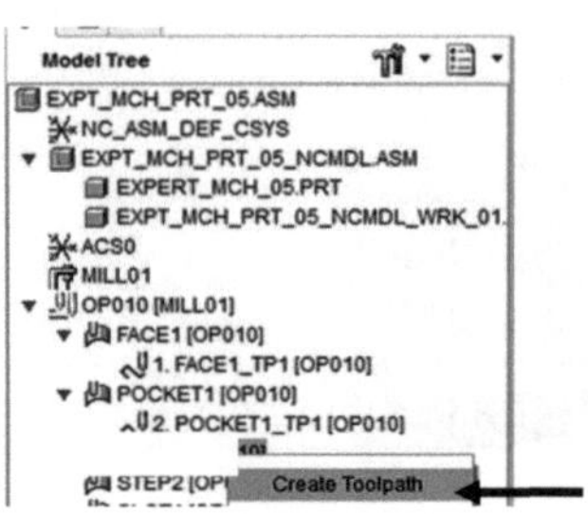

Fig. 8.115 Activating Create Toolpath STEP1 [OP010] operation

The Step Milling dialogue window is activated on the main graphic window ≫ Click on the Tool Path Name section box and type STEP1_TP1 ≫ Click on the downward pointing arrow on the Cutting Tool icon section box; and on the drop-down list, click on T0004 as the tool for the Step Milling operation. See Fig. 8.116.

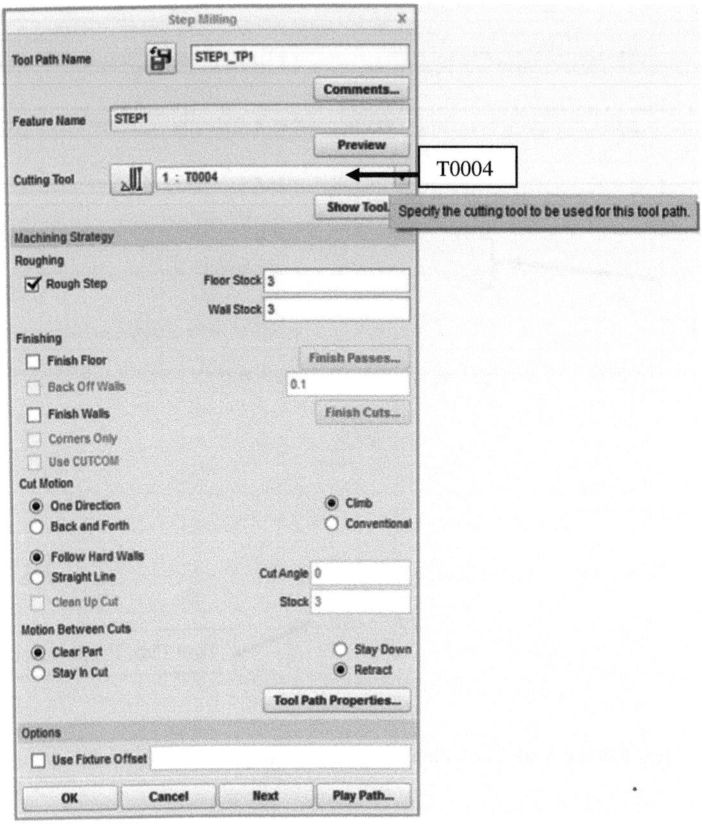

Fig. 8.116 Activating the Step Milling operation Tool

Add Cutting Tool

Click on the Tool Path Properties tab on the Step Milling dialogue window to activate the Tool Path Properties window for Step Milling process as illustrated in Fig. 8.117.

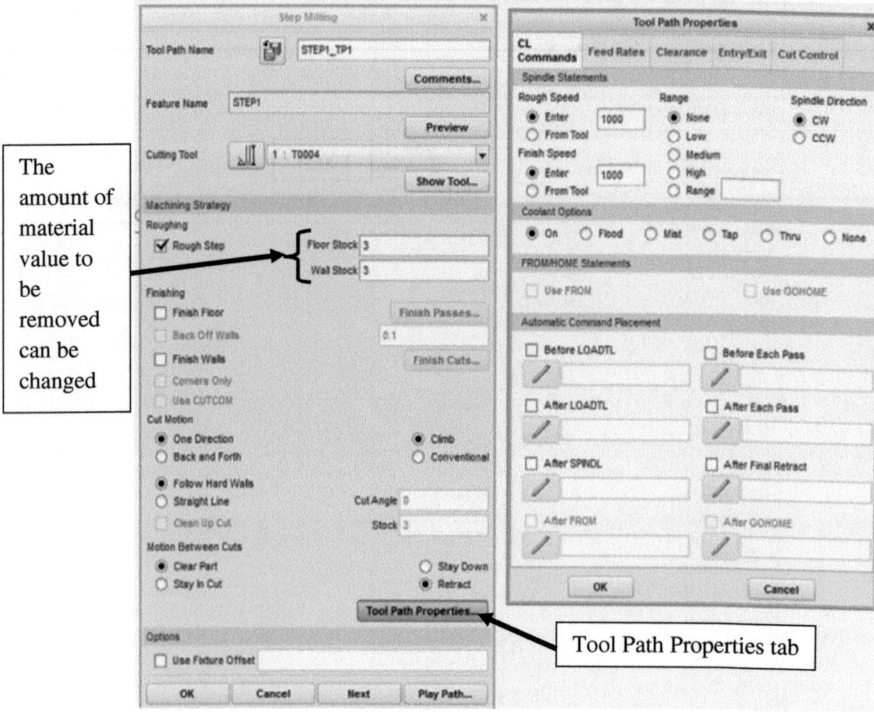

Fig. 8.117 Step Milling and Tool Path Properties dialogue windows on the main dialogue window

Add Spindle Speed

Click on the CL Commands tab to activate its content ≫ On the Spindle Statements group, click on the Enter radio button on the Rough Speed section, enter a value in the Rough Speed section box ≫ Click on the Enter radio button on the Finish Speed section and enter a value for the finish Speed. See Fig. 8.118 in concise form.

Fig. 8.118 Creating the Spindle Speed parameters

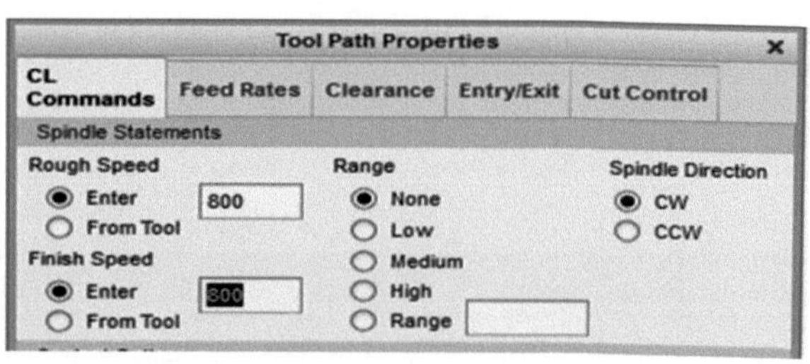

Add Feed Rates

Click on the Feed Rates tab to activate its content ≫ On the Feed Rates group, go to Rough Cutting section box and enter the Rough Cutting Feed Rate value ≫ Go to the Finish Cutting section box, and enter a value for the Finish Cutting Feed rate as indicated by the arrows in Fig. 8.119.

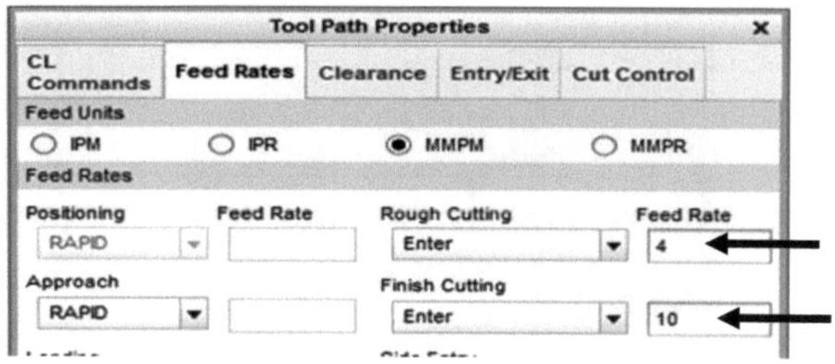

Fig. 8.119 Creating the Feed Rates parameters

Add Depth of Cut

Click on the Cut Control tab to activate its content ≫ On the Maximum Cut group, go to Depth of Cut section, and click on the Enter radio button to make it active and enter a value in the value section box ≫ Click on the Enter radio button on the Stepover section and enter a value for the Stepover. See Fig. 8.120.

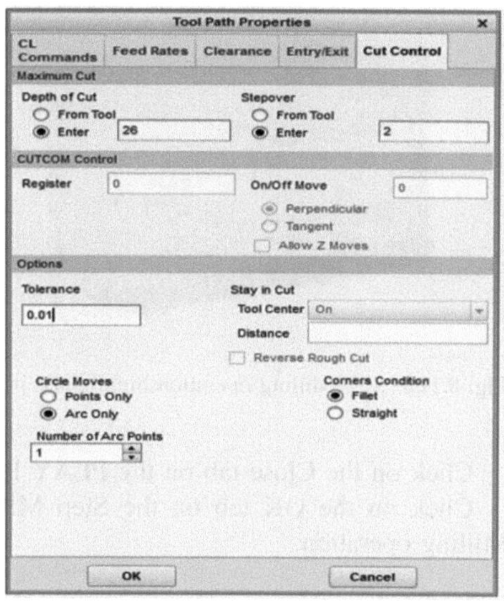

Fig. 8.120 Creating the Cut Control parameters

Click on the OK tab in Fig. 8.120 to exit the Tool Path Properties dialogue window.

Activate the Step Milling operation

Click on Play Path tab on the Step Milling dialogue window.

The cutting Tool and the PLAY PATH dialogue window are activated on the main graphic window as shown in Fig. 8.121.

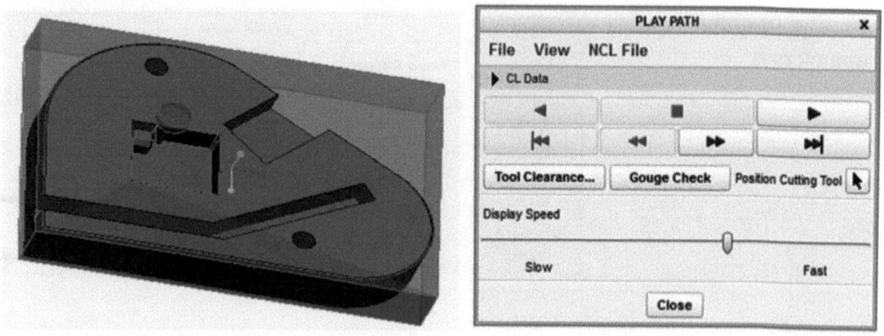

Fig. 8.121 Activated cutting Tool and PLAY PATH dialogue box

Click on the Play tab on the PLAY PATH dialogue box to start the Step Milling operation.

The end of Step Milling operation is as shown in Fig. 8.122.

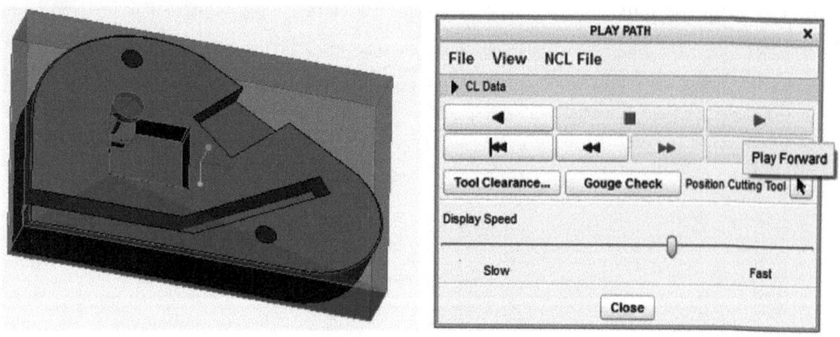

Fig. 8.122 Step Milling operation highlighted in *red* colour

Click on the Close tab on the PLAY PATH dialogue box.

Click on the OK tab on the Step Milling dialogue window to exit the Step Milling operation.

8.13.3 *Create Tool Path for Slot1 Milling*

Go to the Model Tree and right click on SLOT1 [OP010] ≫ Now click on Create Toolpath on the drop-down list as highlighted in Fig. 8.123.

Fig. 8.123 Activating SLOT1 Milling Create Toolpath

The Slot Milling dialogue window is activated on the main graphic window.

Add Cutting Tool

Click on the Tool Path Name section box and type SLOT1_TP1 ≫ Click on Cutting Tool icon to activate its content ≫ Now click on T0002 on the drop-down menu list as the Tool for the Slot1 Milling operation as shown in Fig. 8.124.

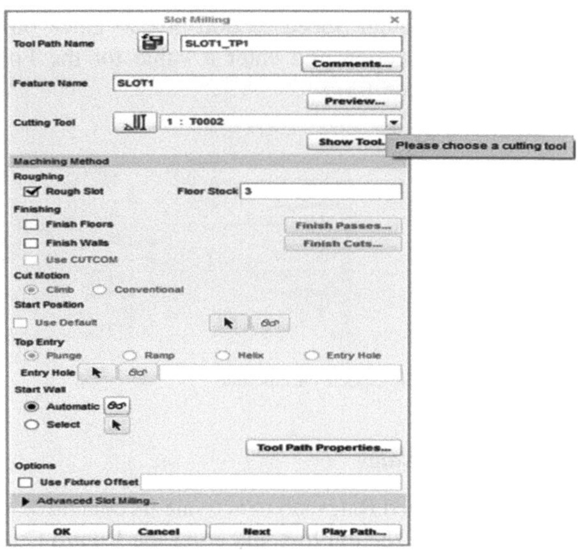

Fig. 8.124 Creating the Slot1 Milling cutting Tool

Click on the Tool Path Properties tab on the Slot Milling dialogue window in Fig. 8.124 to activate the Tool Path Properties dialogue window as shown in Fig. 8.125.

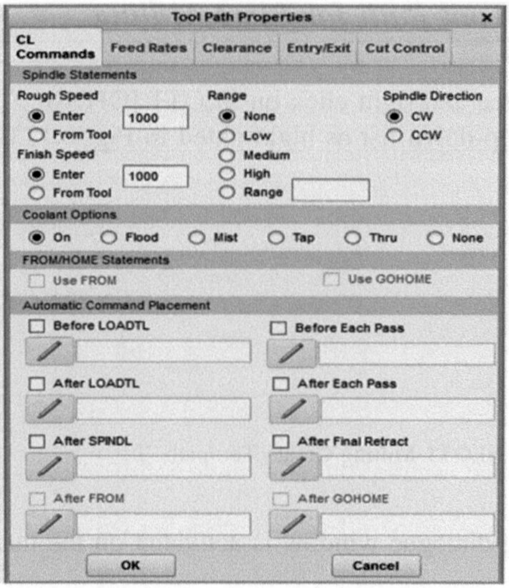

Fig. 8.125 Activated Tool Path Properties dialogue window

Add Spindle Speed

Click on the CL Commands tab to activate its content ≫ On the Spindle Statements group, click on the Enter radio button, on the Rough Speed section, and enter a value in the Rough Speed section box ≫ Click on the Enter radio button on the Finish Speed section and enter a value for the Finish Speed. See Fig. 8.126 as shown in concise form.

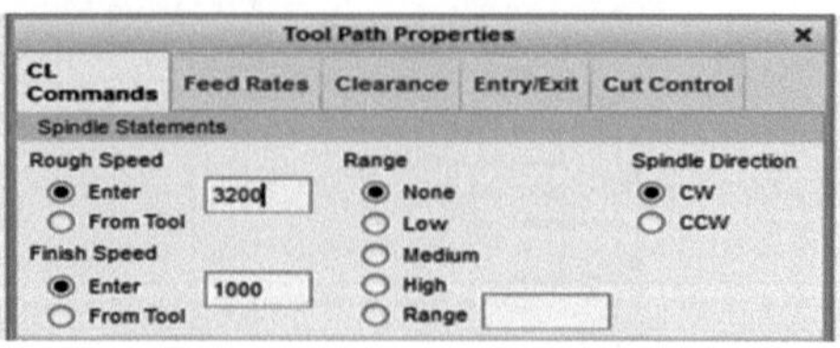

Fig. 8.126 Spindle Speed parameters

Adding Feed Rates

Click on the Feed Rates tab to activate its content ≫ On the Feed Rates group, go to Rough Cutting section box and enter the Rough Cutting Feed Rate value ≫ Go to the Finish Cutting section box, and Enter a value for the Finish Cutting Feed rate. See Fig. 8.127.

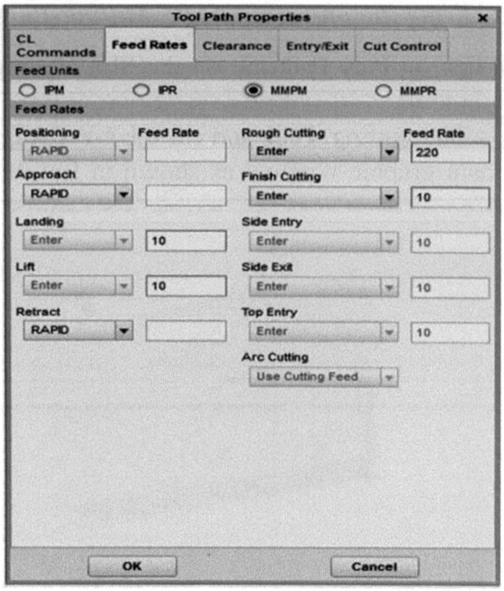

Fig. 8.127 Feed Rates parameters

Add Depth of Cut

Click on the Cut Control tab to activate its content ≫ On the Maximum Cut group, go to Depth of Cut section, click on the Enter radio button to make it active and enter a value in the value section box ≫ Click on the Enter radio button on the Stepover section and enter a value for the Stepover. See Fig. 8.128.

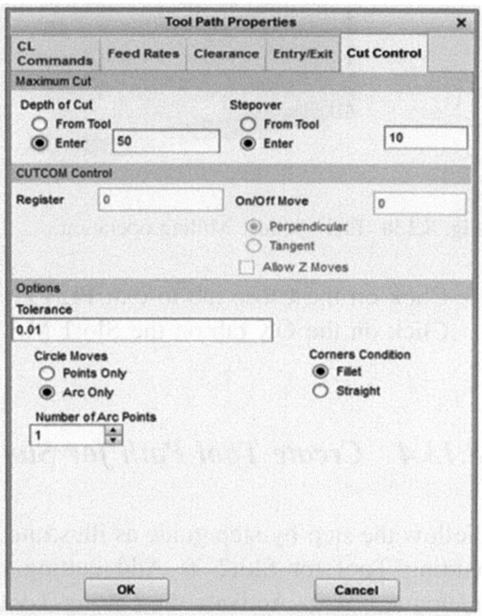

Fig. 8.128 Cut Control parameters

Click on the OK tab to exit the Tool Path Properties dialogue window.

Activate the on-screen Slot1 Milling Play Path

Click on Play Path tab on the Slot Milling dialogue window on the main graphic window.

The cutting Tool and the PLAY PATH dialogue window are activated on the main graphic window as shown in Fig. 8.129.

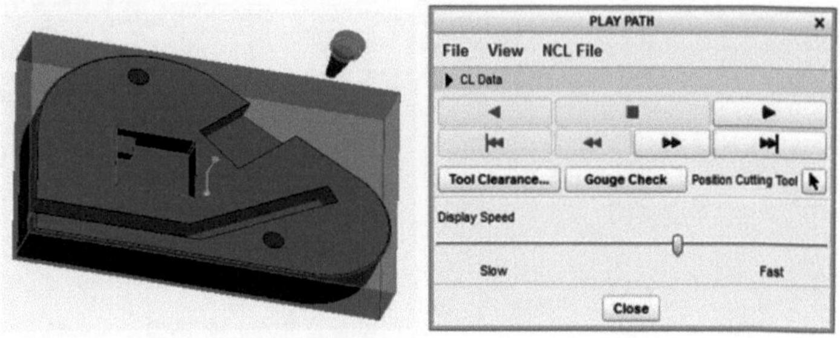

Fig. 8.129 Activated cutting Tool and PLAY PATH dialogue box

The end of the Slot1 Milling operation is as shown in Fig. 8.130.

Fig. 8.130 End of Slot1 Milling operation

Click on the Close tab to exit PLAY PATH dialogue box.
Click on the OK tab on the Slot1 Milling dialogue window to exit.

8.13.4 Create Tool Path for Slot2 Milling

Follow the step-by-step guide as illustrated in Sect. 8.13.3 ≫ Select the appropriate cutting Tool for Slot2 ≫ Add cutting parameters on the Tool Path Properties dialogue box ≫ Activate the cutting Tool and the PLAY PATH dialogue box for the on-screen operation.

8.13.5 Create Tool Drill Group 2

Go to the Model Tree and right click on DRILL_GROUP_2 [OP010] ≫ Click on Create Toolpath on the activated drop-down menu list. See Fig. 8.131.

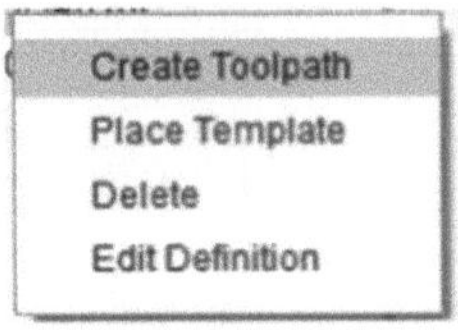

Fig. 8.131 Activating Create Toolpath for Drilling operation

The Drilling Strategy dialogue window is activated on the main graphic window ≫ Type the Name in the Tool Path Name section box ≫ Add T0001 as the Drilling Tool ≫ On the Holemaking Method group, click on the Drill radio button on the Cycle Type group ≫ Click on the Standard radio button on the Cycle Modifier group ≫ Click on Auto radio button on the Cycle Depth section. See Fig. 8.132.

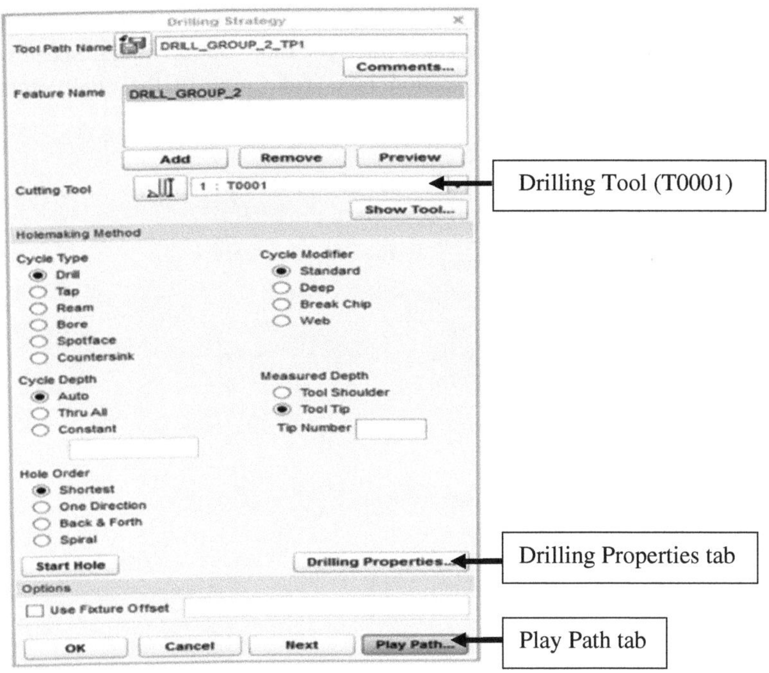

Fig. 8.132 Drilling Strategy dialogue window

Now click on Drilling Properties tab on the Drilling Strategy dialogue window.

The Drilling Properties dialogue window is activated on the main graphic window as shown in Fig. 8.133.

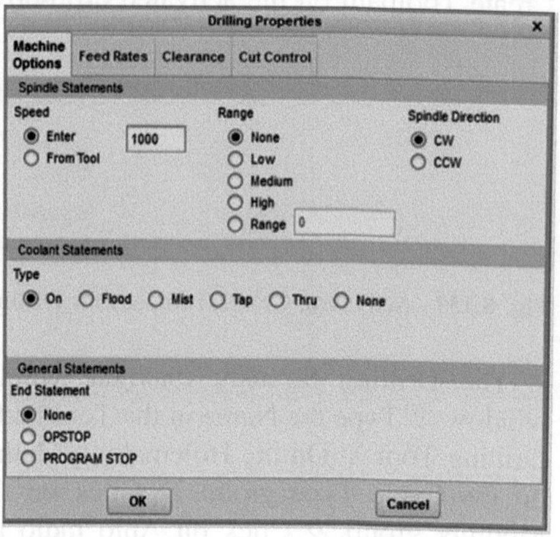

Fig. 8.133 Activated Drilling Properties dialogue window

Add spindle Speed

Click on the Machine Options tab to make active its content ≫ Go to the Spindle Statements group and click on the Enter dialogue radio button on the Speed section, and enter a value in the Speed section box. Take note of the other sections in Fig. 8.134.

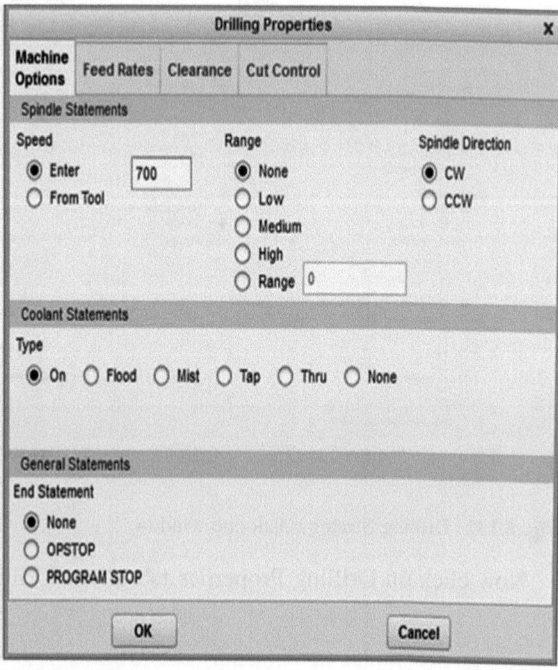

Fig. 8.134 Spindle Speed parameters

Add Feed Rates

Click on the Feed Rates tab to activate its content ≫ On the Feed Rates group, go to Cutting section and enter the Cutting Feed Rate value. Take note of the active radio buttons as shown in Fig. 8.135.

Note: No depth of cut will be added to the drilling process.

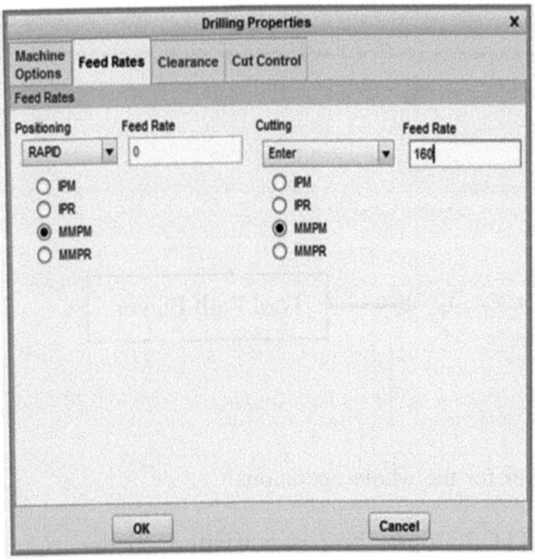

Click on the OK tab to exit.

Fig. 8.135 Feed Rates parameters

Activate the on-screen Drilling operation

Click on Play Path tab on the Drilling Strategy dialogue window.

The drilling Tool and the PLAY PATH dialogue box are activated on the main graphic window as shown in Fig. 8.136.

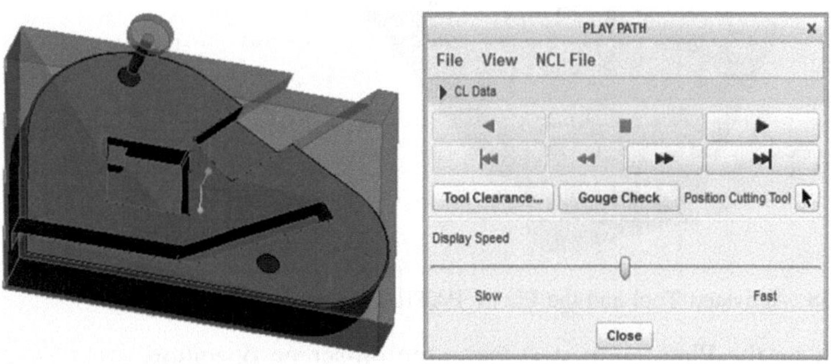

Fig. 8.136 Activated drilling Tool and the PLAY PATH dialogue box

Click on the Play tab button to start the on-screen Drilling operation.
At the end of the Drilling operation, click on the Close tab.
Click on the OK tab on the Drilling Strategy dialogue window to exit.

8.13.6 Run the Whole Operations as Created

To run the whole operations:

Go to Model Tree ≫ Right click on OP010 (MILL01) to activate the drop-down list ≫ Click on Tool Path Player on the drop-down menu list as indicated by the arrow in Fig. 8.137.

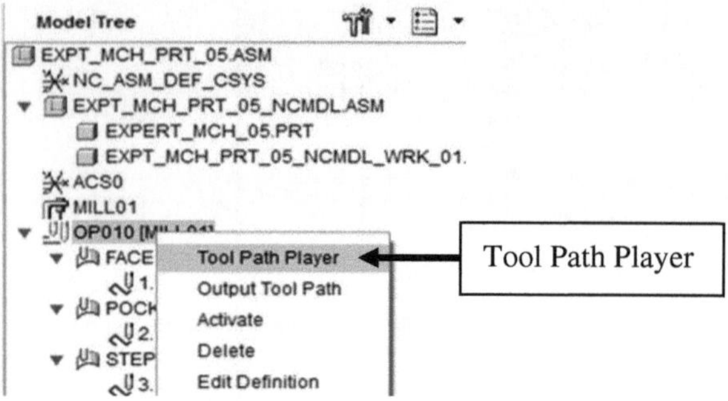

Fig. 8.137 Activating Tool Path Player for the whole operations

The Tool and the PLAY PATH dialogue box is activated on the main graphic window as shown in Fig. 8.138.

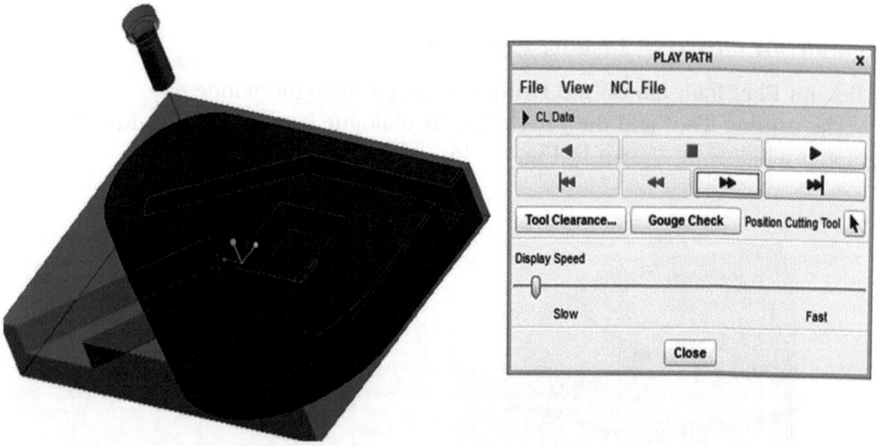

Fig. 8.138 Activated Tool and the PLAY PATH dialogue box

Click on the Play tab to start the entire on-screen operation.

The end of the Face, Pocket, Step, Slot and Drilling operations is shown in Fig. 8.139.

Click on the Close tab to exit the whole operation.

Fig. 8.139 End of Pocket, Step, Slots and Drilling operations

Note: All the parameters values used in this Expert Machinist tutorial are experimental values. To calculate the correct parameters values using the correct formulas, it is highly recommended that you consult relevant handbooks and textbooks.

8.14 Cutter Location (CL) Data

To create the Cutter Location (CL) data output file in Expert Machist, follow the step-by-step guide below.

Go to Model Tree ≫ Right click on OP010 (MILL01) to activate the drop-down list ≫ Click on Output Tool Path on the drop-down menu list as indicated below by the arrow in Fig. 8.140.

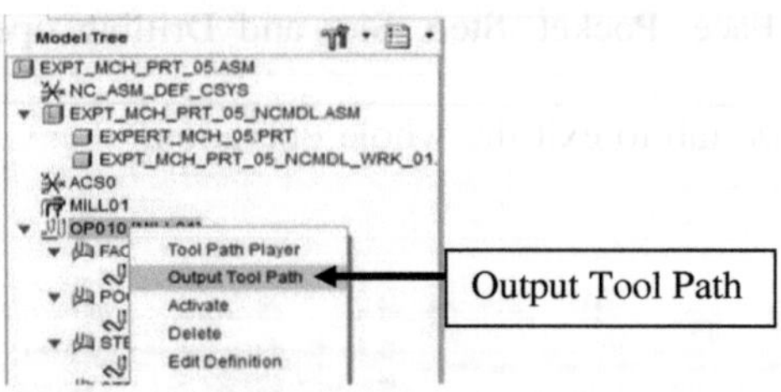

Fig. 8.140 Activating Output Tool Path on the Model Tree

The Save a Copy dialogue window is activated on the main graphic window. See Fig. 8.141.

Fig. 8.141 Activated Save a Copy dialogue window in concise form

Now click on the OK tab to save CL data file in the selected working directory.

8.15 Create Post Processor

Click on the Applications menu bar to activate its content ≫ Now click on NC Process icon to activate the NC Process application as indicated by the arrow below.

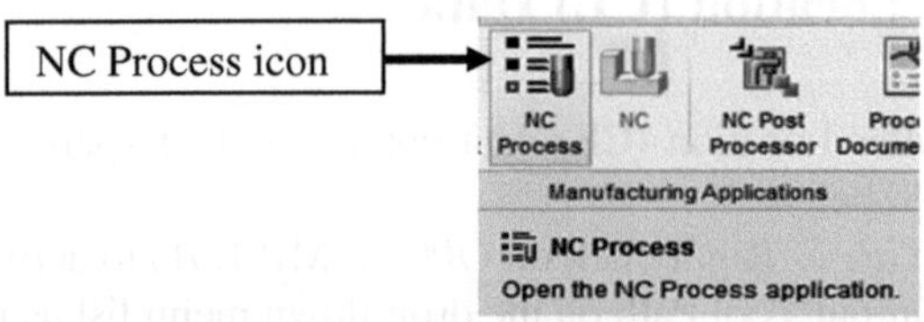

The Save a Copy dialogue window is activated on the main graphic window see Fig. 8.142.

Note: In the File Name section box, the CL file name already saved in the directory is automatically added. The file extension in the Type section box is CL File (*.ncl).

Fig. 8.142 Activated Save a Copy dialogue window in concise form

Now click on the Open tab and the PP OPTIONS Menu Manager dialogue box is activated on the main window ≫ Check Mark Verbose and Trace as shown in Fig. 8.143, if not Checked Marked by the system automatically.

Fig. 8.143 Activated PP OPTIONS Menu Manager dialogue box

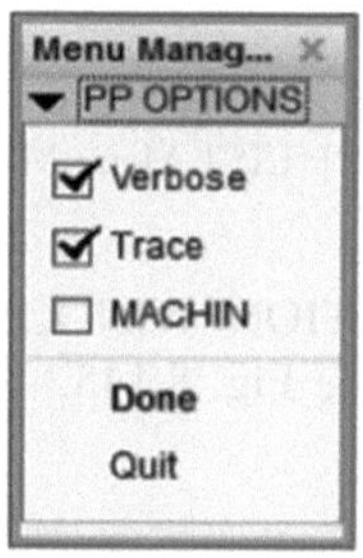

Click on Done to exit.

The PP LIST Menu Manager dialogue box is activated as shown in Fig. 8.144.

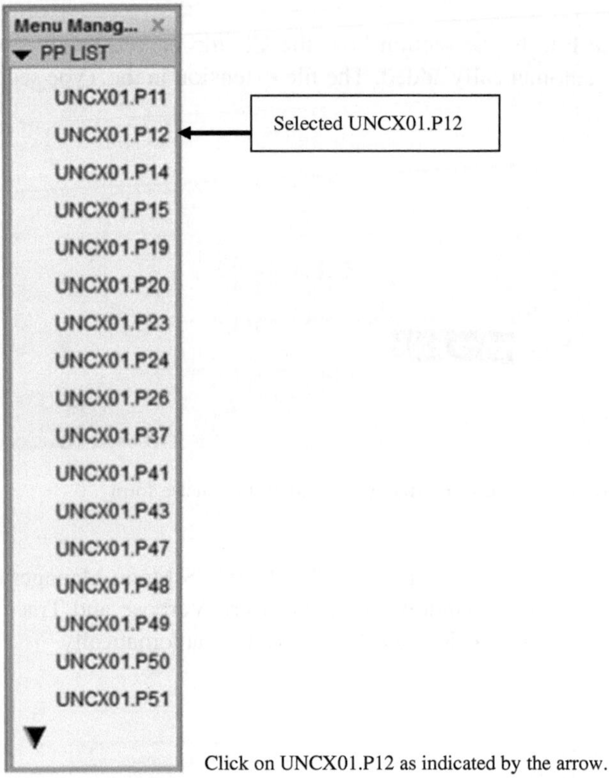

Fig. 8.144 Activated PP LIST Menu Manager dialogue box

The INFORMATION WINDOW dialogue window is activated on the main graphic window. See Fig. 8.145.

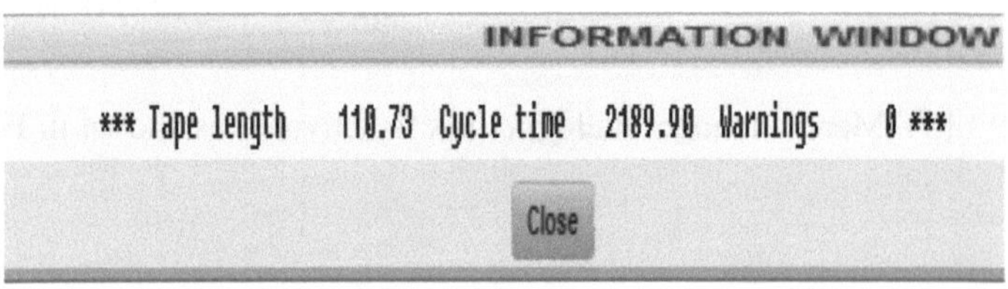

Fig. 8.145 Activated INFORMATION WINDOW in concise form

Click on the Close tab to exit the INFORMATION WINDOW dialogue window.

The G Codes will be created and stored as a TAP file in the working directory. Use notepad to open the TAP file.

Congratulation for completing this tutorial on Expert Machinist.

Some of the G-codes generated for this process is shown below and on the proceeding pages.

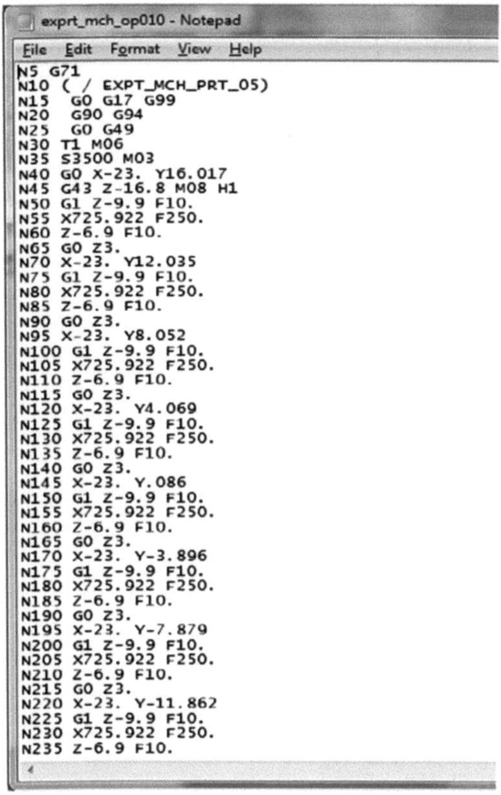
exprt_mch_op010 - Notepad

File Edit Format View Help

```
N5 G71
N10 ( / EXPT_MCH_PRT_05)
N15  G0 G17 G99
N20  G90 G94
N25  G0 G49
N30 T1 M06
N35 S3500 M03
N40 G0 X-23. Y16.017
N45 G43 Z-16.8 M08 H1
N50 G1 Z-9.9 F10.
N55 X725.922 F250.
N60 Z-6.9 F10.
N65 G0 Z3.
N70 X-23. Y12.035
N75 G1 Z-9.9 F10.
N80 X725.922 F250.
N85 Z-6.9 F10.
N90 G0 Z3.
N95 X-23. Y8.052
N100 G1 Z-9.9 F10.
N105 X725.922 F250.
N110 Z-6.9 F10.
N115 G0 Z3.
N120 X-23. Y4.069
N125 G1 Z-9.9 F10.
N130 X725.922 F250.
N135 Z-6.9 F10.
N140 G0 Z3.
N145 X-23. Y.086
N150 G1 Z-9.9 F10.
N155 X725.922 F250.
N160 Z-6.9 F10.
N165 G0 Z3.
N170 X-23. Y-3.896
N175 G1 Z-9.9 F10.
N180 X725.922 F250.
N185 Z-6.9 F10.
N190 G0 Z3.
N195 X-23. Y-7.879
N200 G1 Z-9.9 F10.
N205 X725.922 F250.
N210 Z-6.9 F10.
N215 G0 Z3.
N220 X-23. Y-11.862
N225 G1 Z-9.9 F10.
N230 X725.922 F250.
N235 Z-6.9 F10.
```

exprt_mch_op010 - Notepad

File Edit Format View Help

```
N240 G0 Z3.
N245 X-23. Y-15.845
N250 G1 Z-9.9 F10.
N255 X725.922 F250.
N260 Z-6.9 F10.
N265 G0 Z3.
N270 X-23. Y-19.827
N275 G1 Z-9.9 F10.
N280 X725.922 F250.
N285 Z-6.9 F10.
N290 G0 Z3.
N295 X-23. Y-23.81
N300 G1 Z-9.9 F10.
N305 X725.922 F250.
N310 Z-6.9 F10.
N315 G0 Z3.
N320 X-23. Y-27.793
N325 G1 Z-9.9 F10.
N330 X725.922 F250.
N335 Z-6.9 F10.
N340 G0 Z3.
N345 X-23. Y-31.776
N350 G1 Z-9.9 F10.
N355 X725.922 F250.
N360 Z-6.9 F10.
N365 G0 Z3.
N370 X-23. Y-35.758
N375 G1 Z-9.9 F10.
N380 X725.922 F250.
N385 Z-6.9 F10.
N390 G0 Z3.
N395 X-23. Y-39.741
N400 G1 Z-9.9 F10.
N405 X725.922 F250.
N410 Z-6.9 F10.
N415 G0 Z3.
N420 X-23. Y-43.724
N425 G1 Z-9.9 F10.
N430 X725.922 F250.
N435 Z-6.9 F10.
N440 G0 Z3.
N445 X-23. Y-47.707
N450 G1 Z-9.9 F10.
N455 X725.922 F250.
N460 Z-6.9 F10.
N465 G0 Z3.
N470 X-23. Y-51.689
N475 G1 Z-9.9 F10.
N480 X725.922 F250.
N485 Z-6.9 F10.
N490 G0 Z3.
N495 X-23. Y-55.672
N500 G1 Z-9.9 F10.
N505 X725.922 F250.
N510 Z-6.9 F10.
N515 G0 Z3.
N520 X-23. Y-59.655
N525 G1 Z-9.9 F10.
N530 X725.922 F250.
N535 Z-6.9 F10.
N540 G0 Z3.
N545 X-23. Y-63.638
N550 G1 Z-9.9 F10.
N555 X725.922 F250.
N560 Z-6.9 F10.
N565 G0 Z3.
N570 X-23. Y-67.62
N575 G1 Z-9.9 F10.
N580 X725.922 F250.
N585 Z-6.9 F10.
N590 G0 Z3.
N595 X-23. Y-71.603
N600 G1 Z-9.9 F10.
N605 X725.922 F250.
N610 Z-6.9 F10.
N615 G0 Z3.
N620 X-23. Y-75.586
N625 G1 Z-9.9 F10.
N630 X725.922 F250.
N635 Z-6.9 F10.
N640 G0 Z3.
N645 X-23. Y-79.569
N650 G1 Z-9.9 F10.
N655 X725.922 F250.
N660 Z-6.9 F10.
N665 G0 Z3.
N670 X-23. Y-83.551
N675 G1 Z-9.9 F10.
N680 X725.922 F250.
N685 Z-6.9 F10.
N690 G0 Z3.
N695 X-23. Y-87.534
N700 G1 Z-9.9 F10.
N705 X725.922 F250.
```

exprt_mch_op010 - Notepad

File Edit Format View Help

```
N710 Z-6.9 F10.
N715 G0 Z3.
N720 X-23. Y-91.517
N725 G1 Z-9.9 F10.
N730 X725.922 F250.
N735 Z-6.9 F10.
N740 G0 Z3.
N745 X-23. Y-95.5
N750 G1 Z-9.9 F10.
N755 X725.922 F250.
N760 Z-6.9 F10.
N765 G0 Z3.
N770 X-23. Y-99.482
N775 G1 Z-9.9 F10.
N780 X725.922 F250.
N785 Z-6.9 F10.
N790 G0 Z3.
N795 X-23. Y-103.465
N800 G1 Z-9.9 F10.
N805 X725.922 F250.
N810 Z-6.9 F10.
N815 G0 Z3.
N820 X-23. Y-107.448
N825 G1 Z-9.9 F10.
N830 X725.922 F250.
N835 Z-6.9 F10.
N840 G0 Z3.
N845 X-23. Y-111.431
N850 G1 Z-9.9 F10.
N855 X725.922 F250.
N860 Z-6.9 F10.
N865 G0 Z3.
N870 X-23. Y-115.413
N875 G1 Z-9.9 F10.
N880 X725.922 F250.
N885 Z-6.9 F10.
N890 G0 Z3.
N895 X-23. Y-119.396
N900 G1 Z-9.9 F10.
N905 X725.922 F250.
N910 Z-6.9 F10.
N915 G0 Z3.
N920 X-23. Y-123.379
N925 G1 Z-9.9 F10.
N930 X725.922 F250.
N935 Z-6.9 F10.
N940 G0 Z3.
N945 X-23. Y-127.362
N950 G1 Z-9.9 F10.
N955 X725.922 F250.
N960 Z-6.9 F10.
N965 G0 Z3.
N970 X-23. Y-131.344
N975 G1 Z-9.9 F10.
N980 X725.922 F250.
N985 Z-6.9 F10.
N990 G0 Z3.
N995 X-23. Y-135.327
N1000 G1 Z-9.9 F10.
N1005 X725.922 F250.
N1010 Z-6.9 F10.
N1015 G0 Z3.
N1020 X-23. Y-139.31
N1025 G1 Z-9.9 F10.
N1030 X725.922 F250.
N1035 Z-6.9 F10.
N1040 G0 Z3.
N1045 X-23. Y-143.293
N1050 G1 Z-9.9 F10.
N1055 X725.922 F250.
N1060 Z-6.9 F10.
N1065 G0 Z3.
N1070 X-23. Y-147.275
N1075 G1 Z-9.9 F10.
N1080 X725.922 F250.
N1085 Z-6.9 F10.
N1090 G0 Z3.
N1095 X-23. Y-151.258
N1100 G1 Z-9.9 F10.
N1105 X725.922 F250.
N1110 Z-6.9 F10.
N1115 G0 Z3.
N1120 X-23. Y-155.241
N1125 G1 Z-9.9 F10.
N1130 X725.922 F250.
N1135 Z-6.9 F10.
N1140 G0 Z3.
N1145 X-23. Y-159.224
N1150 G1 Z-9.9 F10.
N1155 X725.922 F250.
N1160 Z-6.9 F10.
N1165 G0 Z3.
N1170 X-23. Y-163.206
N1175 G1 Z-9.9 F10.
```

exprt_mch_op010 - Notepad

File Edit Format View Help

```
N1180 X725.922 F250.
N1185 Z-6.9 F10.
N1190 G0 Z3.
N1195 X-23. Y-167.189
N1200 G1 Z-9.9 F10.
N1205 X725.922 F250.
N1210 Z-6.9 F10.
N1215 G0 Z3.
N1220 X-23. Y-171.172
N1225 G1 Z-9.9 F10.
N1230 X725.922 F250.
N1235 Z-6.9 F10.
N1240 G0 Z3.
N1245 X-23. Y-175.155
N1250 G1 Z-9.9 F10.
N1255 X725.922 F250.
N1260 Z-6.9 F10.
N1265 G0 Z3.
N1270 X-23. Y-179.137
N1275 G1 Z-9.9 F10.
N1280 X725.922 F250.
N1285 Z-6.9 F10.
N1290 G0 Z3.
N1295 X-23. Y-183.12
N1300 G1 Z-9.9 F10.
N1305 X725.922 F250.
N1310 Z-6.9 F10.
N1315 G0 Z3.
N1320 X-23. Y-187.103
N1325 G1 Z-9.9 F10.
N1330 X725.922 F250.
N1335 Z-6.9 F10.
N1340 G0 Z3.
N1345 X-23. Y-191.086
N1350 G1 Z-9.9 F10.
N1355 X725.922 F250.
N1360 Z-6.9 F10.
N1365 G0 Z3.
N1370 X-23. Y-195.068
N1375 G1 Z-9.9 F10.
N1380 X725.922 F250.
N1385 Z-6.9 F10.
N1390 G0 Z3.
N1395 X-23. Y-199.051
N1400 G1 Z-9.9 F10.
N1405 X725.922 F250.
N1410 Z-6.9 F10.
N1415 G0 Z3.
N1420 X-23. Y-203.034
N1425 G1 Z-9.9 F10.
N1430 X725.922 F250.
N1435 Z-6.9 F10.
N1440 G0 Z3.
N1445 X-23. Y-207.017
N1450 G1 Z-9.9 F10.
N1455 X725.922 F250.
N1460 Z-6.9 F10.
N1465 G0 Z3.
N1470 X-23. Y-210.999
N1475 G1 Z-9.9 F10.
N1480 X725.922 F250.
N1485 Z-6.9 F10.
N1490 G0 Z3.
N1495 X-23. Y-214.982
N1500 G1 Z-9.9 F10.
N1505 X725.922 F250.
N1510 Z-6.9 F10.
N1515 G0 Z3.
N1520 X-23. Y-218.965
N1525 G1 Z-9.9 F10.
N1530 X725.922 F250.
N1535 Z-6.9 F10.
N1540 G0 Z3.
N1545 X-23. Y-222.948
N1550 G1 Z-9.9 F10.
N1555 X725.922 F250.
N1560 Z-6.9 F10.
N1565 G0 Z3.
N1570 X-23. Y-226.93
N1575 G1 Z-9.9 F10.
N1580 X725.922 F250.
N1585 Z-6.9 F10.
N1590 G0 Z3.
N1595 X-23. Y-230.913
N1600 G1 Z-9.9 F10.
N1605 X725.922 F250.
N1610 Z-6.9 F10.
N1615 G0 Z3.
N1620 X-23. Y-234.896
N1625 G1 Z-9.9 F10.
N1630 X725.922 F250.
N1635 Z-6.9 F10.
N1640 G0 Z3.
N1645 X-23. Y-238.879
```

exprt_mch_op010 - Notepad

File Edit Format View Help

```
N1650 G1 Z-9.9 F10.
N1655 X725.922 F250.
N1660 Z-6.9 F10.
N1665 G0 Z3.
N1670 X-23. Y-242.861
N1675 G1 Z-9.9 F10.
N1680 X725.922 F250.
N1685 Z-6.9 F10.
N1690 G0 Z3.
N1695 X-23. Y-246.844
N1700 G1 Z-9.9 F10.
N1705 X725.922 F250.
N1710 Z-6.9 F10.
N1715 G0 Z3.
N1720 X-23. Y-250.827
N1725 G1 Z-9.9 F10.
N1730 X725.922 F250.
N1735 Z-6.9 F10.
N1740 G0 Z3.
N1745 X-23. Y-254.81
N1750 G1 Z-9.9 F10.
N1755 X725.922 F250.
N1760 Z-6.9 F10.
N1765 G0 Z3.
N1770 X-23. Y-258.792
N1775 G1 Z-9.9 F10.
N1780 X725.922 F250.
N1785 Z-6.9 F10.
N1790 G0 Z3.
N1795 X-23. Y-262.775
N1800 G1 Z-9.9 F10.
N1805 X725.922 F250.
N1810 Z-6.9 F10.
N1815 G0 Z3.
N1820 X-23. Y-266.758
N1825 G1 Z-9.9 F10.
N1830 X725.922 F250.
N1835 Z-6.9 F10.
N1840 G0 Z3.
N1845 X-23. Y-270.741
N1850 G1 Z-9.9 F10.
N1855 X725.922 F250.
N1860 Z-6.9 F10.
N1865 G0 Z3.
N1870 X-23. Y-274.723
N1875 G1 Z-9.9 F10.
N1880 X725.922 F250.
N1885 Z-6.9 F10.
N1890 G0 Z3.
N1895 X-23. Y-278.706
N1900 G1 Z-9.9 F10.
N1905 X725.922 F250.
N1910 Z-6.9 F10.
N1915 G0 Z3.
N1920 X-23. Y-282.689
N1925 G1 Z-9.9 F10.
N1930 X725.922 F250.
N1935 Z-6.9 F10.
N1940 G0 Z3.
N1945 X-23. Y-286.672
N1950 G1 Z-9.9 F10.
N1955 X725.922 F250.
N1960 Z-6.9 F10.
N1965 G0 Z3.
N1970 X-23. Y-290.654
N1975 G1 Z-9.9 F10.
N1980 X725.922 F250.
N1985 Z-6.9 F10.
N1990 G0 Z3.
N1995 X-23. Y-294.637
N2000 G1 Z-9.9 F10.
N2005 X725.922 F250.
N2010 Z-6.9 F10.
N2015 G0 Z3.
N2020 X-23. Y-298.62
N2025 G1 Z-9.9 F10.
N2030 X725.922 F250.
N2035 Z-6.9 F10.
N2040 G0 Z3.
N2045 X-23. Y-302.603
N2050 G1 Z-9.9 F10.
N2055 X725.922 F250.
N2060 Z-6.9 F10.
N2065 G0 Z3.
N2070 X-23. Y-306.585
N2075 G1 Z-9.9 F10.
N2080 X725.922 F250.
N2085 Z-6.9 F10.
N2090 G0 Z3.
N2095 X-23. Y-310.568
N2100 G1 Z-9.9 F10.
N2105 X725.922 F250.
N2110 Z-6.9 F10.
N2115 G0 Z3.
```

exprt_mch_op010 - Notepad

File Edit Format View Help

```
N2120 X-23. Y-314.551
N2125 G1 Z-9.9 F10.
N2130 X725.922 F250.
N2135 Z-6.9 F10.
N2140 G0 Z3.
N2145 X-23. Y-318.534
N2150 G1 Z-9.9 F10.
N2155 X725.922 F250.
N2160 Z-6.9 F10.
N2165 G0 Z3.
N2170 X-23. Y-322.516
N2175 G1 Z-9.9 F10.
N2180 X725.922 F250.
N2185 Z-6.9 F10.
N2190 G0 Z3.
N2195 X-23. Y-326.499
N2200 G1 Z-9.9 F10.
N2205 X725.922 F250.
N2210 Z-6.9 F10.
N2215 G0 Z3.
N2220 X-23. Y-330.482
N2225 G1 Z-9.9 F10.
N2230 X725.922 F250.
N2235 Z-6.9 F10.
N2240 G0 Z3.
N2245 X-23. Y-334.464
N2250 G1 Z-9.9 F10.
N2255 X725.922 F250.
N2260 Z-6.9 F10.
N2265 G0 Z3.
N2270 X-23. Y-338.447
N2275 G1 Z-9.9 F10.
N2280 X725.922 F250.
N2285 Z-6.9 F10.
N2290 G0 Z3.
N2295 X-23. Y-342.43
N2300 G1 Z-9.9 F10.
N2305 X725.922 F250.
N2310 Z-6.9 F10.
N2315 G0 Z3.
N2320 X-23. Y-346.413
N2325 G1 Z-9.9 F10.
N2330 X725.922 F250.
N2335 Z-6.9 F10.
N2340 G0 Z3.
N2345 X-23. Y-350.395
N2350 G1 Z-9.9 F10.
```

Chapter 9
Electric Discharge Machining (EDM)

Wire EDM is mainly used for cutting hard metals.

EDM is a two-axis Machining process.

In this tutorial, the following will be covered.

- Activate Manufacturing application
- Parts will be imported and Constrained
- Stock is created via Automatic Workpiece Method
- Machine Coordinate System is created
- Tool is created
- Create Cut Line Motion for Tool

9.1 Activate Manufacturing Application

Start Creo Parametric either from your computer desktop or from programme ≫ Click on the New icon as shown below.

The New dialogue box is activated on the main window ≫ On the Type group, click on the Manufacturing radio button to make it active ≫ On the Sub-type group,

P.O. Kanife, *Computer Aided Virtual Manufacturing Using Creo Parametric*,
DOI 10.1007/978-3-319-23359-8_9

click on the NC Assembly radio button ≫ In the Name section box, type Wire_EDM_mfg01 (type whatever name you want). Click on the Use default template square box to clear the Check Mark icon. See Fig. 9.1.

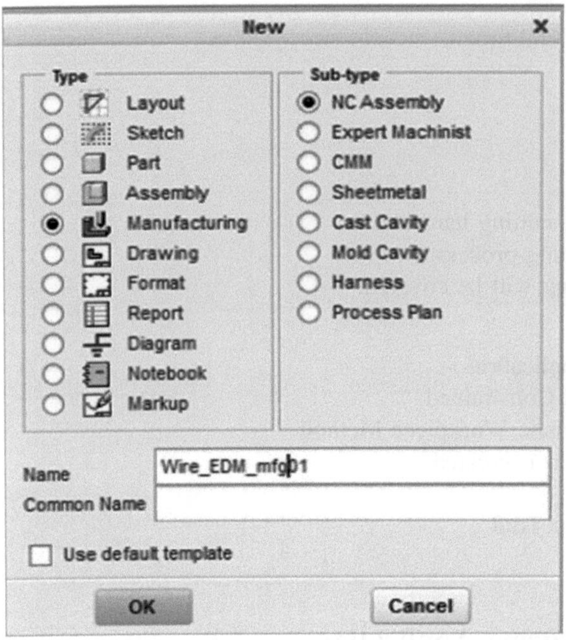

Click on the OK tab to proceed.

Fig. 9.1 Activated New dialogue box

The New File Options dialogue window is activated on the main window ≫ Click on the "mmns_mfg_nc" on the Template group ≫ In this tutorial, nothing will be typed in the MODELLED BY and in the DESCRIPTION section boxes. See Fig. 9.2.

Note: You can type whatever you want in the MODELLED BY and in the DESCRIPTION section boxes.

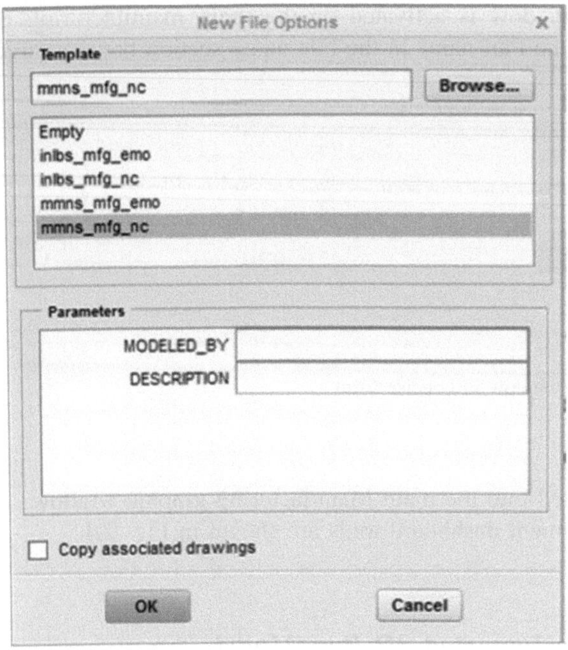

Click on the OK tab.

Fig. 9.2 Selecting "mmns_mfg_nc" as the SI unit

The Manufacturing application main graphic user interface window is activated.

9.2 Import and Constrain the 3D Model

9.2.1 Import the Reference Model

To import the 3D Part, click on the Reference Model icon and on the drop-down menu list, click on Assemble Reference Model as shown below.

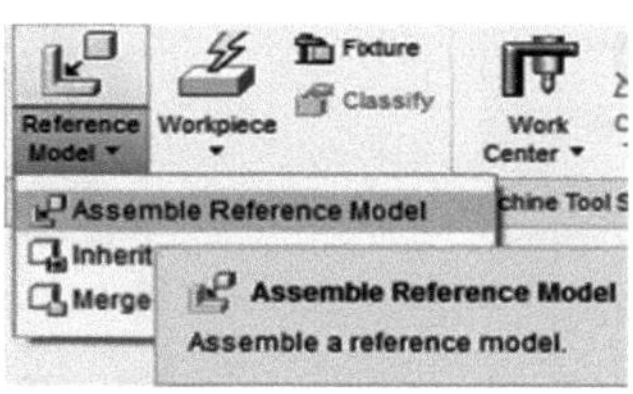

The Open dialogue window is activated on the main manufacturing graphic window. Make sure that the Part name in the File name section box is correct. See Fig. 9.3.

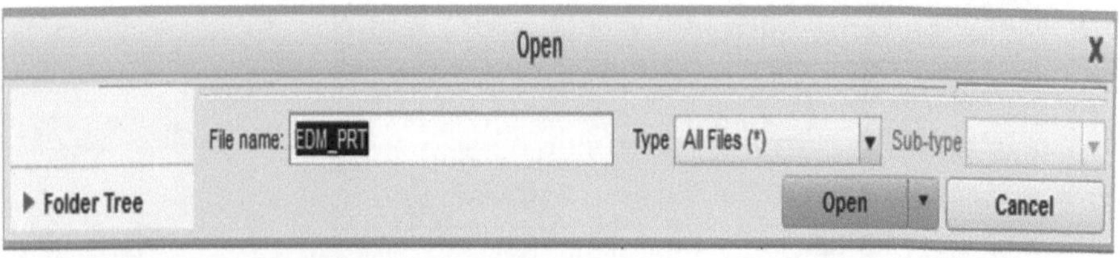

Fig. 9.3 The Open dialogue window in concise form

Click on the Open tab.
The 3D Part is imported into the main Manufacturing graphic window.
The Component Placement dashboard tools are shown in Fig. 9.4.

9.2.2 *Constrain the Imported 3D Part/Model*

Once the Component Placement ribbon toolbar is activated ≫ Click on the Automatic downward arrow and on the drop-down menu list, click on Default as indicated by the arrow in Fig. 9.4.

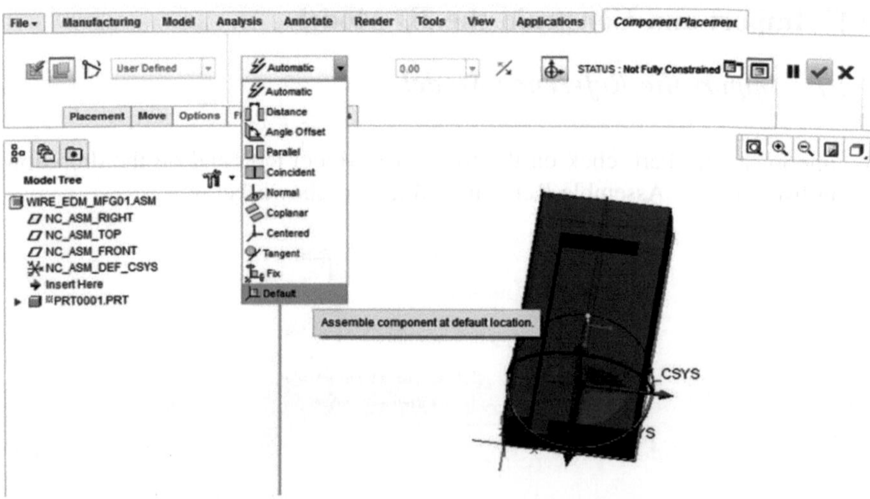

Fig. 9.4 Constraining the imported 3D Part

Once Default is clicked on the drop-down menu list, the imported 3D Part is constrained on the main Manufacturing graphic window.

Now click on the Check Mark icon.

9.3 Create Automatic Stock for the Part

Click on the Workpiece icon to activate its drop-down menu list ≫ Now click on Automatic Workpiece on the activated drop-down menu list as shown below.

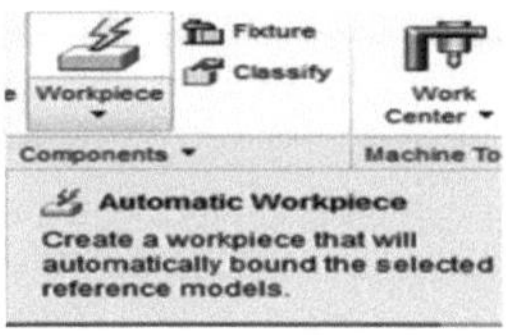

The Auto Workpiece Creation dashboard tools are activated as shown in Fig. 9.5.

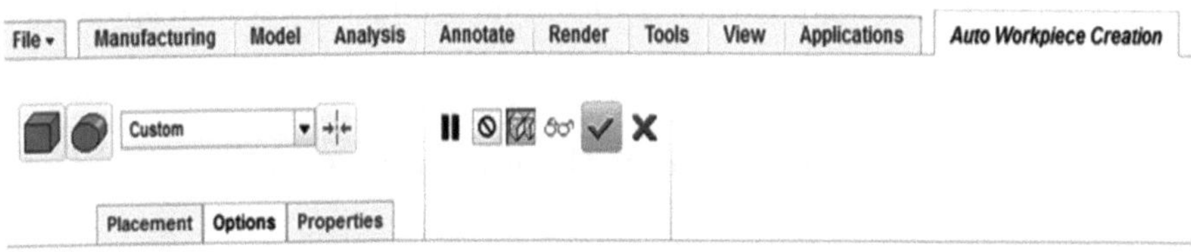

Fig. 9.5 Activated Auto Workpiece Creation dashboard in concise form

Click on the Options tab to activate its content ≫ On the Linear offsets group, click on the Current Offsets tab ≫ Add offset values in the ± X and Z section boxes. Take note of the values in the Rotation Offsets section, as illustrated in Fig. 9.6.

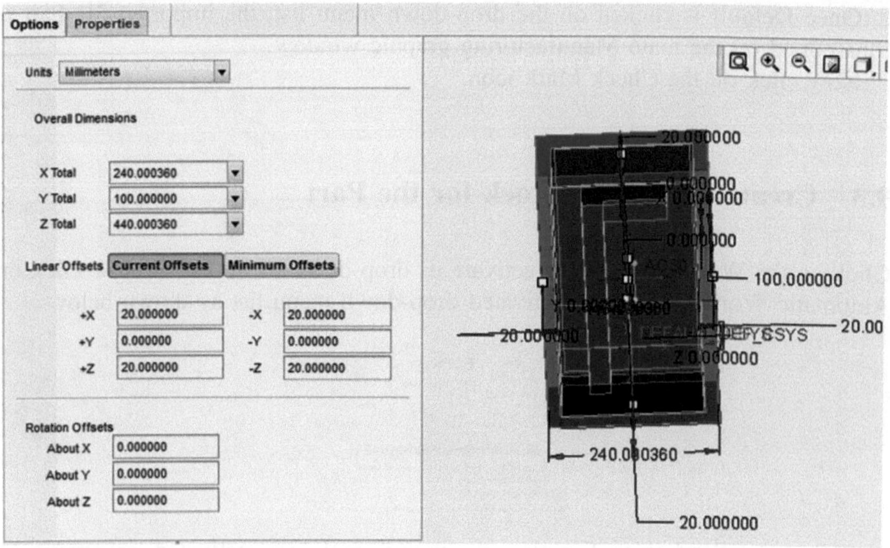

Fig. 9.6 Activated Options panel and the Automatic Workpiece dimensions

Click on the Check Mark icon to exit the Auto Workpiece Creation application.

9.4 Create Coordinate System (Programme Zero)

Click on the Coordinate System icon tab as shown below.

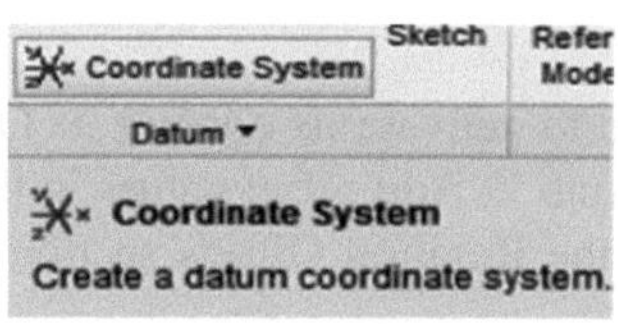

The Coordinate System dialogue box is activated on the main graphic window.

Now click on the Workpiece top surface and also on the two side surfaces, while holding down the Ctrl key on the key board. These selected top and side surfaces are added automatically into the References section box on the Origin tab group of the Coordinate System dialogue box. See Fig. 9.7.

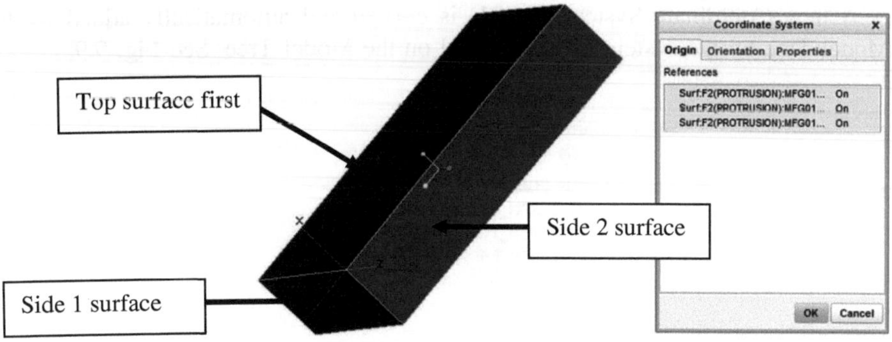

Fig. 9.7 Selected Workpiece references used in creating the Coordinate System

To orient the X, Y and Z coordinate axes to the correct orientation

Click on the Orientation tab on the Coordinate System dialogue box. On the "Orient by" group, click on "References selection" radio button on ≫ Click on the downward pointing arrow on the "to determine" section box, and then click on *Z* axis on the activated drop-down menu list ≫ Click on the downward pointing arrow on the "to project" section box and then click on the *X* axis on the activated drop-down menu list. Now click on the Flip tab to flip *X* axis orientation. The arrows as illustrated in Fig. 9.8 indicate the correct orientations for the created Coordinate System axes.

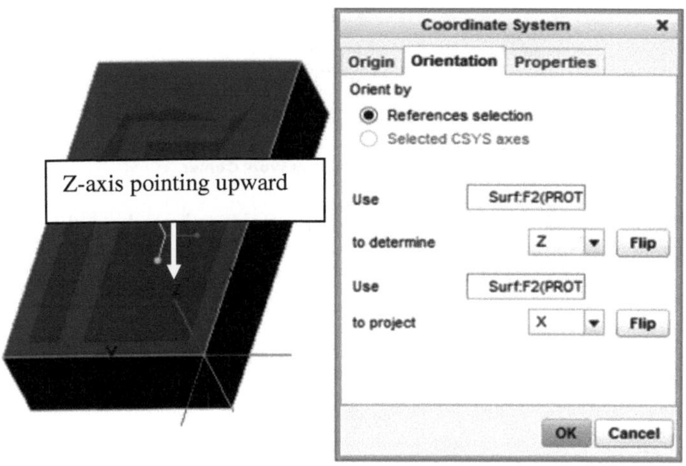

Fig. 9.8 Corrcet Orientaions of the *X*, *Y* and *Z* axes on the Workpiece

Click on the OK tab on the Coordinate System dialogue box to exit.

A new Coordinate System (ACS1) is created and automatically added to the Model Tree by the system as highlighted on the Model Tree. See Fig. 9.9.

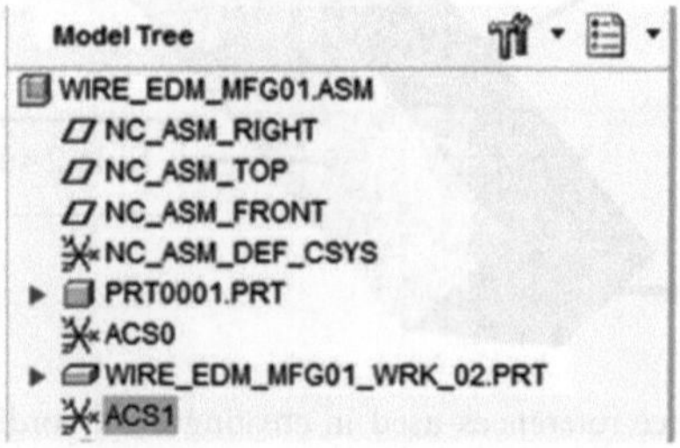

Fig. 9.9 Created ACS1 (Coordinate System) on the Model Tree

9.5 Create Work Centre

Click on the Work Centre icon to activate its drop-down menu list.

Click on Wire EDM on the drop-down list as indicated by the arrow below.

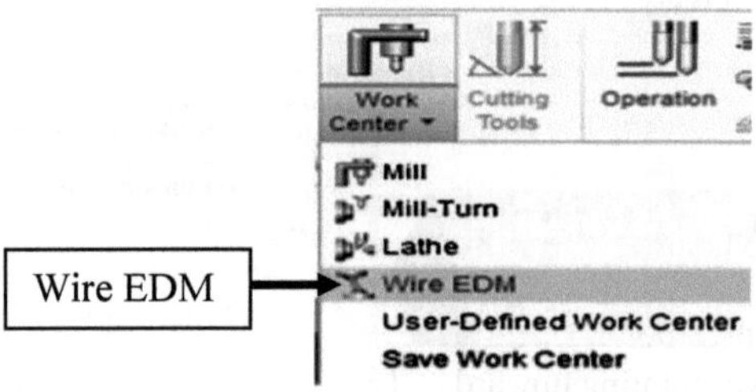

The WEDM Work Centre dialogue window is activated on the main graphic window as shown below. Take note of the Name, Type, Number of Axes section boxes as shown in Fig. 9.10.

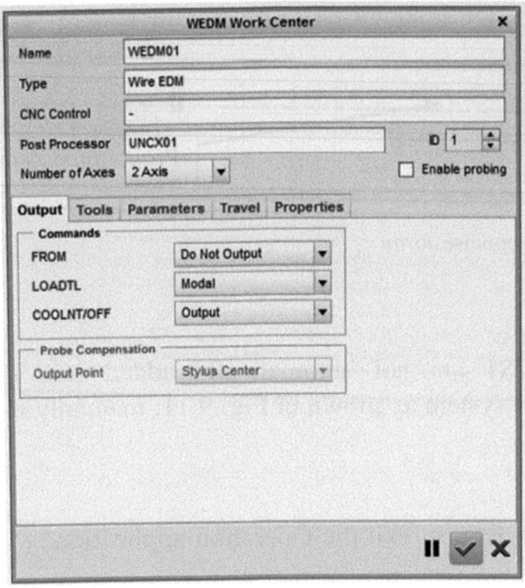

Click on the Check Mark icon.

Fig. 9.10 Activated WEDM Work Centre dialogue window

9.6 Create Operation

Click on the Operation icon as shown below.

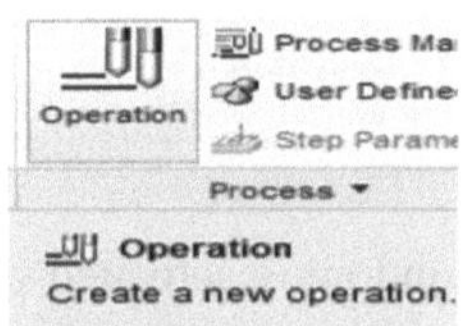

The Operation dashboard is activated as shown in Fig. 9.11. WEDM01 and ACS1 are automatically added by the system into their respective section boxes as indicated by the arrows in Fig. 9.11.

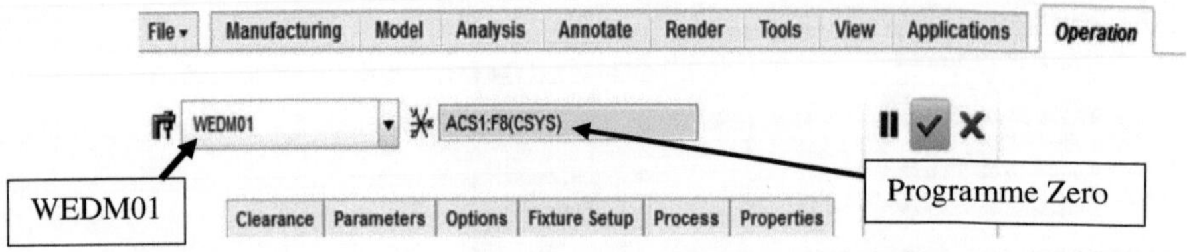

Fig. 9.11 Operation ribbon toolbar in concise form

> Note: If WEDM01 and ACS1 are not automatically added into their respective section boxes by the system as shown in Fig. 9.11, manually input them yourself

Now click on the Check Mark icon to exit the Operation application.

9.7 Create Cutting Tool

Click on the Cutting Tools icon as shown below.

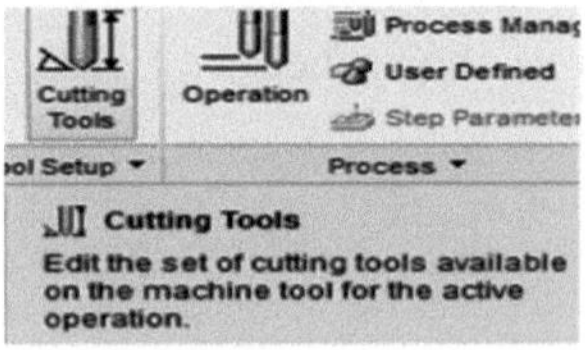

The Tools Setup dialogue window is activated on the main graphic window as shown in Fig. 9.12.

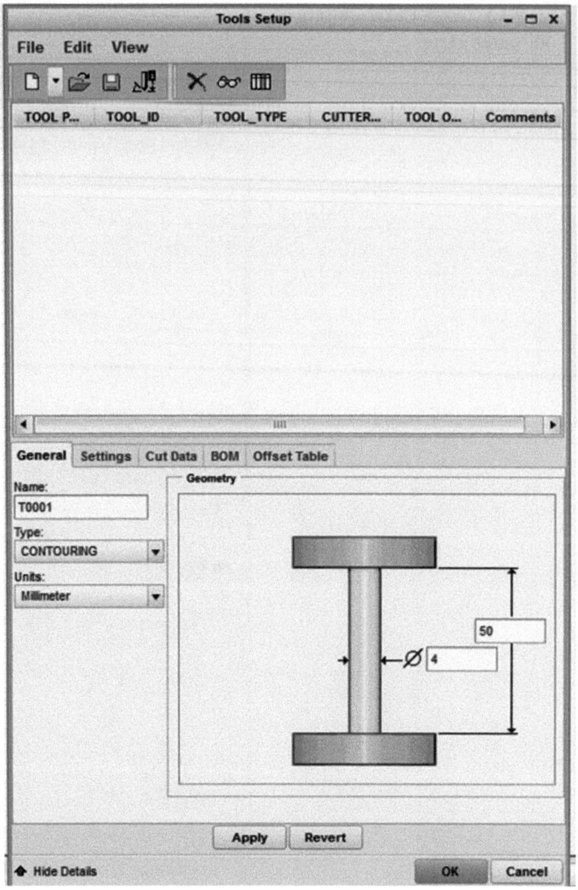

Fig. 9.12 Activated Tools Setup dialogue window

Click on the General tab to activate its content ≫ In the Name section box, type T01 ≫ In the Type section box, click on the downward pointing arrow and now click on CONTOURING on the drop-down menu list ≫ Check to make sure that the Unit is correct ≫ Add dimensions to the CONTOURING Tool. See Fig. 9.13.

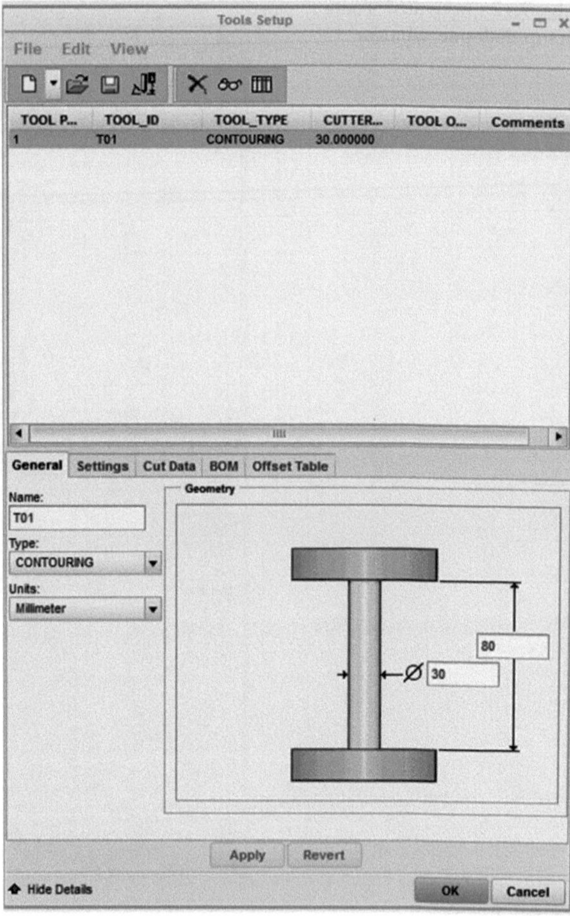

Fig. 9.13 Wire EDM cutting Tool parameters

Now click on the Apply tab.
Click on the OK tab to exit the Tools Setup dialogue window.

9.8 Create the Contouring Sequences

Click on the Wire EDM menu bar ≫ On the activated menu list/ribbon, click on the Contour icon as indicated by the arrow as shown below.

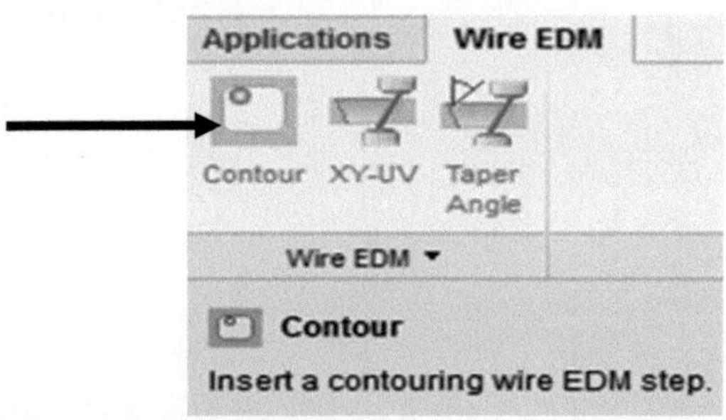

The NC SEQUENCE Menu Manager dialogue box is activated on the main graphic window. Make sure that Tool and Parameters are Checked Marked as shown in Fig. 9.14.

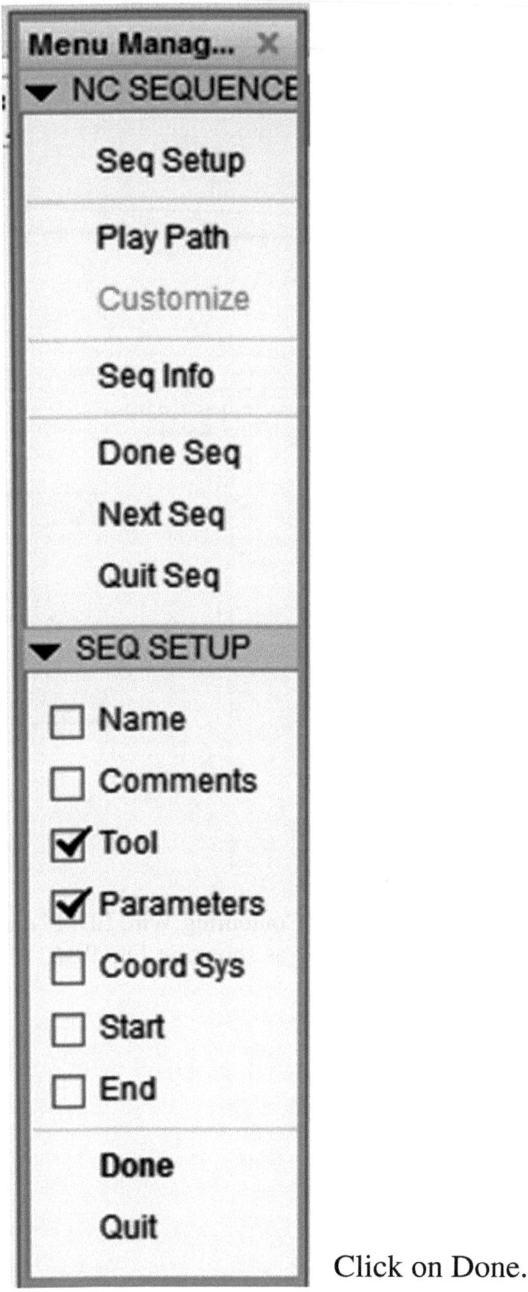

Fig. 9.14 Checked Marked Tool and Parameters on SEQ SETUP group

The Tools Setup dialogue window is activated again as shown in Fig. 9.15.

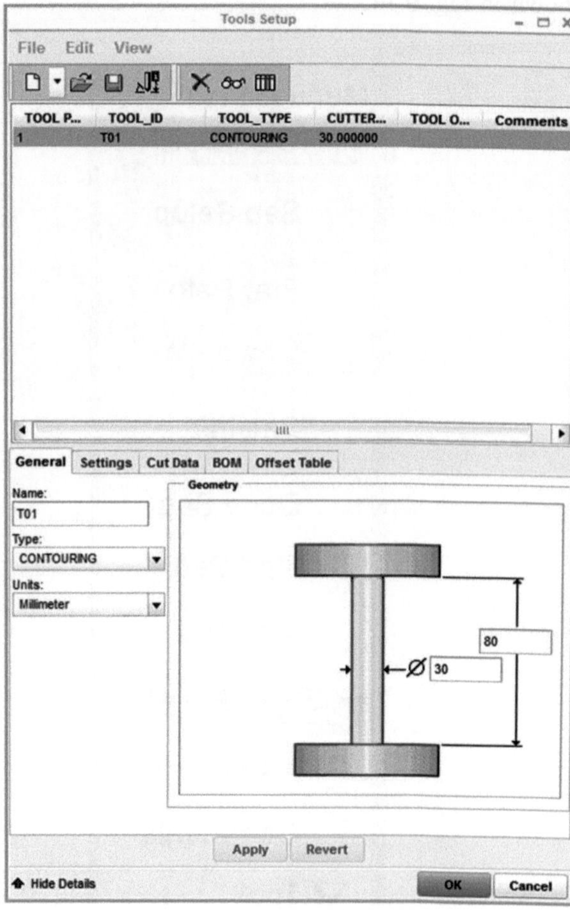

Fig. 9.15 Activated Wire EDM cutting Tool

The Edit Parameters of Sequence "Contouring Wire EDM" dialogue window is activated on the main graphic window as shown in Fig. 9.16.

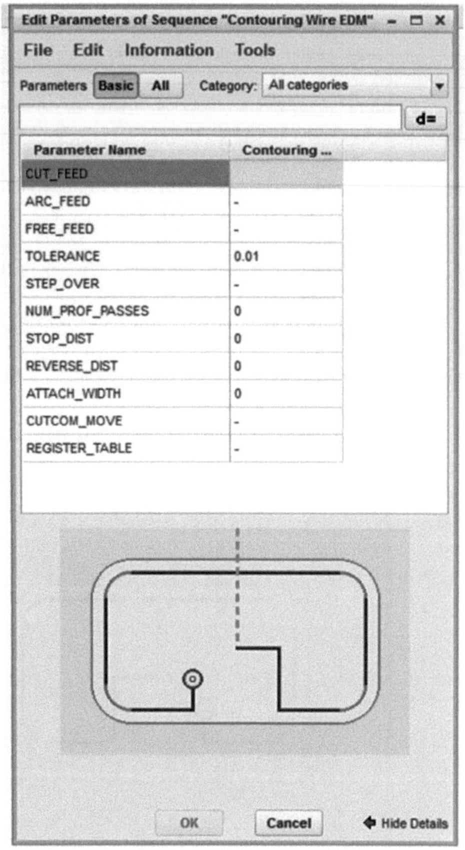

Fig. 9.16 Activated Edit Parameters of Sequence "Contouring Wire EDM" dialogue window

Add the Wire EDM parameters as shown in Fig. 9.17.

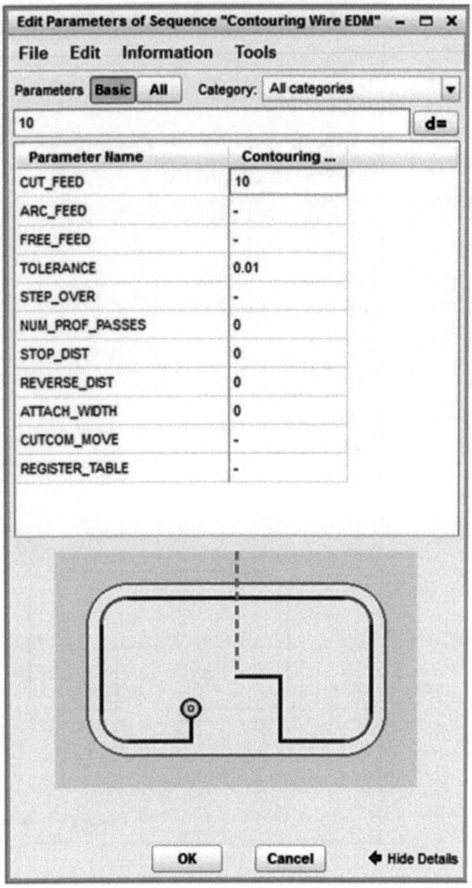

Click on OK tab to proceed.

Fig. 9.17 Wire EDM parameters

Create Tool Motion Path Sequence

The Customize dialogue box is activated on the main graphic window ≫ Click on Insert tab to create Tool motion as indicated by the arrow in Fig. 9.18.

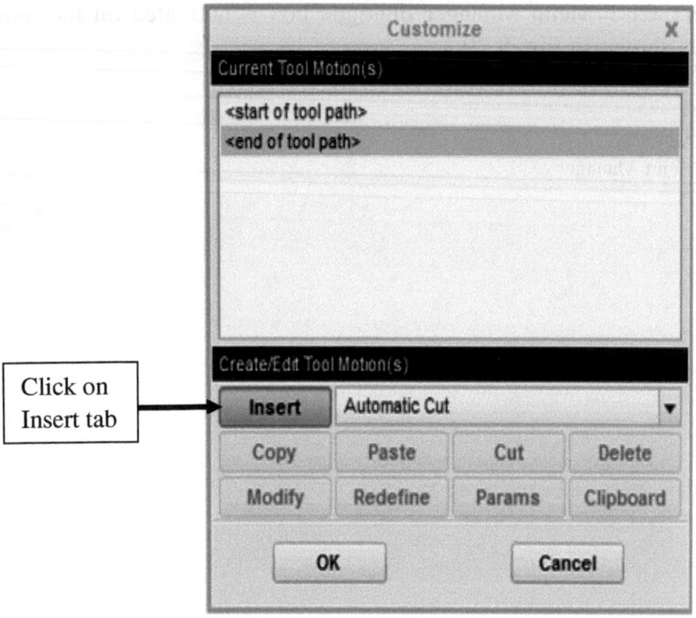

Fig. 9.18 Activated Customize dialogue box

Once Insert tab is clicked, the WEDM OPT Menu Manager dialogue box is activated on the main graphic window » Check Mark Rough on the WEDM OPT group and make sure that Sketch is highlighted on the WEDM OPT group as shown in Fig. 9.19.

Fig. 9.19 WEDM OPT Menu Manager dialogue box with Sketch highlighted and Rough check marked

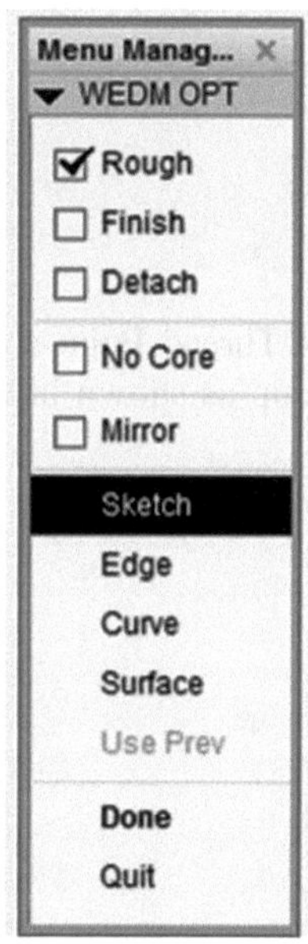

Now click on Done.

The INT CUT Menu Manager dialogue box is activated on the main graphic window as shown in Fig. 9.20.

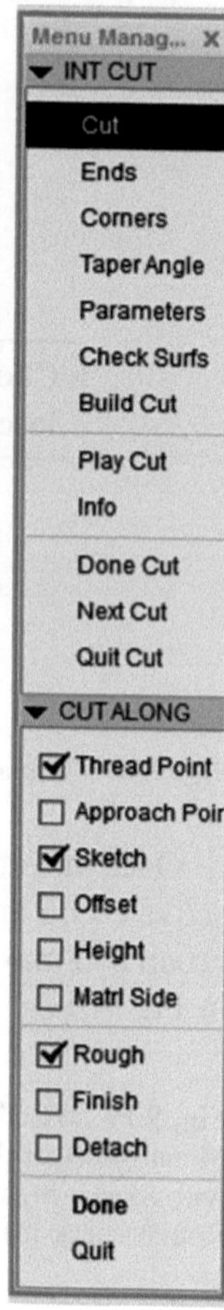

Fig. 9.20 Activated INT CUT Menu Manager dialogue box

Now un-Check Thread Point and Check Mark only Sketch and Rough on the CUT ALONG group as shown in Fig. 9.21.

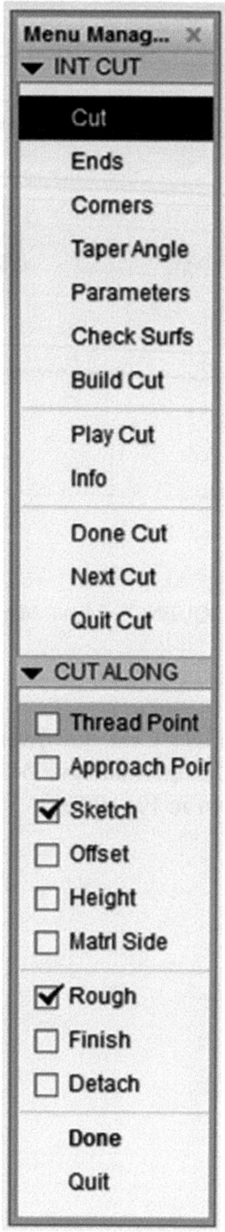

Fig. 9.21 Check Marked Sketch and Rough on the CUT ALONG group

Click on Done.

The NC SEQUENCE, Menu Manager and Customize dialogue boxes remains active on the main graphic window as illustrated in Fig. 9.22.

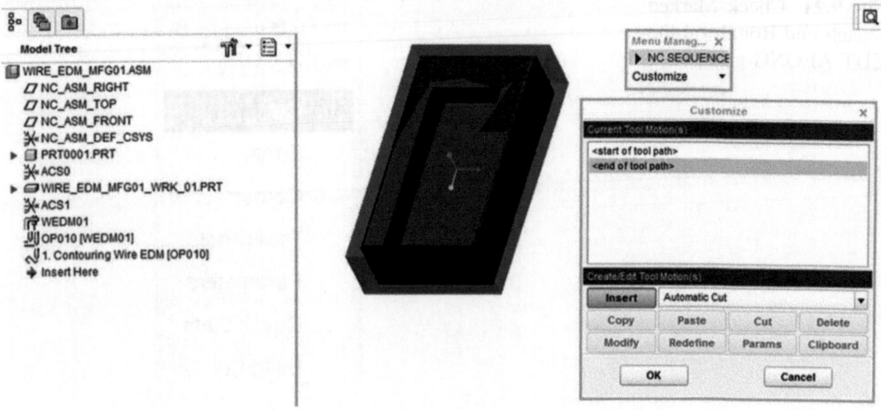

Fig. 9.22 NC SEQUENCE Menu Manager and the Customize dialogue boxes on the main graphic window

Click on the INT CUT downward pointing arrow to expand the INT CUT group on the Menu Manager dialogue box ≫ The SETUP SK PLN group is added by the system as shown in Fig. 9.23.

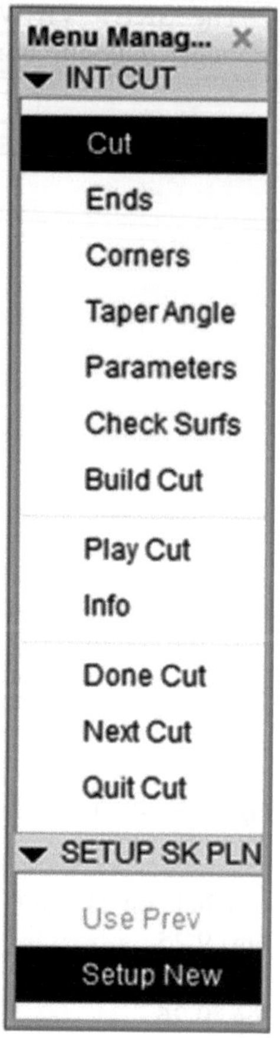

Fig. 9.23 SETUP SK PLN group on the INT CUT Menu Manager dialogue box

System is asking for new sketch plane to be selected on the message bar.

Now click on the Workpiece top surface as the new sketch plane as indicated by the arrow in Fig. 9.24.

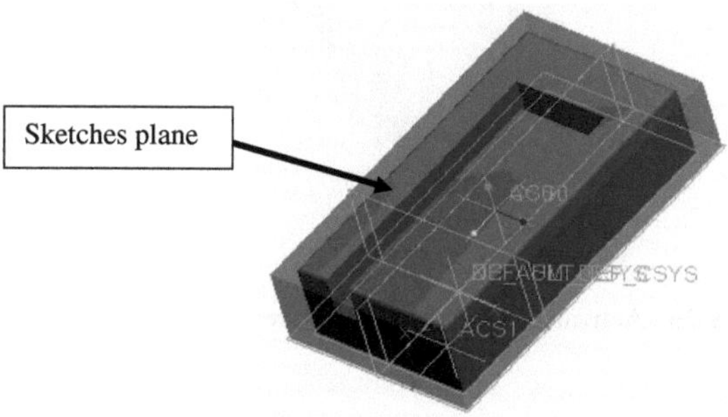

Fig. 9.24 Selected Workpiece top surface as the sketch plane

The SKET VIEW group is automatically added into the INT CUT Menu Manager dialogue box as indicated by the arrow in Fig. 9.25.

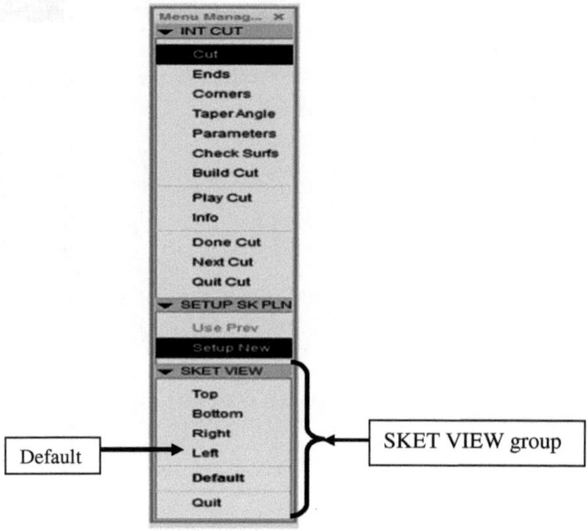

Fig. 9.25 SKET VIEW group automatically added by the system

Click on Default on SKET VIEW group as indicated above by the arrow in Fig. 9.25.

The References dialogue box is activated on the main graphic window. See Fig. 9.26.

Fig. 9.26 Activated References dialogue box on the main graphic window

Click on the Sketch view icon  to orient the Workpiece to the correct Sketch Plane as shown in Fig. 9.27.

Now click on the edges of the Workpiece as indicated by the arrows in Fig. 9.27. Manually click on NC_ASM_RIGHT, NC_ASM_FRONT and ASC1 (Coordinate System) axis if they are not automatically selected and added by the system.

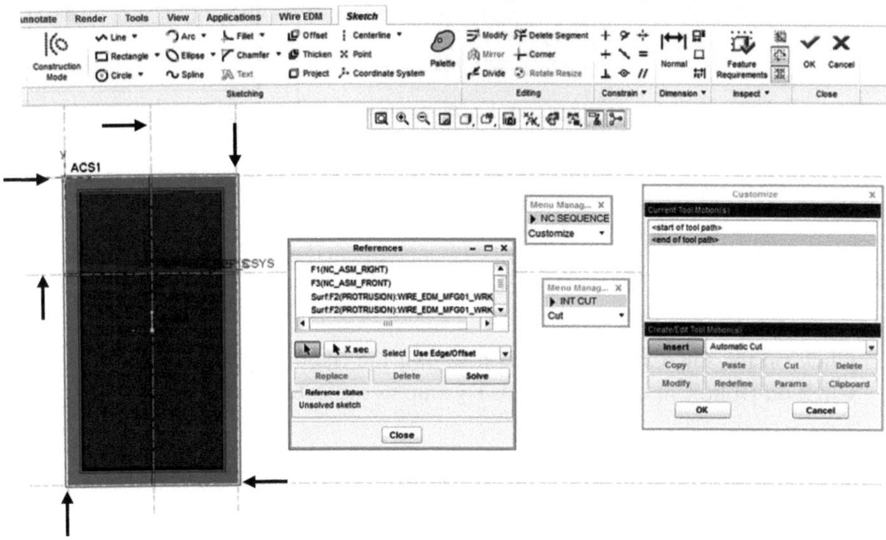

Fig. 9.27 Creating the References

Click on the Close tab on the References dialogue box to exit.

Sketch the Cut Motion Line/Tool Motion Path

Sketch a line that will touch all the edges of the reference Part, starting from ACS1 and back to ACS1 as illustrated by the arrows in Fig. 9.28.

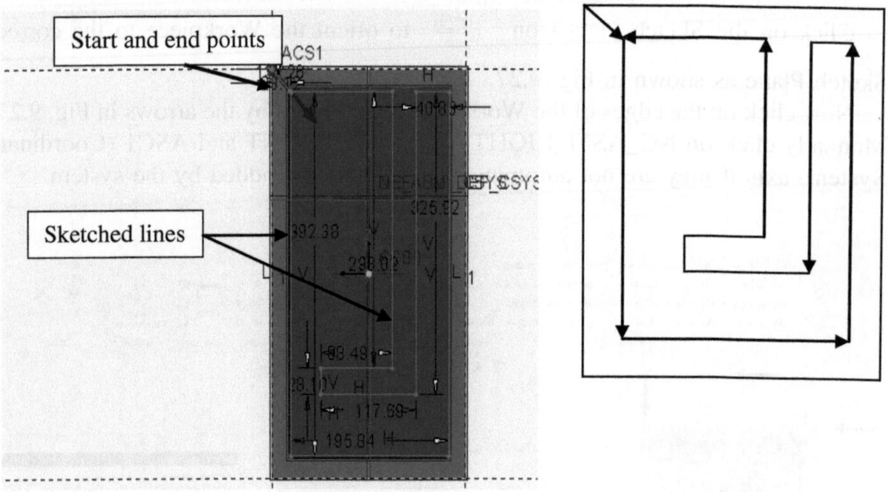

Fig. 9.28 Sketched Tool Motion Path

Click on the OK/Check Mark icon to exit the Sketch application.
Click on Done Cut on the INT CUT group as indicated by the arrow in Fig. 9.29.

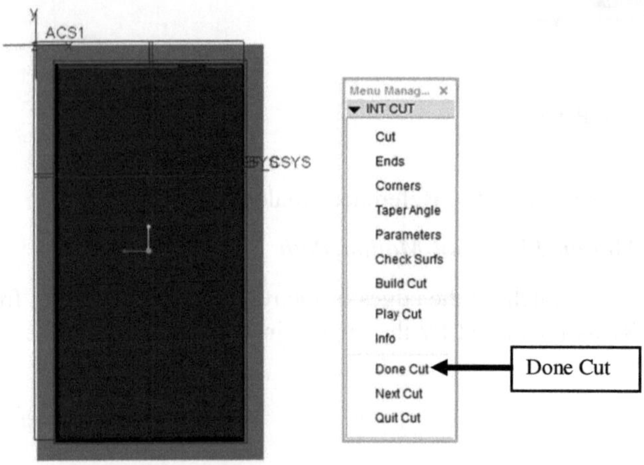

Fig. 9.29 Workpiece and INT CUT group on the Menu Manager dialogue box

The Follow Cut dialogue box is activated on the main graphic window as shown in Fig. 9.30.

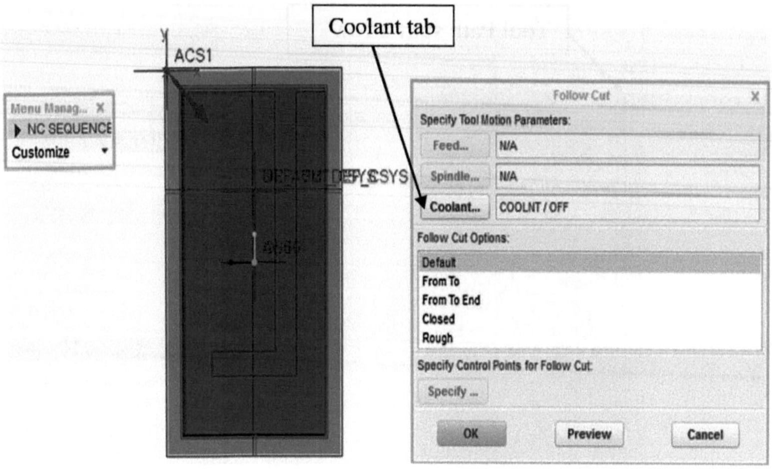

Fig. 9.30 Activated Follow Cut dialogue box and Workpiece

Click on Coolant tab on the Follow Cut dialogue box as indicated by the arrow on Fig. 9.30.

The Coolant Parameters dialogue box is activated on the main graphic window ≫ Click on the Coolant Option downward pointing arrow to activate the menu list ≫ Click on "ON" on the drop-down menu list as shown in Fig. 9.31.

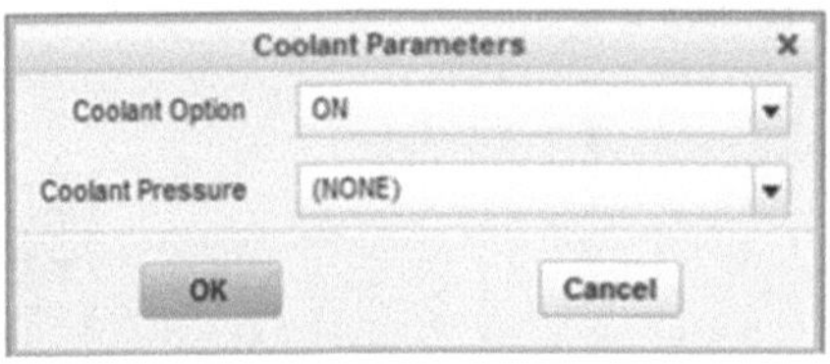

Click on OK tab to exit.

Fig. 9.31 Activating "ON" on the Coolant Parameters dialogue box

Make sure that Default is selected on the Follow Cut Options group and then Click on OK tab on the Follow Cut dialogue box to exit.

Automatic Cut is added automatically by the system into the Current Tool Motion section on the Customize dialogue box. The Tool Path start arrow changes to orange on the Workpiece surface as illustrated in Fig. 9.32.

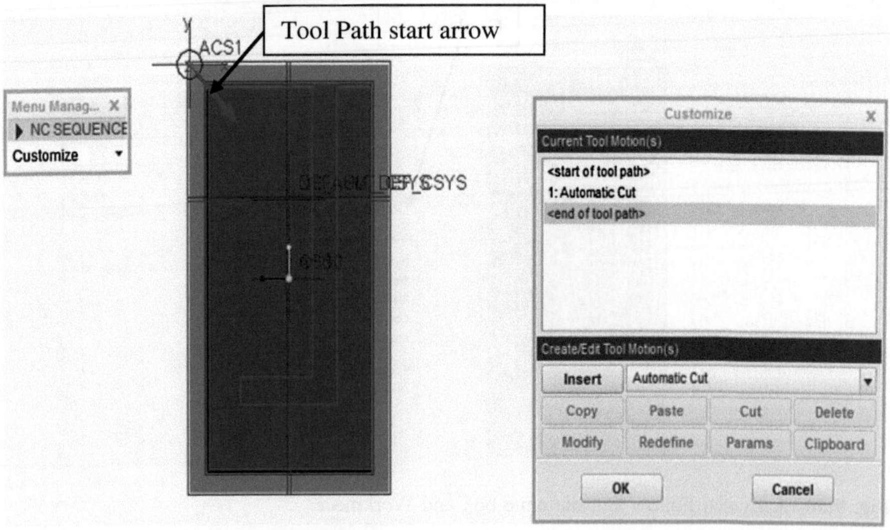

Fig. 9.32 Customize dialogue box indicating cut motion parameters

Click on the OK tab to exit the Customize dialogue box.

Activate Play Path

Click on Play Path on the NC SEQUENCE group and the PLAY PATH group is automatically added ≫ Check Mark Compute CL and click on Screen Play on the PLAY PATH group as shown in Fig. 9.33.

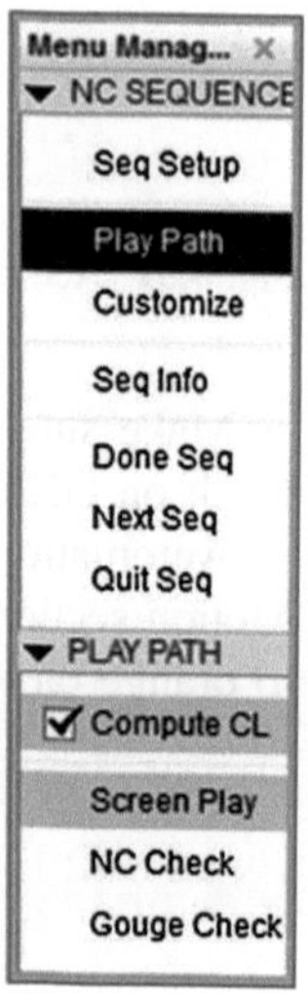

Fig. 9.33 Activating Screen Play Path

The cutting Tool and PLAY PATH dialogue box is activated on the main graphic window as shown in Fig. 9.34.

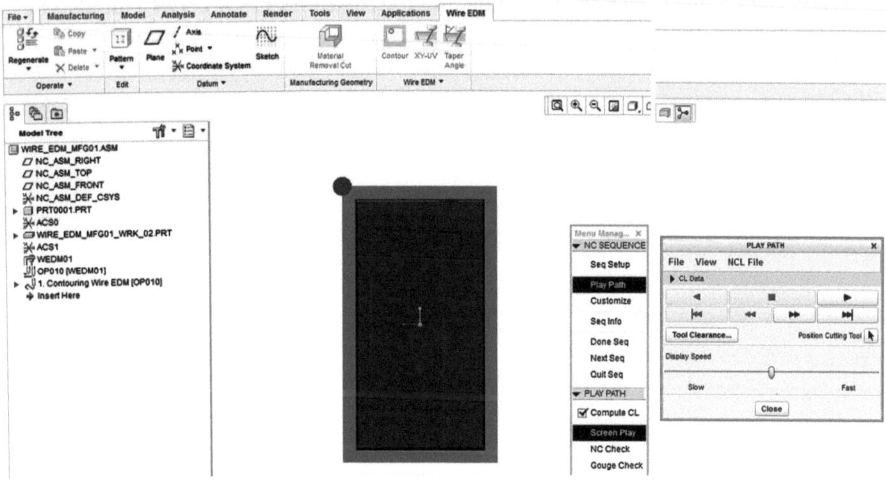

Fig. 9.34 Activated cutting Tool and the PALY PATH dialogue box

Click on the CL Data arrow to view the CL data file if you want to view the cutter location data.

Click on the Play tab to start the on Screen Play of the Electric Discharge Machining (EDM) operation as shown in Fig. 9.35.

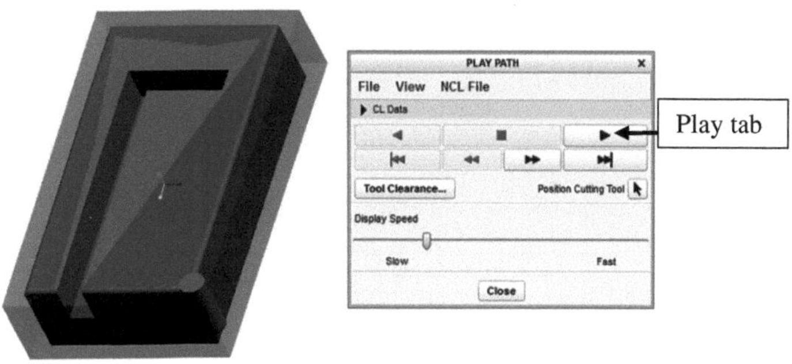

Fig. 9.35 EDM machining Screen simulation

At the end of the EDM process, click on the Close tab to exit PLAY PATH dialogue box » Now click on Done Seq on the NC SEQUENCE group to exit.

Note: The Cutter Location (CL) data and the Post Processor codes can also be created by following the step-by-step guide as explained in the Expert Machinist tutorial.

Chapter 10
CNC Area Lathe Turning

Area Turning is used to remove excess material from stock.

Area Turning is a two-axis machining process.

The following steps will be covered in this tutorial.

- Activate the Manufacturing application
- Import the Part into the Manufacturing environment
- Constrain the Part into the Manufacturing environment
- Create Automatic Workpiece (Stock)
- Create Programme Zero (Machine Coordinate System)
- Create Work Centre
- Create Operation
- Create Cutting Tool
- Create Area Turning using Profile Method
- Generate the Cutter Location (CL) Data File
- Generate the G-code Data

10.1 Activate the Manufacturing Application

Start Creo Parametric either from your Desktop or from Programme.

Click on the New icon as shown below.

The New dialogue box is activated on the main graphic window ≫ On the Type group, click on the Manufacturing radio button ≫ On the Sub type group, click on the NC Assembly radio button ≫ In the Name section box, type "Lathe_Area_mfg01". Click on the Use default template square box to clear the Check Mark icon as shown in Fig. 10.1.

P.O. Kanife, *Computer Aided Virtual Manufacturing Using Creo Parametric*,
DOI 10.1007/978-3-319-23359-8_10

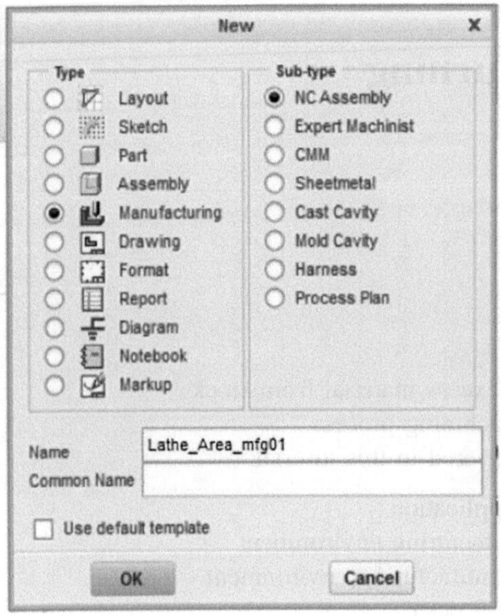

Click on the OK tab to proceed.

Fig. 10.1 New dialogue box

The New File Options dialogue box is activated on the main graphic window ≫ Click on the “mmns_mfg_nc” on the Template group ≫ On the Parameters group, type your name in the MODELLED BY section box and in the DESCRIPTION section box, type “Area Turning” as shown in Fig. 10.2.

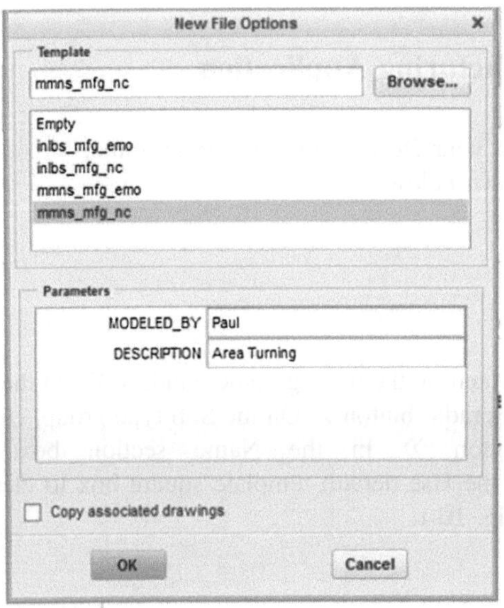

Click on OK tab to proceed.

Fig. 10.2 New File Options dialogue box

Creo Parametric Manufacturing main Graphic User Interface window is activated.

10.1.1 Import Reference Model

Click on the Reference Model icon ≫ Now click on Assemble Reference Model.

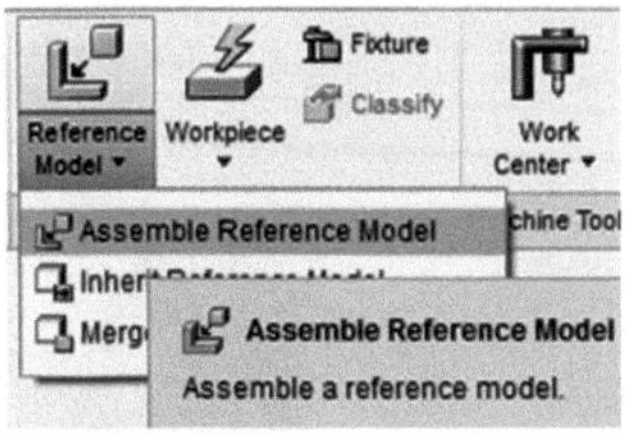

The Open dialogue window is activated on the main graphic window. Make sure that the name in the File name section box is the correct Part that you want to open (Fig. 10.3).

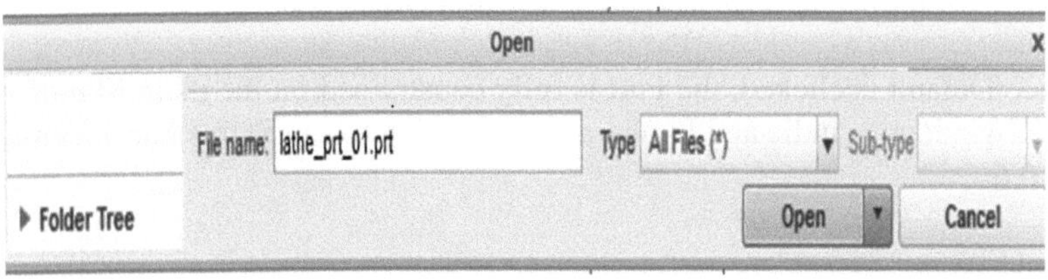

Fig. 10.3 Activated Open dialogue window in concise form

Click on Open tab to exit.

The Reference Model is imported into the Manufacturing main graphic window.

The Component Placement dashboard is activated. See Fig. 10.4.

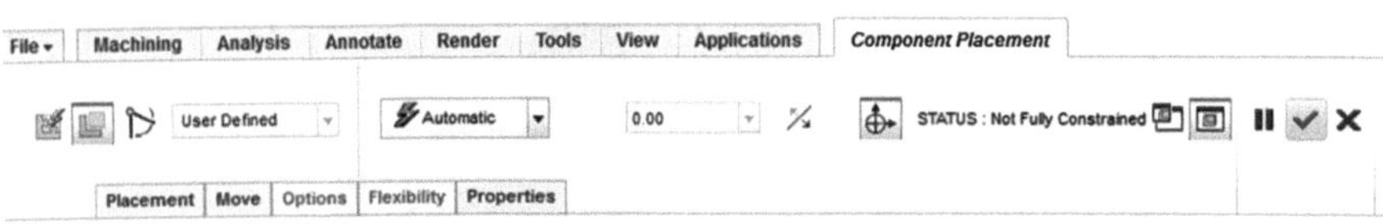

Fig. 10.4 Component Placement dashboard in concise form

10.1.2 Constrain the Model

Once the Component Placement tools are activated ≫ Click on the Automatic downward pointing arrow and on the drop-down list, and click on Default as indicated by the arrow in Fig. 10.5.

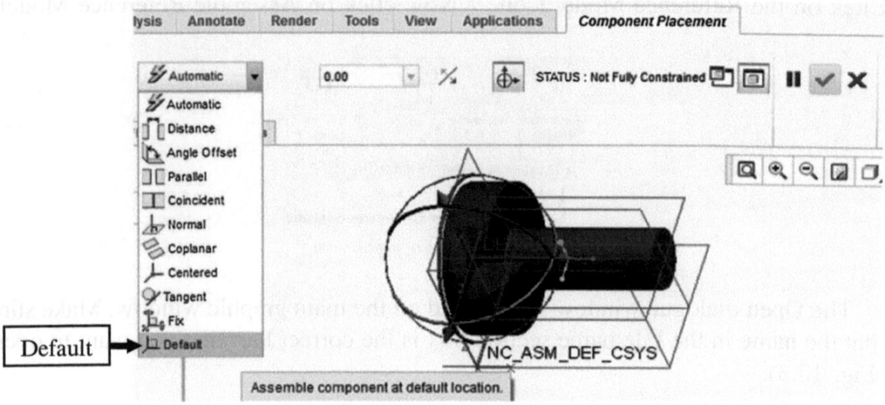

Fig. 10.5 Constraining the imported Part on the main graphic window

Once Default is clicked, the Part is fully constrained on the main Manufacturing graphic window as indicated by the arrow pointing to the Status bar information in Fig. 10.6.

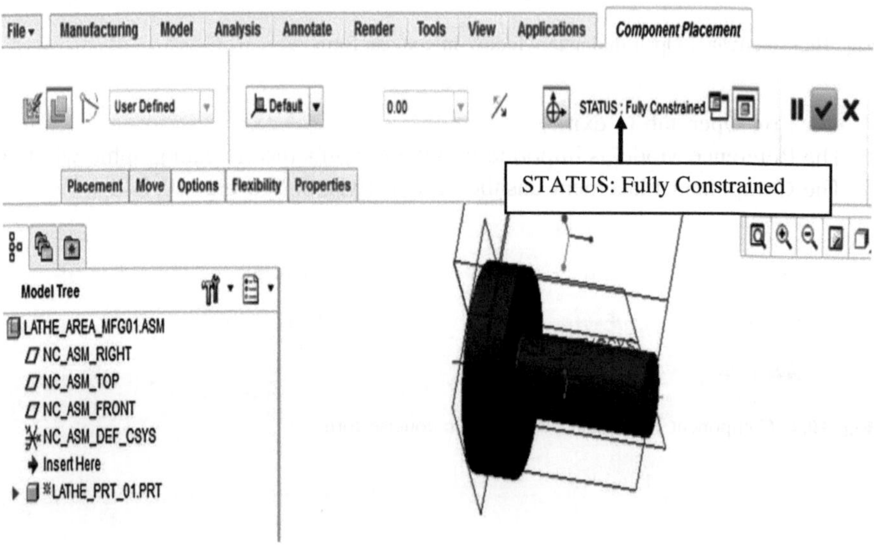

Fig. 10.6 Part is constrained on the main graphic window

Now click on the Check Mark icon.

10.2 Create Automatic Workpiece

Click on Workpiece icon ≫ Now select Automatic Workpiece on the drop-down menu list as shown below.

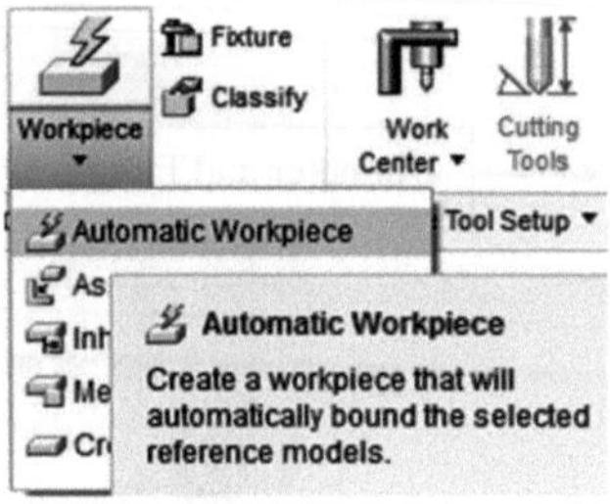

The Auto Workpiece Creation dashboard tools are activated. See Fig. 10.7.

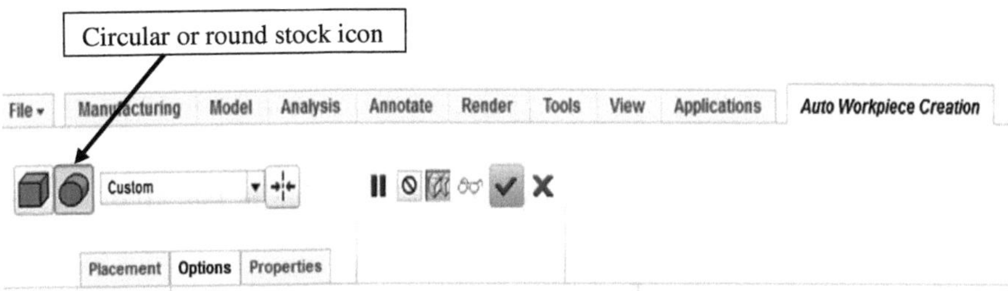

Fig. 10.7 Activated Auto Workpiece Creation dashboard

Click on the Round or Circular icon  in Fig. 10.7 to activate it ≫ Click on the Options tab to activate its panel ≫ Take note of the Units and values in the Overall Dimensions group ≫ On the Linear Offsets group, click on the Current Offsets tab and add values in the Diameter and Length section boxes ≫ On the Rotation Offset group, add value in the About Y section box as illustrated on the Option panel in Fig. 10.8.

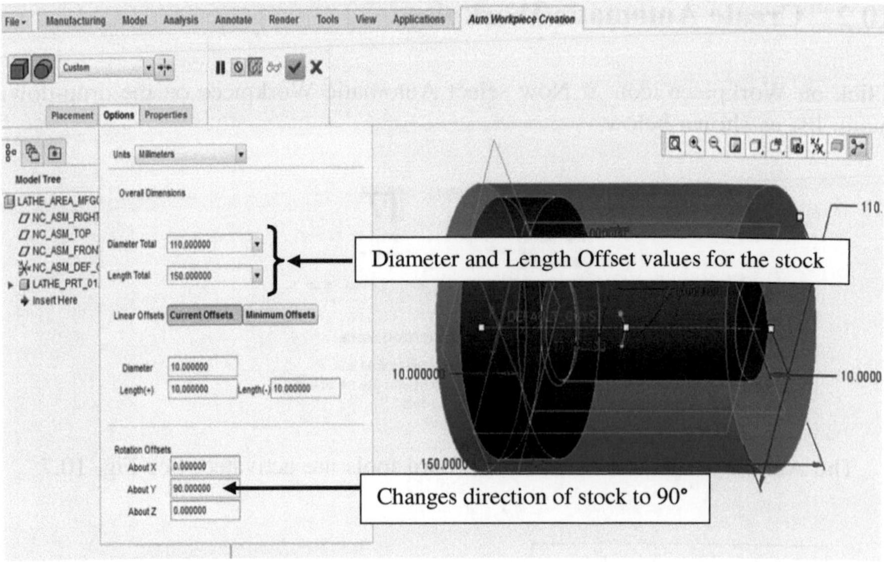

Fig. 10.8 Activated Options panel

Click on the Check Mark icon.
Stock is added automatically to the Workpiece as shown in Fig. 10.9.

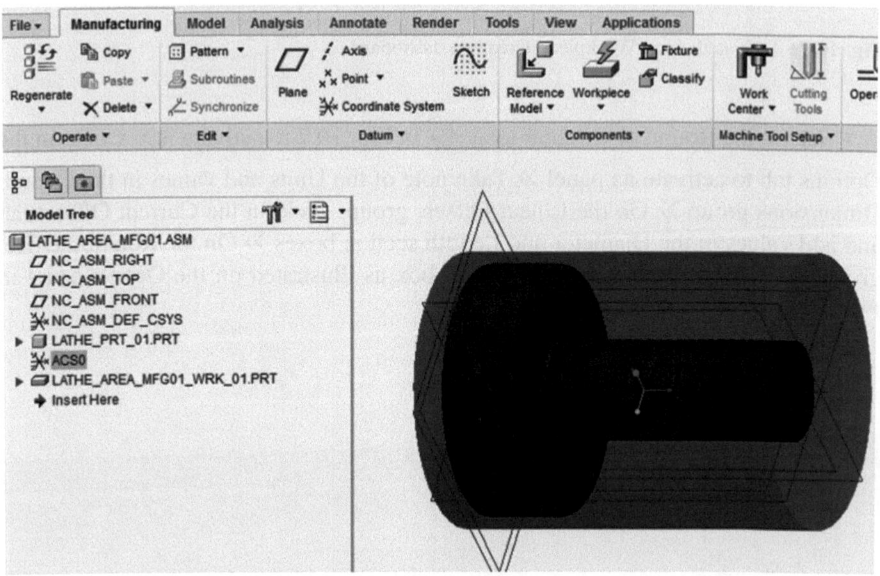

Fig. 10.9 Round Workpiece added to Part

10.3 Create Programme Zero (Coordinate System)

Click on the Coordinate System icon as shown below.

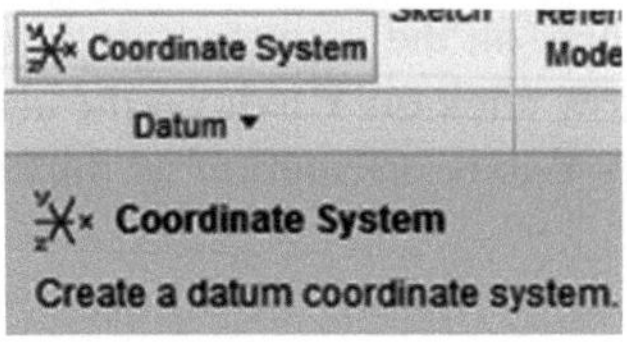

The Coordinate System dialogue box is activated on the main graphic window ≫ Now click on the Origin tab on the Coordinate System dialogue box ≫ Click on the Workpiece surface, NC_ASM_TOP and NC_ASM_FRONT while holding down the Ctrl key as indicated by the arrows in Fig. 10.10.

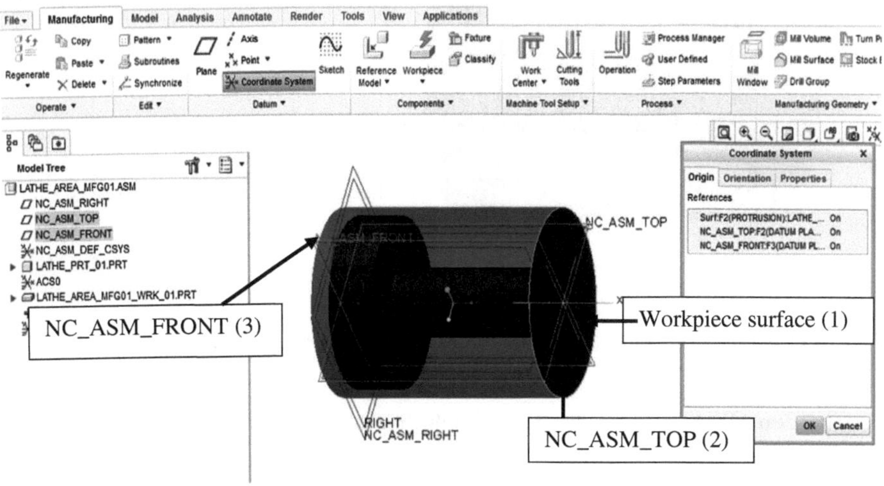

Fig. 10.10 Adding references for creating the Coordinate System

The Orientation of the created Coordinate System is not correct as shown in Fig. 10.11.

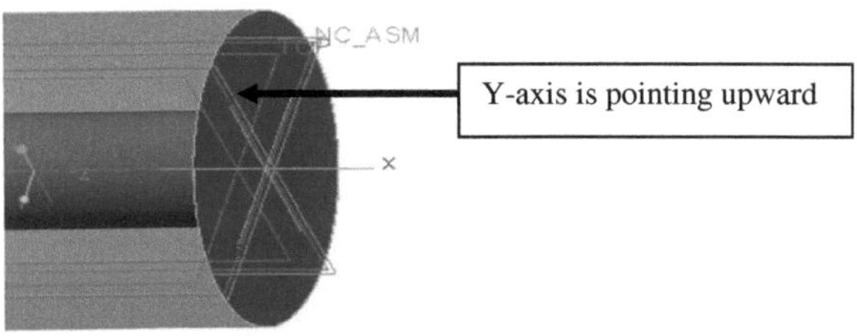

Fig. 10.11 Wrong orientations of *X*, *Y* and *Z* axes

To orient the X, Y and Z coordinate axes to the correct orientation

Click on the Orientation tab on the Coordinate System dialogue box to activate its content ≫ On the "Orient by" group, click on the radio button of "References selection" ≫ Click on the "to determine" section box to activate its drop-down menu list, now click on Z axis on the activated drop-down list ≫ Click on the "to project" section box, and now click on Y axis on the activated drop-down list. Now click on the Flip tab to flip Y axis orientation as illustrated in Fig. 10.12.

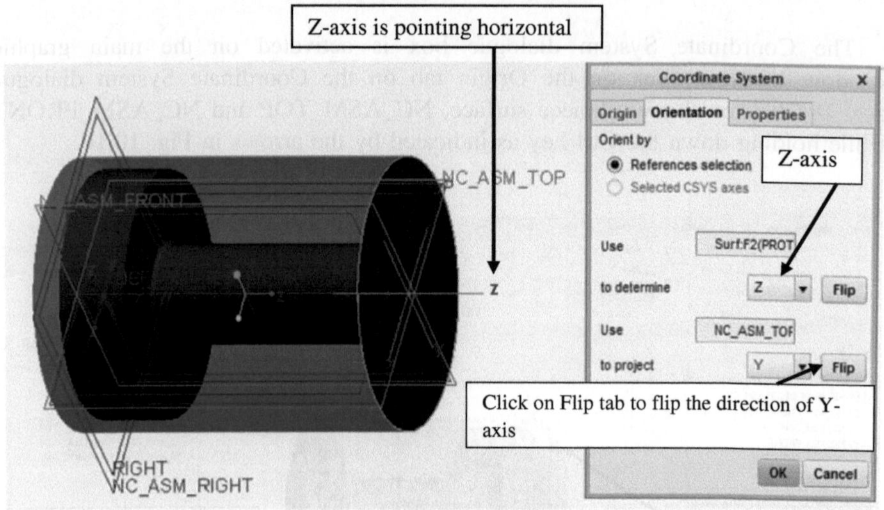

Fig. 10.12 Correct orientation of *X*, *Y* and *Z* axes of the Coordinate System

Click on OK tab to exit the Coordinate System dialogue box.

10.4 Create Work Centre

Click on the Work Centre icon to activate its content.

Click on Lathe on the activated drop-down menu list as shown below.

The Lathe Work Centre dialogue window is activated on the main graphic window as shown in Fig. 10.13.

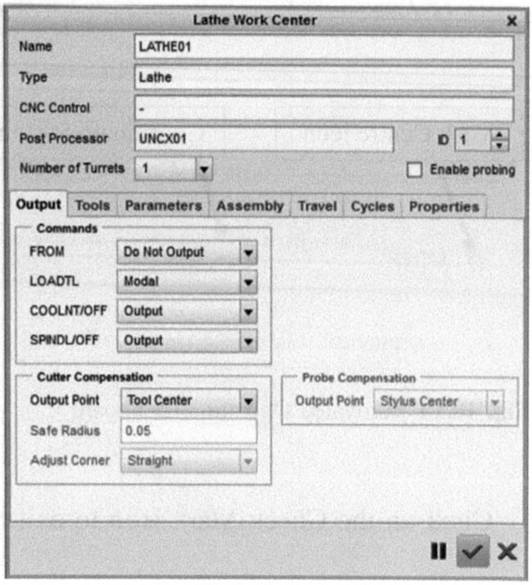

Fig. 10.13 Activated Lathe Work Centre dialogue window

Click on the Check Mark icon.

10.5 Create Operation

Click on the Operation icon as shown below.

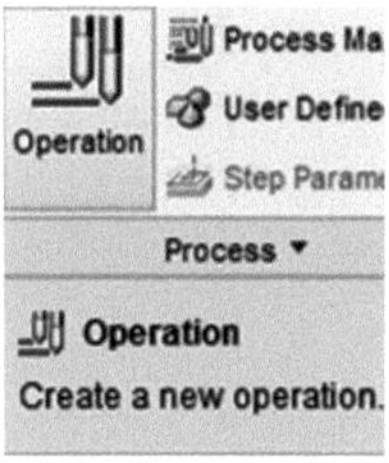

The Operation application tools are activated ≫ LATHE01 and ACS1 are automatically generated and added by the system as indicated by the arrows in Fig. 10.14.

Note: If LATHE01 and ACS1 are not automatically generated by the system in their respective section boxes as indicated by the *arrows*, manually select and input them yourself.

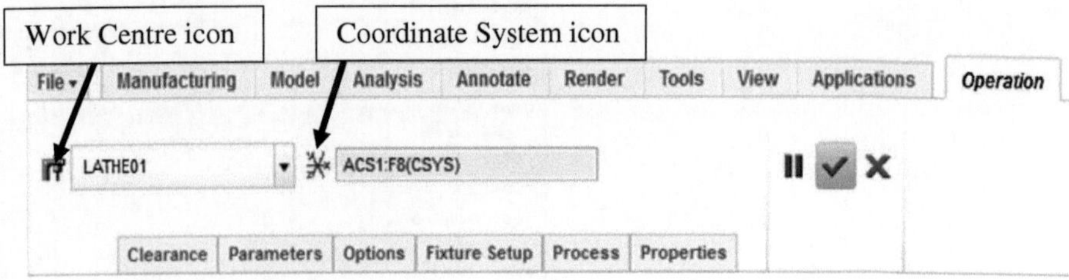

Fig. 10.14 Activated Operation dashboard

Click on the Check Mark icon to exit the Operation application.

10.6 Create Cutting Tool

Click on the Cutting Tools icon 

.

The Tools Setup dialogue window is activated on the main graphic window as shown in Fig. 10.15.

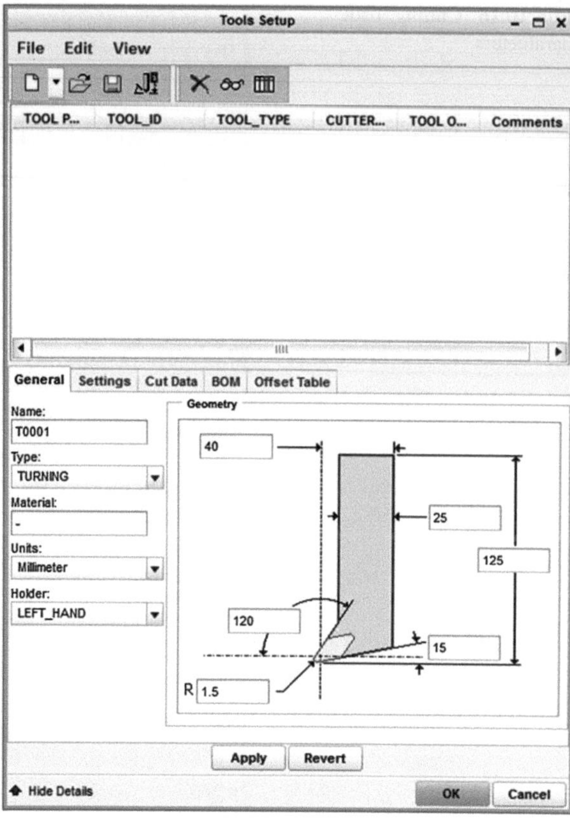

Fig. 10.15 Activated Tools Setup dialogue window

Add cutting Tool parameters

Click on the General tab on the Tools Setup dialogue window to activate its content ≫ In the Name section box, type T0001 ≫ Click on the downward pointing arrow on the Type section box, and click on TURNING on the activated drop-down menu list ≫ In the Material section box, type TUNSTEIN CARBIDE ≫ Check to make sure that the Unit is correct ≫ Click on the Holder section box downward pointing arrow, click on LEFT_ HAND on the menu list. Add dimensions to the TURNING Tool ≫ Now click on the Apply tab. The created TURNING Tool is shown in Fig. 10.16.

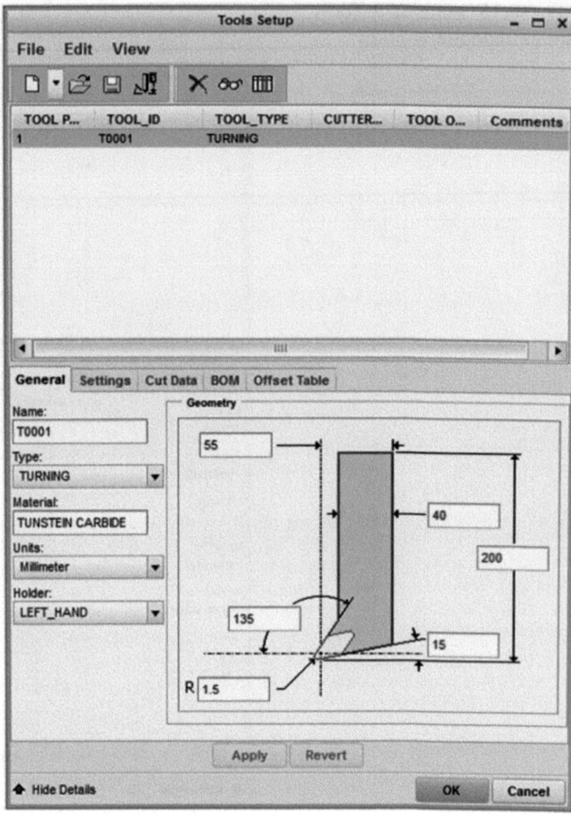

Fig. 10.16 Cutting Tool parameters

Click on the OK tab to exit.

10.7 Create Turn Profile Using Method 1

Click on the Turn menu bar to activate its content ≫ Now click on the Turn Profile icon on the Manufacturing Geometry group as indicated by the arrow as shown below.

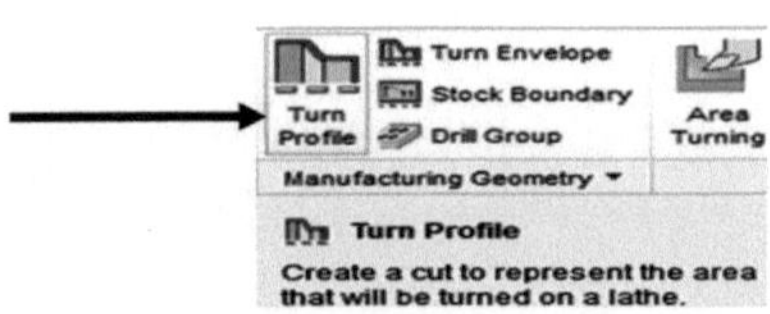

The Turn Profile application tools are activated as shown in Fig. 10.17.

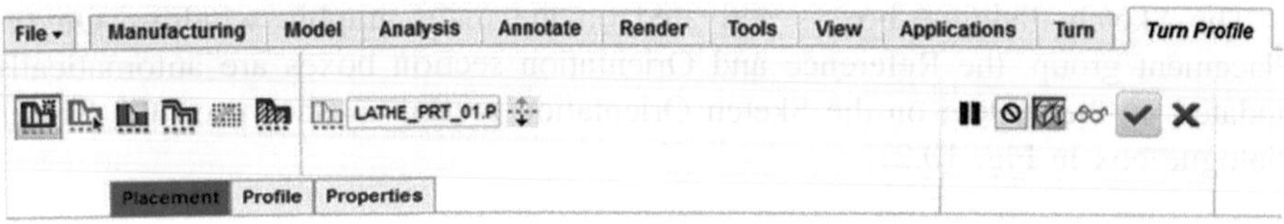

Fig. 10.17 Activated Turn Profile dashboard

Click on the Placement tab to activate its content ≫ Click on the Placement Csys section box, now go to the Model Tree and click on the ACS1 as the Coordinate System as indicated by the arrows in Fig. 10.18.

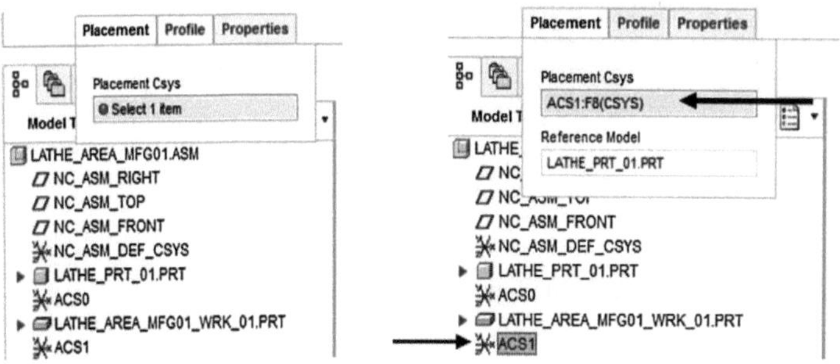

Fig. 10.18 Adding the Plament parameters

Click on the Use sketch to define turn profile icon in Fig. 10.17 as indicated by the arrow in Fig. 10.19.

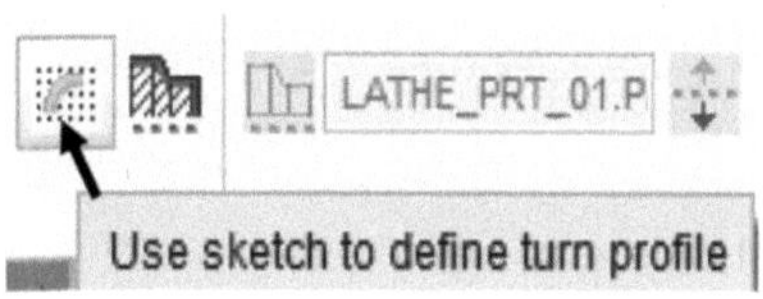

Fig. 10.19 Activating Turn Profile sketch

Click on the Define an Internal Sketch icon as indicated by the arrow in Fig. 10.20.

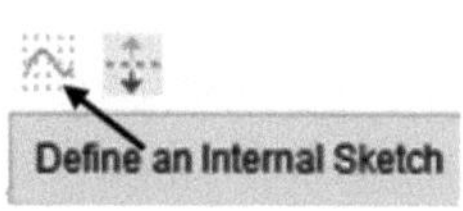

Fig. 10.20 Activating Define an Internal Sketch

The Sketch dialogue box is activated on the main graphic window. On the Placement group, the Reference and Orientation section boxes are automatically updated by the system on the Sketch Orientation section as shown on the Sketch dialogue box in Fig. 10.21.

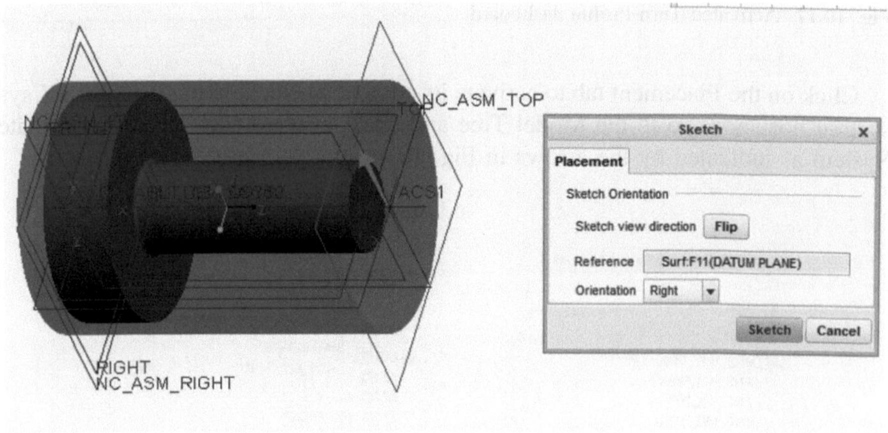

Fig. 10.21 Activated Sketch dialogue box

Click on the Sketch tab on the Sketch dialogue box.
The Sketch tools are activated as shown in Fig. 10.22.

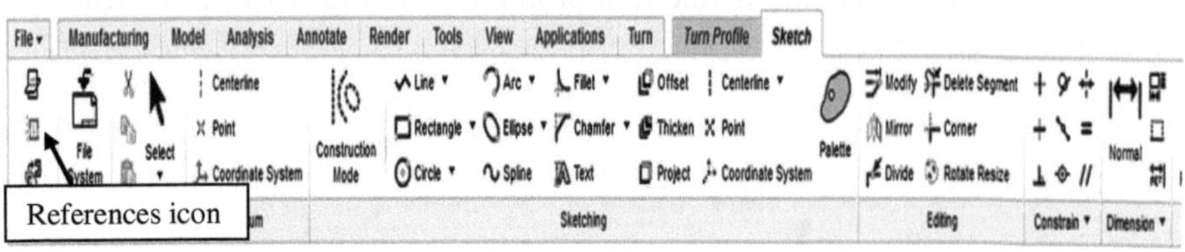

Fig. 10.22 Activated Sketch tools

Note: If the Reference dialogue box is not activated, activate it yourself by following the steps in Sect. 10.7.1.

10.7.1 Define References

Click on the References icon

The References dialogue box is activated on the main graphic window » Click on the Part as indicated by the arrows and the clicked references are automatically added into the reference section on the Reference dialogue box as illustrated in Fig. 10.23.

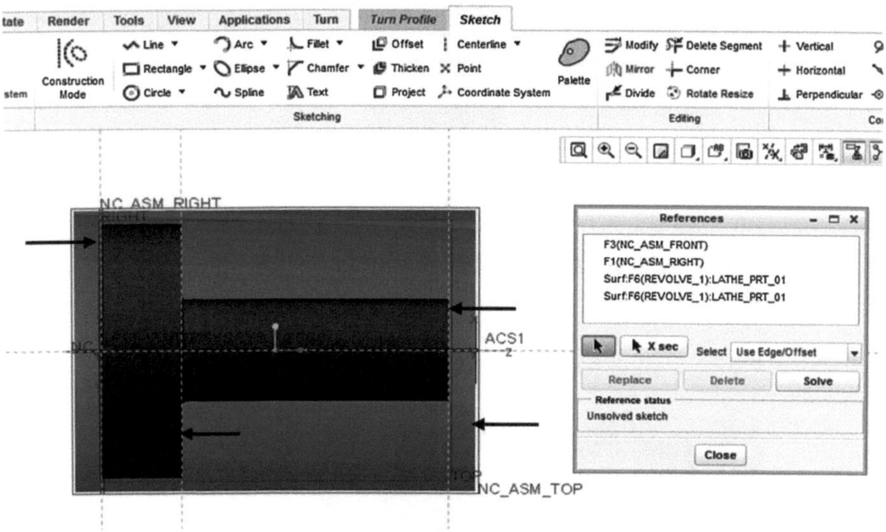

Fig. 10.23 Adding references for the Turn Profile sketch

Click on the Line tool icon Line in Fig. 10.23 on the Sketch tools.

Now draw a line that start from ACS1 to the edge of reference Part as indicated by the arrow in Fig. 10.24. Middle click the mouse button to exit line sketching.

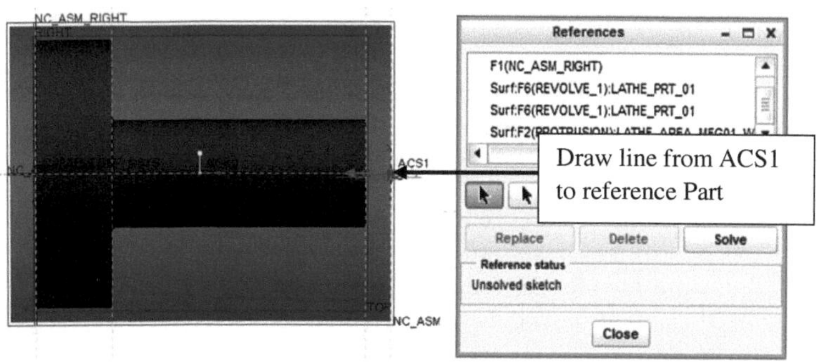

Fig. 10.24 Sketching start point for the cutting Tool

10.7.2 Project Lines and Curve

Click on the Project icon 

as shown in Fig. 10.23.

The Type dialogue box is activated on the main graphic window as shown in Fig. 10.25.

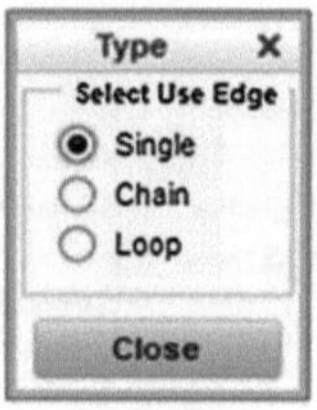

Fig. 10.25 Activated Type dialogue box

Click on Lines and Curves on the Part to be projected as indicated by the arrows as illustrated in Fig. 10.26.

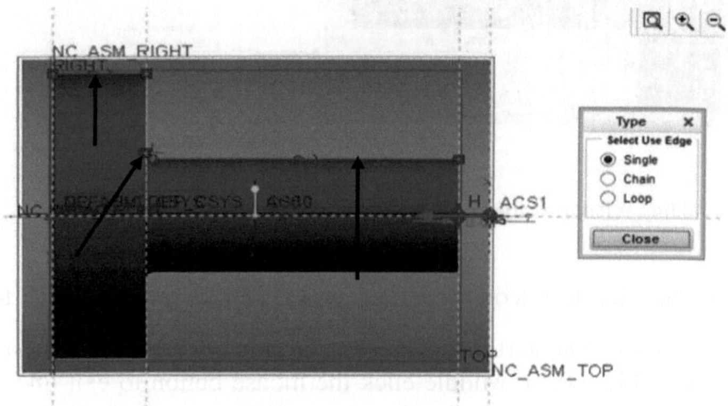

Fig. 10.26 Projecting Lines and Curve into the sketch plane

Click on the Line icon tool again >> Draw the Lines as indicated by the arrows below to join the projected Lines and Curve together as illustrated in Fig. 10.27.

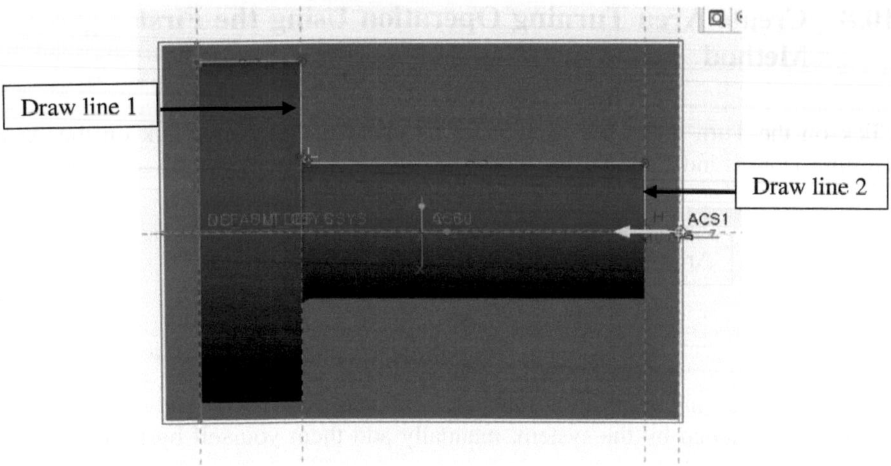

Fig. 10.27 Connecting the projected Lines and Curve

Click on the Check Mark icon to exit the Sketch application.

The Turn Profile application tools are activated again. As illustrated in Fig. 10.28, the cut lines will start from the right to left (Start to End). The Start and End direction can be reversed by clicking on either Start or End point.

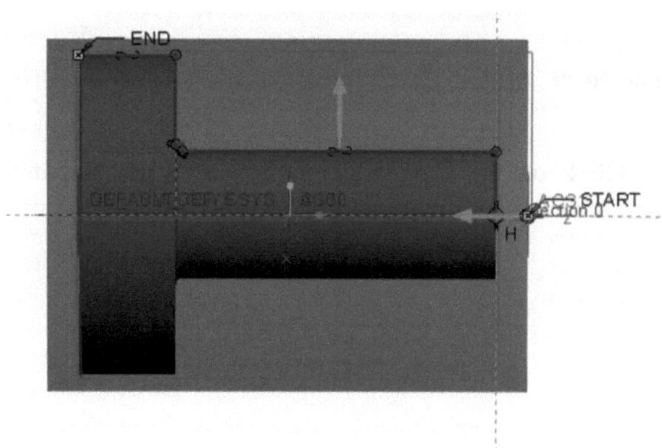

Fig. 10.28 Turn Profile lines and direction activated

Click on the Check Mark icon to exit the Turn Profile application.

10.8 Create Area Turning Operation Using the First Method

Click on the Turn menu bar to activate its menu list ≫ Now click on the Area Turning icon as indicated by the arrow below.

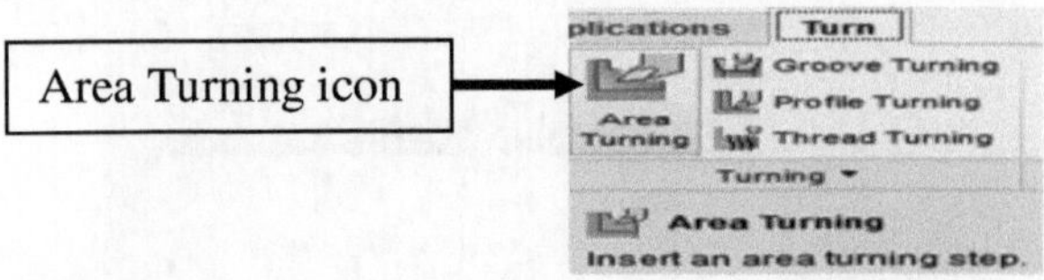

The Area Turning application tools are activated ≫ If T0001 and ACS1 are not automatically added by the system, manually add them yourself by following the step-by-step guide below.

Click on the Tool icon section box to activate its menu bar, now select T0001 on the activated drop-down list ≫ Click on the Coordinate System section box, and go to the Model Tree and click on ACS1. See Fig. 10.29.

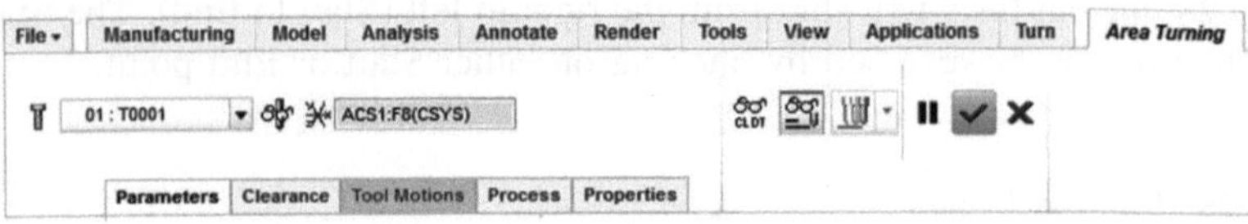

Fig. 10.29 Activated Area Turning dashboard indicating T0001 and ACS1

Click on the Parameters tab to activate its panel ≫ Now Input values to the parameters as shown in Fig. 10.30.

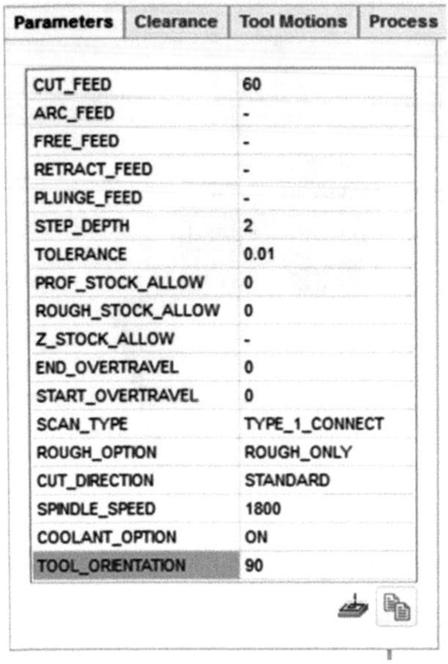

Parameters | Clearance | Tool Motions | Process

CUT_FEED	60
ARC_FEED	-
FREE_FEED	-
RETRACT_FEED	-
PLUNGE_FEED	-
STEP_DEPTH	2
TOLERANCE	0.01
PROF_STOCK_ALLOW	0
ROUGH_STOCK_ALLOW	0
Z_STOCK_ALLOW	-
END_OVERTRAVEL	0
START_OVERTRAVEL	0
SCAN_TYPE	TYPE_1_CONNECT
ROUGH_OPTION	ROUGH_ONLY
CUT_DIRECTION	STANDARD
SPINDLE_SPEED	1800
COOLANT_OPTION	ON
TOOL_ORIENTATION	90

Fig. 10.30 Activated Parameters panel

Click on the Check Mark icon to exit Area Turning application

10.8.1 Activate Play Path

Go to the Model Tree ≫ Right click on Area Turning 1(OP010) to activate the drop-down menu list ≫ Click on Play Path on the drop-down list.

The cutting Tool and PLAY PATH dialogue box are activated on the main graphic window as shown in Fig. 10.31.

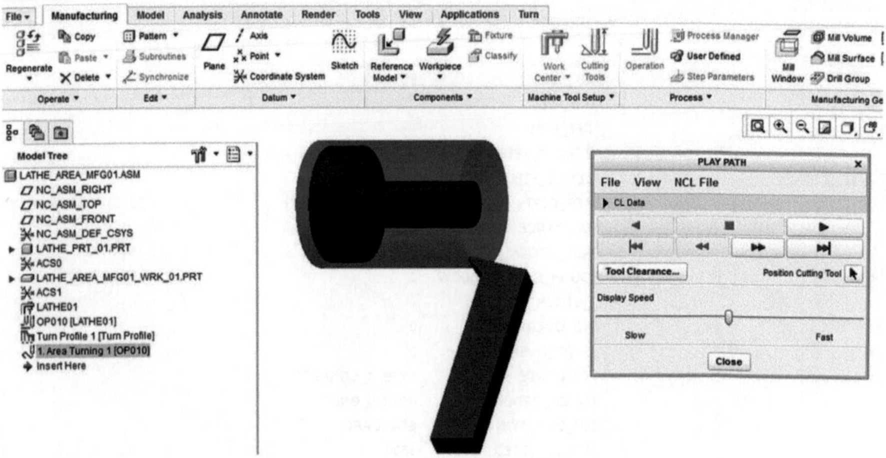

Fig. 10.31 Activated Turning Tool and PLAY PATH dialogue box

Click on the Play tab to start the on-screen Area Turning operation. The end of the Area Turning operation is shown in Fig. 10.32.

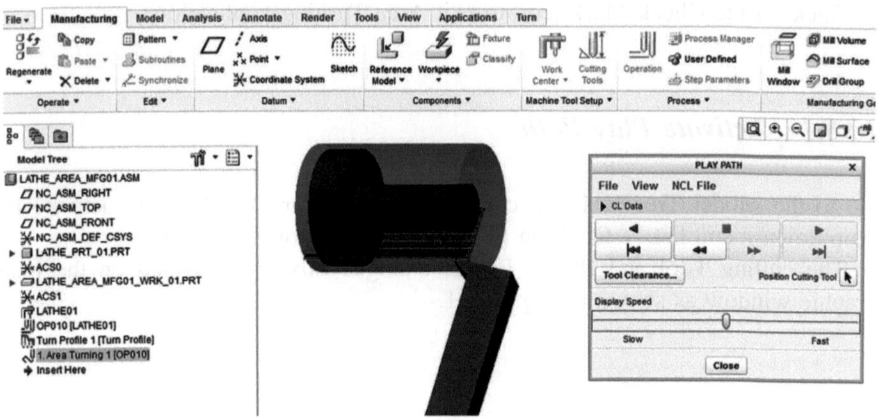

Fig. 10.32 End of on-screen Area Turning operation

Click on Close tab to exit.

10.9 Create Area Turning Operation Using the Second Method

Click on Turn menu bar to activate its content, and now click on Area Turning icon as indicated by the arrow below.

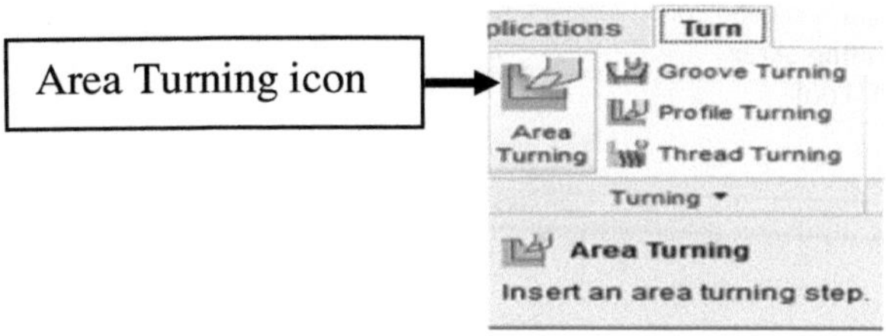

The Area Turning application tools are activated ≫ Input manually the created Cutting Tool and Coordinate System if they are not automatically generated by the system. Click on the Parameters bar to activate its panel ≫ Now add values to the Area Turning Parameters as shown in Fig. 10.33.

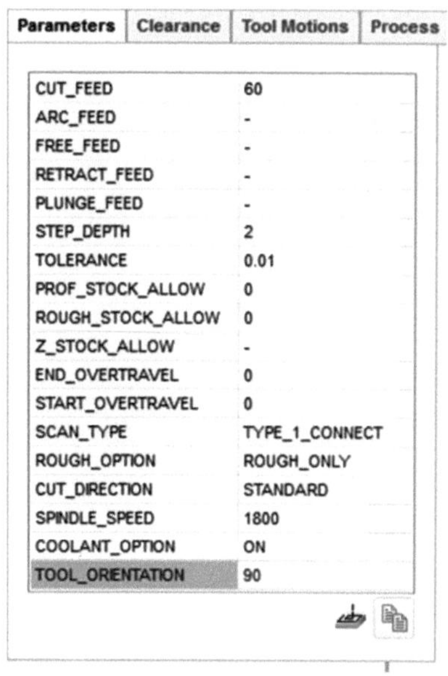

Fig. 10.33 Area Area Turning parameters

10.9.1 *Create Tool Motion for Area Turning*

Click on Tool Motions tab to activate its panel ≫ Click on Area Turning Cut tab to activate its drop-down menu list, and now click on Area Turning Cut on the activated drop-down menu list as illustrated in Fig. 10.34.

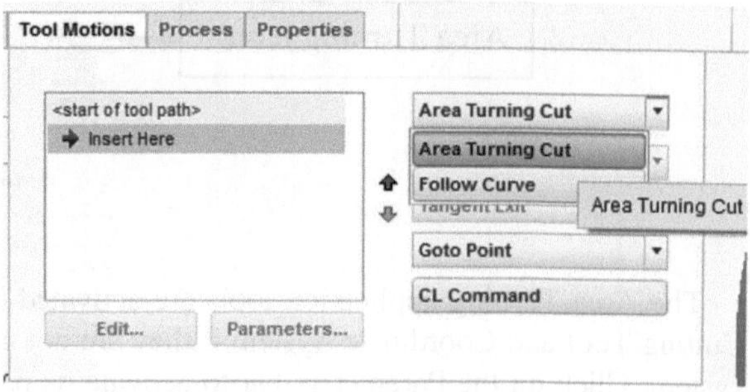

Fig. 10.34 Activating Area Turning Cut from Tool Motions panel

The Area Turning Cut dialogue box is activated on the main graphic window as shown in Fig. 10.35.

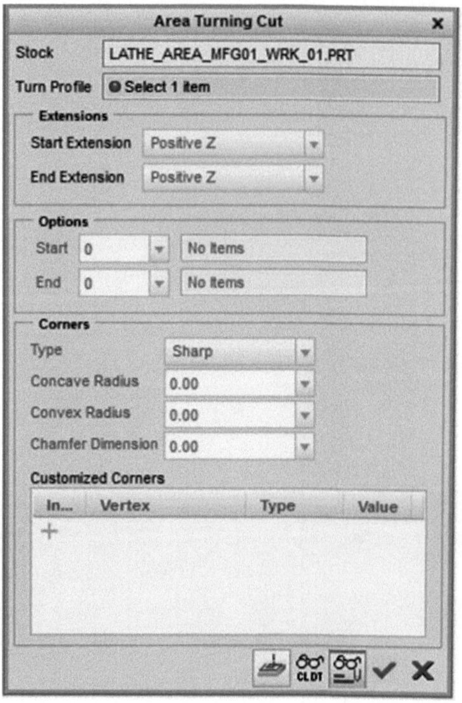

Fig. 10.35 Activated Area Turning dialogue box

Click on the Turn Profile section box on the Area Turning Cut dialogue box ≫ Now click on the created Turn Profile on the Model Tree or on the Workpiece as indicated by the arrows in Fig. 10.36.

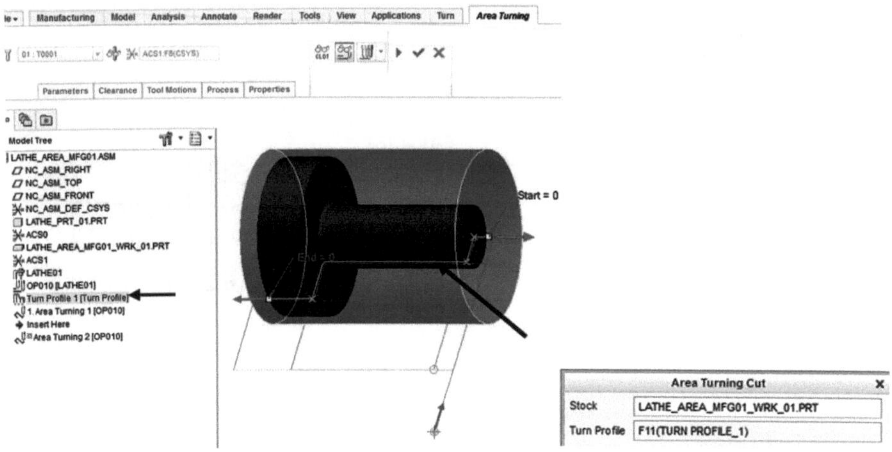

Fig. 10.36 Adding the created Turning Profile into the Area Turning Cut dialogue box

On the Extensions group, make sure that the Positive Z and X are selected on the Start and End Extension section boxes as shown in the Area Turning Cut dialogue window as indicated by the arrow in Fig. 10.37.

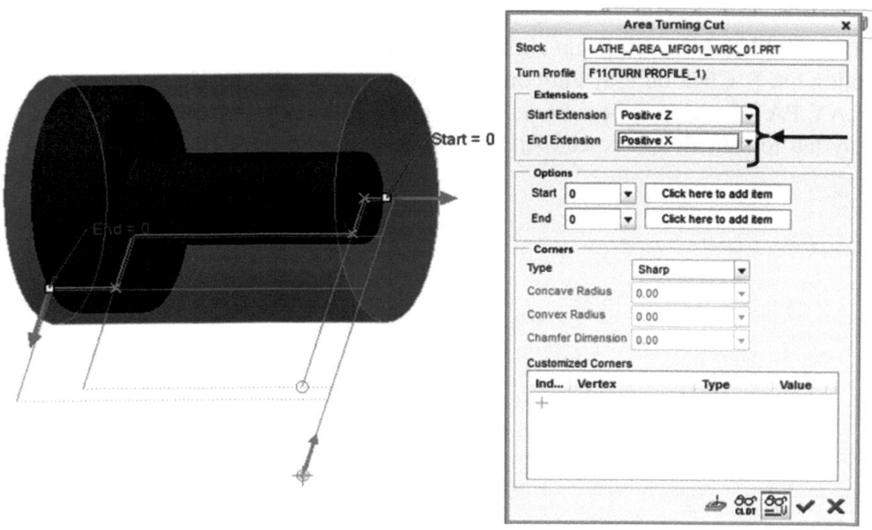

Fig. 10.37 Area Turning Cut parameters indicating positive Z and X

Click on the Check Mark icon to exit the Area Turning Cut dialogue box (Fig. 10.38).

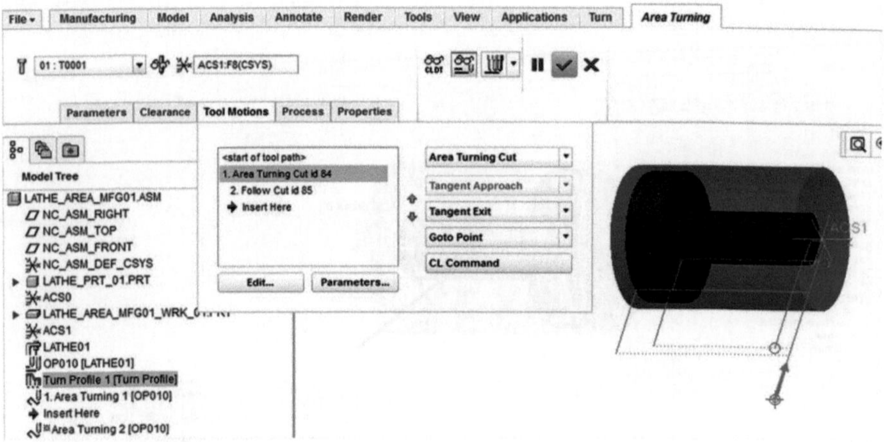

Fig. 10.38 Area Turning indicating the created Tool Motions parameters

Click on Check Mark icon to exit the Area Turning application.

10.9.2 Activate Play Path

Go to the Model Tree ≫ Right click on Area Turning 2(OP010) to activate its menu option list ≫ Click on Play Path on the activated option list ≫ The cutting Tool and PLAY PATH dialogue box are activated on main graphic window ≫ Click on the Play tab to start the on-screen Area Turning operation (Fig. 10.39).

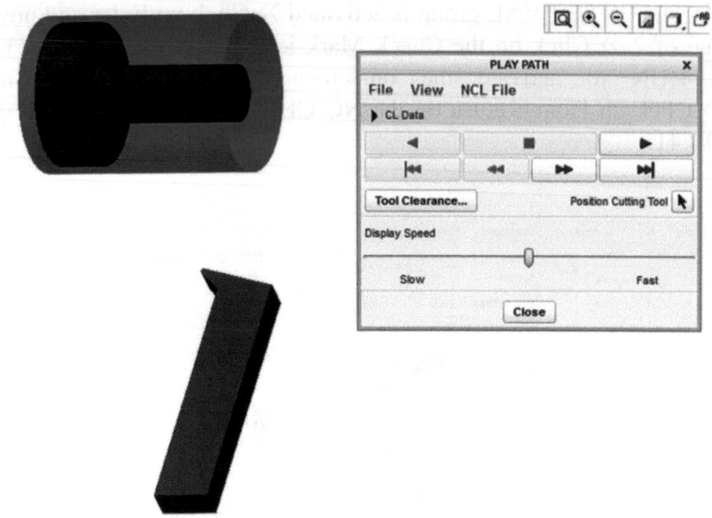

Fig. 10.39 Activated cutting Tool and PLAY PATH dialogue box

End of Area Turning operation is shown in Fig. 10.40.

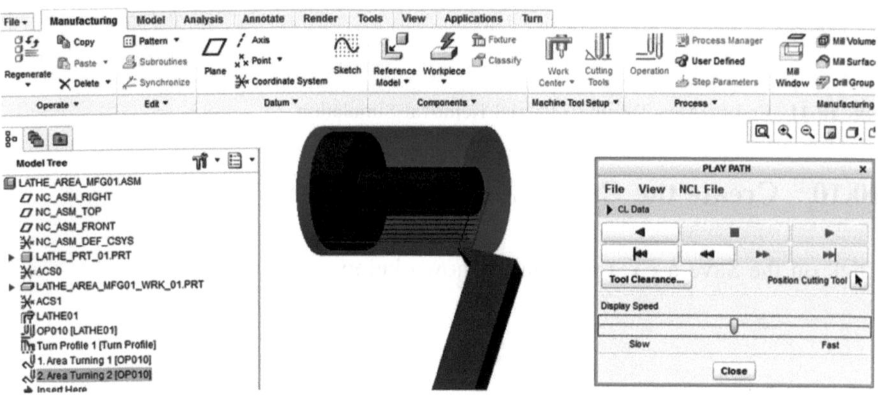

Fig. 10.40 End of Area Turning operation

10.9.3 Activate the Material Removal Simulation

To activate Material Removal Simulation process ≫ Right click on 'Area Turning 2 [OP010]' ≫ Now select Material Removal Simulation on the menu list ≫ The NC CHECK Menu Manager dialogue box is activated on the main graphic window ≫ Click on Step Size ≫ The STEP SIZE group is activated ≫ Now click on Enter ≫ The ENTER VAL group is activated ≫ Click on Enter and now enter a lower value of 2 ≫ Click on the Check Mark icon (Press the Enter key) ≫ Now click on RUN to activate the on-screen Material Removal Simulation process ≫ Click on Done/Return on the NC CHECK group to exit the operation. See Fig. 10.41.

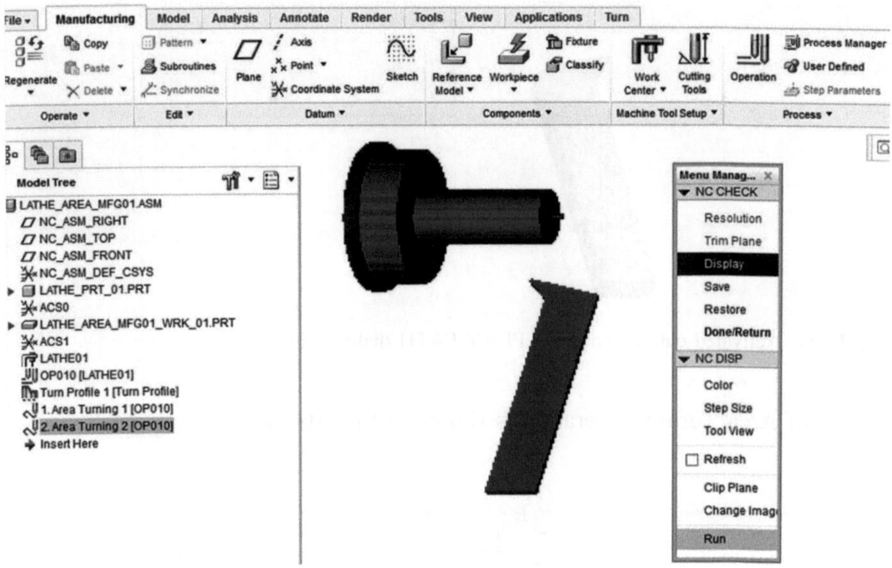

Fig. 10.41 End of Area Turning Material Removal Simulation

10.10 Create the Cuter Location (CL) Data

Click on the Save a CL File icon as shown below.

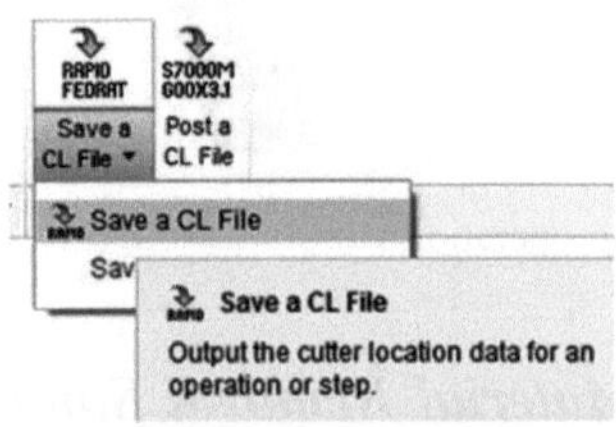

The SELECT FEAT Menu Manager dialogue box is activated on the main graphic window ≫ Click on Operation, and now click on OP010 on the SEL MENU group as highlighted on the Menu Manager dialogue box shown in Fig. 10.42.

Fig. 10.42 Activated SELECT FEAT Menu Manager dialogue box

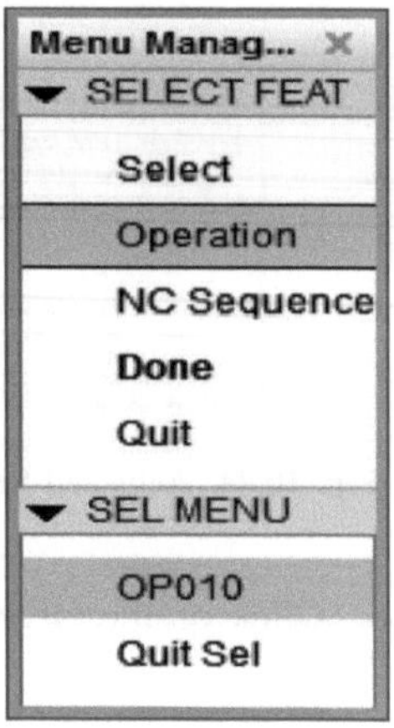

Once OP010 is clicked, the PATH Menu Manager dialogue box is activated on the main graphic window ≫ Click on File on the PATH group and Check Mark CL File and Interactive square boxes on the OUTPUT TYPE group as shown in Fig. 10.43.

Fig. 10.43 Activated PATH Menu Manager dialogue box

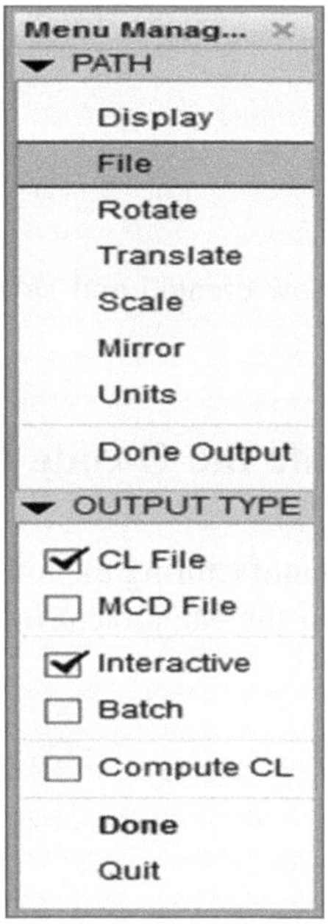

Click on Done.

The Save a Copy dialogue window is activated on the main graphic window. Make sure that the correct Part name is on the New Name section box. See Fig. 10.44.

Fig. 10.44 Activated Save a Copy dialogue window in concise form

Click on the OK tab to save the CL File data in your already chosen directory folder.

Click on Done Output on the PATH group on the Menu Manager dialogue box as indicated by the arrow in Fig. 10.45.

Fig. 10.45 Activating Done Output on the PATH group

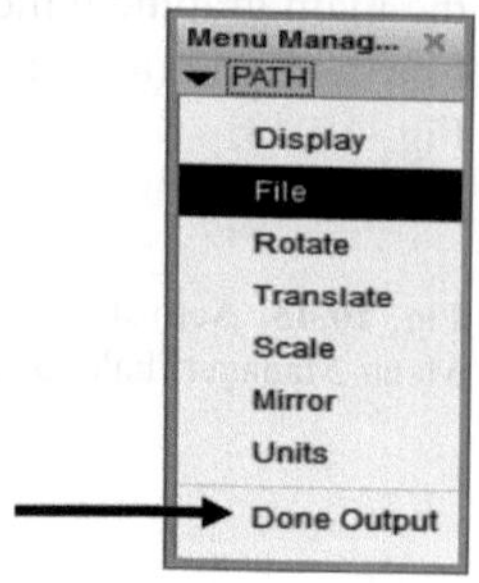

CL File is now created and saved in the chosen directory.

10.11 Create the G-code Data

Click on the Manufacturing menu bar to activate its menu list ≫ Click on the Post a CL File icon on the Manufacturing menu list as shown below.

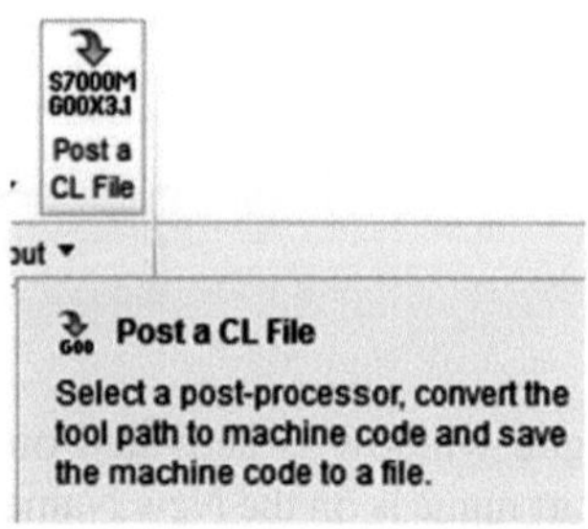

The Open dialogue window is activated on the main graphic window with the correct CL File name. See Fig. 10.46.

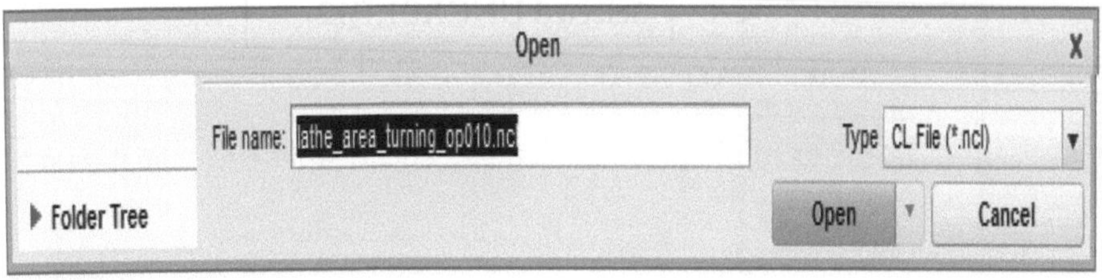

Fig. 10.46 Activating Open dialogue window

Click on the Open tab to exit.

The PP OPTIONS Menu Manager dialogue box is activated on the main graphic window ≫ Check Mark the square boxes of Verbose and Trace if they are not Checked Marked automatically by the system as shown in Fig. 10.47.

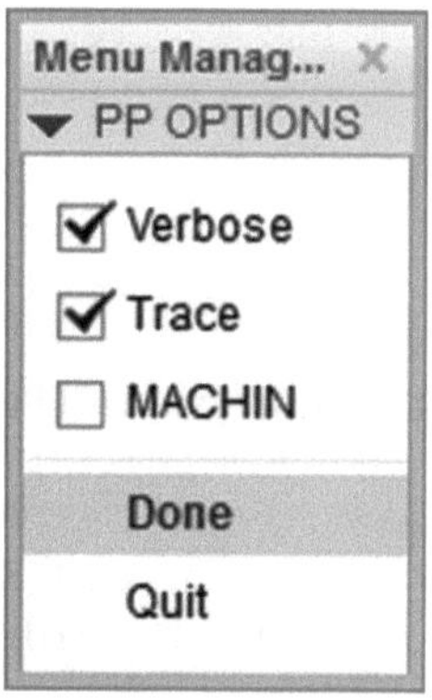

Fig. 10.47 Activated PP OPTIONS Menu Manager dialogue box

Click on Done.

The PP LIST Menu Manager is automatically generated by the system as shown in Fig. 10.48.

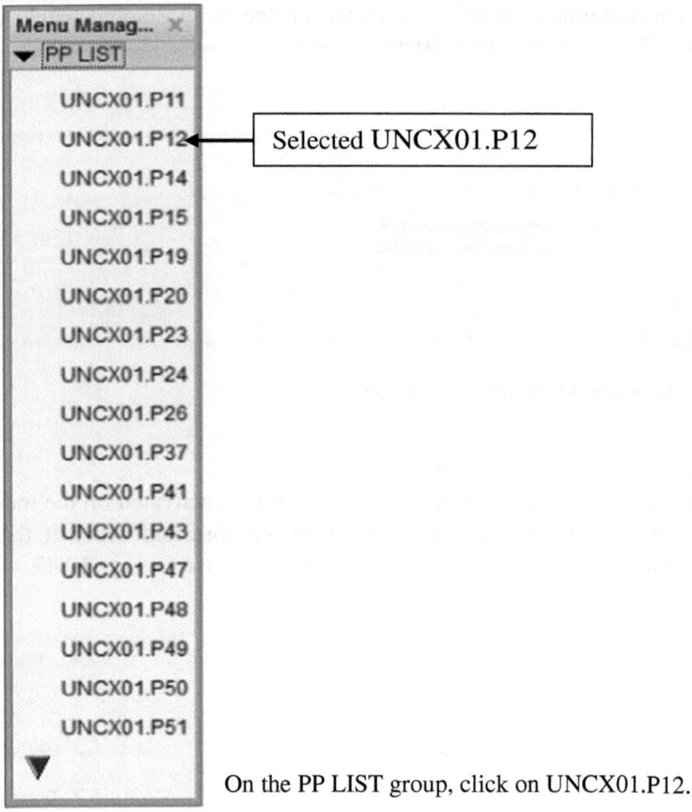

Fig. 10.48 Activated PP LIST Menu Manager dialogue box

Note: The UNCX01.P12 is the post process code for NIIGATAHN50A-FANUC-B TABLE Lathe machine.

The INFORMATION WINDOW dialogue window is activated on the main graphic window as shown in concise form in Fig. 10.49.

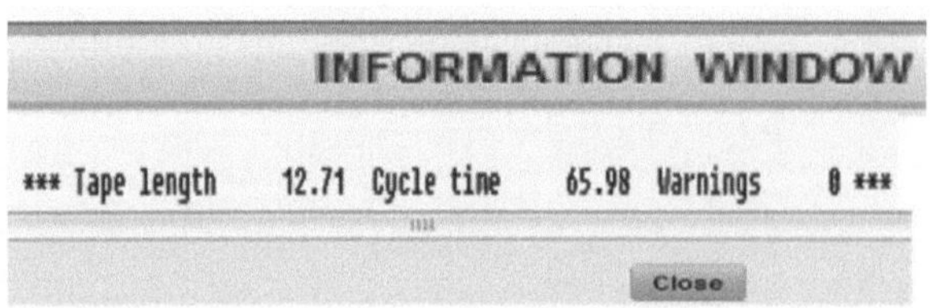

Fig. 10.49 Activated INFORMATION WINDOW in concise form

The generated G-code data is stored as a TAP file in the chosen work directory.

Now go to where the TAP file was saved and Open the TAP file using Notepad option to view the generated G-codes.

Chapter 11
Area Lathe Turning, Drilling, Boring and Volume Milling

The following steps will be used as a guide to achieve the set objectives in this chapter.

- Activate the New Manufacturing application
- Import the 3D Part into the main Manufacturing Graphic window
- Constrain the imported 3D Part into the main Manufacturing Graphic window
- Add Automatic Workpiece (Stock)
- Create Two Programme Zeroes (Coordinate Systems)
- Add Work Centre
- Create Operation
- Create Cutting Tool
- Create Area Turning using Turn Profile Method
- Create Point that will be used as Start and End points for Tool
- Drilling operation
- Boring operation
- Create Volume Rough using Mill Volume operation
- Generate Cutter Location (CL) Data
- Generate the Post Process data

11.1 Start the Manufacturing Application

Start Creo Parametric either on your Computer desktop or from Programme.

Once Creo Parametric is activated ≫ Click on the New icon as shown below.

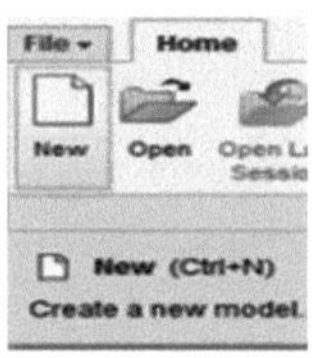

The New dialogue box is activated on the main window ≫ On the Type group, click on the Manufacturing radio button ≫ On the Sub-type group, click on NC Assembly radio button ≫ In the Name section box type

P.O. Kanife, *Computer Aided Virtual Manufacturing Using Creo Parametric*,
DOI 10.1007/978-3-319-23359-8_11

"Turn_Drill_Boring_Mill_V_mfg01". Click on the Use default template square box to clear the Check Mark icon. The New dialogue box changes as shown in Fig. 11.1.

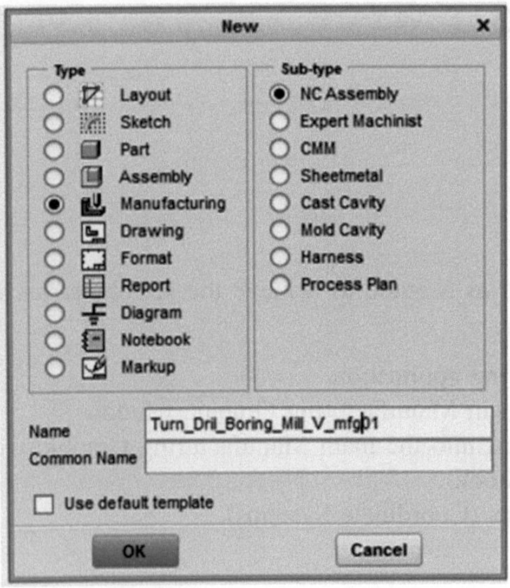

Click on OK tab to proceed.

Fig. 11.1 Activating Manufacturing application on the New dialogue box

The New File Options dialogue box is activated on the main window ≫ Click on "mmns_mfg_nc" on the Template group ≫ On the Parameters group, type your name in the MODELLED BY section box and in the DESCRIPTION section box, type "Area Turning_Drill_Boring_Volume" as shown in Fig. 11.2.

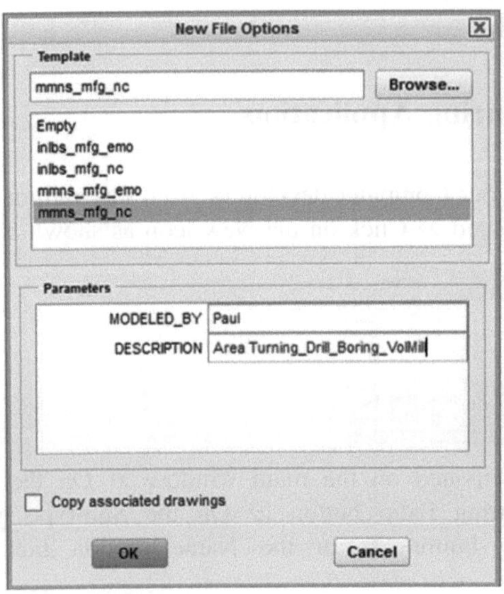

Click on OK tab to proceed

Fig. 11.2 New File Options dialogue box

Creo Parametric Manufacturing application main graphic window is activated.

11.1.1 Import Reference Part

Click on the Reference Model icon tab to activate the drop menu-down list ≫ On the drop-down menu list, click on Assemble Reference Model as shown below.

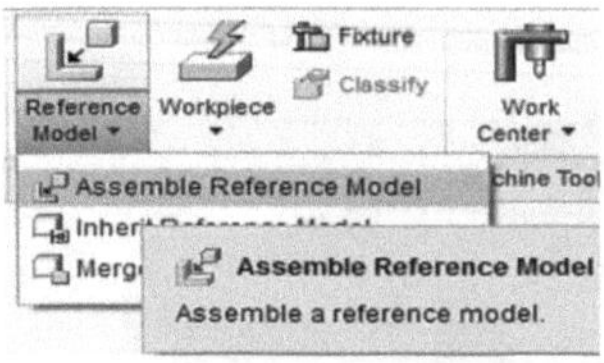

The Open dialogue window is activated on the main graphic window. Make sure that the name in the File name section box is the correct Part name that you want to import as shown in Fig. 11.3.

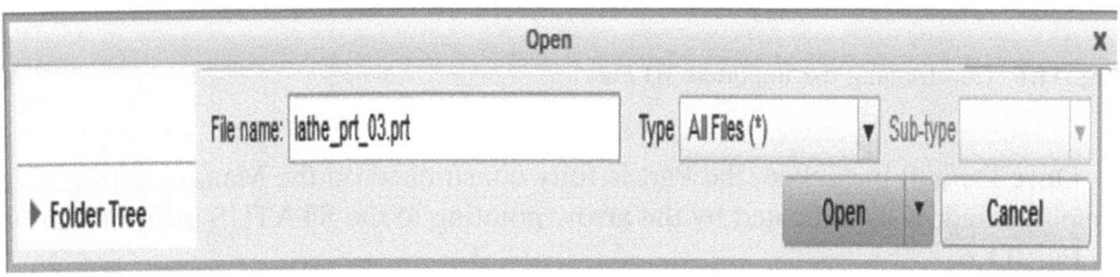

Fig. 11.3 Activated Open dialogue window in concise form

Click on Open tab to proceed.
The Reference Model is imported into the Manufacturing main graphic window.
The Component Placement dashboard tools are activated. See Fig. 11.4.

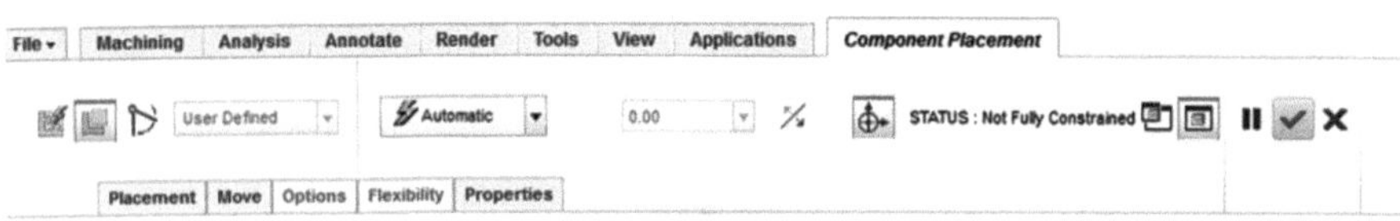

Fig. 11.4 Activated Component Placement dashboard

11.1.2 Constrain the Reference Model

Once the Component Placement ribbon toolbar is activated ≫ Click on the Automatic section box downward pointing arrow and on the drop-down list, click on Default as highlighted in Fig. 11.5.

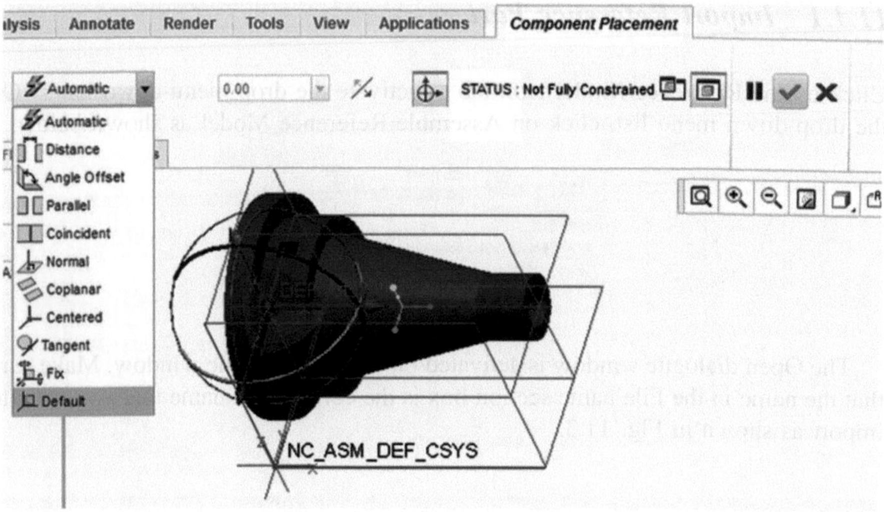

Fig. 11.5 Constraining the imported 3D Part

Once Default is clicked, the Part is fully constrained on the Manufacturing main graphic window as indicated by the arrow pointing to the STATUS information bar in Fig. 11.6.

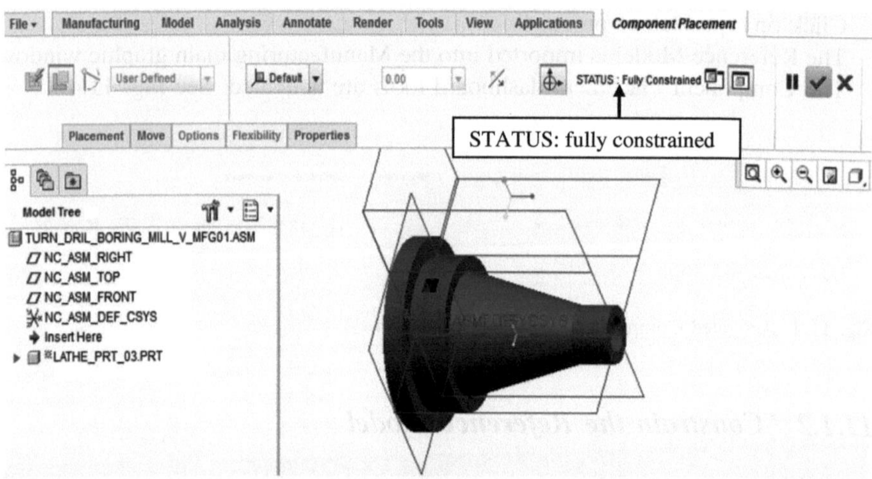

Fig. 11.6 Imported Part is fully constrained

Click on the Check Mark icon to proceed.

11.1.3 *Create Automatic Workpiece*

To create automatic Workpiece, click on the Workpiece icon tab and on the drop-down list, click on Automatic Workpiece as shown below.

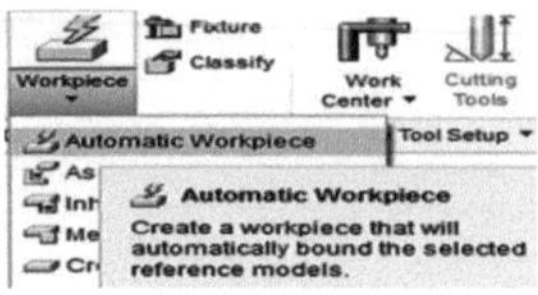

The Auto Workpiece Creation dashboard is activated ≫ Click on the Round or Circular icon  tab to make it active as shown in Fig. 11.7.

Fig. 11.7 Activated Auto Workpiece Creation ribbon toolbar

Click on the Options tab to activate its panel ≫ Take note of the Units, values in the Overall Dimensions group ≫ On the Linear Offsets group, click on the Current Offsets tab, add values in the Diameter and Length section boxes ≫ On the Rotation Offset group, add value (90°) in the About Y section box as illustrated below on the Option panel as shown in Fig. 11.8.

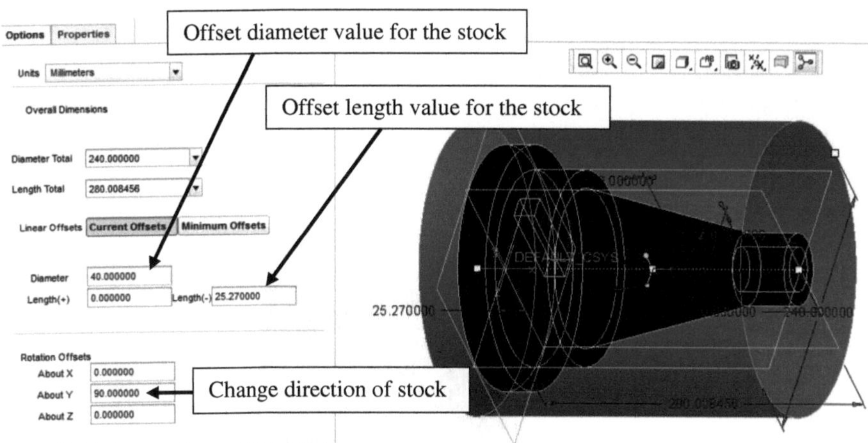

Fig. 11.8 Activated Options Panel showing Workpiece dimensions

Click on the Check Mark icon.

Stock is added to the Workpiece as shown in Fig. 11.9.

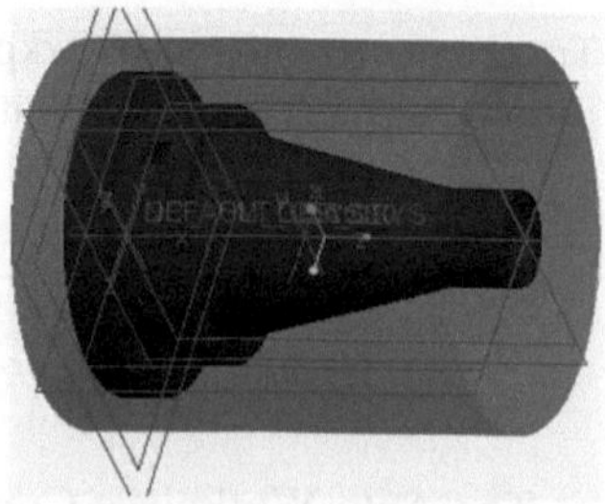

Fig. 11.9 Round Workpiece added to 3D Part

11.1.4 Create Programme Zero (Coordinate System)

Click on the Coordinate System icon tab as shown below.

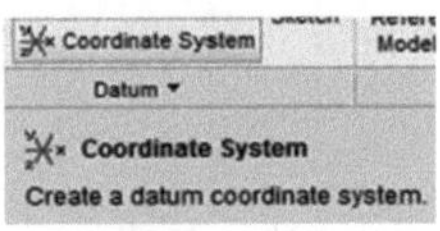

The Coordinate System dialogue box is activated on the main graphic window ≫ Now click on the Origin tab on the Coordinate System dialogue box ≫ Click on the Workpiece surface, NC_ASM_TOP and NC_ASM_FRONT datum planes while holding down the Ctrl key as indicated by the arrows in Fig. 11.10.

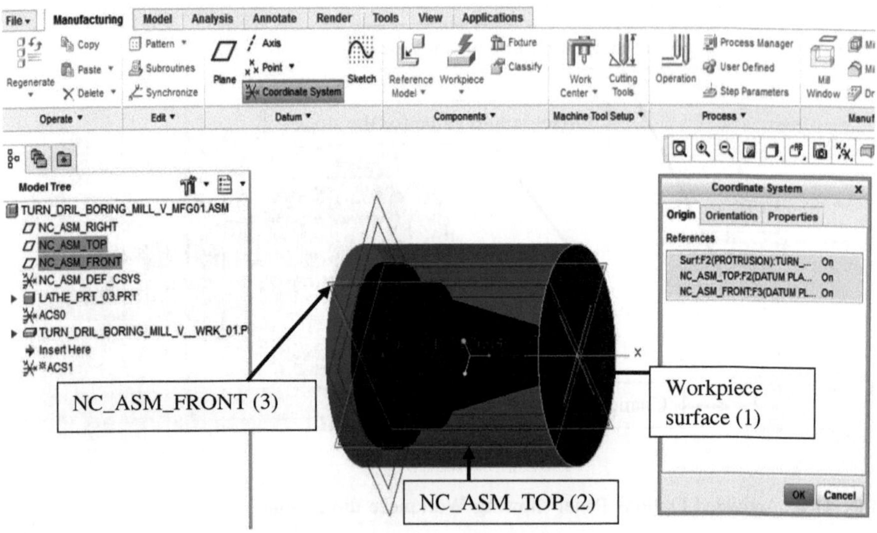

Fig. 11.10 Adding references for the Coordinate System creation

The orientation of the created Coordinate System (*X*, *Y* and Zaxes) is not correct as illustrated in Fig. 11.11.

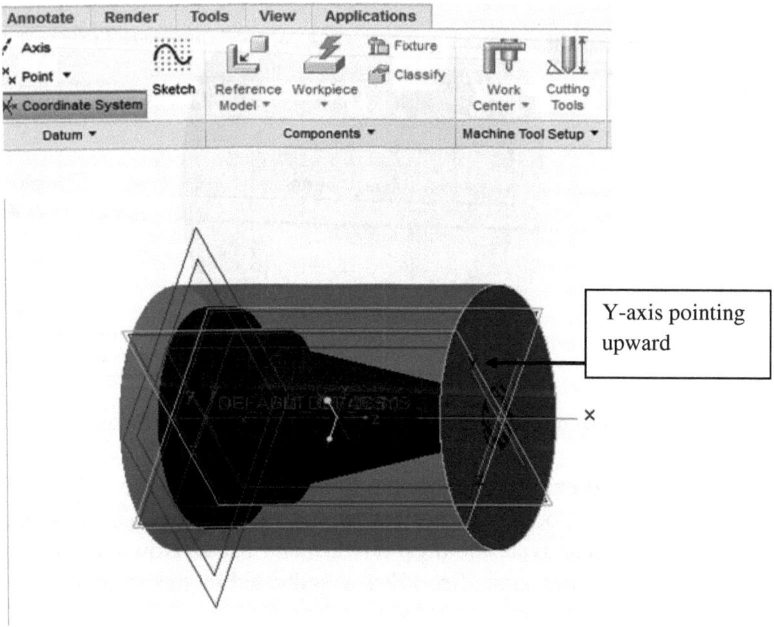

Fig. 11.11 Wrong orientations of the *X*, *Y* and *Z* axes of the Coordinate System

To orient the X, Y and Z coordinate axes to the correct orientation

Click on the Orientation tab on the Coordinate System dialogue box to activate its content ≫ On the "Orient by" group, click on the radio button of "References selection" ≫ Click on the downward arrow on the "to determine" section box to activate its drop-down menu list, now click on *Z* axis on the drop-down list ≫ Click on the downward arrow on the "to project" section box, now click on *Y*-axis on the activated drop-down list. Now click on the Flip tab, to flip *Y*-axis orientation as illustrated in Fig. 11.12.

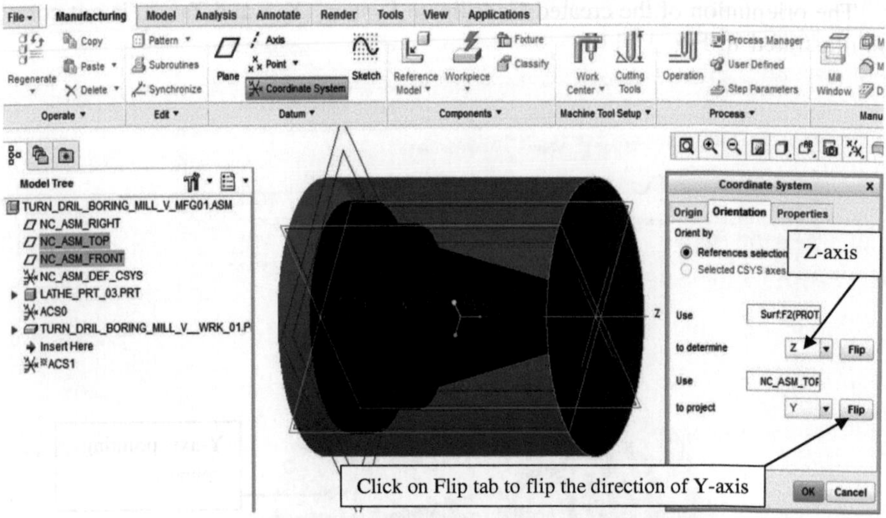

Fig. 11.12 Correct orientation of the X, Y and Z axes of the Coordinate System

Click on OK tab to exit the Coordinate System dialogue box.

Let change the new Coordinate System name to PZ01 by pause click on ACS1 and then right click to activate the drop-down menu list ≫ Now click on Rename and Type PZ01 as the new name. See PZ01 as indicated by the arrow in Fig. 11.13.

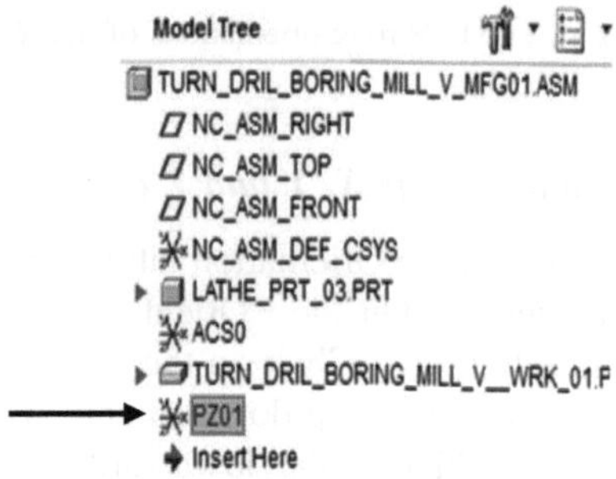

Fig. 11.13 ACS1 renamed as PZ01 on the Model Tree

11.1.5 Create Work Centre

Click on the Work Centre icon to activate its drop-down menu list ≫ Now select Lathe on the drop-down list as illustrated below.

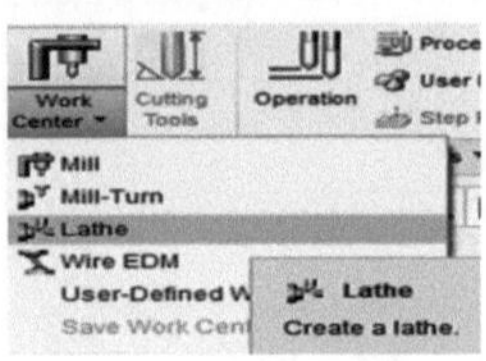

The Lathe Work Centre dialogue window is activated as shown in Fig. 11.14.

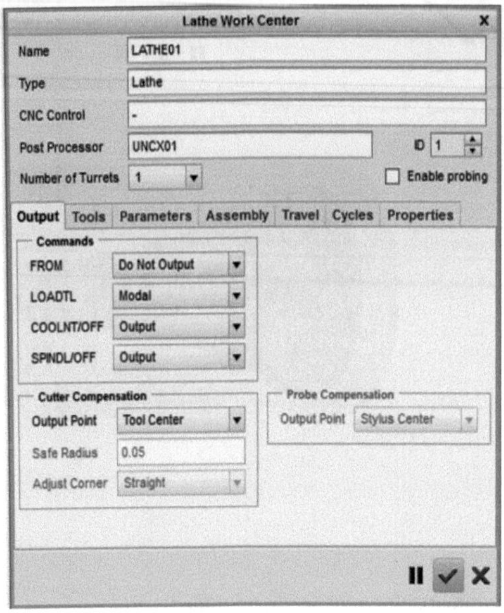

Click on Check Mark icon.

Fig. 11.14 Activated Lathe Work Centre dialogue window

11.1.6 Create Operation

Click on the Operation icon tab as shown below.

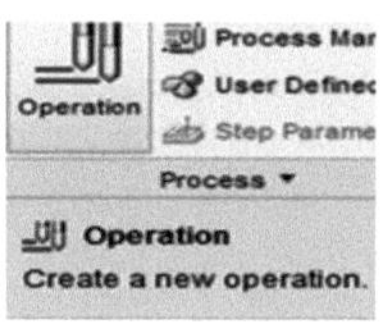

The Operation ribbon dashboard is activated. LATHE01 and PZ01 are automatically generated by the system as indicated by the arrow in Fig. 11.15.

Note: If LATHE01 and PZ01 are not automatically generated by the system in their respective section boxes as indicated by the arrows, manually select and input them yourself.

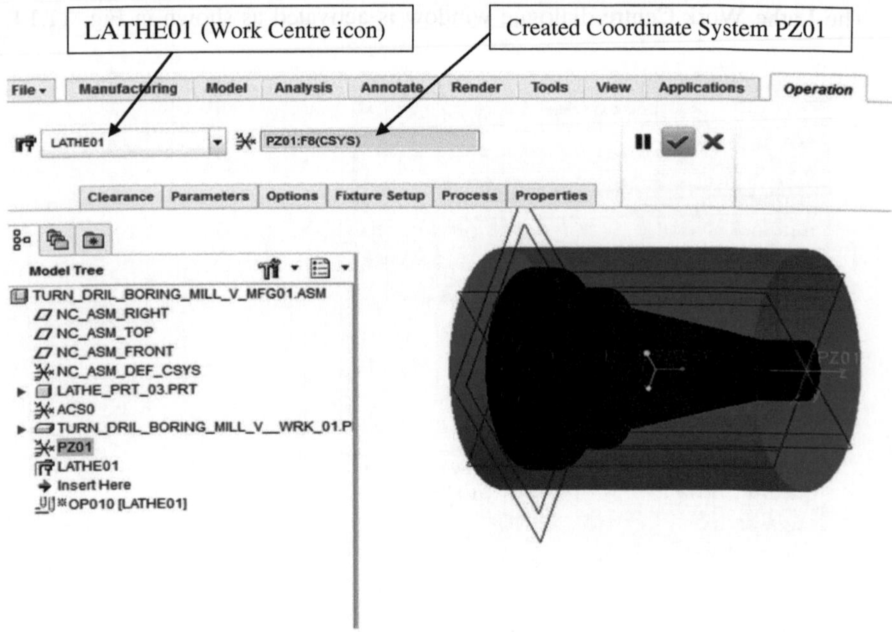

Fig. 11.15 LATHE01 and PZ01 are automatically generated and added

Create Retract Plane

Click on the Clearance tab to activate its panel ≫ On the Retract group, click on the Type section box and click on Plane on the drop-down list ≫ On the Reference section box, click on the Workpiece surface as indicated by the arrow in Fig. 11.17 ≫ In the Value section box, type 10 mm as the Retract distance as shown in Fig. 11.16.

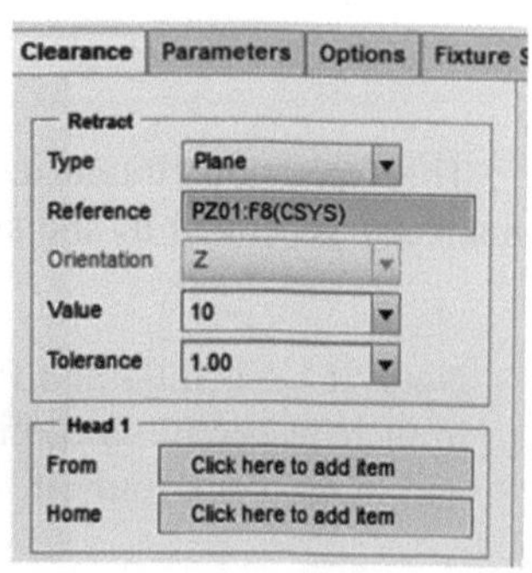

Fig. 11.16 Clearance panel with Retract parameters

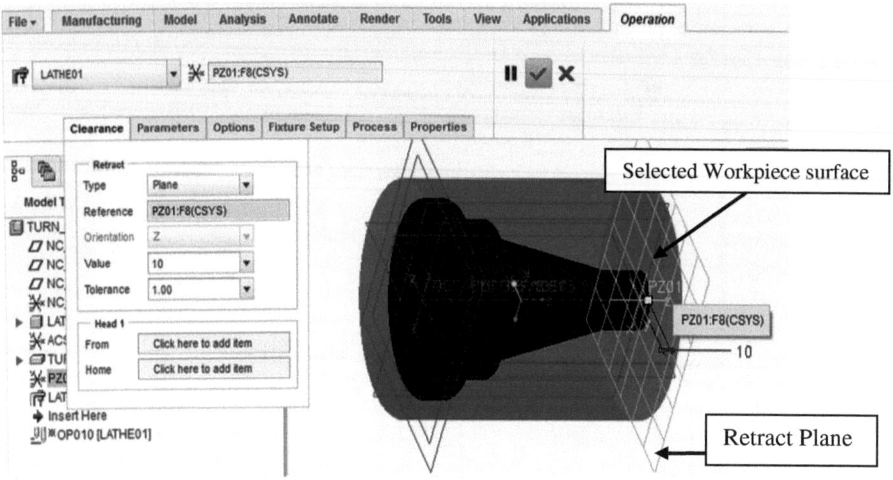

Fig. 11.17 Activating Clearance panel and Retract plane parameter

Click on the Check Mark icon to exit Operation.

11.1.7 Create Cutting Tools

Click on the Cutting Tools icon as shown below.

The Tools Setup dialogue window is activated on the main window as shown in Fig. 11.18.

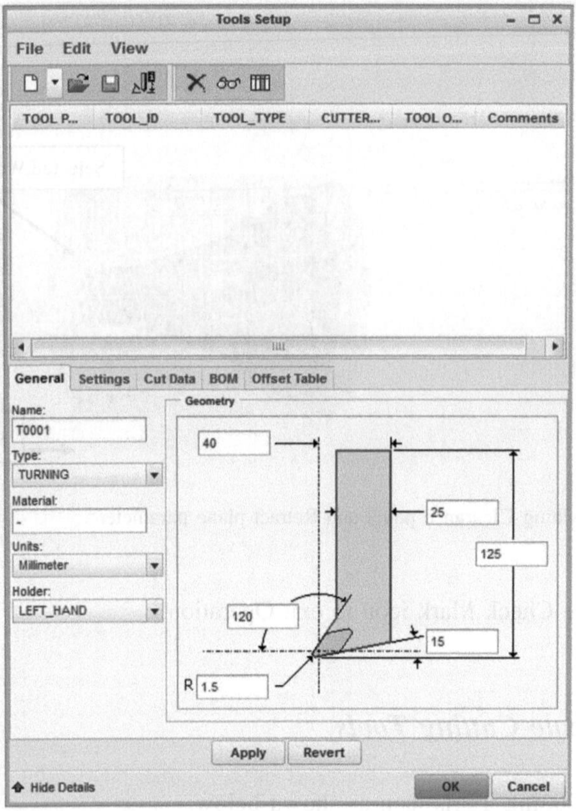

Fig. 11.18 Activated Tools Setup dialogue window

Add TURNING Tool

Click on the General tab to activate its content, only when it is not active ≫ In the Name section box, type T0001 ≫ On the Type section box, click on the downward pointing arrow to activate its content, now click on TURNING Tool on the drop-down list ≫ Check to make sure that the Units and Holder section boxes are correect ≫ Add dimensions to the TURNING Tool ≫ Now click on the Apply tab and the TURNING Tool is added. As shown in Fig. 11.19.

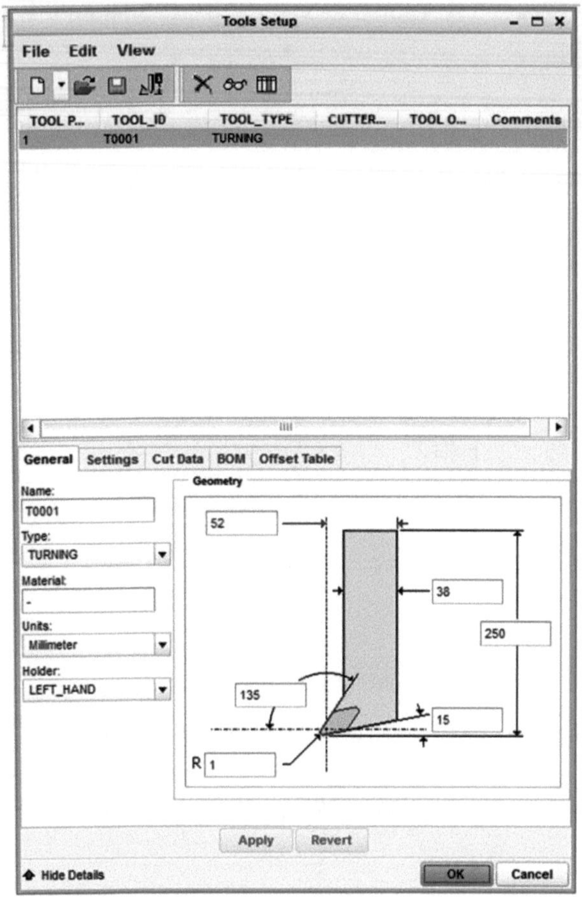

Fig. 11.19 Turning Tool parameters

Add DRILLING Tool

In the Name section box, type T0002 ≫ In the Type section box, click on the the downward pointing arrow, now click on DRILLING Tool on the drop-down list ≫ Check to make sure that the Units section box is correect ≫ Add dimensions to the DRILLING Tool ≫ Now click on the Apply tab and the DRILLING Tool is added. Repeat the same procedure for T0003 (Turning Tool). See Fig. 11.20.

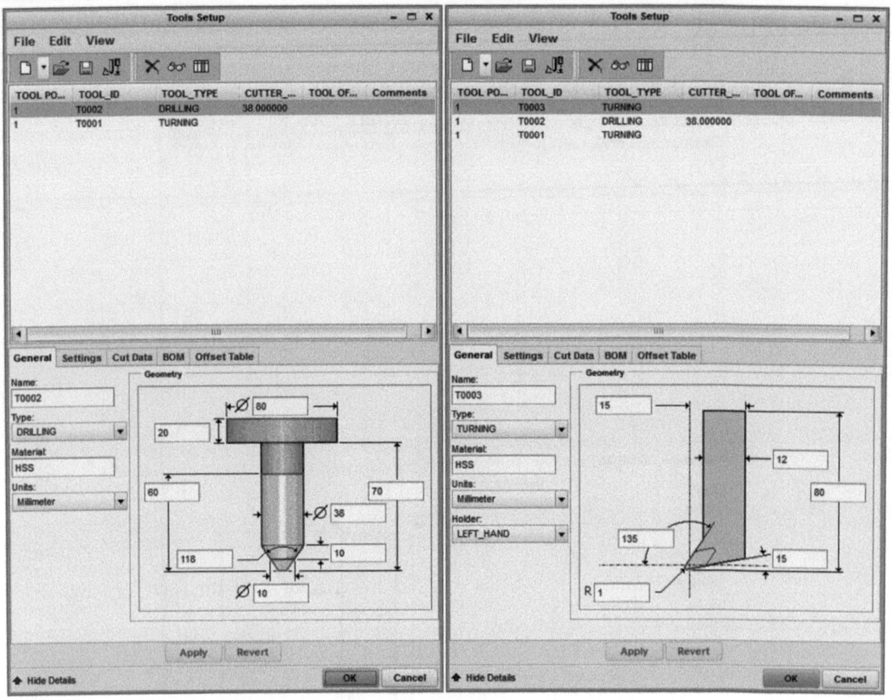

Fig. 11.20 Drilling Tool (T0002) and Turning Tool (T0003) parameters

Click on the OK tab to exit Tools Setup dialogue window

11.2 Create a Datum Point

The Datum Point to be created is a point where Tool will move from and after the machining process (From and To). Just like the Retract Plane.

Click on the Point icon to activate its content ≫ Now select Offset Coordinate System from the menu drop-down list as shown below.

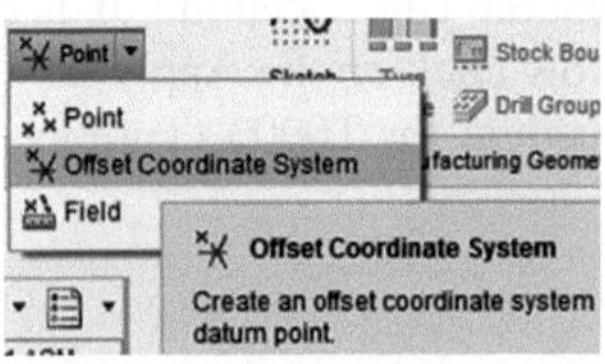

The Datum Point dialogue box opens up on the main graphic window see Fig. 11.21.

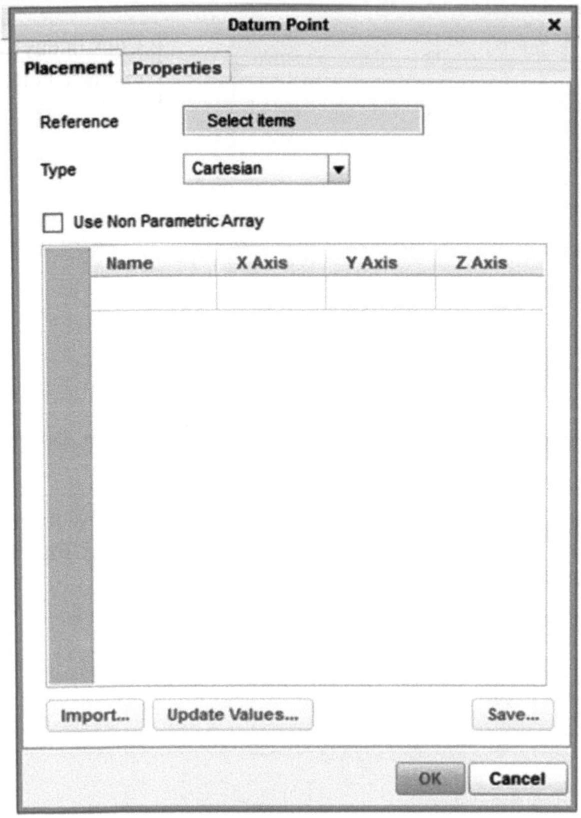

Fig. 11.21 Activated Datum Point dialogue box

Click on the Placement tab to activate its content ≫ Click on the Reference section box, now go to the created Coordinate System (PZ01) on the Model Tree and click on it ≫ Click on the downward arrow on the Type section box to activate its drop-down menu, now click on Cartesian on the drop-down menu list as shown in Fig. 11.21.

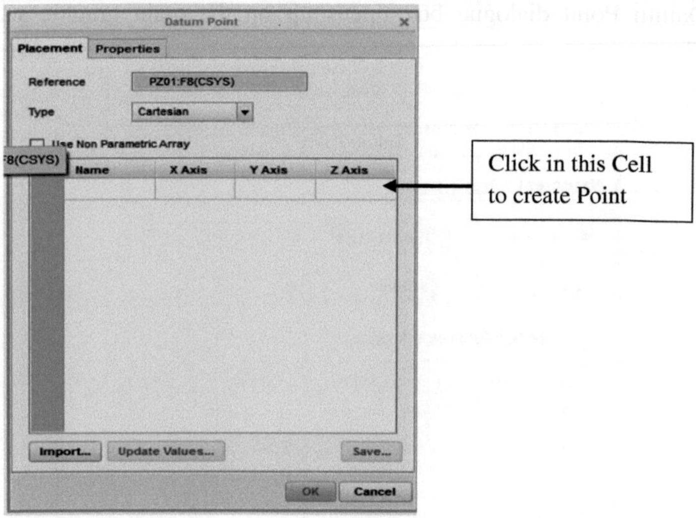

Fig. 11.22 Reference and Type parameters added to the Datum Point dialogue box

Click on the *X*, *Y* and *Z* axes Cell as indicated by the arrow in Fig. 11.22 ≫ Add values to the *X*, *Y* and *Z* axes as indicated by the arrow in Fig. 11.23. Tool will move from and to this Point after machining operation

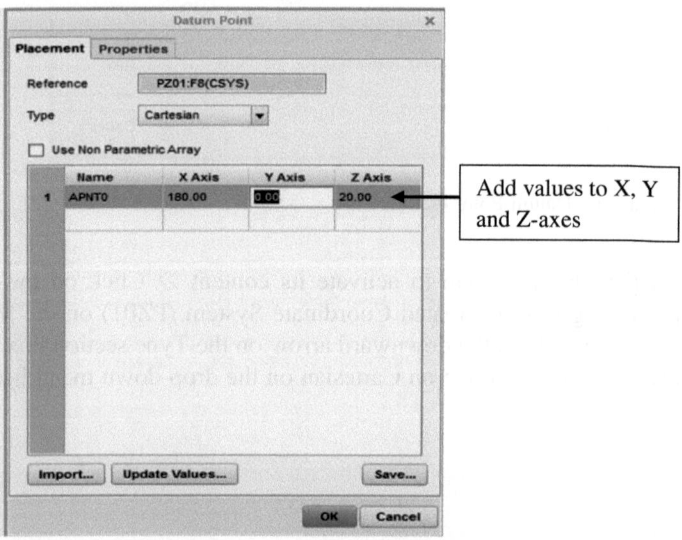

Fig. 11.23 Name and values added to the *X*, *Y* and *Z* axes cells

Click on the OK tab to exit Datum Point dialogue box.

Datum Point is now created on the graphic window as shown in Fig. 11.24.

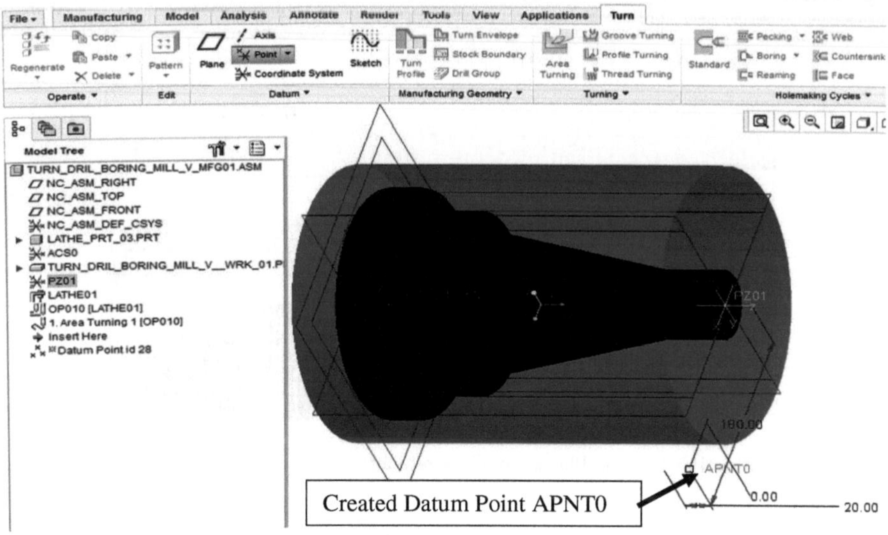

Fig. 11.24 Created Datum Point

11.3 Create Turn Profile

Click on the Turn Profile icon tab as indicated by the arrow shown below.

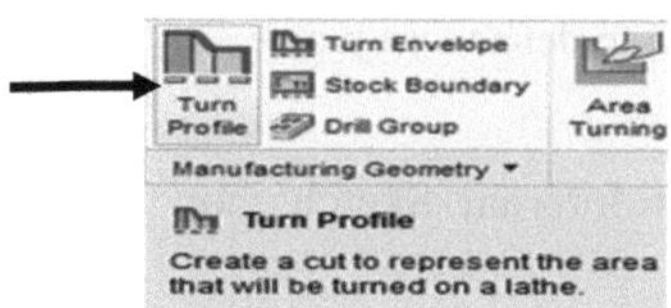

The Turn Profile dashboard tools are activated as shown in Fig. 11.25.

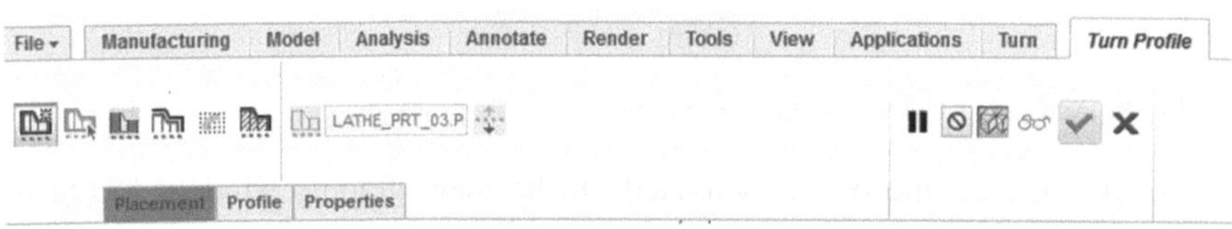

Fig. 11.25 Activated Turn Profile dashboard

Click on the Placement tab to activate its content ≫ Click on the Placement Csys section box, now click on the PZ01 on the Model Tree as indicated by the arrow in Fig. 11.26.

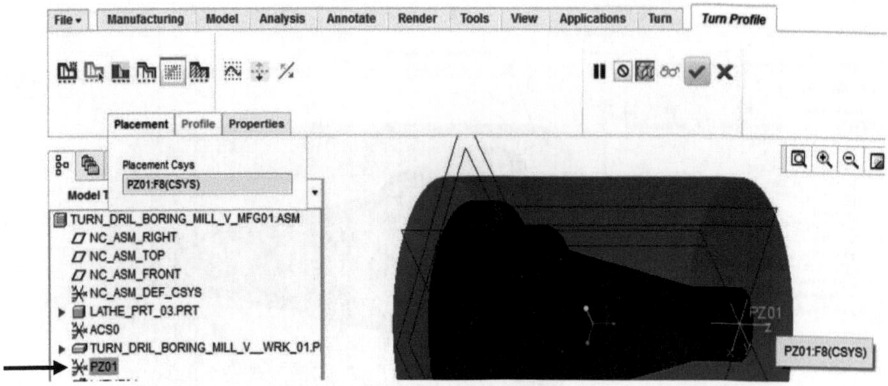

Fig. 11.26 Adding PZ01 as the Placement Csys reference

Click on the Use sketch to define turn profile icon in Fig. 11.26 as indicated by the arrow in Fig. 11.27.

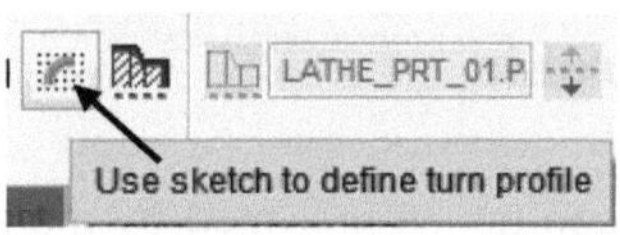

Fig. 11.27 Activating Use sketch to define turn profile

Now click on Define an Internal Sketch icon as indicated by the arrow in Fig. 11.28.

Fig. 11.28 Activating Define an Internal Sketch

The Sketch dialogue box is activated on the main graphic window ≫ On the Sketch Orientation group, the parameter in the Reference section box is automatically generated and added ≫ Make sure that orientation is set to Right in the Orientation section box as illustrated in Fig. 11.29.

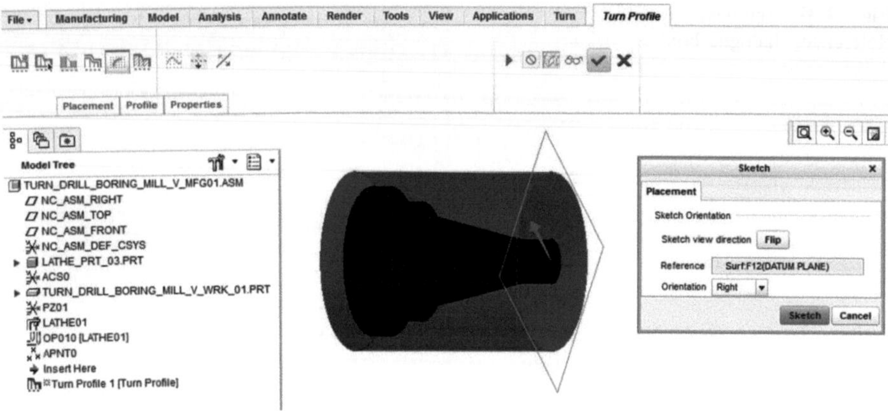

Fig. 11.29 Activated Sketch dialogue box with all Sketching parameters

The Sketch application tools are activated. See Fig. 11.30.

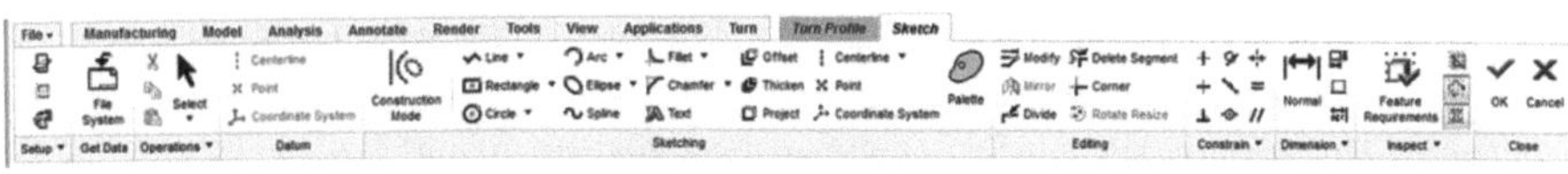

Fig. 11.30 Activated Sketch tools in concise form

The References dialogue box is activated automatically by the system. If not, activate it yourself as described in create reference section.

Create References

Click on the References icon  on the Sketch tools.

The References dialogue box is activated on the graphic window as shown in Fig. 11.31.

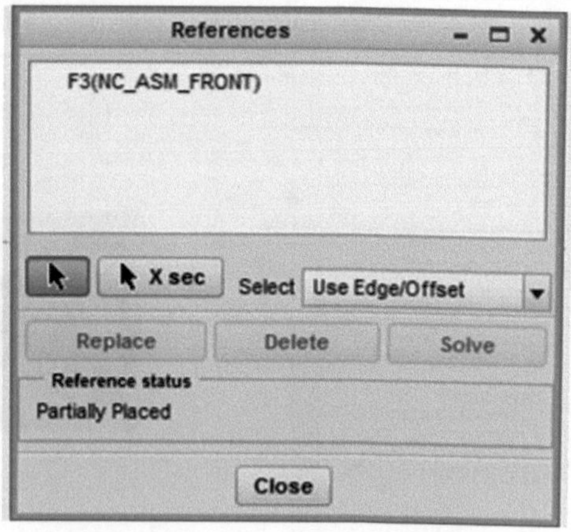

Fig. 11.31 Activated References dialogue box

Click on Part to create the references as illustrated by the arrows in Fig. 11.32.

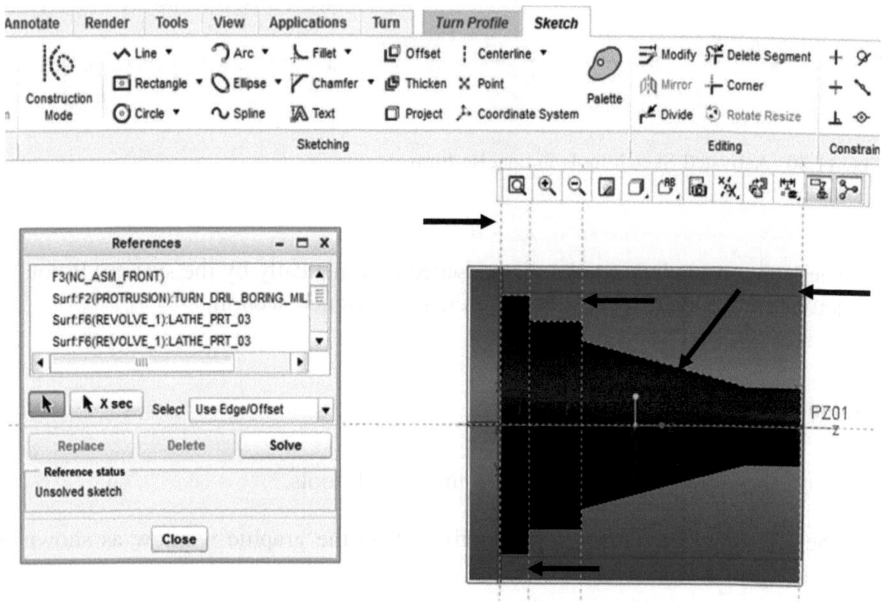

Fig. 11.32 Adding References

Click on the Close tab to exit the References dialogue box.

Project Line and Curve

Now click on the Project icon tab **Project** in Fig. 11.32.

The Type dialogue box opens up on the main graphic window as shown in Fig. 11.33.

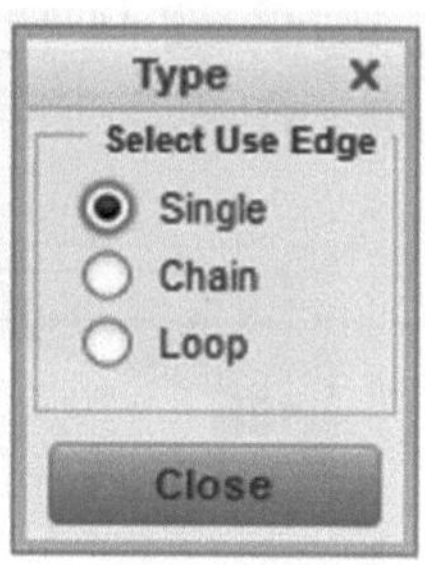

Fig. 11.33 Activated Type dialogue box

Click on Lines on the Part to project them as indicated by the arrows in Fig. 11.34.

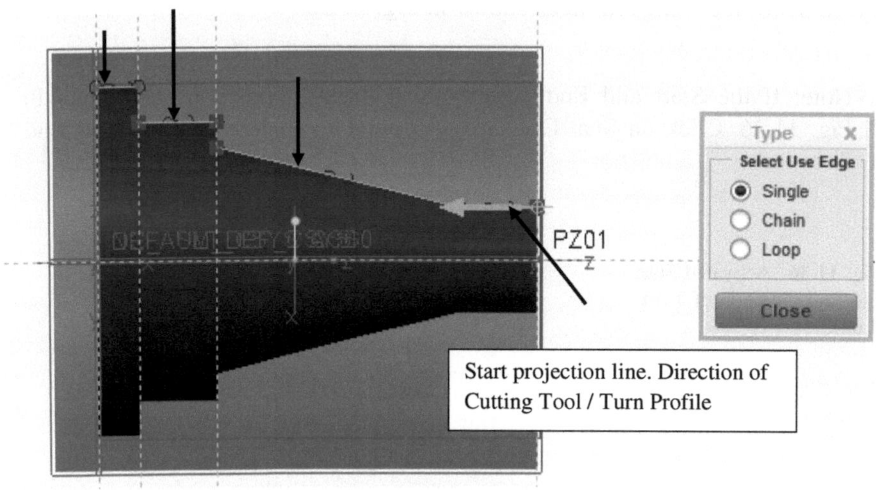

Fig. 11.34 Projecting Lines into the sketch plane

Draw Lines to join with the projected Lines and Curves

Click on the Line tool icon Line on the Sketch tool list.

Draw the Lines as indicated by the arrows below to join with the Projected Lines as illustrated in Fig. 11.35.

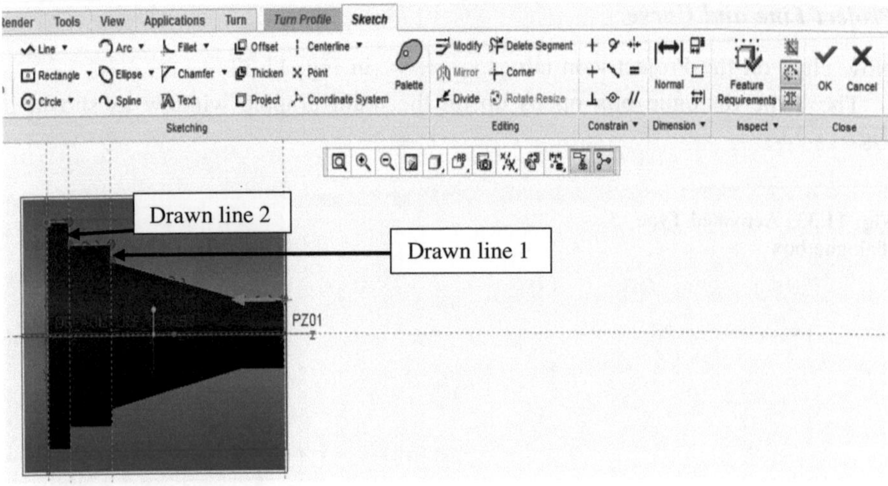

Fig. 11.35 Joining all the projected Lines together

Click on the Check Mark icon to exit the Sketch application.

The Turn Profile ribbon toolbar is activated again indicating the Start and End directions of the Turn Profile as shown in Fig. 11.36.

Note: If the Start and End points are in diffident positions as shown in Fig. 11.36. Click on Start/End arrow or point to change the Start and End directions.

Fig. 11.36 Activated Start and End direction of Turn Profile

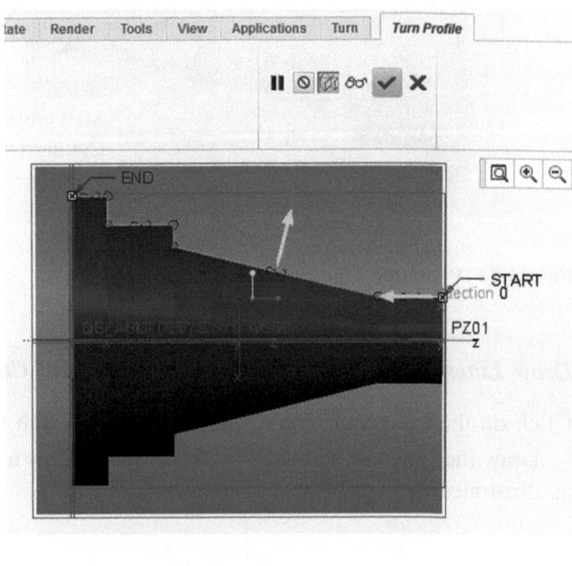

Click on the Check Mark icon to exit Turn Profile.

11.4 Create Area Turning Sequence

Go to the menu bar and click on Turn to activate its menu list >> Click on Area Turning icon as indicated by the arrow below.

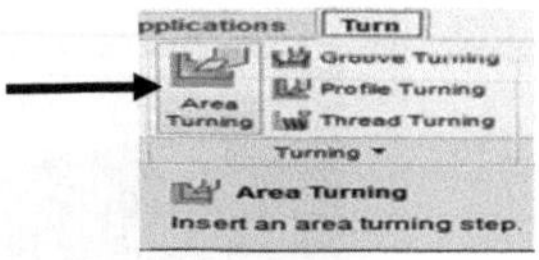

The Area Turning dashboard tools are activated. See Fig. 11.37.

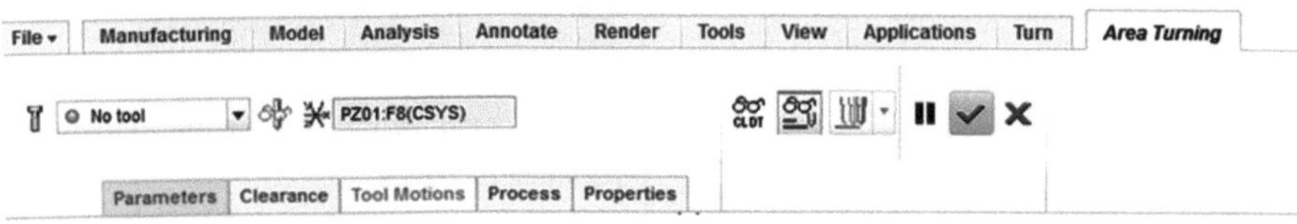

Fig. 11.37 Activated Area Turning dashboard in concise form

Note: Manually, add the correct Cutting Tool and Coordinate System if they are not automatically generated and added by the system.

Click on the Parameters tab to activate its panel >> Now add values to the Area Turning parameters as shown in Fig. 11.38.

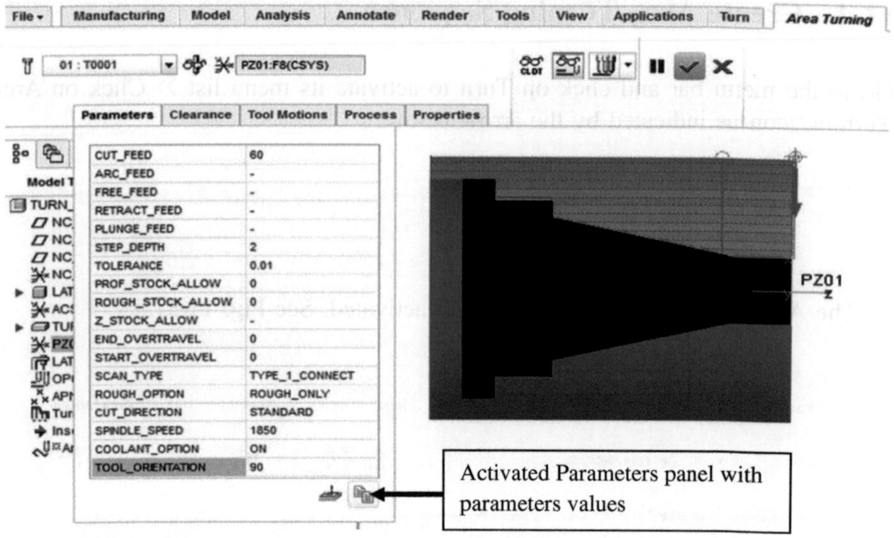

Fig. 11.38 Area Turning parameters and the system activated cutting area

Note: If the area to be turned is not activated automatically by the system, just follow the steps outlined in Sect. 11.4.1 to activate it yourself.

11.4.1 Create Tool Motions for Area Turning

Click on Tool Motions tab to activate its panel ≫ Click on Area Turning Cut downward arrow, now select Area Turning Cut on the drop-down list as shown in Fig. 11.39.

Fig. 11.39 Activated Tool Motions panel

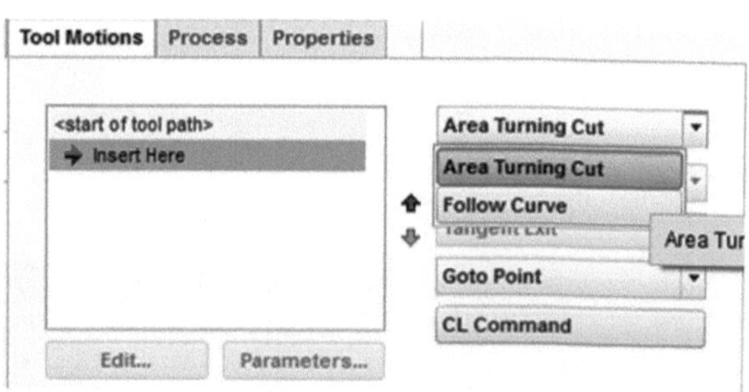

The Area Turning Cut dialogue window is activated on the main graphic window as shown in Fig. 11.40.

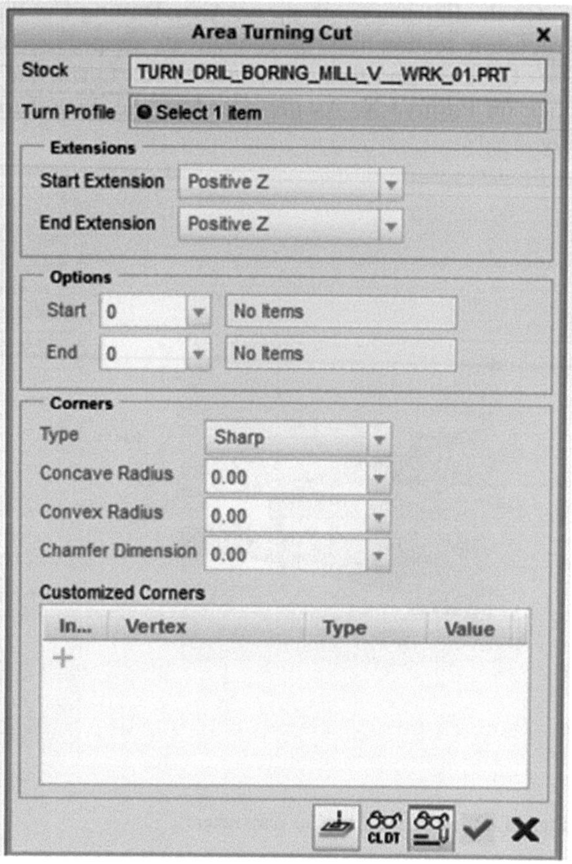

Fig. 11.40 Activated Area Turning Cut dialogue window

Click on the Turn Profile section box on the Area Turning Cut dialogue window ≫ Now click on the Turn Profile highlighted on the Model Tree in Fig. 11.41. Alternatively, click on the sketched (projected sketched) Turn Profile line on the Workpiece.

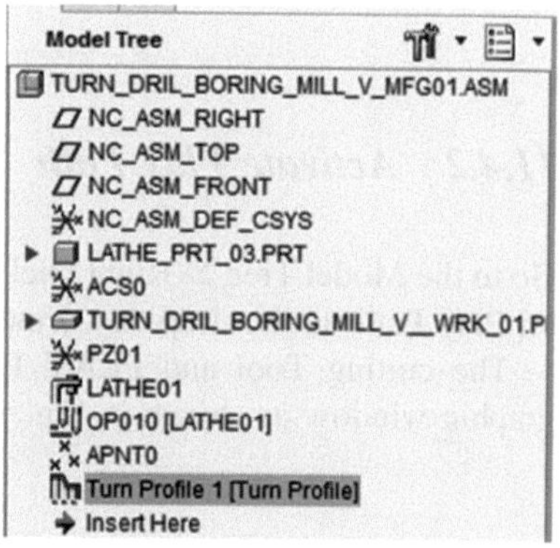

Fig. 11.41 Highlighted Turn Profile Model Tree

On the Extensions group on Area Turning Cut dialogue window, click on Start Extension section box to activate its drop-down menu, now click on Positive Z ≫ Click on End Extension section box to activate its drop-down menu list, now click on Positive X. As illustrated on the Area Turning Cut dialogue window in Fig. 11.42.

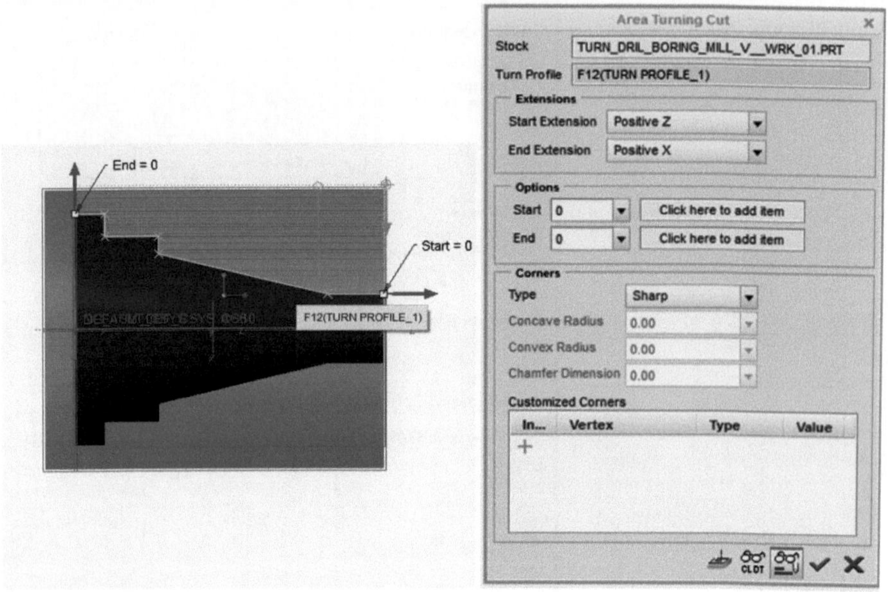

Fig. 11.42 Area Turning Cut parameters

Click on the Check Mark icon to exit Area Turning Cut dialogue window.
Click on the Check Mark icon to exit Area Turning.

Note: Material within the shaded area as shown in Fig. 11.42 will be removed during the Turning process.

11.4.2 Activate Play Path

Go to the Model Tree ≫ Right click the mouse on Area Turning 1 (OP010) ≫ Click on Play Path on the drop-down menu list.

The cutting Tool and PLAY PATH dialogue box are activated on the main graphic window as shown in Fig. 11.43.

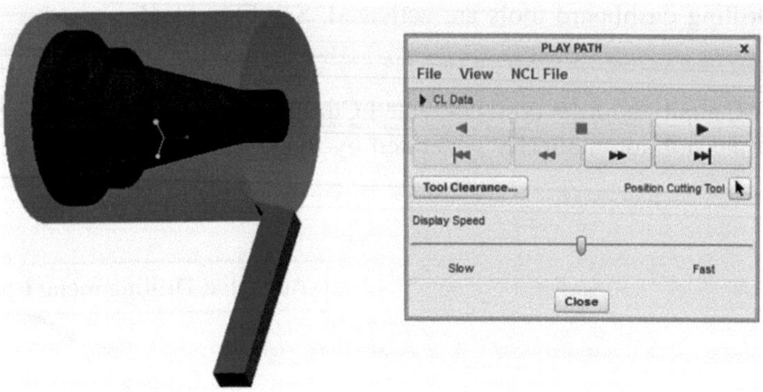

Fig. 11.43 Activated cutting Tool and PLAY PATH dialogue box

Click on the Play tab to start the on-screen the Area Turning operation. The end of the Area Turning operation is as shown in Fig. 11.44.

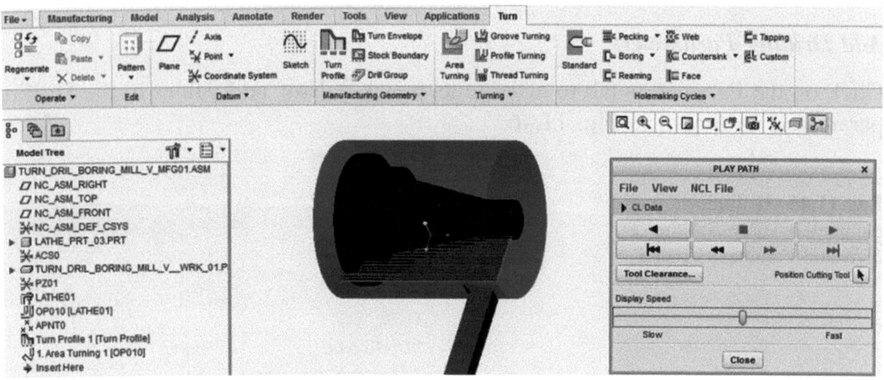

Fig. 11.44 End of the Area Turning operation

Click on Close tab on the PLAY PATH dialogue box to exit the Area Turning operation.

11.5 Create Drilling Sequence

Activate the Turn menu tab ≫ Click on the Standard Drill icon on the Holemaking Cycles group as indicated by the arrow below.

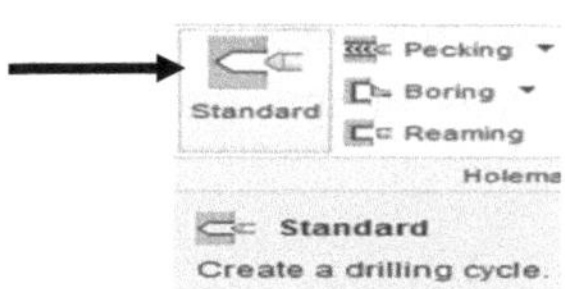

The Drilling dashboard tools are activated. See Fig. 11.45.

Note: Manually add the correct created Cutting Tool and Coordinate System if they are not automatically generated by the system.

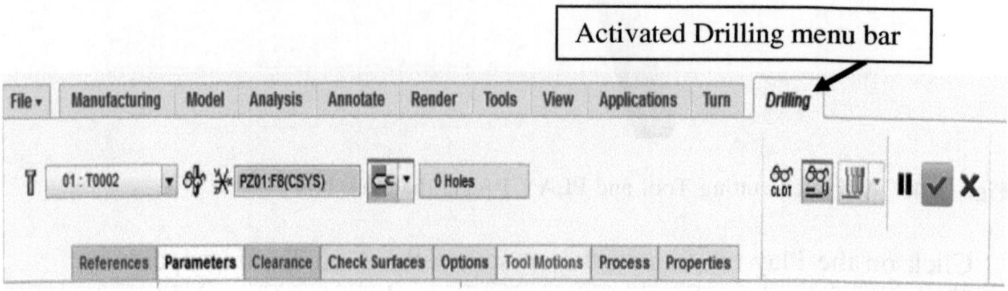

Fig. 11.45 Activated Drilling dashboard in concise form

Add Drilling Parameters

Click on the Parameters tab to activate its panel ≫ Now add values to the Drilling parameters as shown in Fig. 11.46.

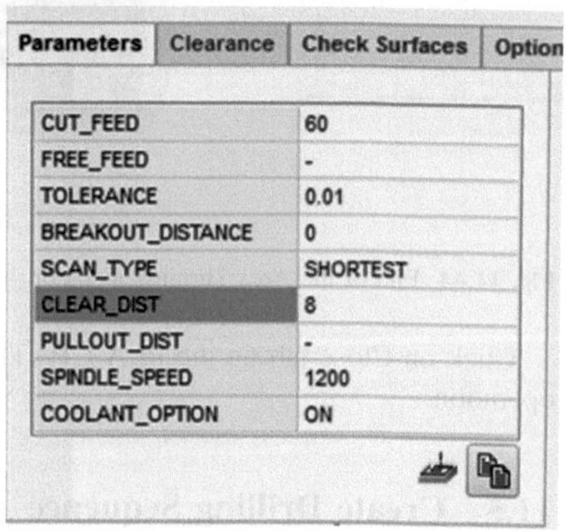

Fig. 11.46 Drilling parameters values on the activated Parameters panel

Add Reference Hole Axis for Drilling

Click on the References tab to activate its panel ≫ Click on the Type section box to activate its drop-down menu list ≫ Click on Axes on the drop-down list ≫ In the Holes section box, click on the axis of the hole to be drilled as illustrated in Fig. 11.47.

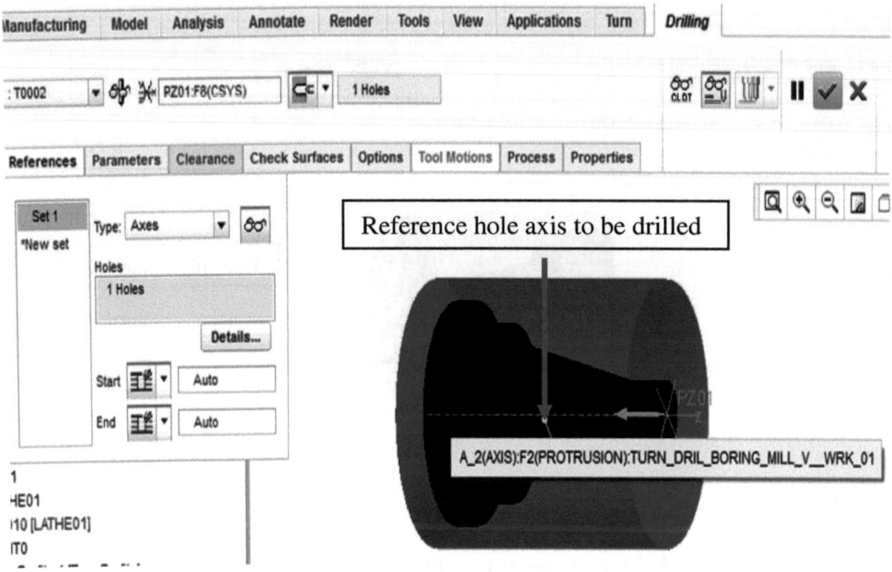

Fig. 11.47 Adding the Drilling reference parameters

Add the Retract parameters

Click on the Clearance tab to activate its panel ≫ On the Retract group, click on the Reference section box ≫ Now click on the surface of the Workpiece as indicated by the arrow in Fig. 11.48. In the value section box, type 10 mm as the value of the Retract distance as shown in Fig. 11.48.

Note: The Start and End Points group are left untouched on the Clearance panel for now.

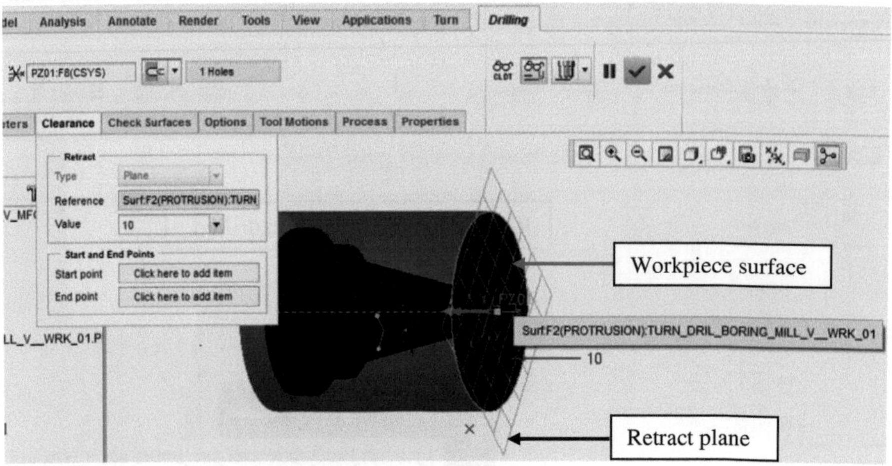

Fig. 11.48 Adding Clearance parameters

Click on the Check Mark icon to exit the Drilling application.

Activate Play Path

Go to the Model Tree ≫ Right click the mouse on Drilling 1(OP010) ≫ Click on Play Path on the drop-down list.

The Drilling Tool and PLAY PATH dialogue box are activated on the main graphic window as shown in Fig. 11.49.

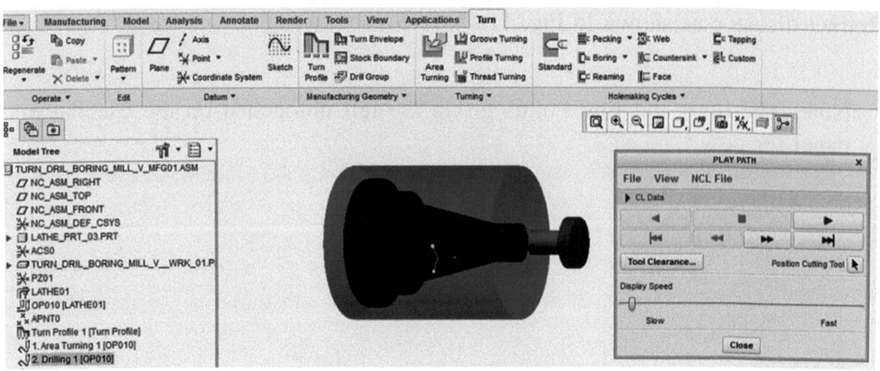

Fig. 11.49 Activated Drilling Tool and PLAY PATH dialogue box

Click on Play tab on the PLAY PATH dialogue box to start.

Change the Start Point

To change the Sart Point to the created Datum Point.

Go to Model Tree >> Right click the mouse on Drilling 1(OP010) >> Click on Edit Definition on the drop-down list and the Drilling ribbon toolbar is activated again.

Now click on the Clearance tab to activate its panel >> On the Start and End Points group, click on the Start point section box, now click on the already created Datum Point (APNT0) on the Model Tree or on the main graphic window as indicated by the arrow in Fig. 11.50.

> Note: You can also add the Datum Point in the End point section box to be the Retract point for Tool after the drilling operation as indicated in this figure.

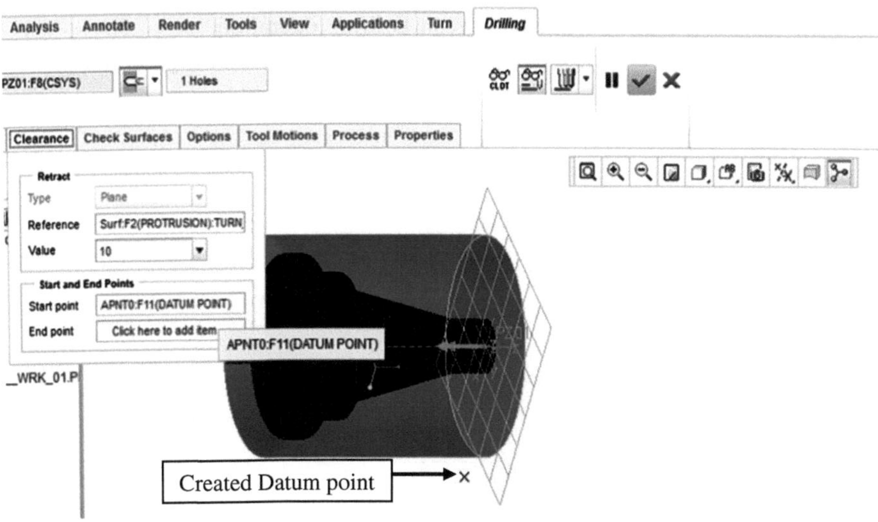

Fig. 11.50 Changing the Retract Start and End Points group parameters

Click on the Check Mark icon to exit.

Go to the Model Tree >> Right click on Area Drilling 1(OP010) >> Click on Play Path on the drop-down list.

The Drilling Tool and the PLAY PATH dialogue box are activated on the main graphic window as shown in Fig. 11.51.

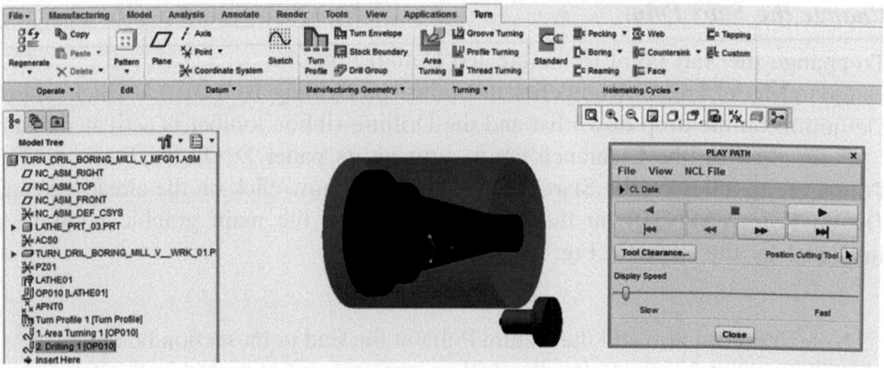

Fig. 11.51 Activated Drilling Tool and the PLAY PATH dialogue box with Tool staring at the created Datum point

Click on Play tab to start the on-Screen Drilling operation.
The end of the on-screen Drilling operation is as shown in Fig. 11.52.

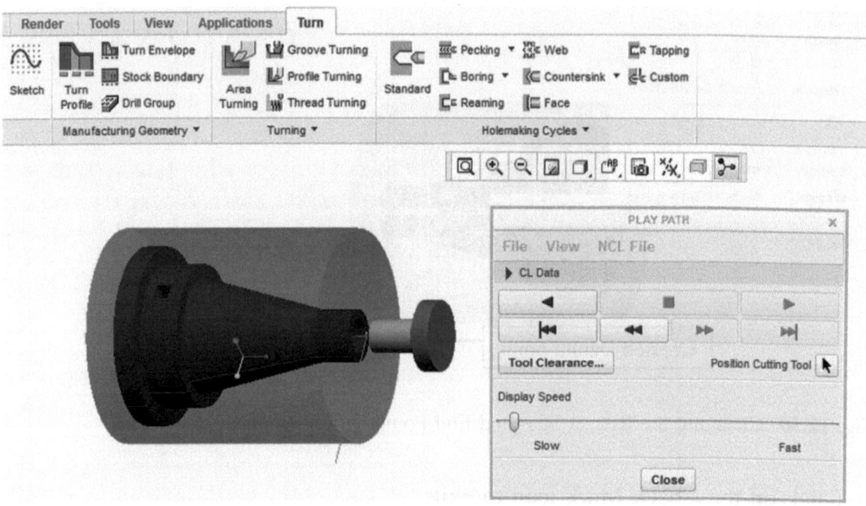

Fig. 11.52 End of on-screen Drilling operation

Click on the Close tab to exit Drilling operation.

11.6 Create Turn Profile for the Internal Hole Diameter

Click on the Turn Profile icon tab as shown below.

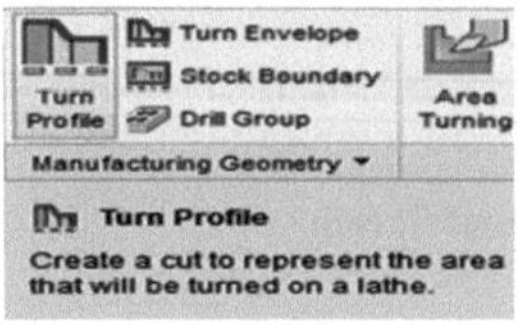

The Turn Profile dashboard tools are activated as shown. See Fig. 11.53.

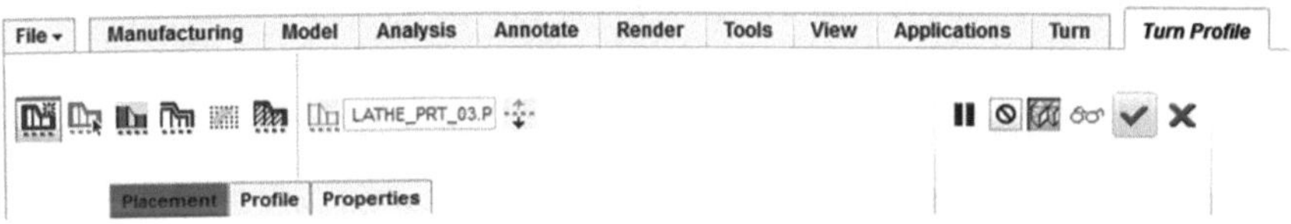

Fig. 11.53 Activated Turn Profile dashboard

Click on the Placement tab to activate its panel ≫ Click on the Placement Csys section box, now click on the PZ01 Coordinate System on the Model Tree as indicated by the arrows in Fig. 11.54.

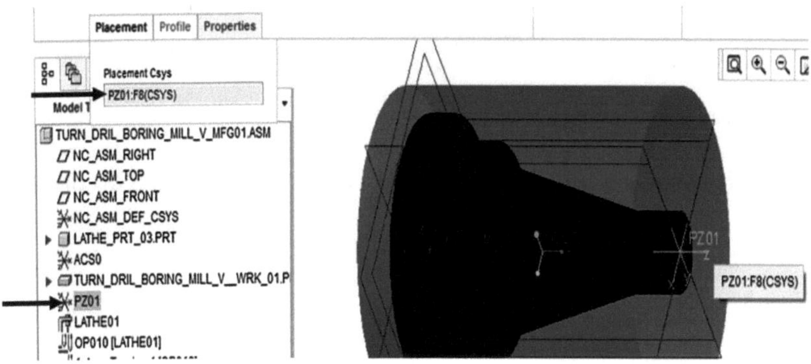

Fig. 11.54 Adding the Placement reference

Click on the Use sketch to define turn profile icon as indicated by the arrow below (Fig. 11.55).

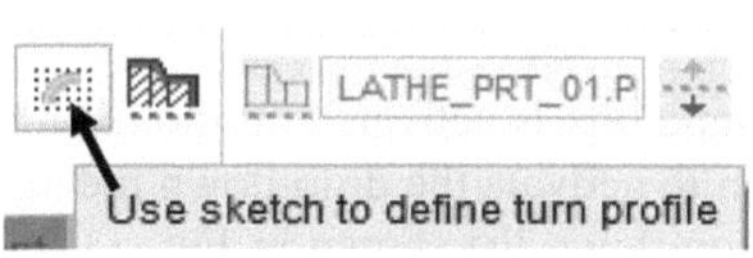

Fig. 11.55 Activating Use sketch to define turn profile

Click on Define an Internal Sketch icon as indicated by the arrow in Fig. 11.56.

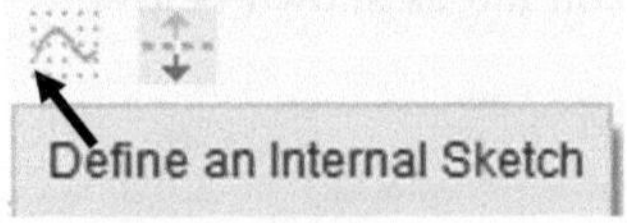

Fig. 11.56 Activating Define an Internal Sketch

The Sketch dialogue box is activated on the main graphic window ≫ On the Sketch Orientation group, the Reference and Orientation section boxes are automatically generated and added by the system as shown in Fig. 11.57.

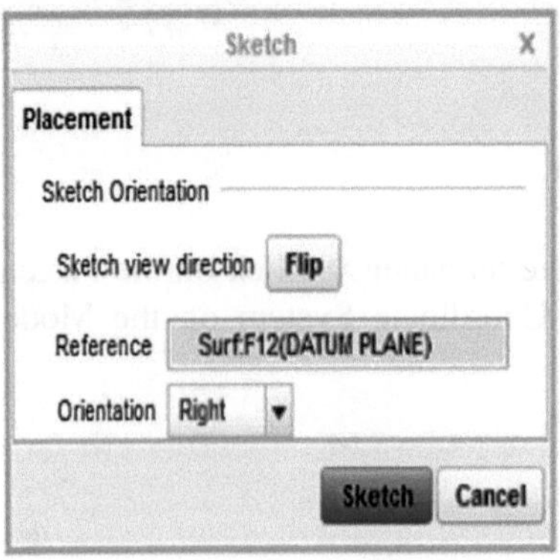

Click on the Sketch tab to exit.

Fig. 11.57 Activated Sketch dialogue box

The Sketch application tools are activated as shown in Fig. 11.58 in concise form.

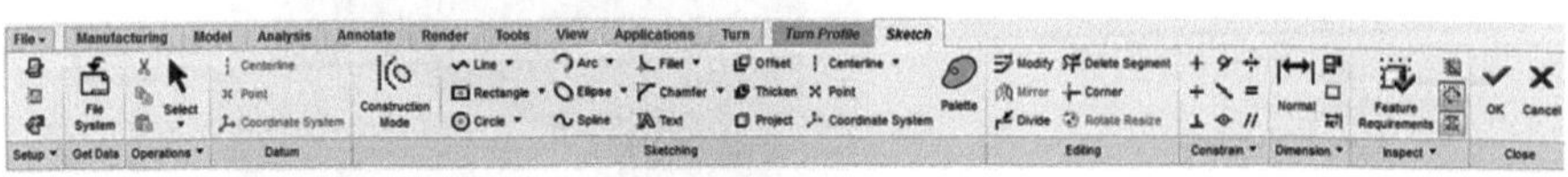

Fig. 11.58 Activated Sketch tools

Create References

Click on the References icon  on the Sketch tool list.

The References dialogue box is activated on the main graphic window ≫ Click on the Display Style icon to activate the drop-down menu, now click on Wireframe as highlighted on the drop-down list shown in Fig. 11.59.

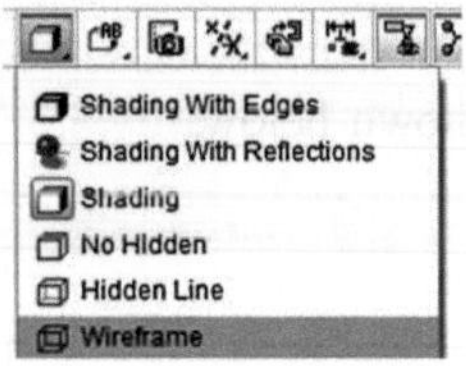

Fig. 11.59 Activating Wireframe on the Display Style drop-down list

Now click on the Sketch View icon  to position the Workpiece on the correct sketch plane /orientation as shown in Fig. 11.60.

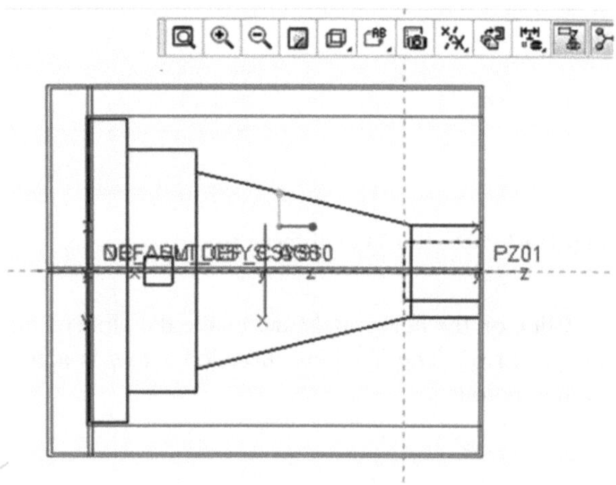

Fig. 11.60 Workpiece in Wireframe display style

Select references on Part as indicated by the arrows as illustrated in Fig. 11.61

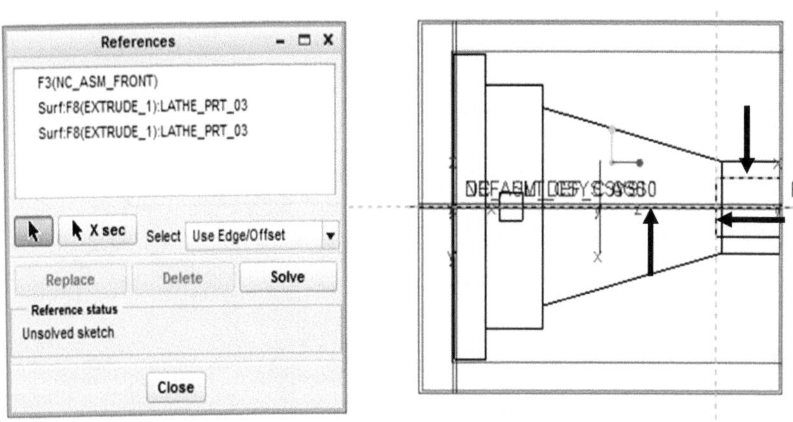

Fig. 11.61 Creating references for the internal diameter

Click on Close tab on the References dialogue box to exit.

Project Sketch

Click on the Project icon as shown below.

The Type dialogue box opens up on the main graphic window as shown in Fig. 11.62.

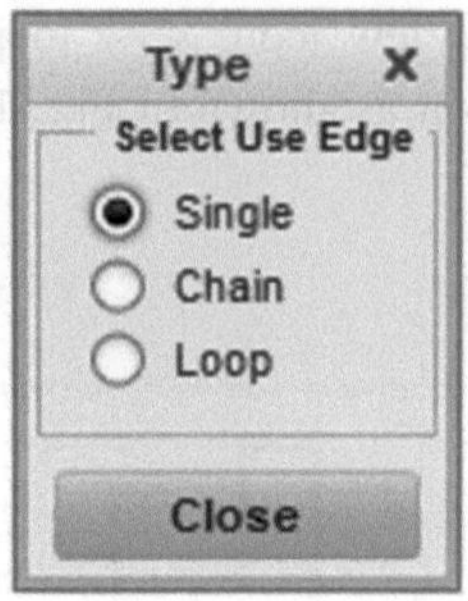

Fig. 11.62 Activated Type dialogue box

Click on the horizontal Line inside the Hole of the Part as indicated by the arrow in Fig. 11.63. The Tool cut direction arrow is activated as shown in Fig. 11.63 in orange colour

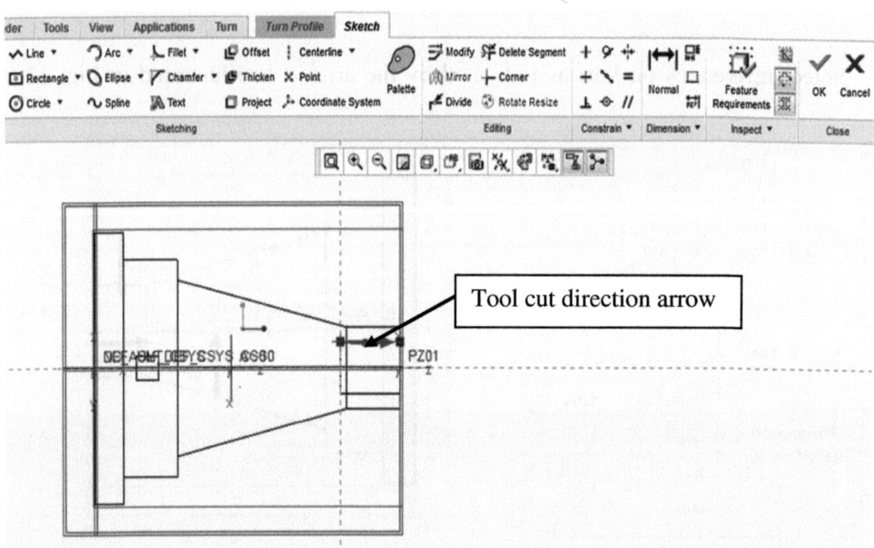

Fig. 11.63 Projecting the internal horizontal line inside the hole

Click on the Close tab on the Type dialogue box to exit.
Click on the Check Mark icon/OK to exit Sketch.
The Turn Profile dashboard tool are activated again.
The Tool cut directions are activated as indicated by the two arrows in purple colour as shown in Fig. 11.64.

Note: The incorrect Tool cut lines will start from the left to right (Start to End) as indicated by the purple coloured arrows.

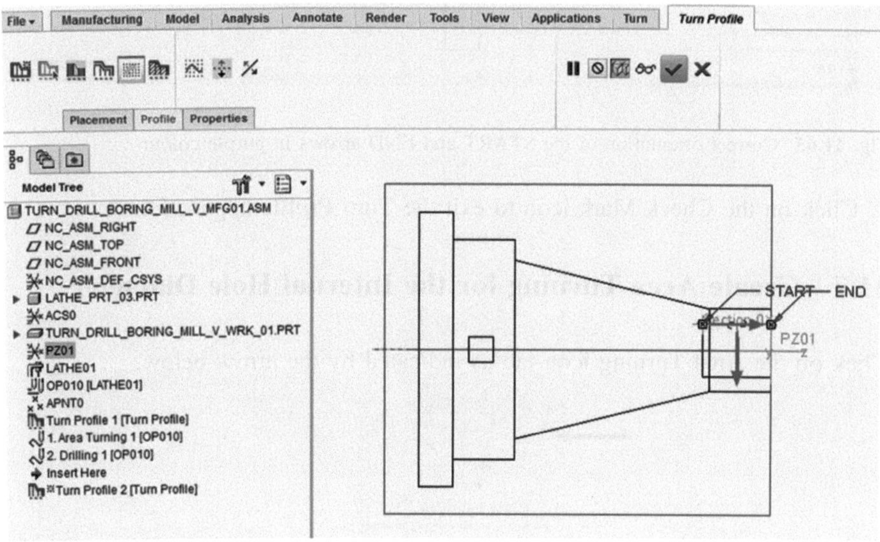

Fig. 11.64 Wrong Tool cut direction

To change the Tool cut direction to the correct direction, click on either the Start or End arrow crossed point in Fig. 11.64 ≫ Now click on the upward pointing arrow to point downwards.

The correct Tool cut direction/cut lines will start from right to left (START to END) as indicated by the arrows in purple colour in Fig. 11.65.

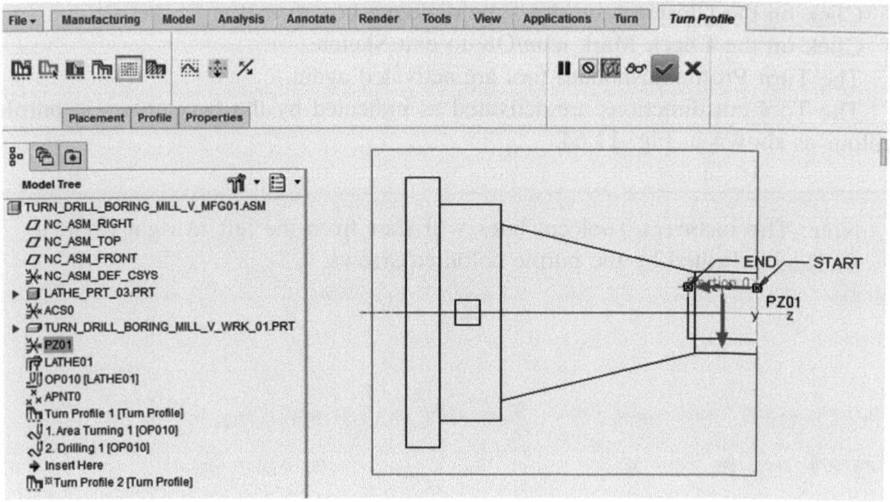

Fig. 11.65 Correct orientation of the START and END arrows in purple colour

Click on the Check Mark icon to exit the Turn Profile application.

11.7 Create Area Turning for the Internal Hole Diameter

Click on the Area Turning icon tab as indicated by the arrow below.

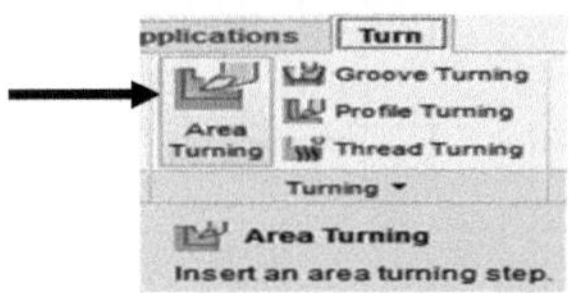

The Area Turning dashboard tools are activated as shown. See Fig. 11.66.

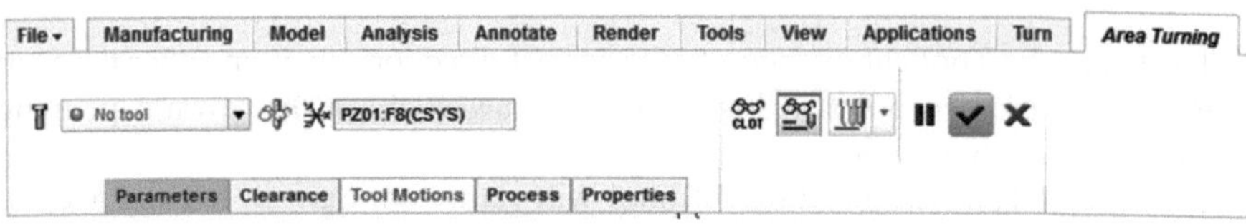

Fig. 11.66 Activated Area Turning dashboard

Note: Add the created Cutting Tool (T0003) as shown in Fig. 11.20 and Coordinate System (PZ01) manually, if they are not automatically generated and added by the system.

Click on the Parameters tab to activate its panel ≫ Now add values to the Parameters as shown in Fig. 11.67.

Note: Make sure Tool_Orientation as indicated by the arrow below is set to zero

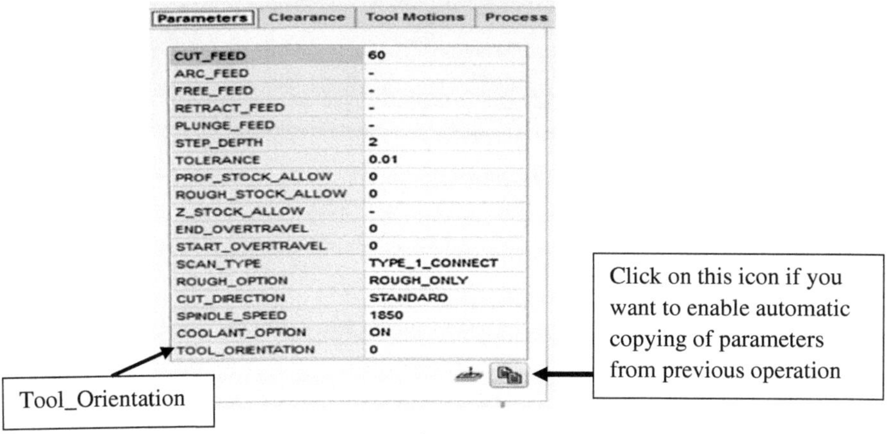

Fig. 11.67 Area Turning parameters

11.7.1 Create Tool Motion for Area Turning

Click on Tool Motions tab to activate its panel ≫ Now click on Area Turning Cut to activate its menu ≫ Select Area Turning Cut on the drop-down menu list as shown in Fig. 11.68.

Note: To delete the created cut area by default tool path motion (Area Turning Cut id 237) ≫ Right click "Area Turning Cut id 237" and then select delete from the drop-down menu list.

Fig. 11.68 Activating Area Turning Cut

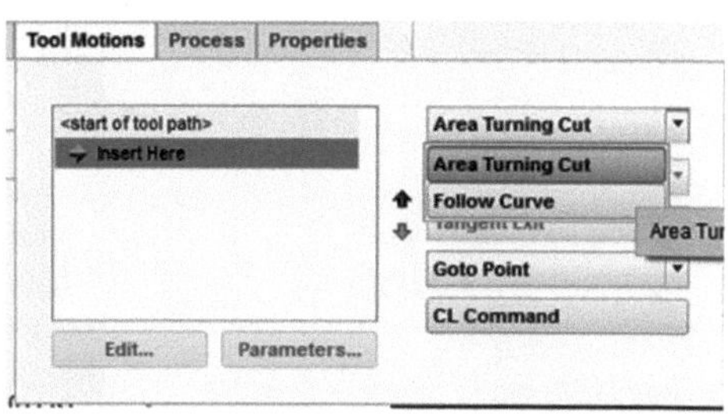

The Area Turning Cut dialogue window is activated on the main graphic window as shown in Fig. 11.69.

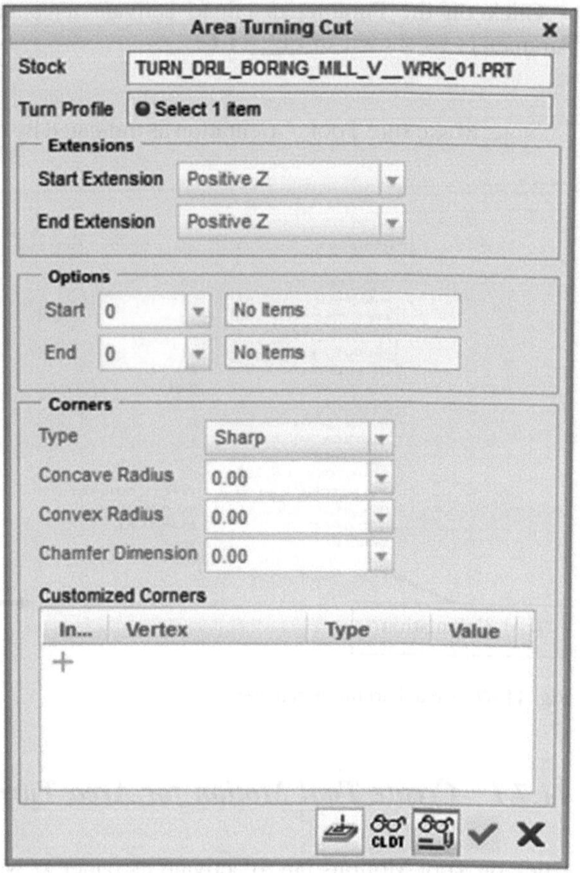

Fig. 11.69 Activated Area Turning Cut dialogue window

Click on the Turn Profile section box on the Area Turning Cut dialogue window ≫ Now click on the created second Turn Profile on the Part. Alternatively, click on the "Turn Profile 2[Turn Profile]" on the Model Tree as highlighted in Fig. 11.70.

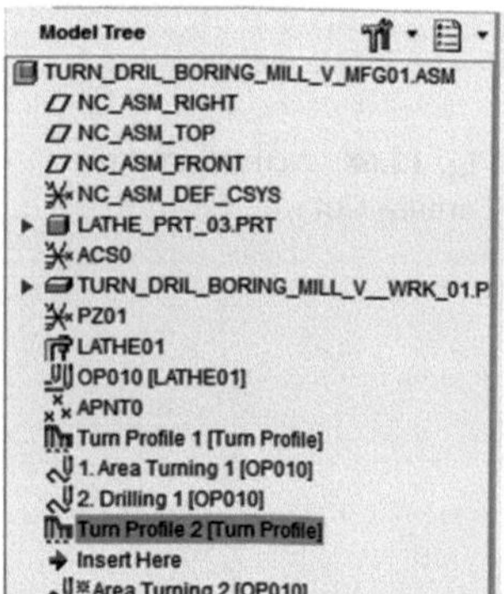

Fig. 11.70 Activating the created Turn Profile 2 [Turn Profile]

On the Extensions group on the Area Turning Cut dialogue window, click on Start Extension section box to activate its drop-down menu, now click on Positive Z >> Click on End Extension section box to activate its drop-down menu list, now click on Negative X on the drop-down list as shown on the Area Turning Cut dialogue window as illustrated in Fig. 11.71.

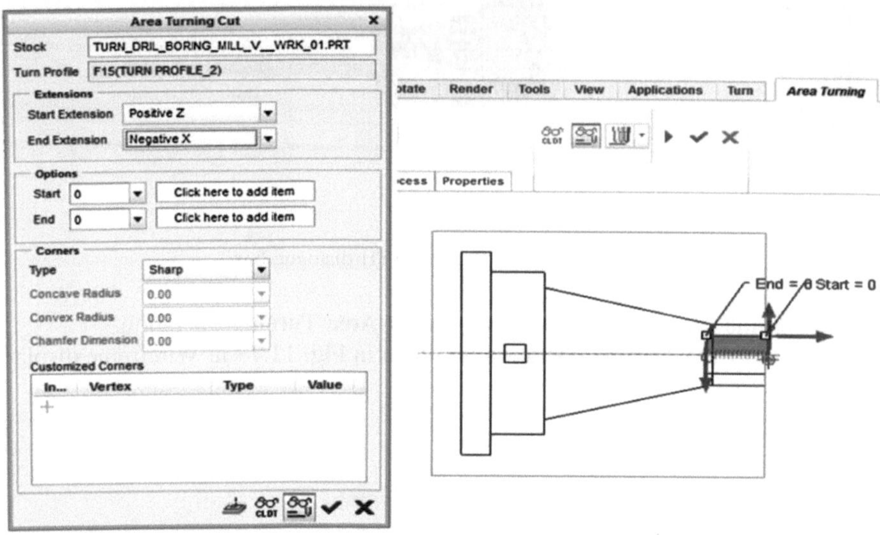

Fig. 11.71 Defining Area Turning Cut parameters

Note: The default Area Cut profile disappears as a result of defining the Extensions limits as shown in Fig. 11.71.

Now click on the Check Mark icon to exit the Area Turning Cut dialogue window.

Click on the Check Mark icon to exit the Area Turning application dashboard tools.

11.7.2 *Activate Play Path*

Go to the Model Tree >> Right click the mouse on Area Turning 2 (OP010) >> Click on Play Path on the drop-down list.

The cutting Tool and the PLAY PATH dialogue box are activated on the main graphic window as shown in Fig. 11.72.

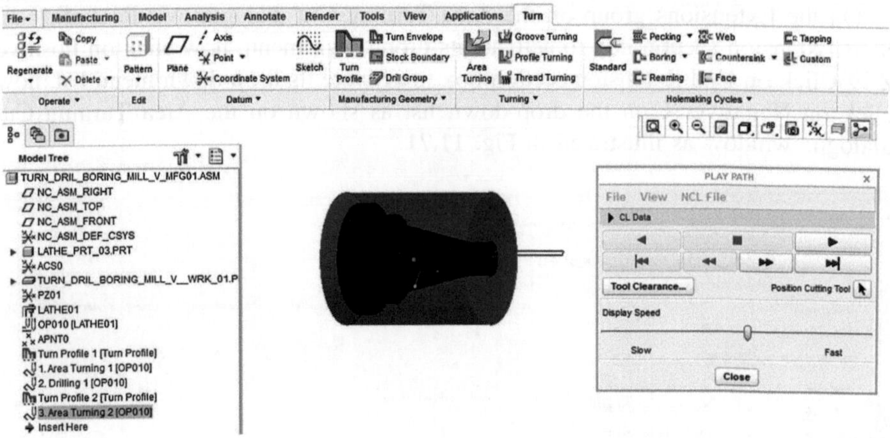

Fig. 11.72 Activated cutting Tool and the PLAY PATH dialogue box

Click on the Play tab to start the on-screen Area Turning operation.

The end of Area Turning operation is shown in Fig. 11.73 in Wireframe display.

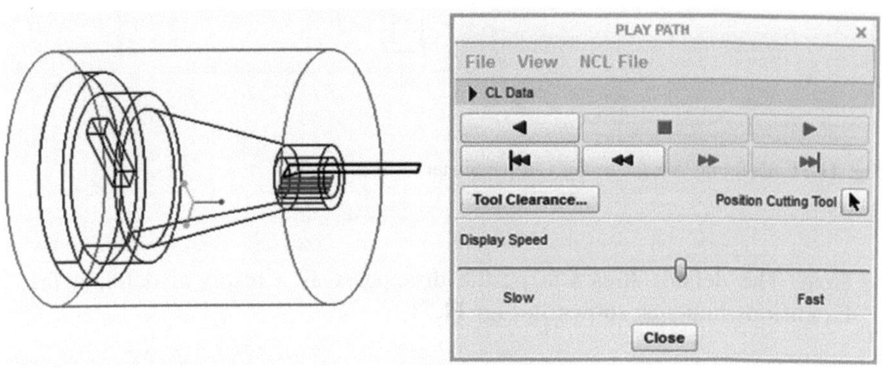

Fig. 11.73 End of Area Turning operation in wireframe display

Click on Close tab to exit the Area Turning operation.

11.8 Create New Coordinate System, Work Centre, Datum and Operation for the Volume Mill Process

11.8.1 Create a New Coordinate System

Another Coordinate System has to be created for the Mill Volume operation ≫ Click on the Coordinate System icon as shown below.

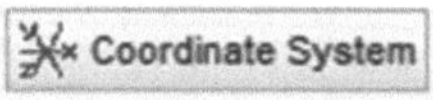

The Coordinate System dialogue box is activated on the main graphic window ≫ Now click on the Origin tab on the Coordinate System dialogue box ≫ Click on the Workpiece surface, NC_ASM_FRONT and NC_ASM_TOP datum planes as indicated by the arrows in Fig. 11.74. The selected references are automatically added into the References section box. Now activate the Orientation tab and click on "Y" Flip tab on the "to project group" to change its direction.

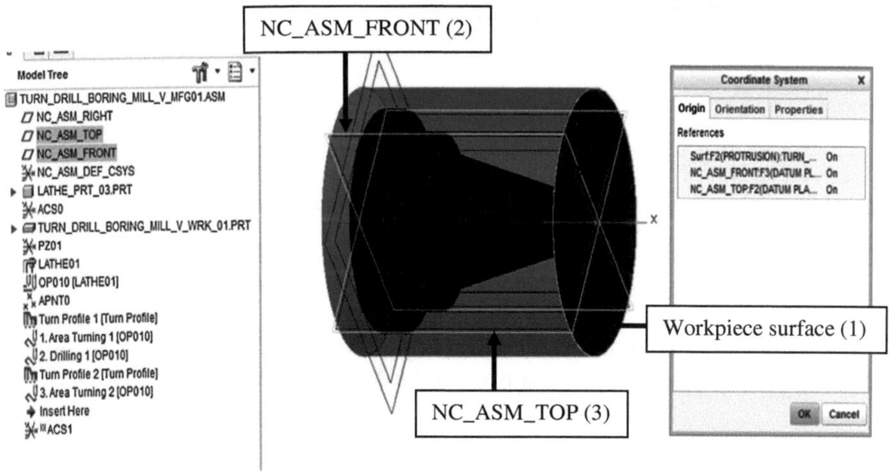

Fig. 11.74 Activating the correct Coordinate System orientation

If your steps of selections are Workpiece surface first, NC_ASM_TOP and NC_ASM_FRONT datum planes. Follow the step-by-step guide below to orient the *X*, *Y* and *Z* axes to the correct orientation.

Click on the Orientation tab on the Coordinate System dialogue box ≫ On the "Orient by" group, click on the radio button on the "References selection" ≫ Click on the downward arrow on the "to determine" section box to activate the drop-down menu, now click on *X*-axis on the drop-down list ≫ Click on the downward arrow on the "to project" section box, and click on *Z* axis on the drop-down list. Now if necessary, click on the Flip tab, to flip the *Z* axis orientation as illustrated in Fig. 11.75.

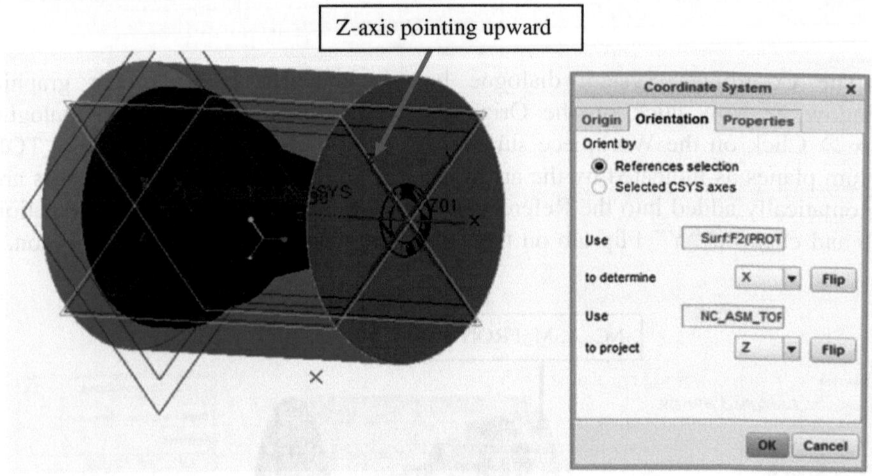

Fig. 11.75 Correct orientation of *X*, *Y* and *Z* axes of the Coordinate System

Click on OK tab to exit the Coordinate System dialogue box.

Let change the new ACS2 Coordinate System to PZ02 ≫ Right click on ACS2 to activate its drop-down menu and then click on Rename ≫ Type PZ02 as the new name as indicated below by the arrow on the Model Tree in Fig. 11.76.

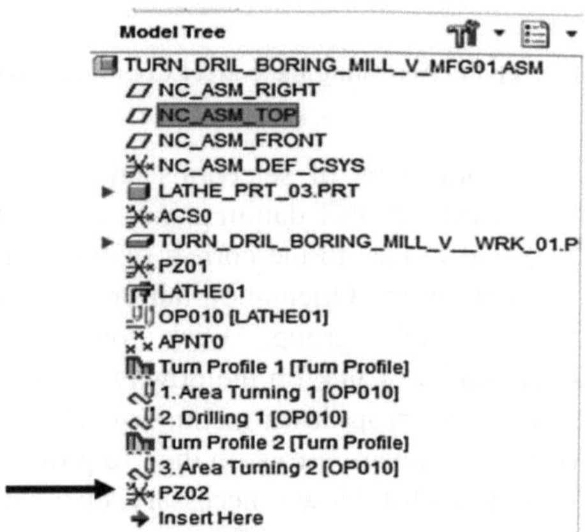

Fig. 11.76 Model Tree indicating the renamed PZ02

11.8.2 Create New Work Centre (MILL02)

Click on the Work Centre icon tab to activate its content ≫ Now click on Mill as indicated by the arrow on the drop-down list as shown below.

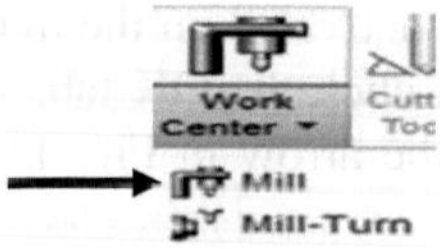

The Milling Work Centre dialogue window is activated on the main graphic window as shown in Fig. 11.77.

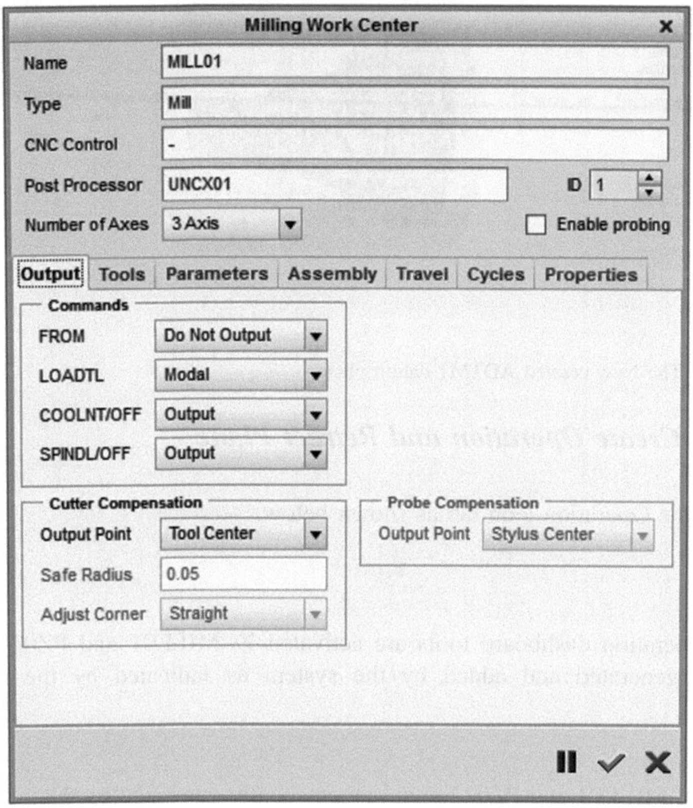

Fig. 11.77 Activated Milling Work Centre dialogue window

Click on the Check Mark icon to exit.

11.8.3 Create a New Datum Plane

Click on NC_ASM_TOP datum plane to highlight it.

Now click on Plane icon 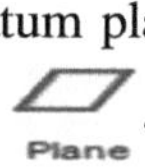
.

The Datum Plane dialogue box is activated on the main graphic window ≫ Add an offset value of 120 mm and then click the OK tab. A new Datum Plane called ADTM1 is created as indicated by the arrow in Fig. 11.78.

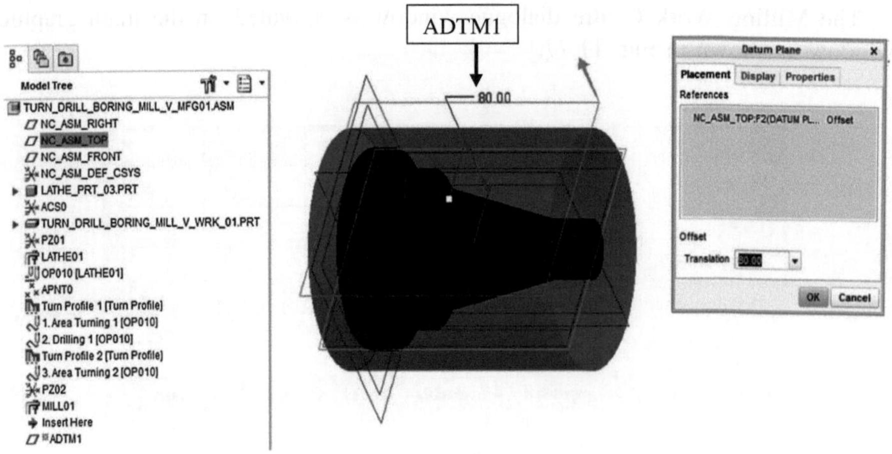

Fig. 11.78 The New created ADTM1 datum plane

11.8.4 Create Operation and Retract Plane

Click on the Operation icon tab as shown below.

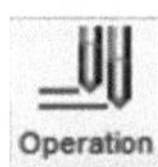

The Operation dashboard tools are activated ≫ MILL01 and PZ02 are automatically generated and added by the system as indicated by the arrows in Fig. 11.79.

> Note: If MILL01 and PZ02 are not automatically generated by the system in their respective section boxes as indicated by the arrows, manually select and input them yourself.

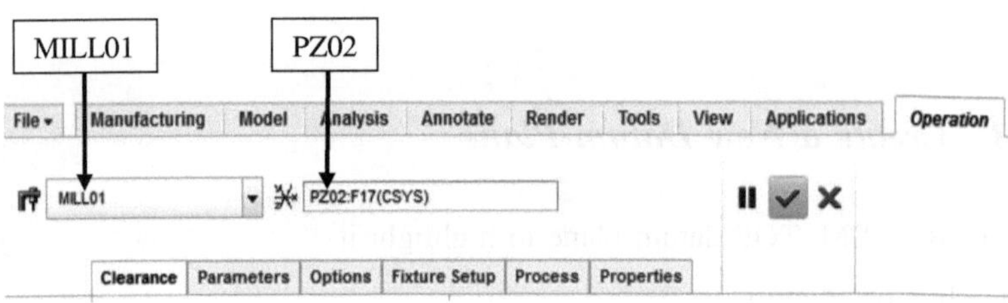

Fig. 11.79 MILL01 and PZ02 in there respective section boxes

Create Retract Plane

Click on the Clearance tab to activate its panel ≫ On the Retract group, click on the Type section box to activate its drop-down menu ≫ Now click on Plane on the drop-down list ≫ Click on the Reference section box, and now click on the newly created ADTM1 datum plane ≫ In the Value section box, type 25 mm as the offset distance as shown in Fig. 11.80.

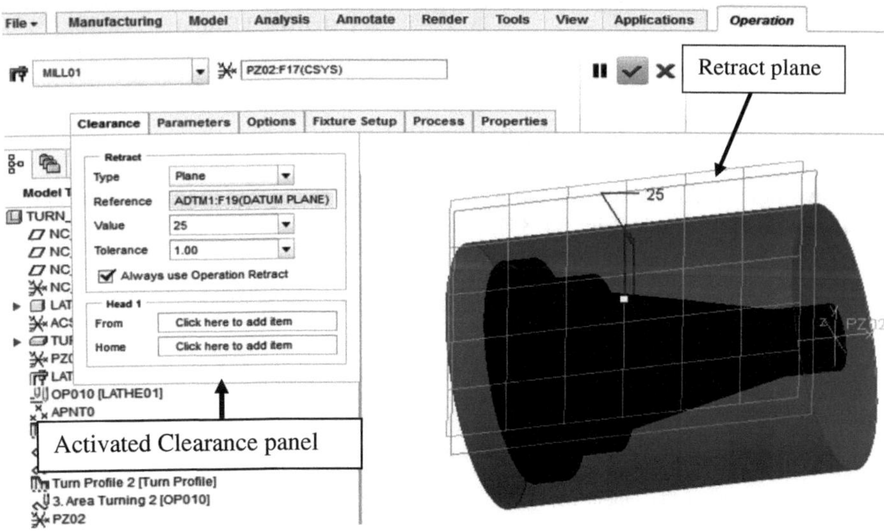

Fig. 11.80 Clearance panel parameters and created retract plane

11.8.5 Create New Mill Volume

Click on Mill Volume icon tab as indicated by the arrow below.

The Mill Volume dashboard tools are activated. See Fig. 11.81 in concise form.

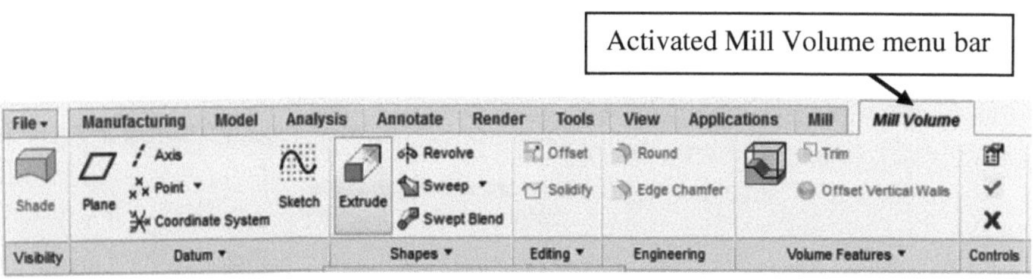

Fig. 11.81 Activated Mill Volume dashboard

Click on the Extrude icon in Fig. 11.81 as shown below.

The Extrude dashboard tools are activated as shown in concise form in Fig. 11.82.

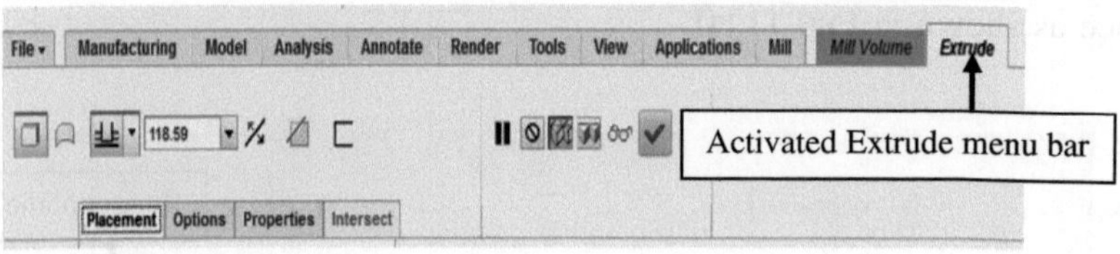

Fig. 11.82 Activated Extrude dashboard

Now click on the Placement tab as to activate its panel ≫ On the Sketch section, click on the Define tab (Fig. 11.83).

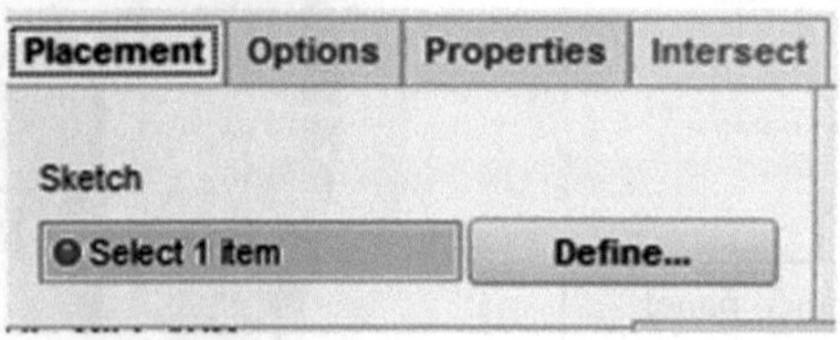

Fig. 11.83 Activated Placement panel

The Sketch dialogue box is activated on the main graphic window ≫ Now click on the Placement tab to activate its content ≫ On the Sketch Plane section, click on the Plane section box; now click on the newly created Datum Plane (ADTM1). Make sure that the Orientation box is set to Right in the Orientation section box. As shown in Fig. 11.84.

Fig. 11.84 Activated Sketch dialogue box

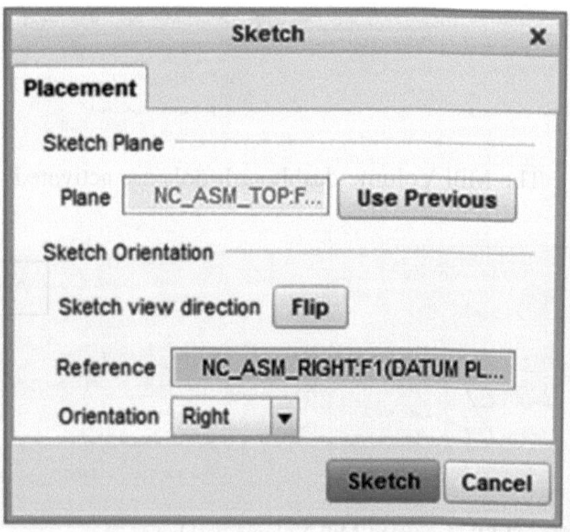

Now click on Sketch tab to exit the Sketch dialogue box.

The Sketch tools are activated as shown. See Fig. 11.85.

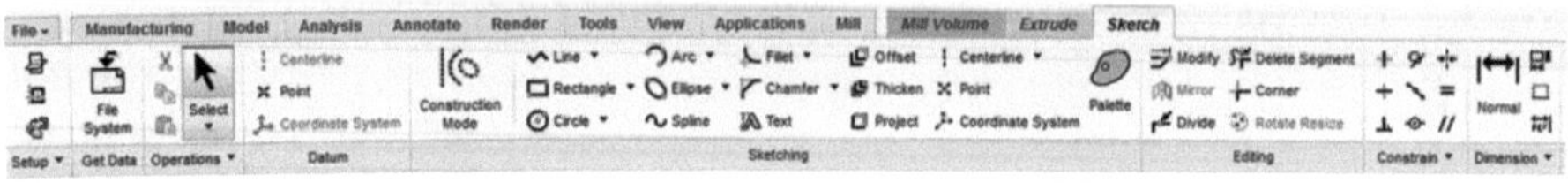

Fig. 11.85 Activated Sketch tools

Now click on the Sketch view icon  on the Sketch toolbar to orient the Workpiece to the correct sketch plane/view.

Project Sketch

Click on Project icon tab 

The Type dialogue box is activated on the main graphic window as shown in Fig. 11.86.

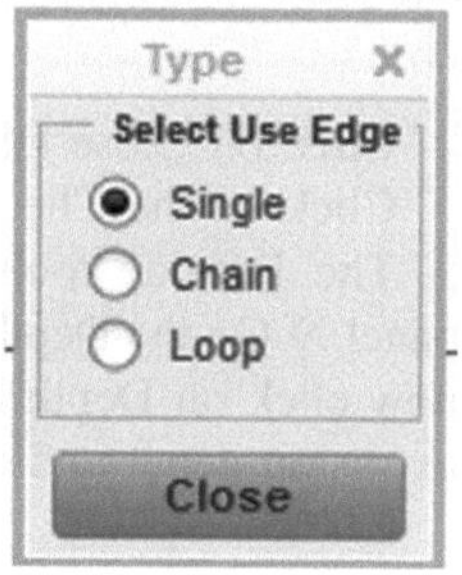

Fig. 11.86 Activated Type dialogue box

Click on the four Lines of the rectangle as indicated by the arrow below in Fig. 11.87.

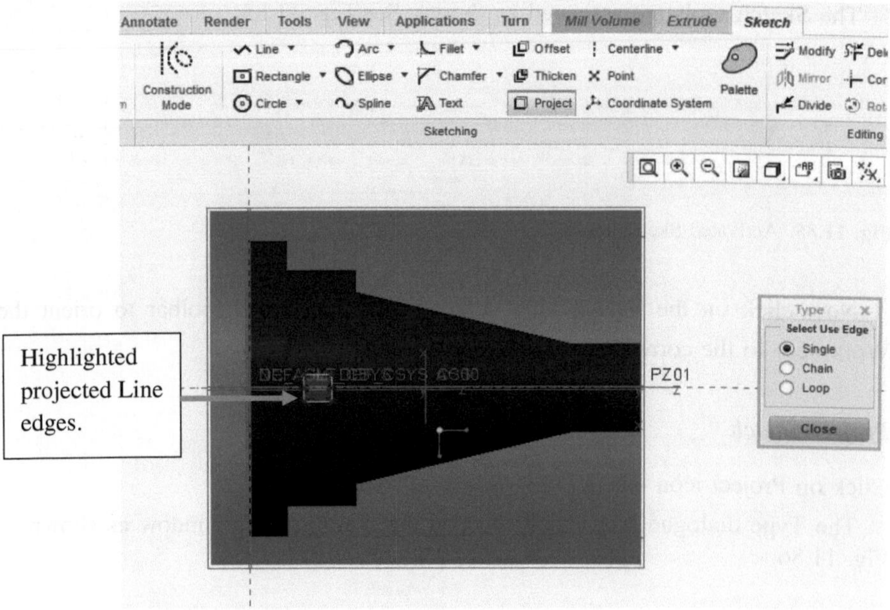

Fig. 11.87 Projecting the lines into the Sketch plane

Click on Close tab on the Type dialogue box.

Click on the Check Mark icon to exit the activated Sketch tools.

The Extrude tools are activated ≫ Click on the Options tab to activate its panel ≫ On the Depth group, click on Side1 section box downward pointing arrow, now click on Depth icon (Blind) on the drop-down list ≫ Type 80 mm to be the extrusion depth and change the extrusion direction if need be. See Fig. 11.88.

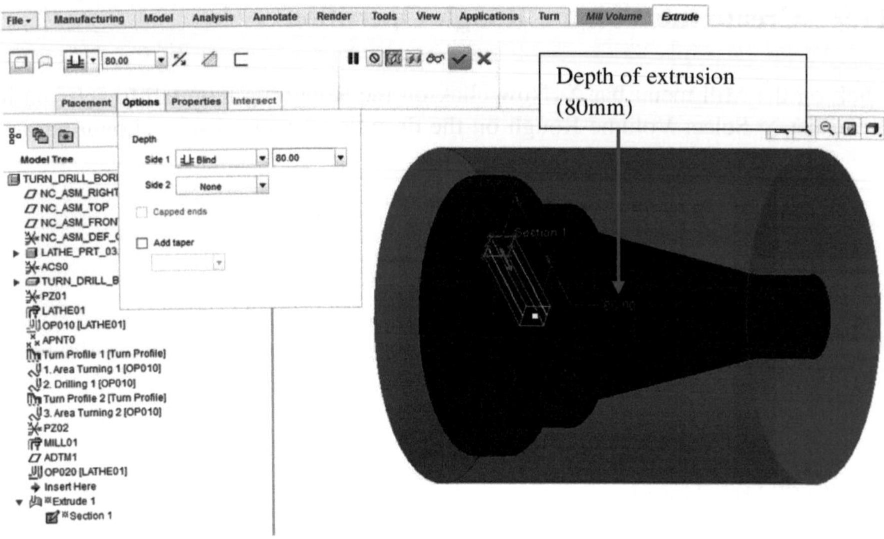

Fig. 11.88 Creating the Mill Volume via extrusion

Click on the Check Mark icon to exit the activated Extrude tools. The created Mill Volume is indicated by the arrow in Fig. 11.89.

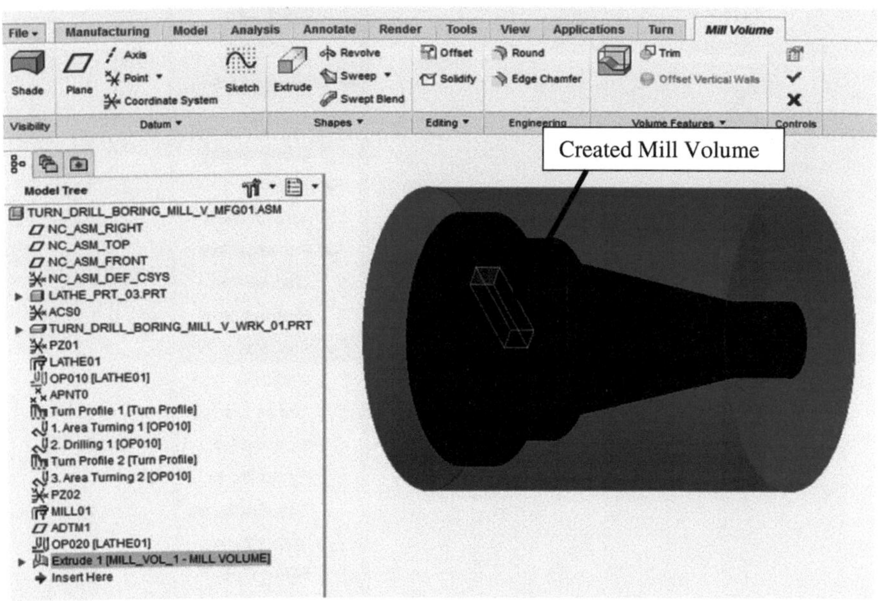

Fig. 11.89 Created Mill Volume

Click on the Check Mark icon to exit the activated Mill Volume tools.

11.9 Create the Volume Rough Operation

Click on the Mill menu bar ≫ Now click on the Roughing icon tab to activate its menu list ≫ Select Volume Rough on the drop-down menu list as shown below.

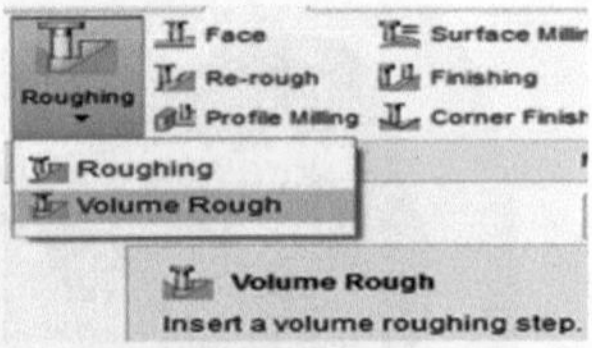

The NC SEQUENCE Menu Manager dialogue box is activated as shown in Fig. 11.90.

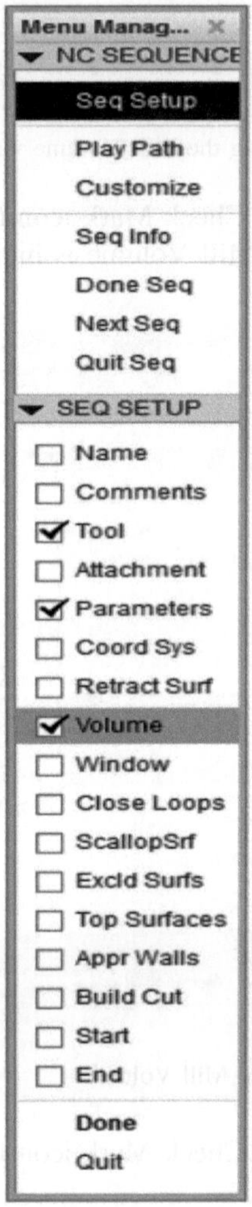

Fig. 11.90 Activated NC SEQUENCE Menu Manager dialogue box

Click on Done.

The Tools Setup dailogue window is activated ≫ To add new End Mill Tool ≫ Click on the General tab ≫ In the Name section box, type T0004 ≫ Click on the Type section box, now click on END MILL on the activated drop-down list ≫ Add dimensions to the END MILL Tool ≫ Now click on the Apply tab. See Fig. 11.91.

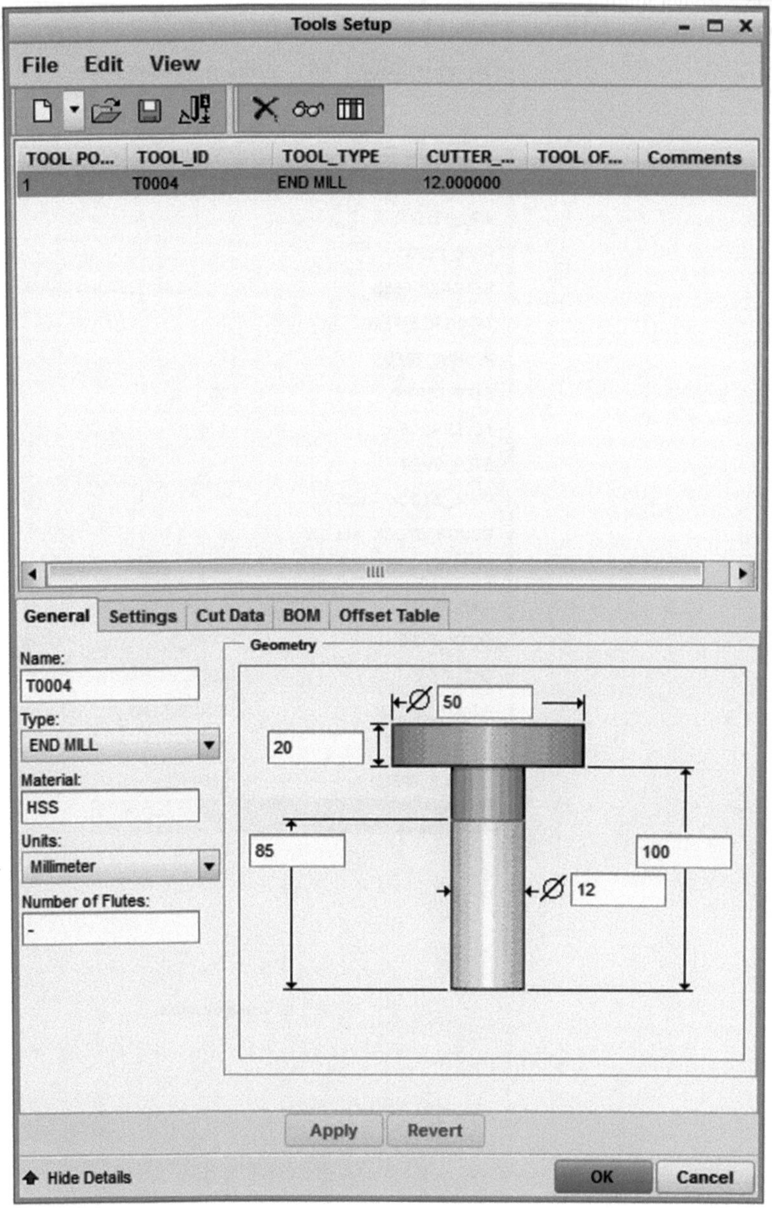

Fig. 11.91 Created END MILL Tool (T0004)

Click on the OK tab to exit.

The Edith Parameters of Sequence ''Volume Milling'' dialogue window is activated on the main graphic window ≫ Input values to the required parameters. See Fig. 11.92.

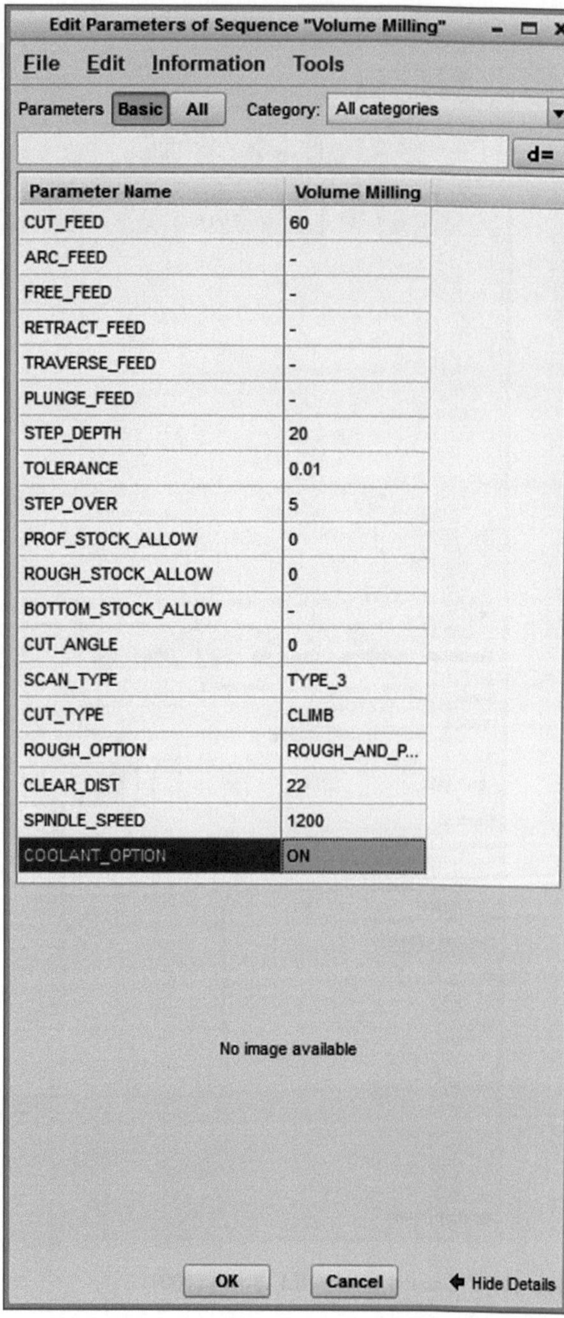

Fig. 11.92 Parameter values for Volume Rough milling operation

Click on OK tab to exit.

System is asking for the created Mill Volume to be added on the message bar at the bottom of the main window ≫ Now click on the created Mill Volume on the part.

Alternatively, click on (EXTRUDE1 [MIL_VOL_1-MILL VOLUME]) on the Model Tree as the volume to be Volume Milled when prompted to do so by the system.

11.9.1 Activate the Play Path Operation

Click on Play Path on the NC SEQUENCE group. Now click on Screen Play on the PLAY PATH group as shown on the Menu Manager dialogue box in Fig. 11.93.

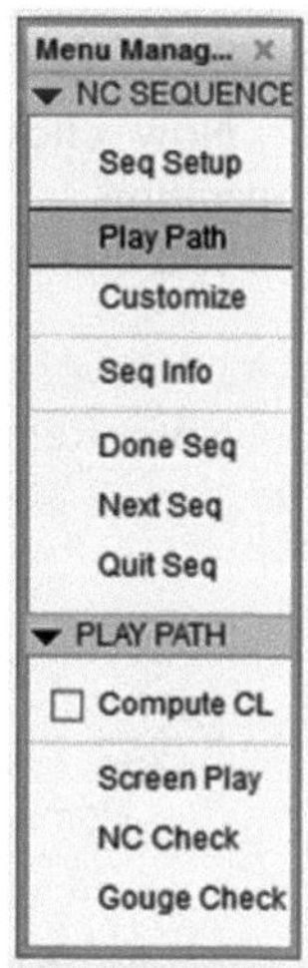

Fig. 11.93 Activating on-screen Volume Rough milling operation

The PLAY PATH dialogue box and the cutting Tool are activated on the main graphic window as shown in Fig. 11.94.

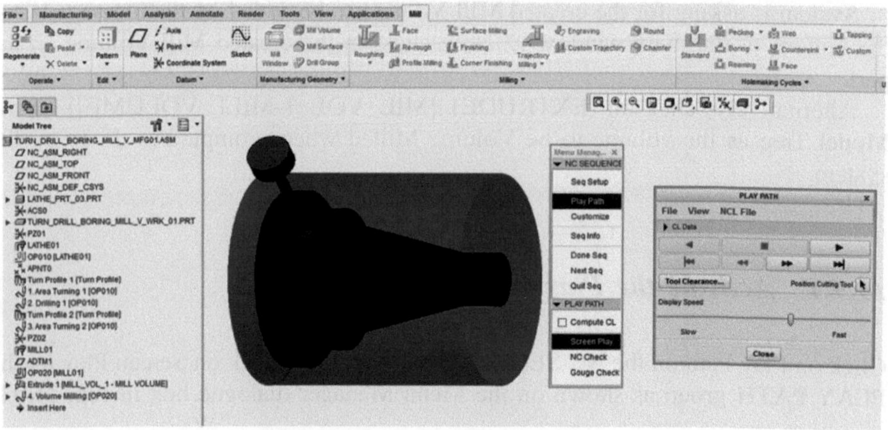

Fig. 11.94 Activated PLAY PATH dialogue box and cutting Tool

Now click on the Play tab to start the on-screen Volume Rough milling operation.

The end of the Volume Rough milling operation is as shown in Fig. 11.95.

Note: Remember to activate the wireframe display.

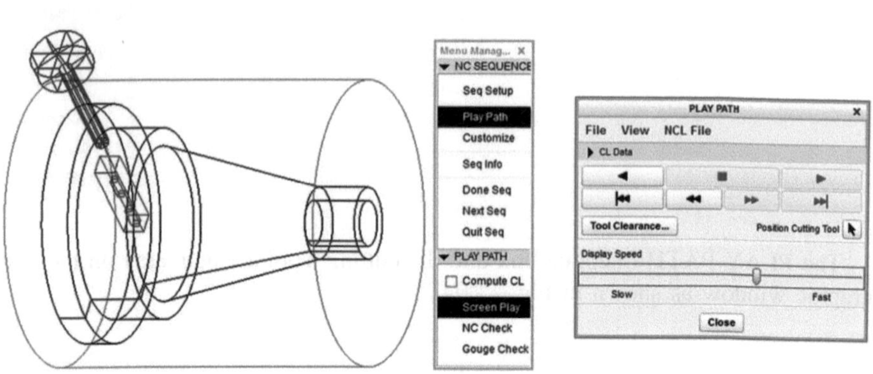

Fig. 11.95 End of the Volume Rough milling operation in Wireframe

Click on the Close tab to exit the PLAY PATH dialogue box.

11.9.2 *Material Removal Simulation for Volume Rough Milling*

Right click the mouse on Volume Milling [OP020] on the Model Tree ≫ Select Material Removal Simulation on the drop-down menu list.

The NC CHECK Menu Manager dialogue box is activated on the main graphic window as shown in Fig. 11.96.

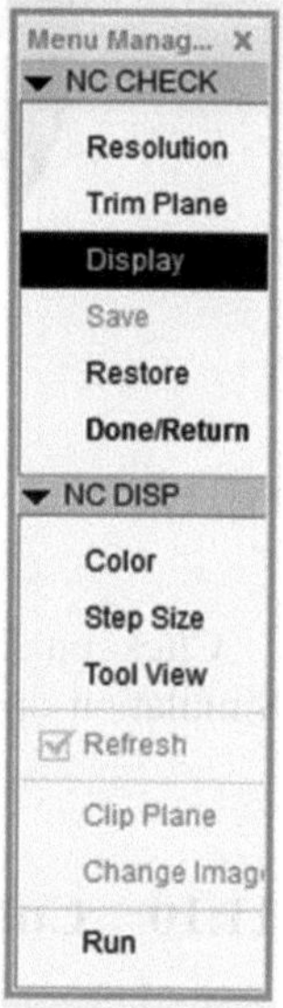

Fig. 11.96 Activated NC CHECK and NC DISP groups

Click on Step Size on the NC DISP group and change the automatically generated value to a lower value, this will slow down the Material Simulation Removal process.

Click on Run on the NC DISP group to start the Material Removal Simulation process as shown in Fig. 11.97.

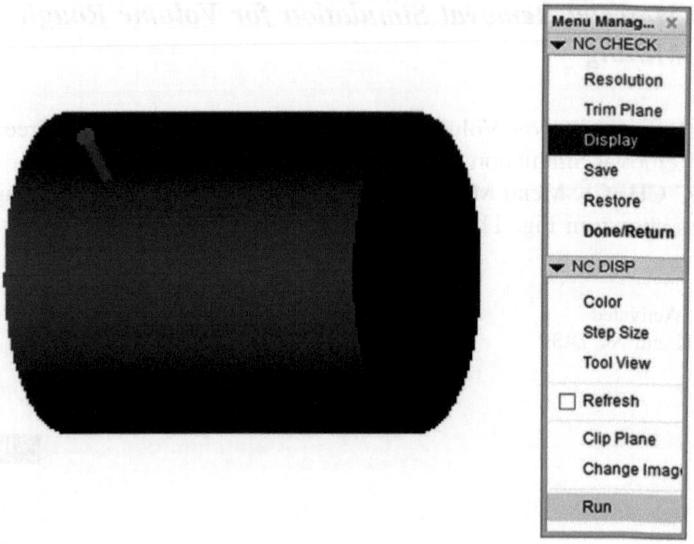

Fig. 11.97 Activated Material Removal Simulation process

Click on Done/Return on NC CHECK group to exit the Material Removal Simulation operation.

11.10 Cuter Location (CL) Data for All Operations

Click on Save a CL File icon to activate the drop-down menu list ≫ Now click on Save a CL File on the menu list as shown below.

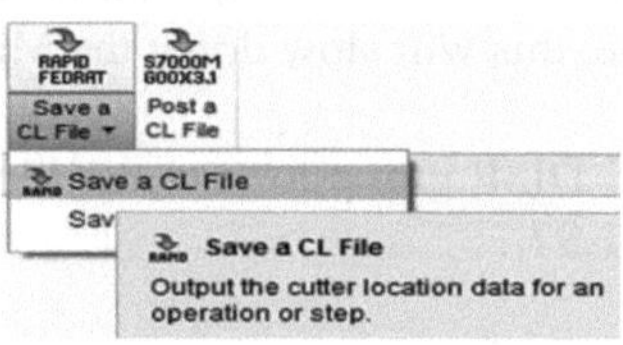

The SELECT FEAT Menu Manager dialogue box is activated on the main graphic window ≫ Click on Operation and then click on OP010 as highlighted in Fig. 11.98.

Note: To generate the Operation G-codes for OP020, repeat the same steps as described in Sects. 11.10 and 11.11.

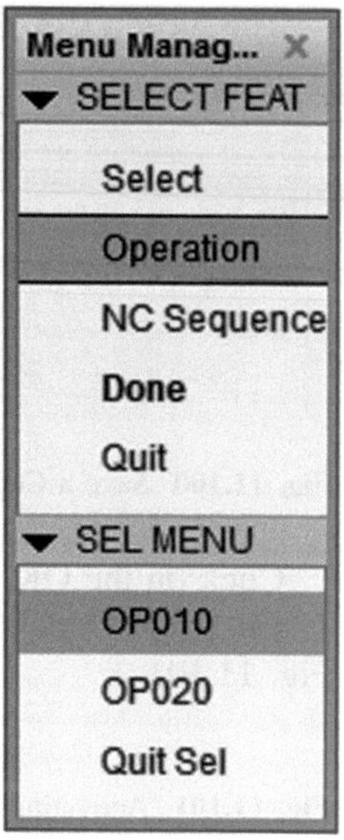

Fig. 11.98 Activated SELECT FEAT and SEL MENU group

Once OP010 is clicked, the PATH Menu Manager dialogue box is activated on the main graphic window ≫ Click on File and Check Mark the CL File and Interactive square boxes, respectively, on the OUTPUT TYPE drop-down list as illustrated in Fig. 11.99.

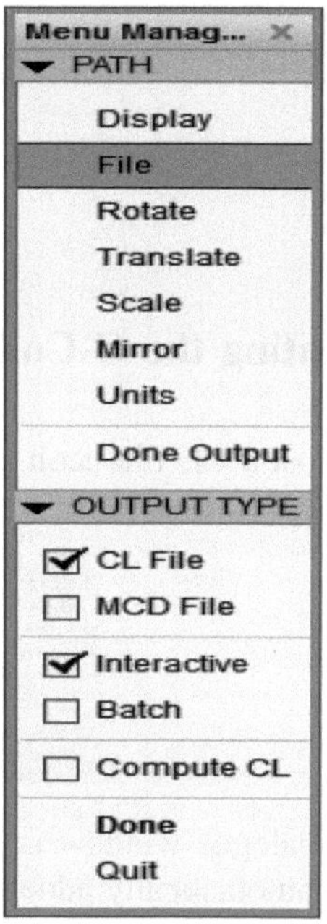

Fig. 11.99 Activated PATH Menu Manager dialogue box

Click on Done.

The Save a Copy dialogue window is activated. See Fig. 11.100.

Fig. 11.100 Save a Copy dialogue window in concise form

Click on the OK tab to exit.

Click on Done Output on the PATH group as indicated by the arrow in Fig. 11.101.

Fig. 11.101 Activating Done Output on the PATH group

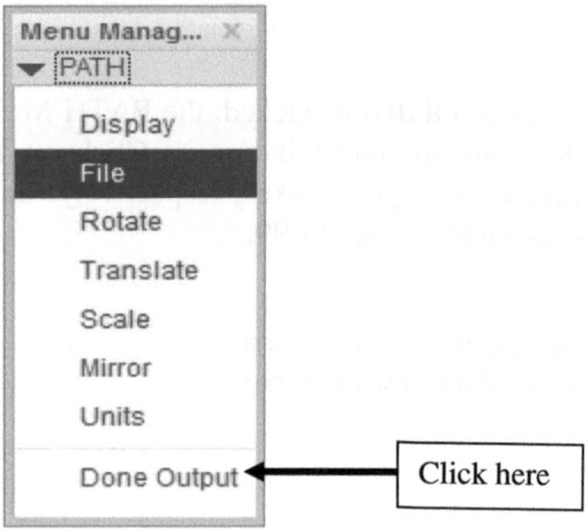

11.11 Creating the G-Code Data

Click on the Post a CL File icon as shown below.

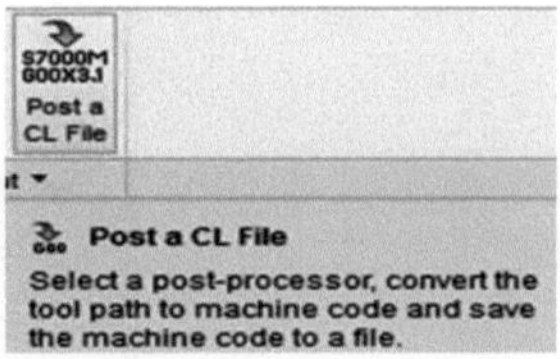

The Open dialogue window is activated on the main graphic window with the CL File name automatically added by the system. See Fig. 11.102.

Fig. 11.102 Activated Open dialogue window in concise form

Click on the Open tab to exit.

PP OPTIONS Menu Manager dialogue box is activated on the main graphic ≫ Check Mark the square boxes of Verbose and Trace if they are not automatically Checked Marked by the system. See Fig. 11.103.

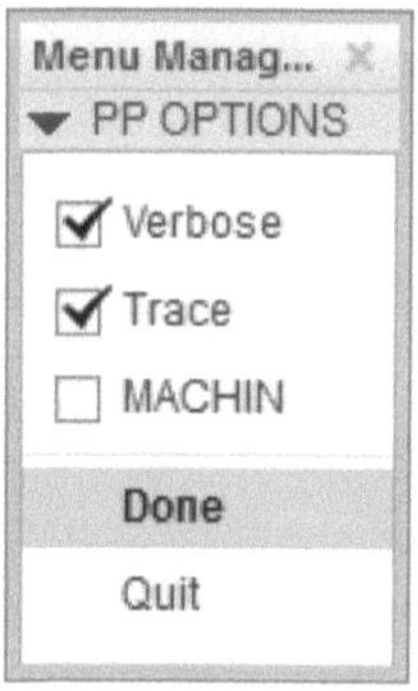

Click on Done.

Fig. 11.103 Activated PP OPTIONS group with Verbose and Trace check marked

The PP LIST Menu Manager dialogue box is activated on the main graphic window as shown in Fig. 11.104.

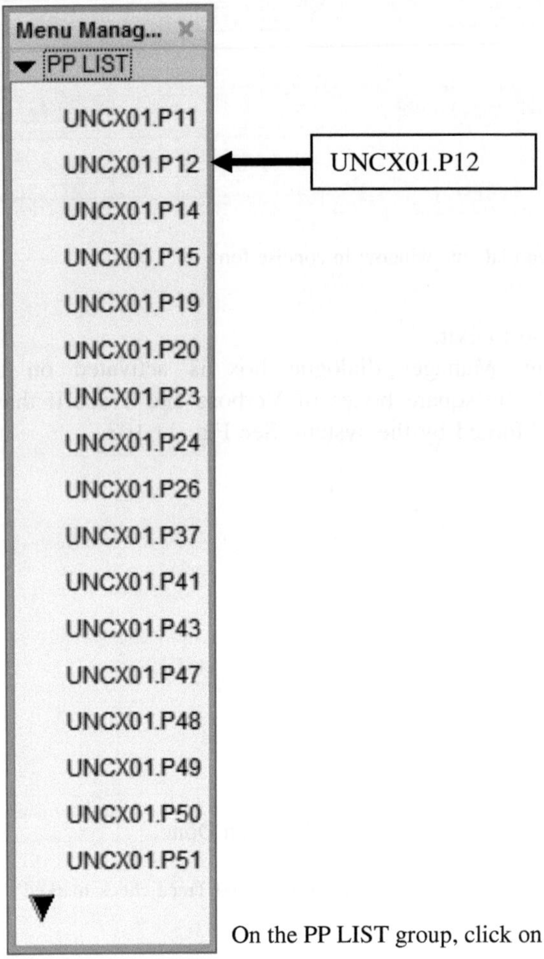

On the PP LIST group, click on UNCX01.P12.

Fig. 11.104 Activated PP LIST Menu Manager dialogue box

Note: The UNCX01.P12 generates post processor for four-axis Milling but it is also capable of generating G-codes for three-axis Milling also.

The INFORMATION WINDOW dialogue Window is activated on the main graphic window. See Fig. 11.105.

Note: Information about the run process is displayed on the INFORMATION WINDOW.

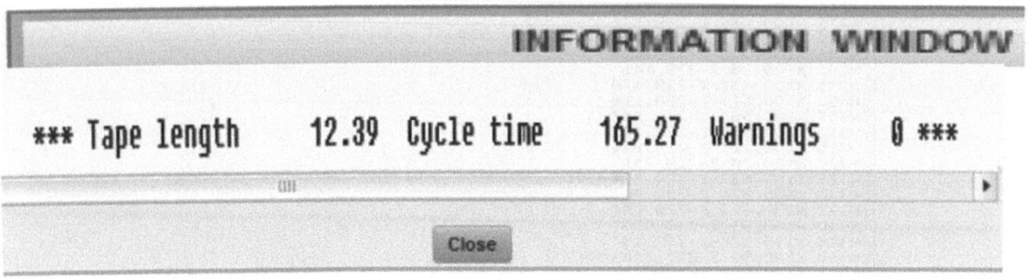

Fig. 11.105 Activated INFORMATION WINDOW in concise form

Click on the Close tab to exit.

The generated G-codes data are stored as a TAP File in the chosen work directory. Open the TAP File using Notepad to view the generated G-codes.

Some of the G-codes data generated in this tutorial are as shown below and on the proceeding pages.

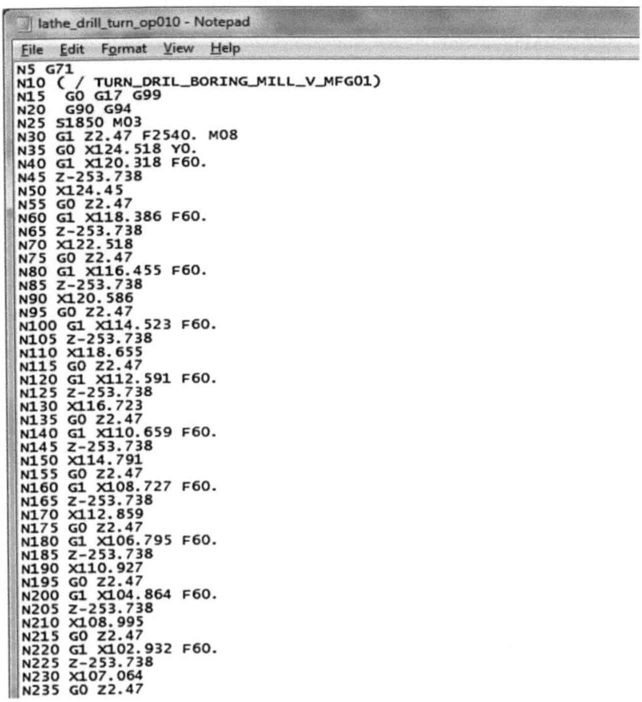
lathe_drill_turn_op010 - Notepad

File Edit Format View Help

```
N5 G71
N10 ( / TURN_DRIL_BORING_MILL_V_MFG01)
N15  G0 G17 G99
N20  G90 G94
N25 S1850 M03
N30 G1 Z2.47 F2540. M08
N35 G0 X124.518 Y0.
N40 G1 X120.318 F60.
N45 Z-253.738
N50 X124.45
N55 G0 Z2.47
N60 G1 X118.386 F60.
N65 Z-253.738
N70 X122.518
N75 G0 Z2.47
N80 G1 X116.455 F60.
N85 Z-253.738
N90 X120.586
N95 G0 Z2.47
N100 G1 X114.523 F60.
N105 Z-253.738
N110 X118.655
N115 G0 Z2.47
N120 G1 X112.591 F60.
N125 Z-253.738
N130 X116.723
N135 G0 Z2.47
N140 G1 X110.659 F60.
N145 Z-253.738
N150 X114.791
N155 G0 Z2.47
N160 G1 X108.727 F60.
N165 Z-253.738
N170 X112.859
N175 G0 Z2.47
N180 G1 X106.795 F60.
N185 Z-253.738
N190 X110.927
N195 G0 Z2.47
N200 G1 X104.864 F60.
N205 Z-253.738
N210 X108.995
N215 G0 Z2.47
N220 G1 X102.932 F60.
N225 Z-253.738
N230 X107.064
N235 G0 Z2.47
```

lathe_drill_turn_op010 - Notepad

File Edit Format View Help

```
N240 G1 X101. F60.
N245 Z-253.738
N250 X105.132
N255 G0 Z2.47
N260 G1 X99. F60.
N265 Z-229.377
N270 X100.
N275 X100.089 Z-229.381
N280 X100.178 Z-229.393
N285 X100.265 Z-229.413
N290 X100.35 Z-229.441
N295 X100.432 Z-229.476
N300 X100.511 Z-229.518
N305 X100.586 Z-229.567
N310 X100.656 Z-229.623
N315 X100.721 Z-229.684
N320 X100.78 Z-229.751
N325 X100.833 Z-229.824
N330 X100.879 Z-229.9
N335 X100.918 Z-229.981
N340 X100.95 Z-230.064
N345 X100.974 Z-230.15
N350 X100.99 Z-230.238
N355 X100.999 Z-230.327
N360 X101. Z-230.377
N365 X103.2
N370 G0 Z2.47
N375 G1 X97. F60.
N380 Z-229.377
N385 X101.2
N390 G0 Z2.47
N395 G1 X95. F60.
N400 Z-229.377
N405 X99.2
N410 G0 Z2.47
N415 G1 X93. F60.
N420 Z-229.377
N425 X97.2
N430 G0 Z2.47
N435 G1 X91. F60.
N440 Z-229.377
N445 X95.2
N450 G0 Z2.47
N455 G1 X89. F60.
N460 Z-229.377
N465 X93.2
N470 G0 Z2.47
N475 G1 X87. F60.
N480 Z-229.377
N485 X91.2
N490 G0 Z2.47
N495 G1 X85. F60.
N500 Z-229.377
N505 X89.2
N510 G0 Z2.47
N515 G1 X83. F60.
N520 Z-229.377
N525 X87.2
N530 G0 Z2.47
N535 G1 X81. F60.
N540 Z-229.377
N545 X85.2
N550 G0 Z2.47
N555 G1 X79. F60.
N560 Z-184.377
N565 X80.
N570 X80.089 Z-184.381
N575 X80.178 Z-184.393
N580 X80.265 Z-184.413
N585 X80.35 Z-184.441
N590 X80.432 Z-184.476
N595 X80.511 Z-184.518
N600 X80.586 Z-184.567
N605 X80.656 Z-184.623
N610 X80.721 Z-184.684
N615 X80.78 Z-184.751
N620 X80.833 Z-184.824
N625 X80.879 Z-184.9
N630 X80.918 Z-184.981
N635 X80.95 Z-185.064
N640 X80.974 Z-185.15
N645 X80.99 Z-185.238
N650 X80.999 Z-185.327
N655 X81. Z-185.377
N660 X83.2
N665 G0 Z2.47
N670 G1 X77. F60.
N675 Z-184.377
N680 X81.2
N685 G0 Z2.47
N690 G1 X75. F60.
N695 Z-184.377
N700 X79.2
N705 G0 Z2.47
```

lathe_drill_turn_op010 - Notepad

File Edit Format View Help

```
N710 G1 X73. F60.
N715 Z-184.377
N720 X77.2
N725 G0 Z2.47
N730 G1 X71. F60.
N735 Z-184.377
N740 X75.2
N745 G0 Z2.47
N750 G1 X69. F60.
N755 Z-184.377
N760 X73.2
N765 G0 Z2.47
N770 G1 X67. F60.
N775 Z-184.377
N780 X71.2
N785 G0 Z2.47
N790 G1 X65. F60.
N795 Z-181.244
N800 X65.781 Z-184.377
N805 X69.2
N810 G0 Z2.47
N815 G1 X63. F60.
N820 Z-173.222
N825 X65. Z-181.244
N830 X67.2
N835 G0 Z2.47
N840 G1 X61. F60.
N845 Z-165.201
N850 X63. Z-173.222
N855 X65.2
N860 G0 Z2.47
N865 G1 X59. F60.
N870 Z-157.179
N875 X61. Z-165.201
N880 X63.2
N885 G0 Z2.47
N890 G1 X57. F60.
N895 Z-149.158
N900 X59. Z-157.179
N905 X61.2
N910 G0 Z2.47
N915 G1 X55. F60.
N920 Z-141.136
N925 X57. Z-149.158
N930 X59.2
N935 G0 Z2.47
N940 G1 X53. F60.
N945 Z-133.114
N950 X55. Z-141.136
N955 X57.2
N960 G0 Z2.47
N965 G1 X51. F60.
N970 Z-125.093
N975 X53. Z-133.114
N980 X55.2
N985 G0 Z2.47
N990 G1 X49. F60.
N995 Z-117.071
N1000 X51. Z-125.093
N1005 X53.2
N1010 G0 Z2.47
N1015 G1 X47. F60.
N1020 Z-109.05
N1025 X49. Z-117.071
N1030 X51.2
N1035 G0 Z2.47
N1040 G1 X45. F60.
N1045 Z-101.028
N1050 X47. Z-109.05
N1055 X49.2
N1060 G0 Z2.47
N1065 G1 X43. F60.
N1070 Z-93.007
N1075 X45. Z-101.028
N1080 X47.2
N1085 G0 Z2.47
N1090 G1 X41. F60.
N1095 Z-84.985
N1100 X43. Z-93.007
N1105 X45.2
N1110 G0 Z2.47
N1115 G1 X39. F60.
N1120 Z-76.963
N1125 X41. Z-84.985
N1130 X43.2
N1135 G0 Z2.47
N1140 G1 X37. F60.
N1145 Z-68.942
N1150 X39. Z-76.963
N1155 X41.2
N1160 G0 Z2.47
N1165 G1 X35. F60.
N1170 Z-60.92
N1175 X37. Z-68.942
```

lathe_drill_turn_op010 - Notepad

File Edit Format View Help

```
N1180 X39.2
N1185 G0 Z2.47
N1190 G1 X33. F60.
N1195 Z-52.899
N1200 X35. Z-60.92
N1205 X37.2
N1210 G0 Z2.47
N1215 G1 X31. F60.
N1220 Z-44.877
N1225 X33. Z-52.899
N1230 X122.518
N1235 S1200
N1240 G0 Z20. M08
N1245 X180.
N1250 X0.
N1255 Z10.
N1260 G81 X0. Y0. Z-60. R10. F60.
N1265 G80
N1270 S1850 M03
N1275 G0 X-3.363
N1280 Z1.345 M08
N1285 G1 X.838 F60.
N1290 Z-49.5
N1295 X-1.363
N1300 G0 Z1.345
N1305 G1 X2.8 F60.
N1310 Z-49.5
N1315 X-1.363
N1320 G0 Z1.345
N1325 G1 X4.763 F60.
N1330 Z-49.5
N1335 X.6
N1340 G0 Z1.345
N1345 G1 X6.725 F60.
N1350 Z-49.5
N1355 X2.563
N1360 G0 Z1.345
N1365 G1 X8.688 F60.
N1370 Z-49.5
N1375 X4.525
N1380 G0 Z1.345
N1385 G1 X10.65 F60.
N1390 Z-49.5
N1395 X6.488
N1400 G0 Z1.345
N1405 G1 X12.613 F60.
N1410 Z-49.5
N1415 X8.45
N1420 G0 Z1.345
N1425 G1 X14.575 F60.
N1430 Z-49.5
N1435 X10.413
N1440 G0 Z1.345
N1445 G1 X16.538 F60.
N1450 Z-49.5
N1455 X12.375
N1460 G0 Z1.345
N1465 G1 X18.5 F60.
N1470 Z-49.5
N1475 X-1.363
N1480 M30
%
```

Chapter 12
Five-Axis Machining of Intricate Part

The aim of this tutorial is to expose learners to the importance and capability of five-axis CNC Milling Machine.

This Tutorial will cover Volume Rough milling operation using Mill Volume and Surface Milling. Drilling and Tapping operation will also be covered.

> Note: In practice, the Reference Model would be cost effective if manufactured through casting process as oppose to machining of a rectangular Stock to the required shape and dimensions.

The steps below will be covered in this tutorial.

- Start the Manufacturing application
- Import and Constrain Reference Model
- Add the Stock
- Create Two Programme Zeroes (Machine Coordinate Systems)
- Create the Work Centres
- Set up Operations
- Create Cutting Tools
- Create Mill Volumes for the Workpiece
- Create Volume Rough Milling operation sequences for each Mill Volume created
- Add Tools, Parameters and Mill Volume for the NC SEQUENCE
- Mill Surface will be created (use for smoothing out rough surfaces after Volume Rough machining)
- Create Cut Lines
- Drilling and Tapping cycle operation are created
- Generate Cutter Location (CL) Data for the whole operation
- Generate G-codes for the whole operation

P.O. Kanife, *Computer Aided Virtual Manufacturing Using Creo Parametric*,
DOI 10.1007/978-3-319-23359-8_12

12.1 Volume Rough Milling Using Mill Volume

12.1.1 Start Creo Parametric

Start Creo Parametric either on your computer desktop or from Programme ≫ Click on the New icon as shown below.

The New dialogue box is activated on the main window ≫ Click on the Manufacturing radio button on the Type group to make it active ≫ On the Sub-type group, click on the NC Assembly radio button to make it active ≫ Click on the Name section box and type Vol_Rough_5Axis_Mill ≫ Click on the square box of Use default template to clear the Check Mark icon. The New dialogue box is as shown in Fig. 12.1.

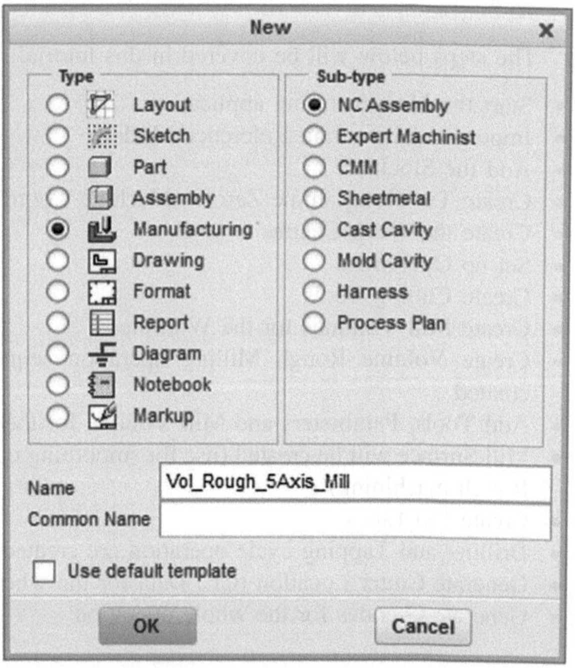

Fig. 12.1 Activating the Manufacturing application

Click on OK tab to proceed.

The New File Options dialogue window is activated on the main window ≫ Click on the "mmns_mfg_nc" on the Template group ≫ On the Parameters group, type your name in the MODELLED BY section box and in the DESCRIPTION section box, type Vol_Rough_5Axis_Mill as shown in Fig. 12.2.

Fig. 12.2 New File Options parameters

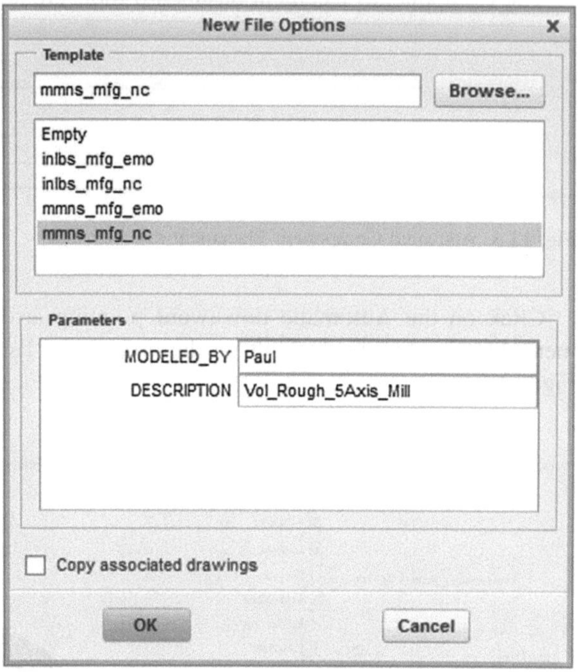

Click on OK tab to proceed.

Creo Parametric Manufacturing application main Graphic User Interface (GUI) window is activated.

12.1.2 *Import and Constrain the Reference Part*

To import the 3D Part into the Manufacturing main graphic window, click on the Reference Model icon tab to activate the drop-down menu list ≫ Now select Assemble Reference Model on the drop-down menu list as shown below.

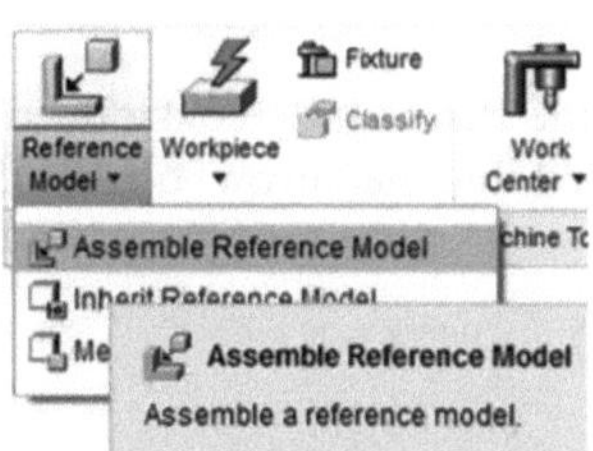

The Open dialogue window is activated on the main graphic window.

Make sure that the File name that you want to open is on the File name section box.

Click on the Open tab and the Reference Model is imported into the Manufacturing main graphic window.

The Component Placement dashboard tools are activated as shown in Fig. 12.3.

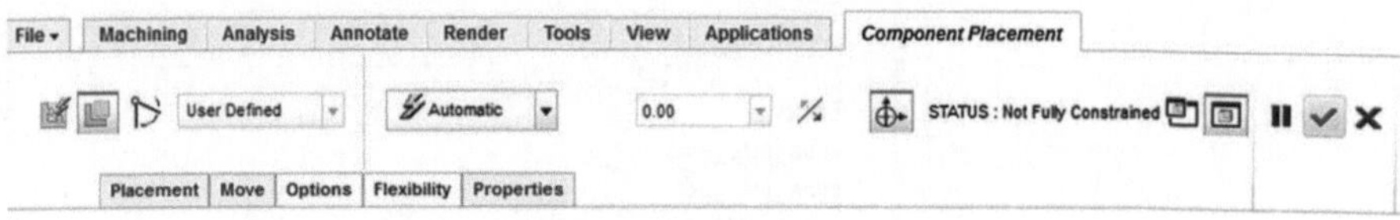

Fig. 12.3 Activated Component Placement dashboard

Click on the Automatic downward pointing arrow to activate the drop-down menu list ≫ Click on Default on the drop-down list as indicated by the arrow in Fig. 12.4.

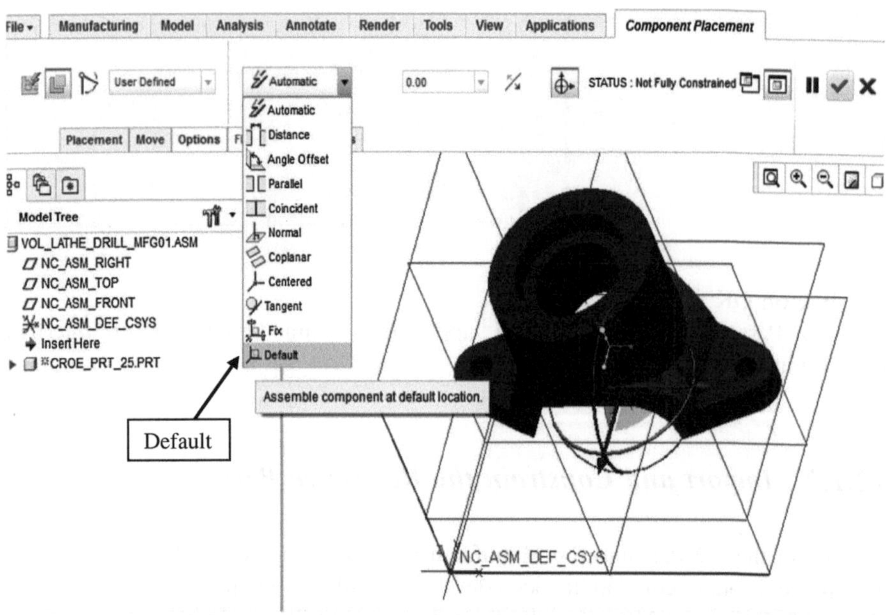

Fig. 12.4 Constraining the imported Part

Once Default is clicked, the Part is fully constrained on the Manufacturing main graphic window as indicated by the arrow on the STATUS bar as shown in Fig. 12.5.

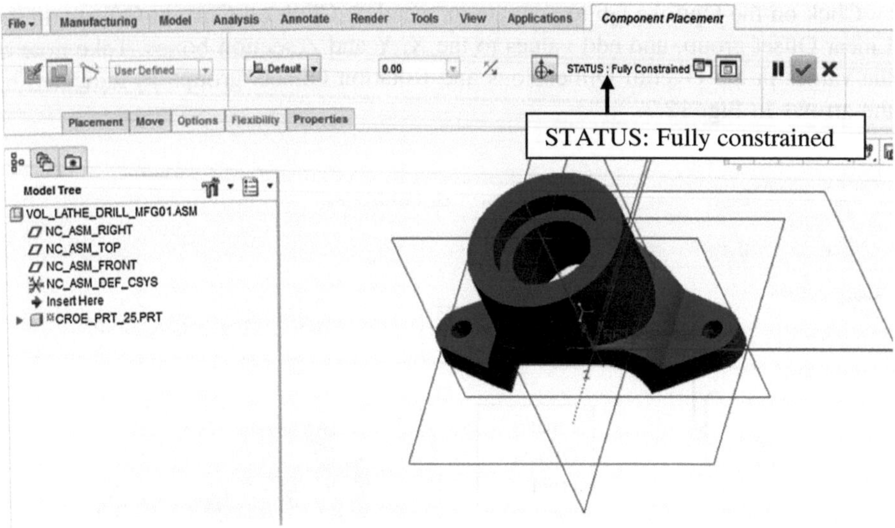

Fig. 12.5 Part now constrained on the main manufacturing GUI window

Click on the Check Mark icon to proceed.

12.1.3 Add Stock to the Part

Click on the Workpiece icon tab to activate the drop-down menu list ≫ Now click on Automatic Workpiece on the drop-down menu list as shown below.

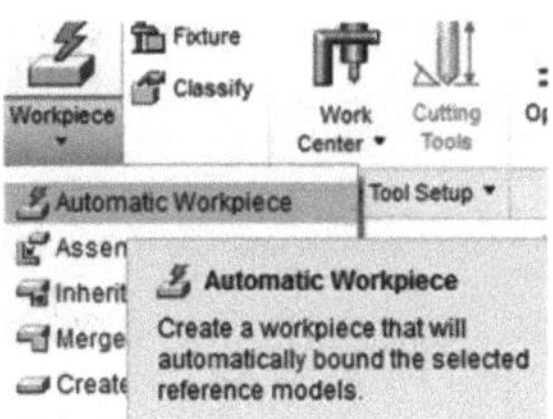

The Auto Workpiece Creation dashboard tools are activated as shown in Fig. 12.6.

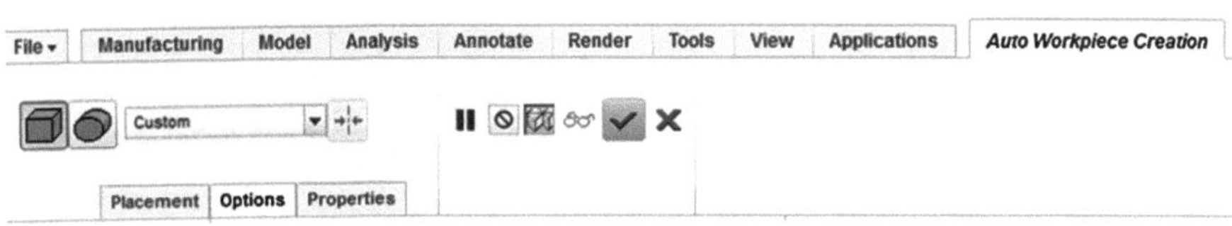

Fig. 12.6 Activated Auto Workpiece Creation dashboard

Click on the Options tab to activate its panel ≫ Click on Current Offsets on the Linear Offset group, and add values to the X, Y and Z section boxes. Take note of the values in the Overall Dimensions and Rotation Offsets groups as indicated by the arrows in Fig. 12.7.

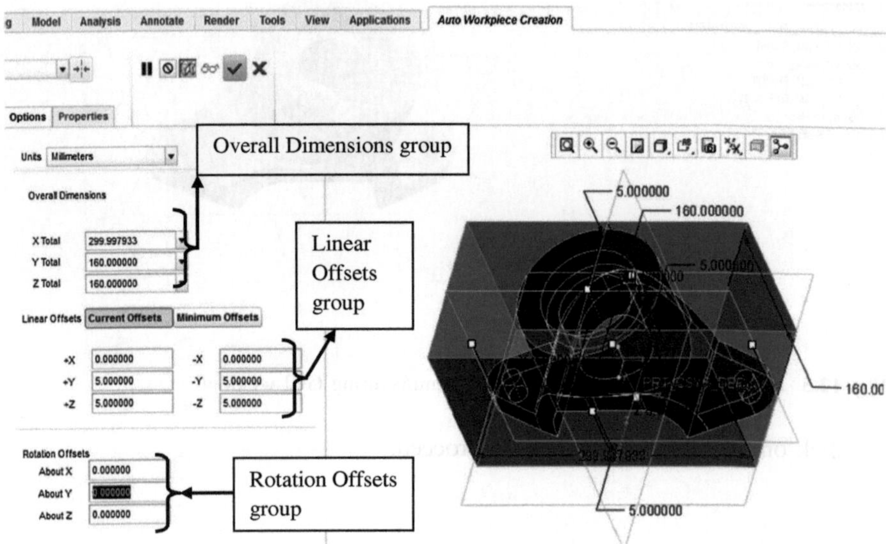

Fig. 12.7 Activated Options Panel showing Wokpiece dimensions

Click on the Check Mark icon to exit the Auto Workpiece Creation. Stock is now added into Part as shown in Fig. 12.8.

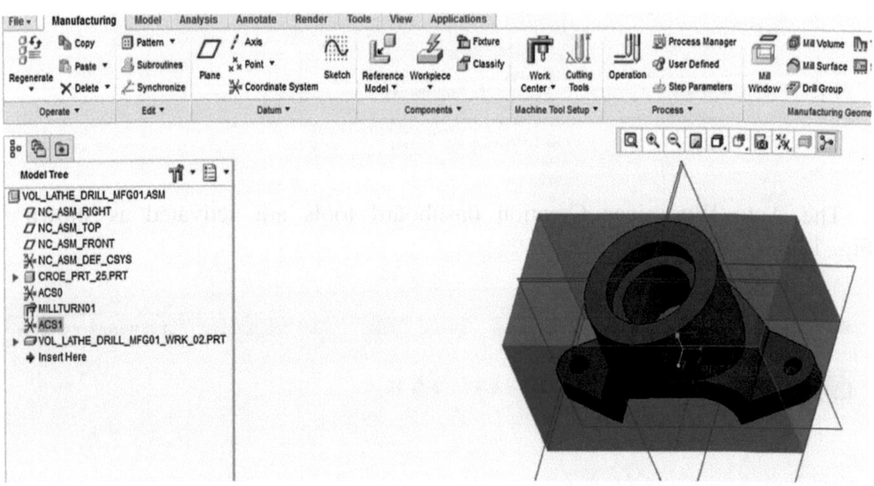

Fig. 12.8 Stock now added to Part

12.1.4 Suppress Holes

Note: The Holes on Part have to be suppressed to avoid Tool gouging.

Go to the Model Tree ≫ Click on the downward arrow on CREO_PRT_25.PRT to expand its content ≫ Right click on the mouse on each Extrude that represent each created Hole individually ≫ Now click on Suppress on the drop-down menu list ≫ The Suppress dialogue box is activated ≫ Click on the Ok tab to completely hide the Hole on the Part. Repeat the same steps to the remaining Extruded Holes.

The Holes on Part are now suppressed as shown in Fig. 12.9.

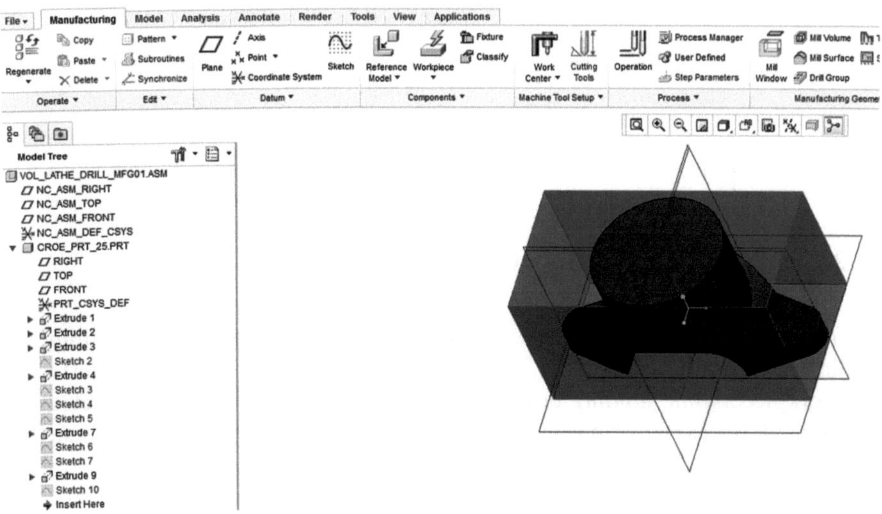

Fig. 12.9 Holes on Part now suppressed

Note: The suppressed Holes will be un-suppressed during the Drilling operation later in this tutorial.

12.2 Create Coordinate Systems for the Workpiece

Click on the Coordinate System icon tab as shown below.

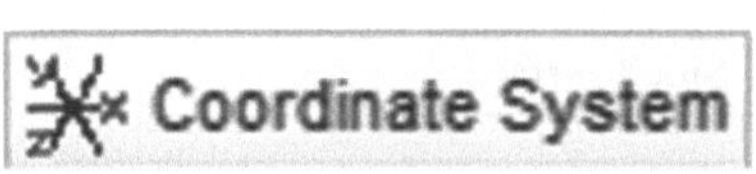

The Coordinate System dialogue box is activated as shown in Fig. 12.10.

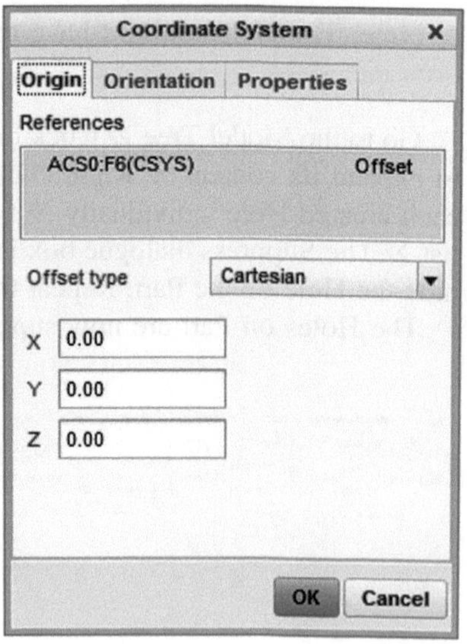

Fig. 12.10 Coordinate System dialogue box

12.2.1 Create Top (First) Coordinate System

To create the First Coordinate System (Programme Zero) for the Top Milling and Drilling operations, click on the Top surface, Side1 surfaces and Side2 surface of the Workpiece while holding down the Ctrl key when clicking. The selected Top and Side surfaces will appear on the Reference section box as illustrated in Fig. 12.11.

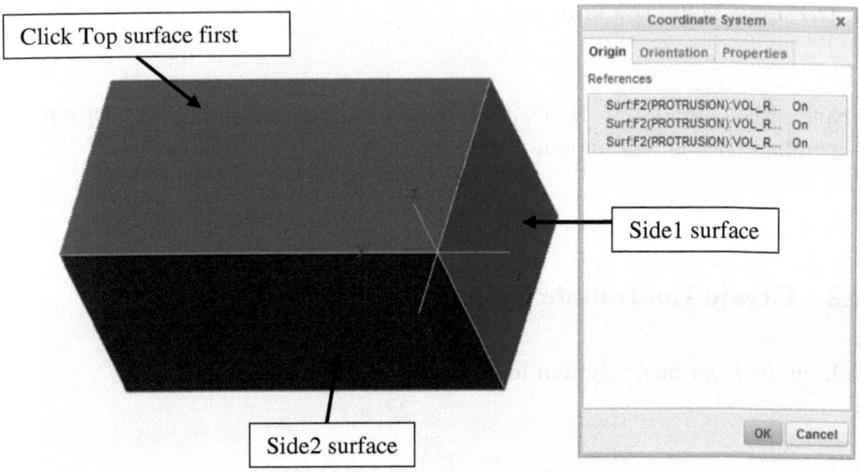

Fig. 12.11 Creating the Coordinate System

To confirm that the *X*, *Y* and *Z* coordinate axes are in their correct orientation, click on the Orientation tab to activate its content. If the axis are not in the correct orientation, change "to determine" and "to project" boxes as shown in Fig. 12.12.

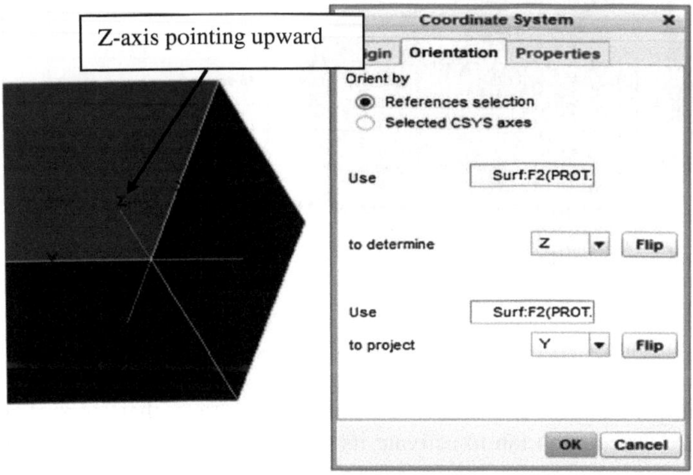

Fig. 12.12 Correct orientations of *X*, *Y* and *Z* axes of Coordinate System

The Coordinate System axis orientation is correct because the *Z* axis is pointing upward. Click on OK tab to exit the coordinate System dialogue box.

12.2.2 Create Bottom (Second) Coordinate System

To create the Second Coordinate System for the Bottom Milling operation, re-orientate the Workpiece to the opposite position using the mouse movement.

Activate the Coordinate System dialogue box again. Now click on the Bottom surface and two Side surfaces of the Workpiece while holding down the Ctrl key when clicking. The selected Bottom and Side surfaces will appear on the Reference section box as shown in Fig. 12.13.

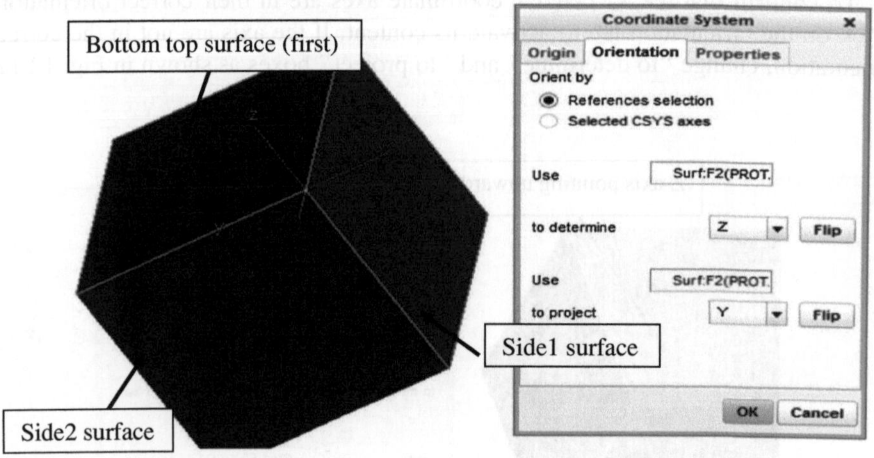

Fig. 12.13 Creating Coordinate System on the Workpiece bottom

To confirm that the *X*, *Y* and *Z* coordinate axes are in their correct orientation, click on the Orientation tab to activate its content. If the axis are not in the correct orientation, and change the "to determine" and the "to project" boxes as shown in Fig. 12.14.

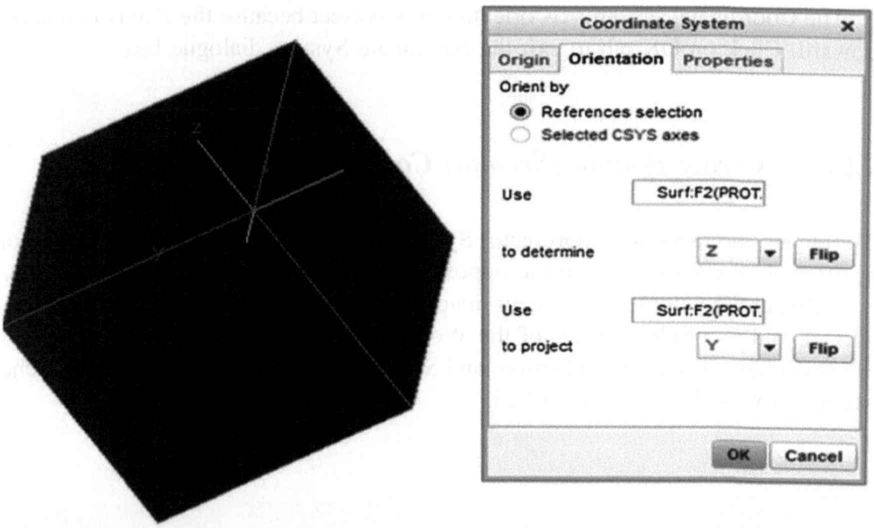

Fig. 12.14 Correct orientation of *X*, *Y* and *Z* axes of the bottom Coordinate System

Click on the OK tab to exit.

Note: New ACS1 and ACS2 Coordinate Systems (Programme Zeroes) are created. The new ACS1 and ACS2 created will be re-named to ACS2 and ACS3, respectively, in this tutorial. Slow click on the ACS1 on the Model Tree and type ACS2 ≫ Highlight ACS2 and right click the mouse and select Rename on the drop-down menu list and type ACS3 as the renamed Coordinate system.

12.3 Set up the Work Centre

Click on the Work Centre icon tab to activate its drop-down menu list ≫ Now click on Mill on the drop-down menu list as shown below.

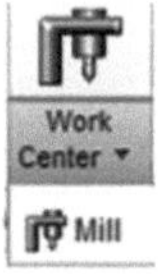

The Milling Work Centre dialogue window is activated on the main graphic window. Click on the Number of Axes section box to activate its drop-down menu list ≫ Now click on 5 Axis on the menu list. Take note of Name, Type and Post Processor section boxes as illustrated in Fig. 12.15.

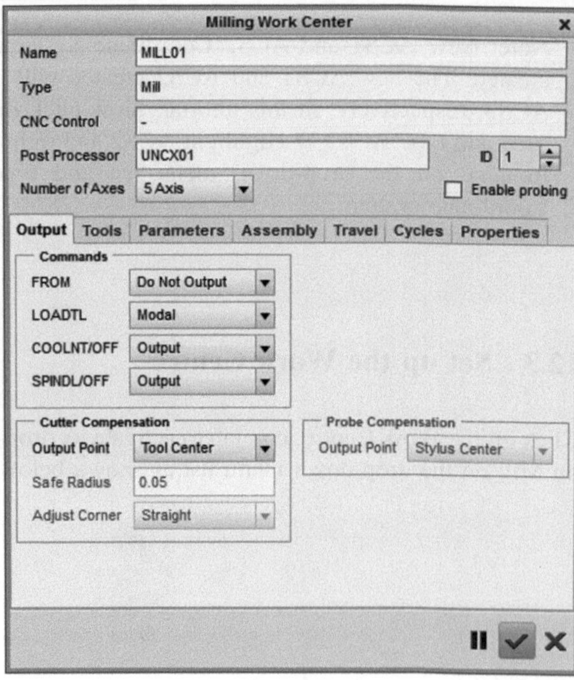

Fig. 12.15 Activated Milling Work Centre dialogue window

Click on the Check Mark icon.

12.4 Set up Operation Using the First Coordinate System Created (ACS2)

Click on the Operation icon as shown below.

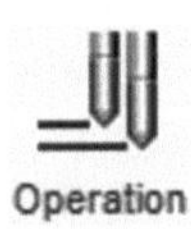

The Operation ribbon toolbar is activated. If MILL01 and ACS2 are not automatically added by the system, maually add them yourself.

To add MILL01 and ACS2, click on the Machine icon section box and then click on MILL01 on the Model Tree ≫ Click on the Coordinate System icon section box, and then click on the created Coordinate System (ACS2) on the Model Tree.

Click on the Clearance tab to activate its panel ≫ On the Retract group, click on the downward pointing arrow on the Type section box, and now click on Plane on the drop-down list ≫ Click on the Reference section box, and then click on the Workpiece surface ≫ Type 15 mm in the Value section box. See the illustration in Fig. 12.16.

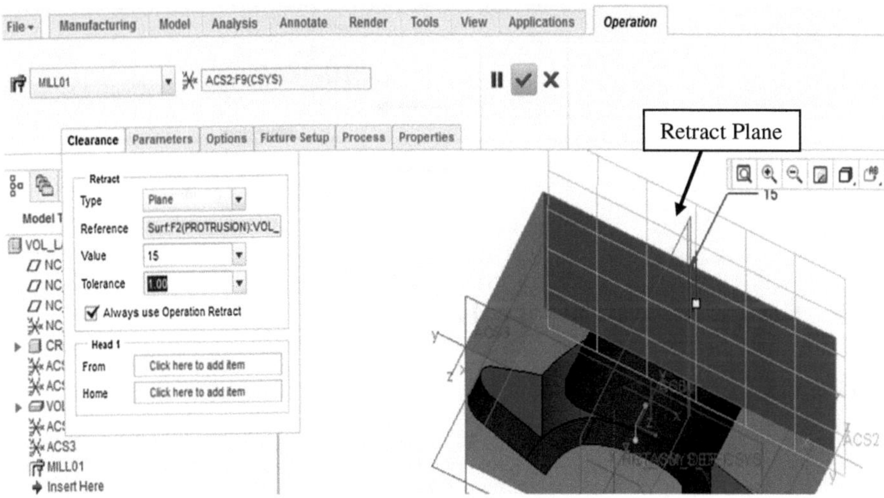

Fig. 12.16 Adding the Clearance parameters for the Retract plane

Click on the Check Mark icon.

Note: Check the Model Tree as shown in Fig. 12.17 to make sure that ACS2, ACS3, MILL01 and OP010 [MILL01] are on the Model Tree as indicated by the arrows.

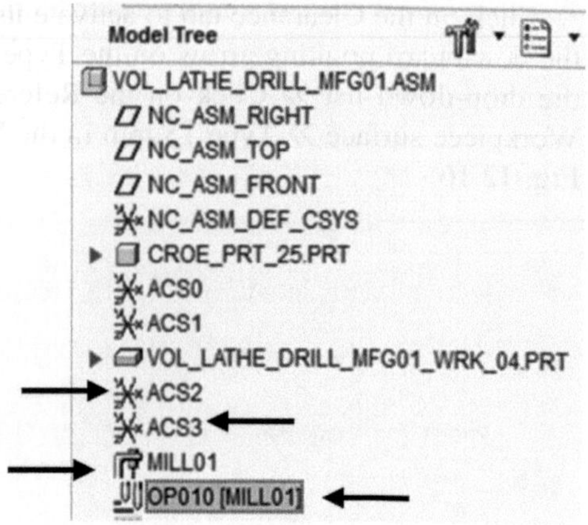

Fig. 12.17 ACS2, ACS3, MILL01 and OP010 [MILL01] on Model Tree

12.5 Set up the Cutting Tools

Click on the Cutting Tools icon

The Tools Setup dailogue window is activated on the main graphic window.

To add new End Mill Tool ⨠ Click on the General tab to activate its content ⨠ In the Name section box, type T01 ⨠ Click on the Type section box downward pointing arrow to activate its drop-down menu, click on END MILL on the drop-down menu list ⨠ Add dimensions to the END MILL Tool ⨠ Now click on the Apply tab and the END MILL Tool is added as shown in Fig. 12.18.

Add T01 as shown in Fig. 12.18.

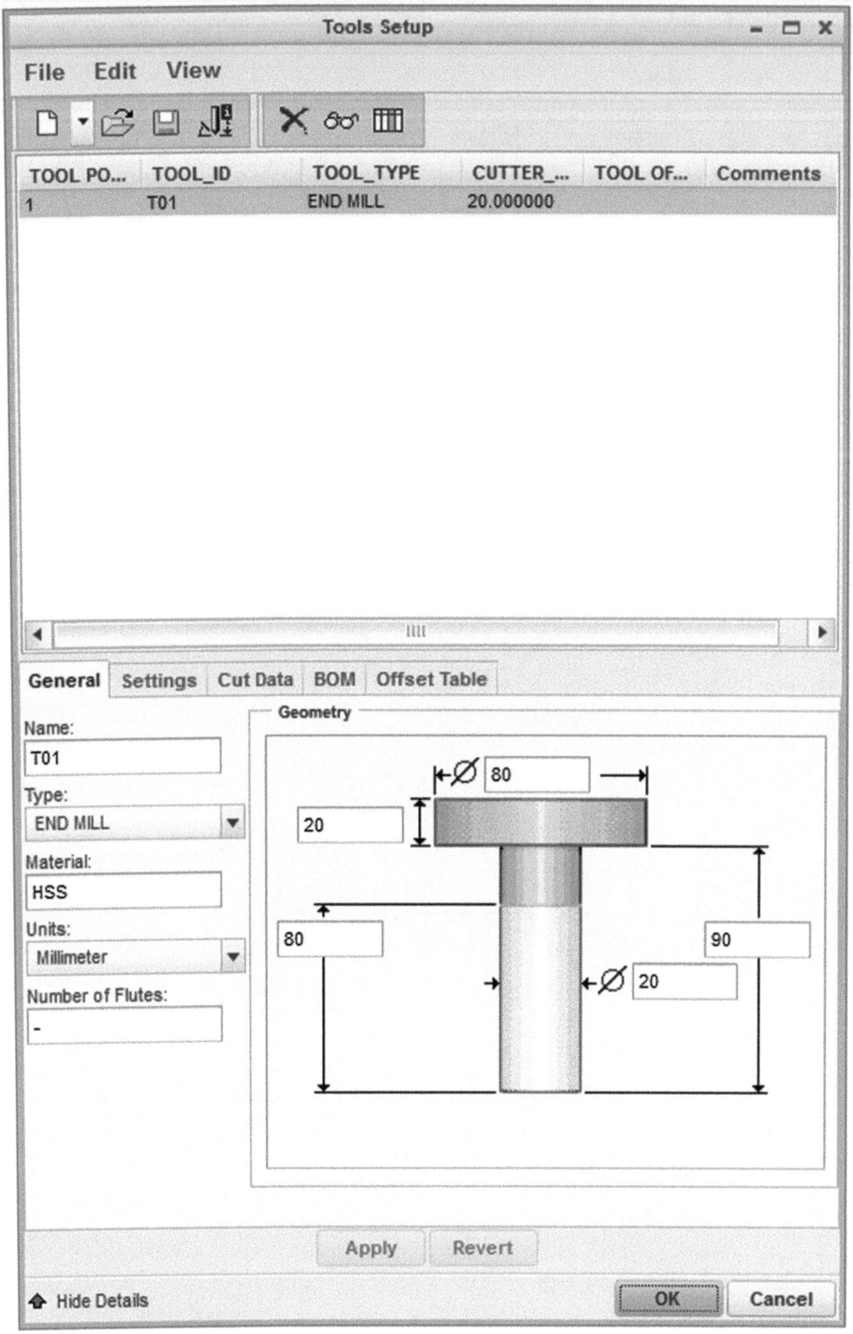

Fig. 12.18 T01 dimensions

***Add T02 as shown in Fig*. 12.19.**

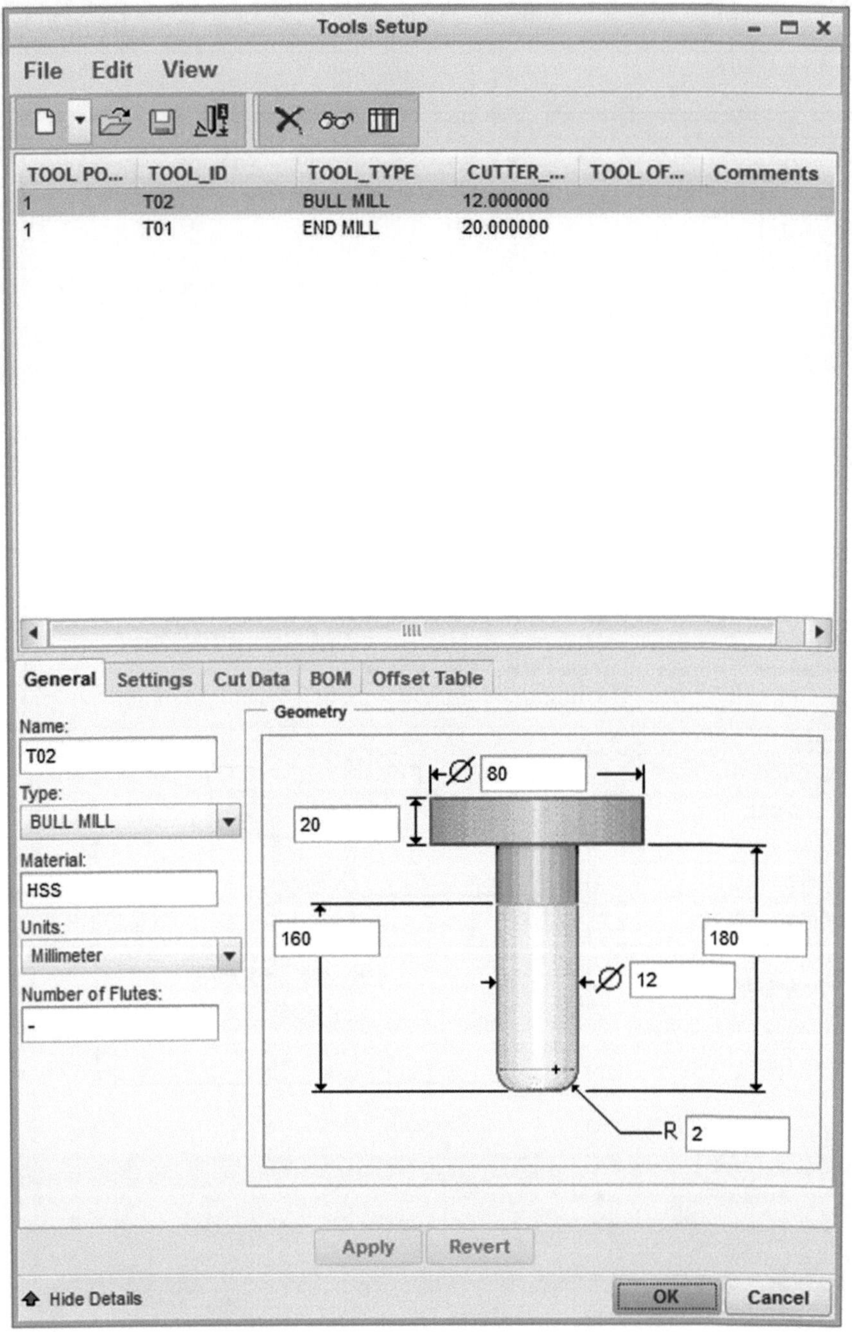

Fig. 12.19 T02 dimensions

***Add T03 as shown in Fig*. 12.20.**

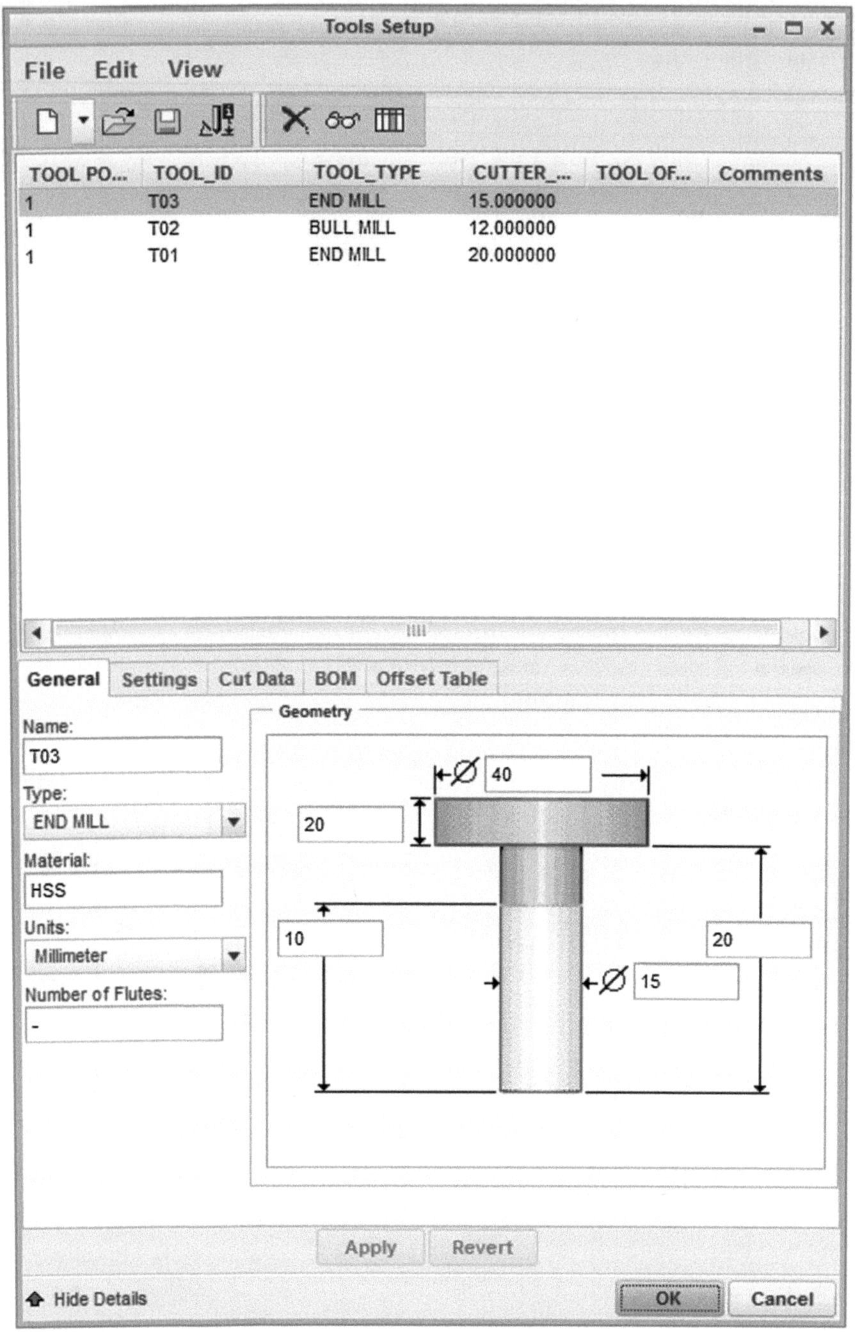

Fig. 12.20 T03 dimensions

***Add T04 as shown in Fig*. 12.21.**

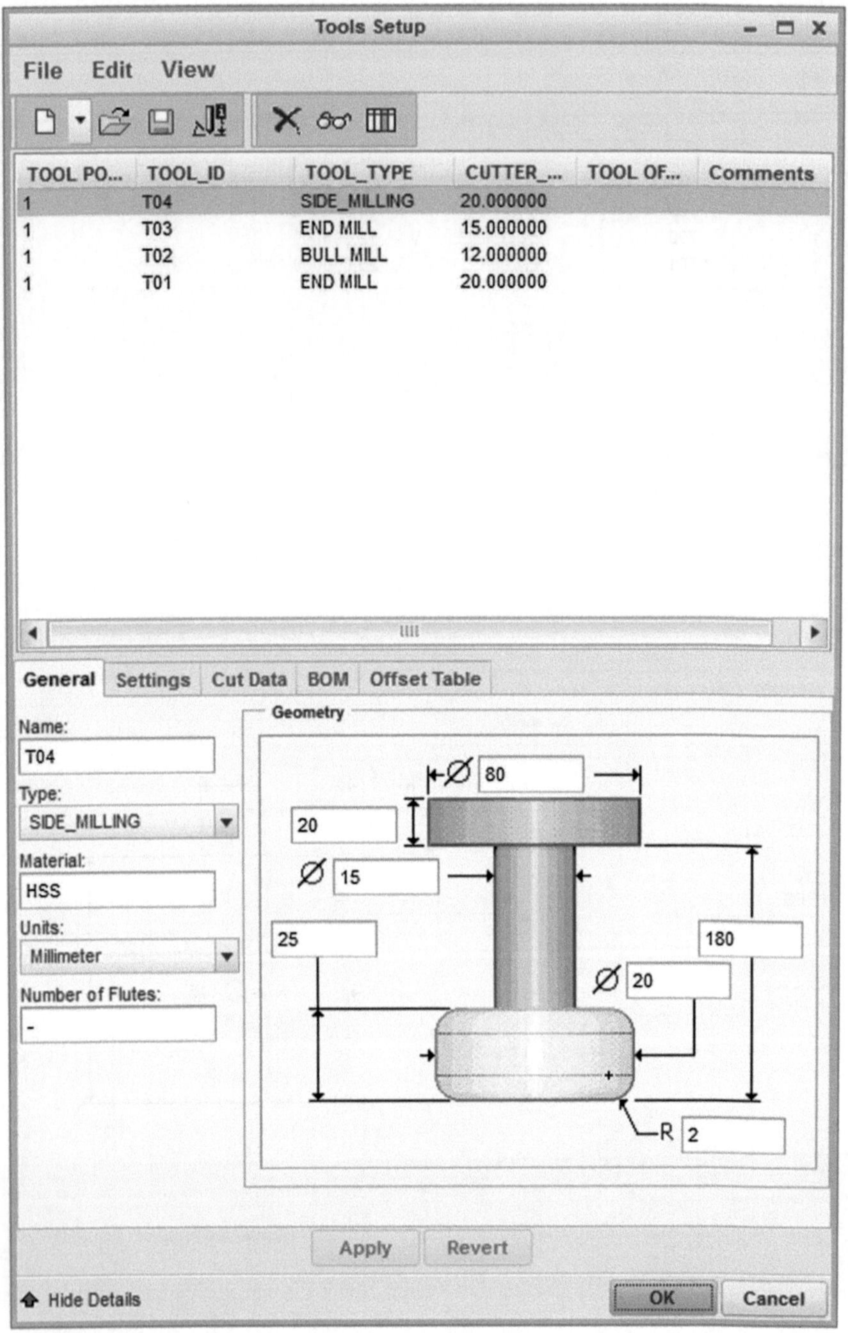

Fig. 12.21 T04 dimensions

Add T05 as shown in Fig. 12.22.

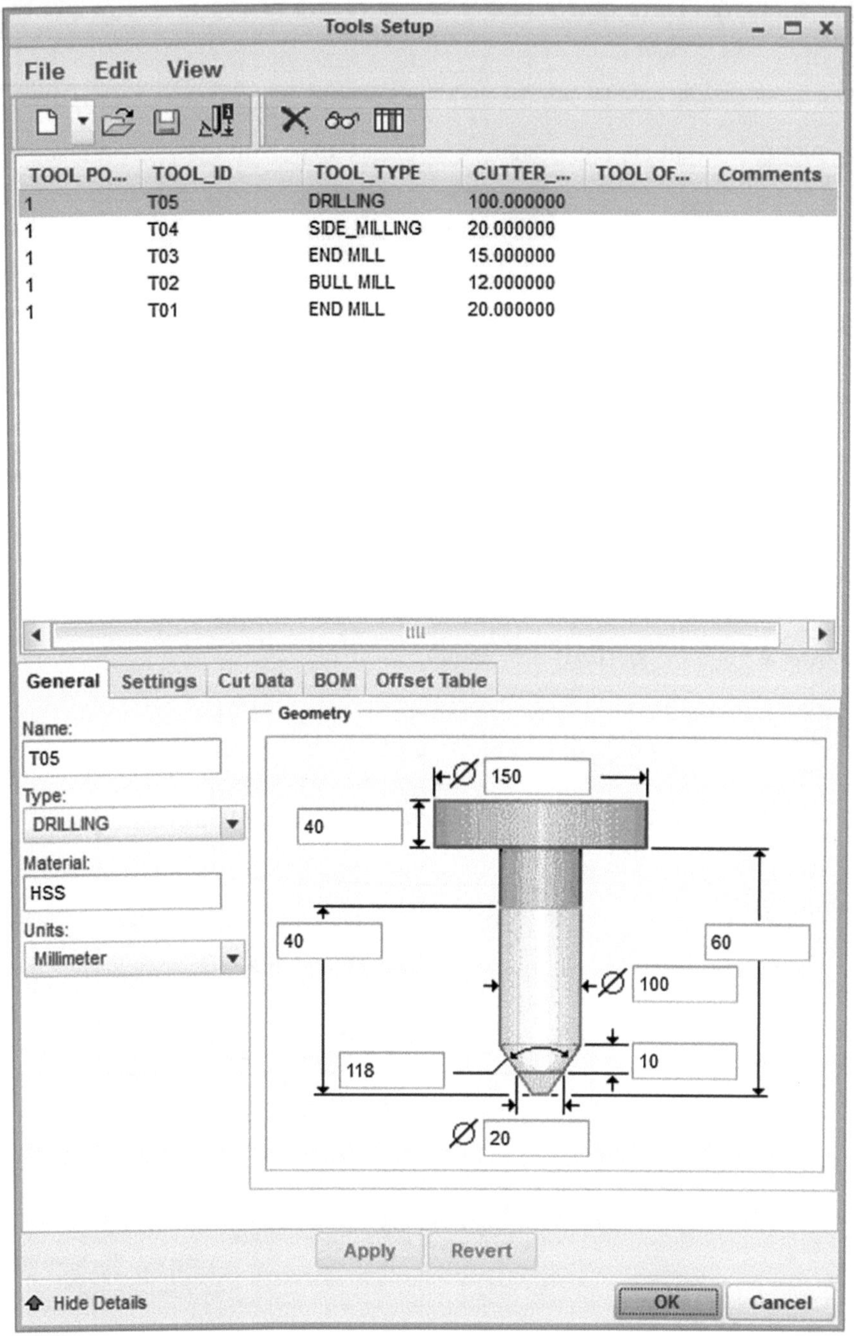

Fig. 12.22 T05 dimensions

Add T06 as shown in Fig. 12.23.

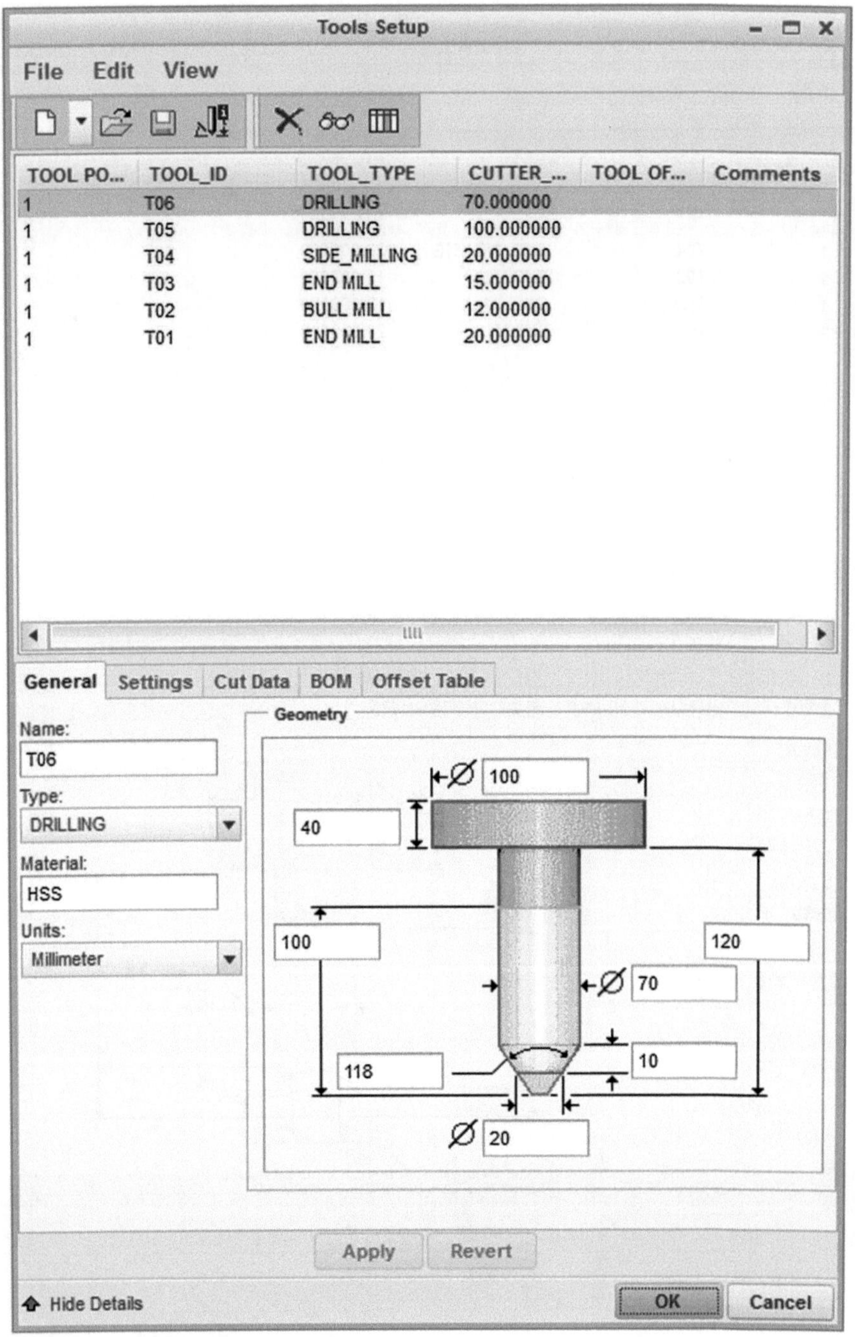

Fig. 12.23 T06 dimensions

Add T07 as shown in Fig. 12.24.

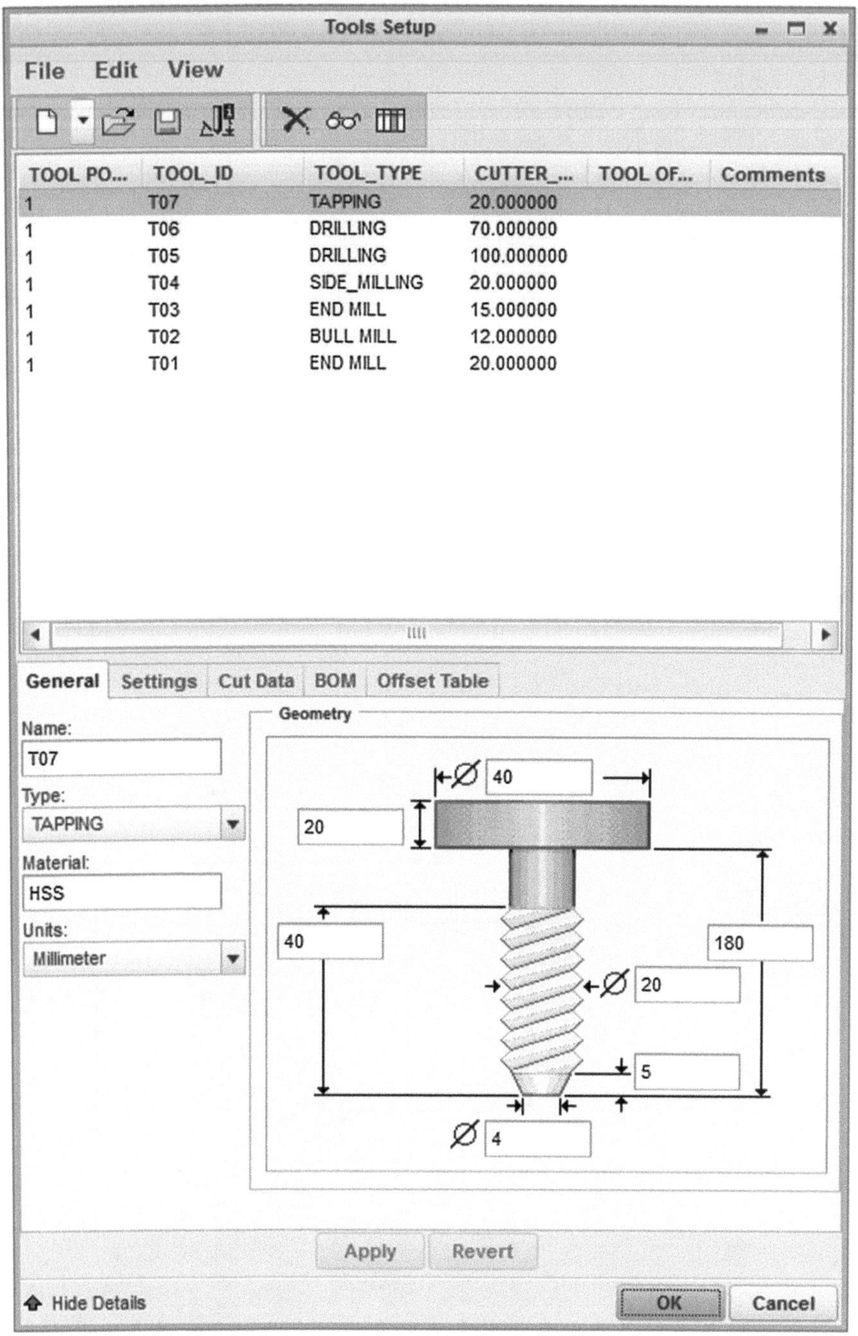

Fig. 12.24 T07 dimensions

Add T08 as shown in Fig**.** 12.25.

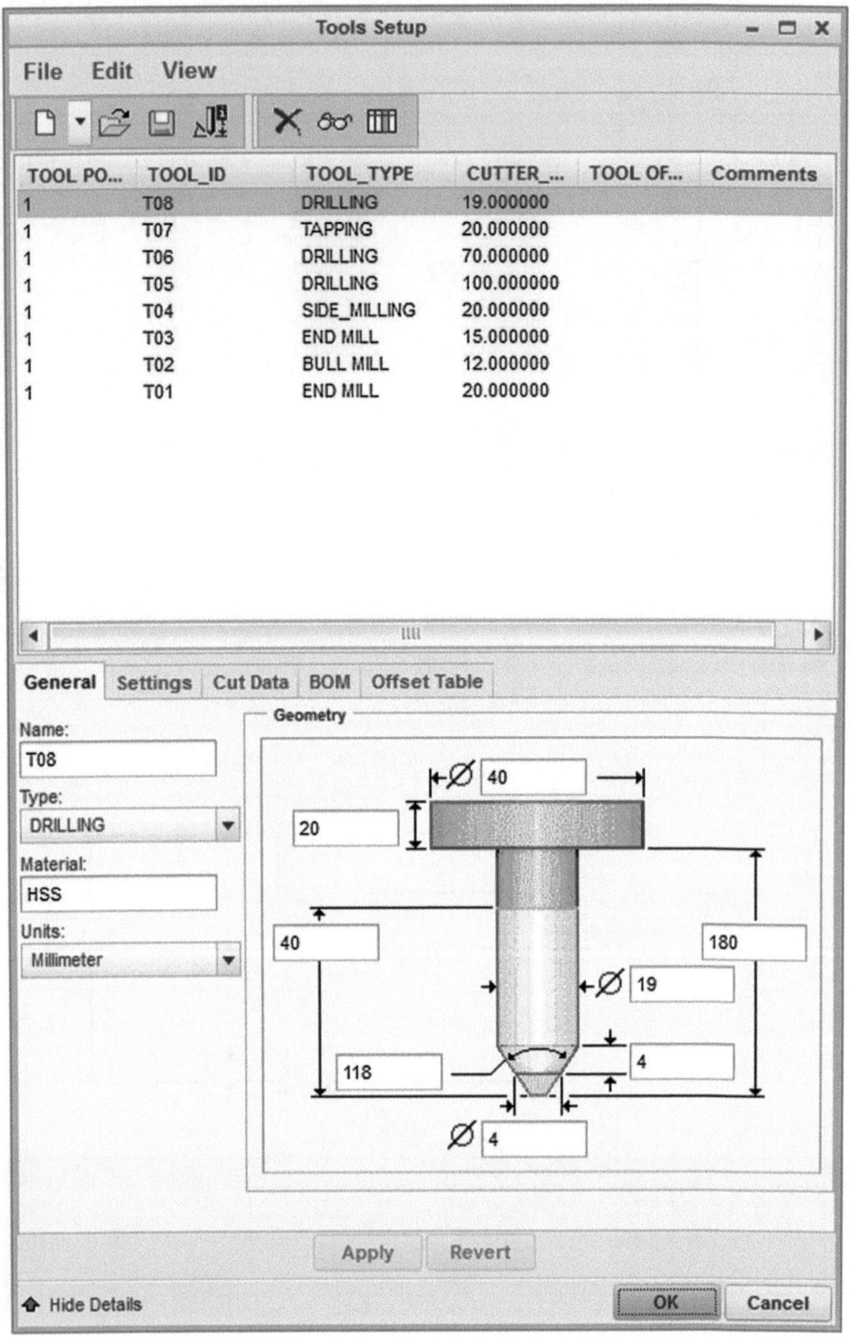

Fig. 12.25 T08 dimensions

Now click on the OK tab to exit Tools Setup dialogue window.

Note: Not all Cutting Tools created during the Tools Setup process will be used in this tutorial.

12.6 Create Mill Volumes

12.6.1 Create First Mill Volume

Click on the Mill menu bar to activate its ribbon ≫ Now click on Mill Volume icon tab as indicated by the arrow below.

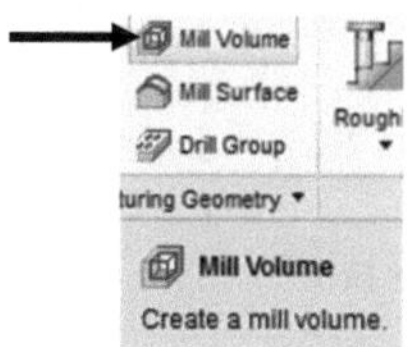

The Mill Volume dashboard tools are activated. See Fig. 12.26.

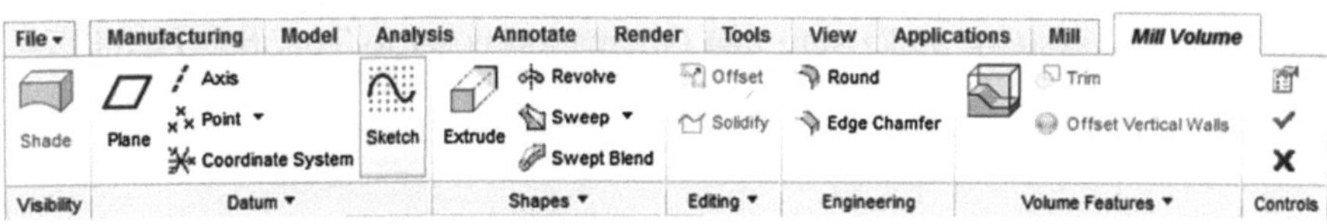

Fig. 12.26 Activated Mill Volume dashboard tools

Click on the Sketch icon on the Mill Volume ribbon toolbar.

The Sketch dialogue box is activated on the main graphic window as shown in Fig. 12.27.

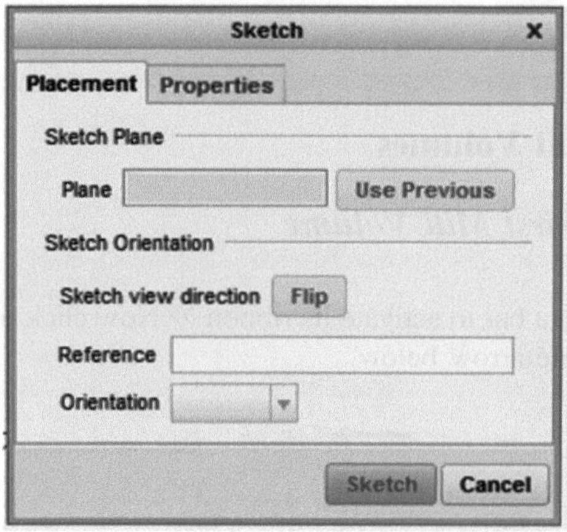

Fig. 12.27 Activated Sketch dialogue box

Click on the Placement tab to activate its content ≫ On the Sketch Plane group, click on the Plane section box and now click on the Top of the Workpiece as the sketch plane ≫ The Reference and Orientation section boxes are automatically updated by the system. Make sure that Orientation is set to Top. See Fig. 12.28.

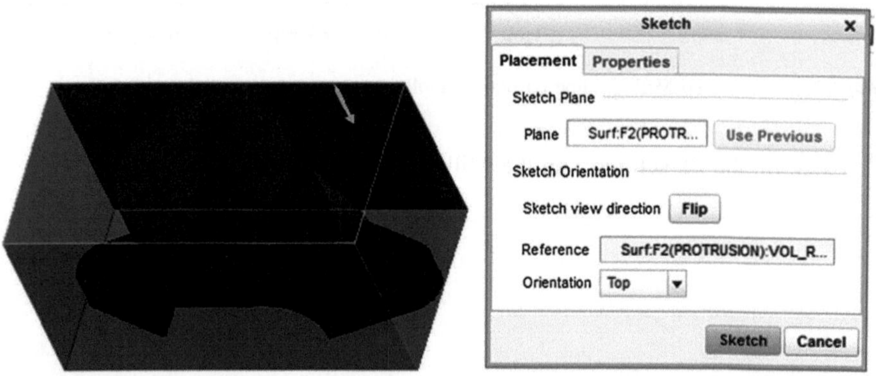

Fig. 12.28 Creating sketch Plane references

Click on Sketch tab to exit.

The Sketch tools are activated as shown in Fig. 12.29.

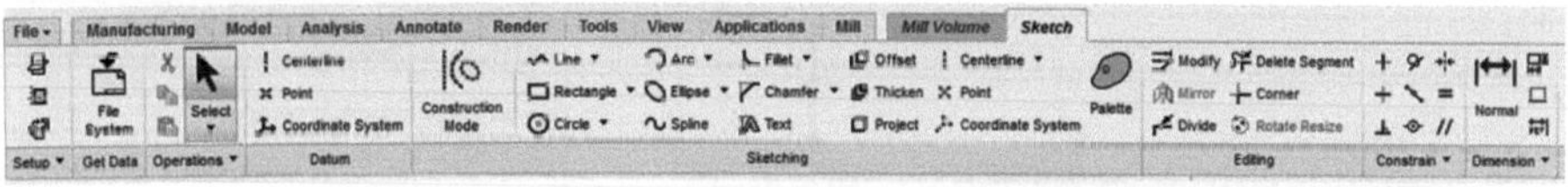

Fig. 12.29 Activated Sketch tools in concise form

Click on Sketch view icon  in Fig. 12.29 on the Sketch tools to orient the Workpiece to the correct sketch view/plane.

The Workpiece is now orientated on the correct sketch plane as shown in Fig. 12.30.

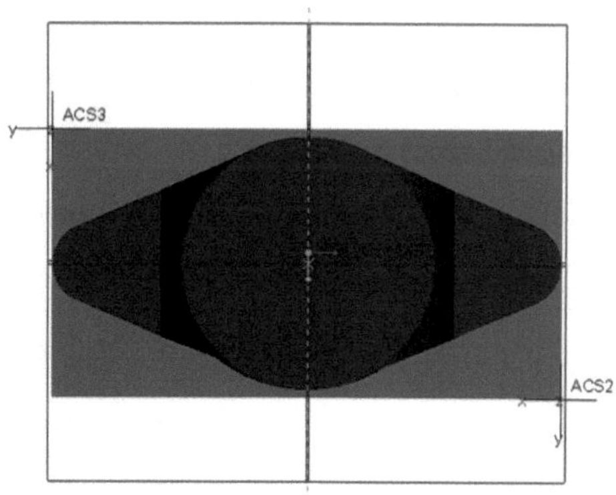

Fig. 12.30 Workpiece in correct orientation on the sketch plane

Create Reference

Click on the Reference icon

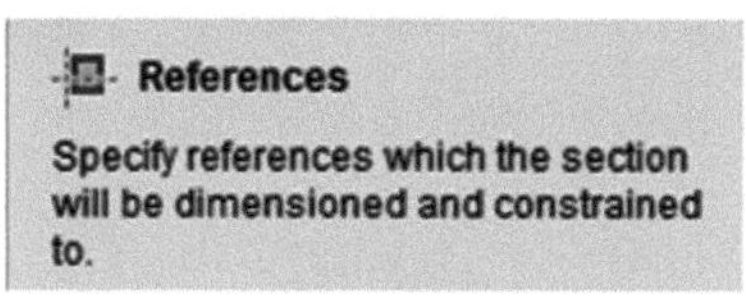

The References dialogue box is activated on the main graphic window ≫ Click on the Workpiece as indicated by the arrows (Fig. 12.31). The selected references are automatically updated and added into the References section box, as illustrated in Fig. 12.31.

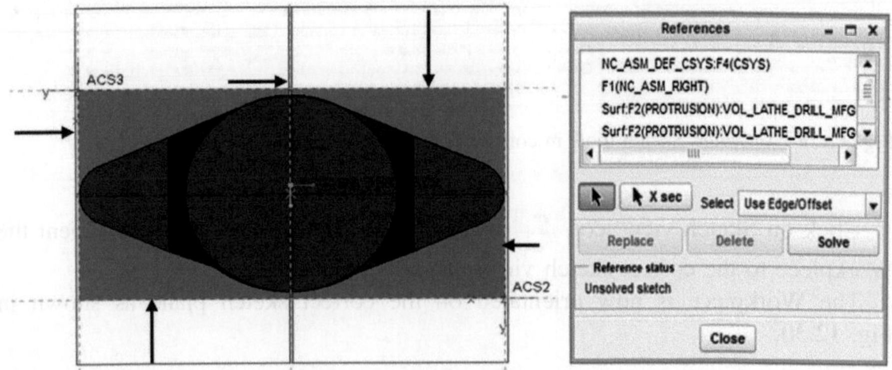

Fig. 12.31 Creating the references

Click on the Close tab to exit the References dialogue box.

Project Sketch

Click on the Project icon as shown below.

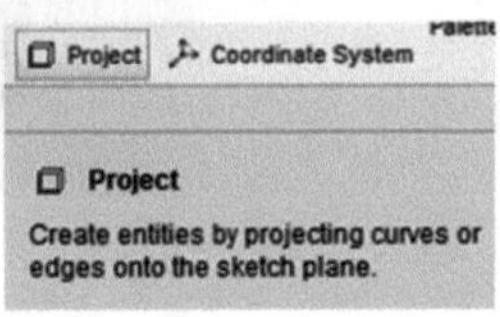

The Type dialogue box is activated on the main graphic window as shown in Fig. 12.32.

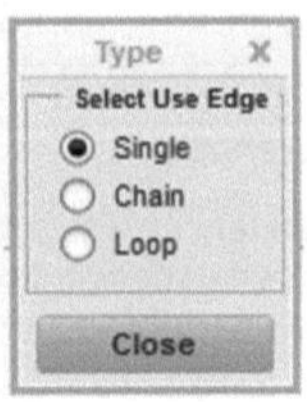

Fig. 12.32 Activated Type dialogue box

Click on the Lines and Curves on the Workpiece as indicated by the arrows. The Lines and Curves are projected into the sketch plane. The Line, Spline, Arc and Delete Segment tools can also be used to accomplish the Sketch as highlighted in Fig. 12.33.

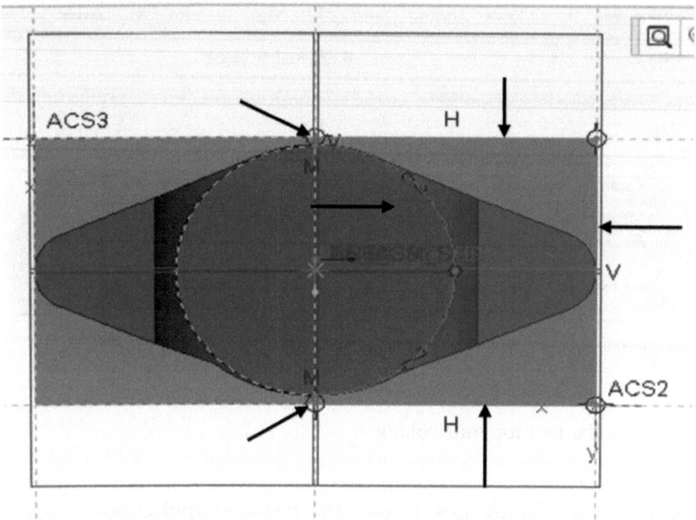

Fig. 12.33 Projecting Lines and Curves into the sketch plane

Click on Close tab of the Type dialogue box if it is still active ≫ Click on the Check Mark icon to exit the Sketch application.

Click on the Extrude icon and the Extrude dashboard is activated as shown in Fig. 12.34.

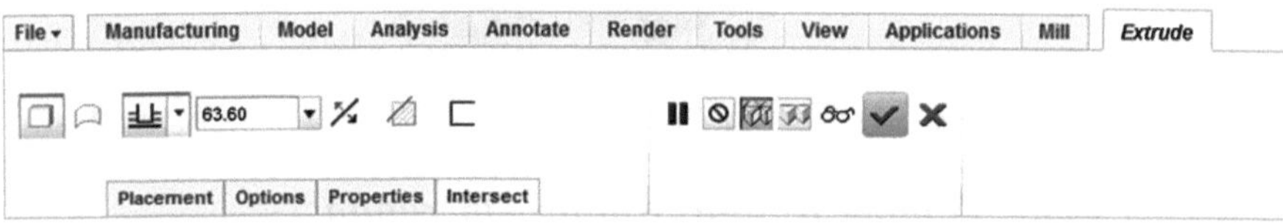

Fig. 12.34 Activated Extrude dashboard

Click on Exclude type icon to activate its drop-down menu list ≫ Click on "Extrude from sketch plane by a specified depth value" icon 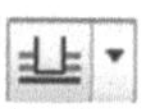 on the activated drop-down menu and type the required depth (63.60 mm) as indicated by the arrow in Fig. 12.35.

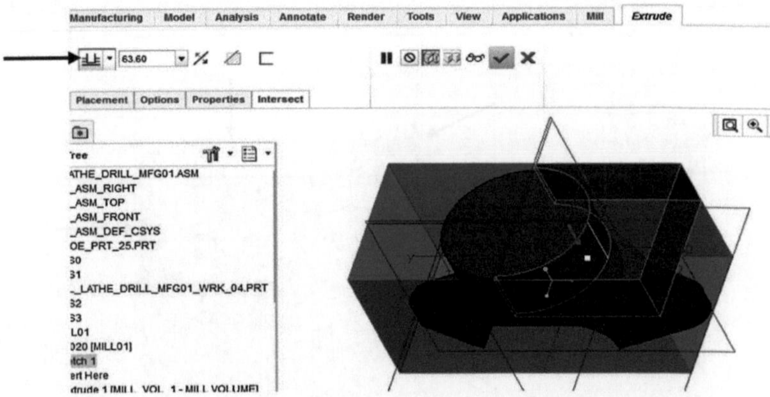

Fig. 12.35 Creating the first top mill volume

Click on the Check Mark icon to exit the Extrude application.
Click on the Check Mark icon to exit the Mill Volume application.
The first Mill Volume is now added to the Workpiece as shown in Fig. 12.36.

Fig. 12.36 First top created mill volume

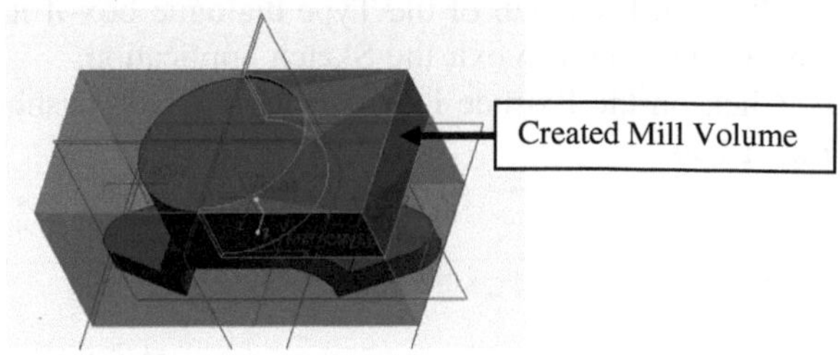

12.6.2 Create Volume Rough Milling Operation for the First Mill Volume

Click on Roughing icon tab to activate its drop-down menu list ≫ Now click on Volume Rough as indicated by the arrow shown below.

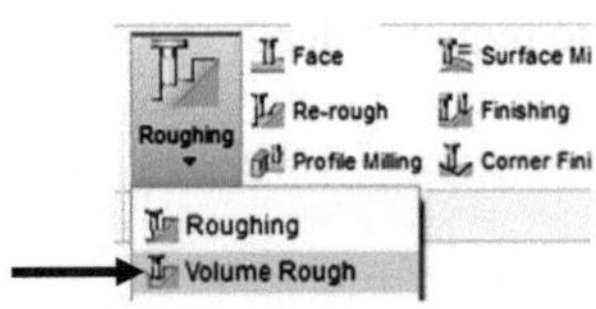

The NC SEQUENCE Menu Manager dialogue box is activated on the main graphic window ≫ Check Mark the Tool, Parameters and Volume square boxes if they are not automatically Check Marked by the system as shown in Fig. 12.37.

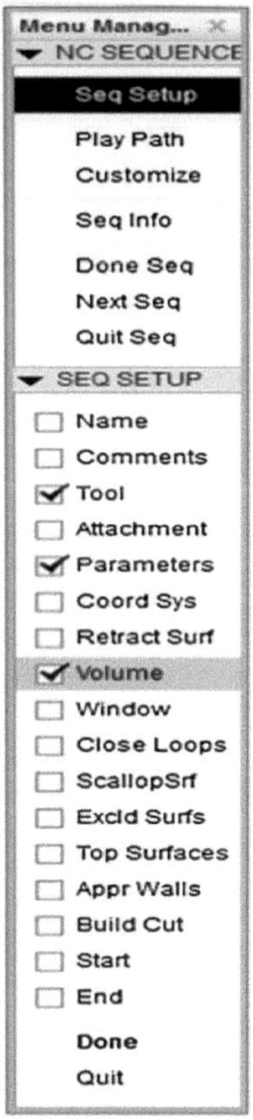

Fig. 12.37 Activated NC SEQUENCE Menu Manager dialogue box

Click on Done.

The Tools Setup dialogue window is activated again on the main graphic window ≫ Click on T01 as the cutting Tool for the first Volume Rough milling operation as highlighted on the Tools Setup dialogue window as shown in Fig. 12.38.

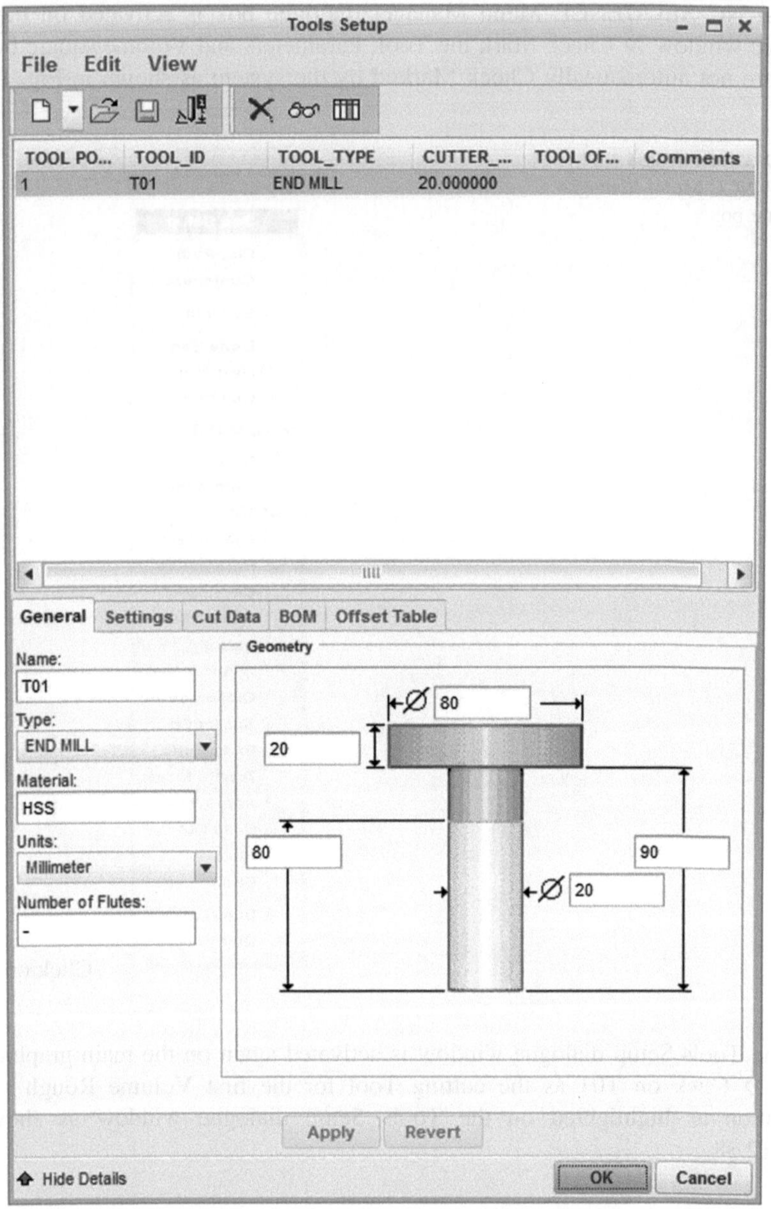

Fig. 12.38 Activating T01 as the milling Tool in concise form

Click on OK tab to exit.

The Edit Parameters of Sequence "Volume Milling" dialogue window is activated on the main graphic window as shown in Fig. 12.39.

Fig. 12.39 Activated Edit Parameters of Sequence "Volume Milling" dialogue window

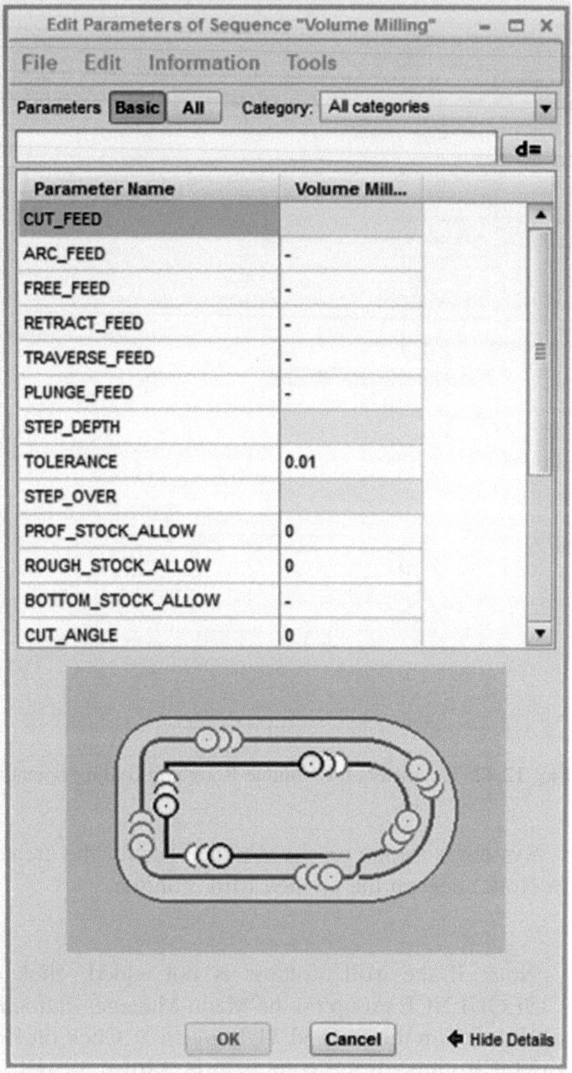

Add values to the activated Edit Parameters of Sequence "Volume Milling" dialogue window as shown in Fig. 12.40.

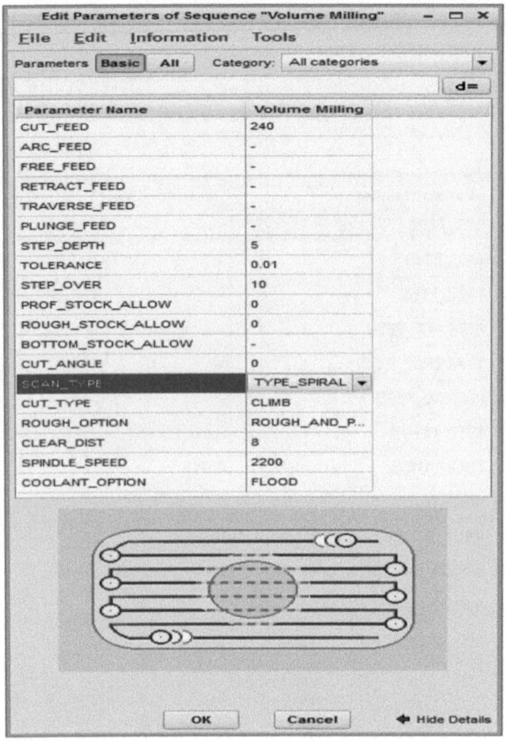

Click on OK tab to exit.

Fig. 12.40 Parameters for Volume Roughing milling operation

System is asking on the message bar for the created Mill Volume to be selected ≫ Now click on the created Mill Volume.

Note: If the Mill Volume is not added, click on Customize on the NC SEQUENCE group on the Menu Manager dialogue box ≫ Now Check Mark Volume on the SEQ SETUP group ≫ Click on Done ≫ Click on the created Mill Volume as the volume to be Milled as indicated by the arrow shown in Fig. 12.41.

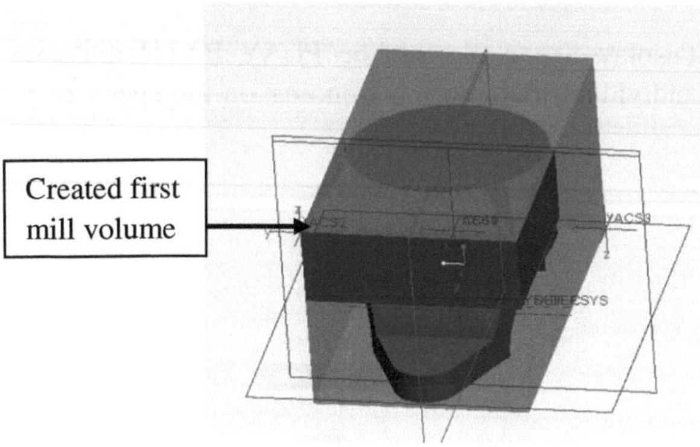

Fig. 12.41 Adding the top first created mill volume

Once Play Path is clicked on the NC SEQUENCE group, the Menu Manager dialogue box expand to include the PLAY PATH group ≫ Now click on Screen Play as shown in Fig. 12.42.

The cutting Tool and PLAY PATH dialogue box are activated on the main graphic window as shown in Fig. 12.42.

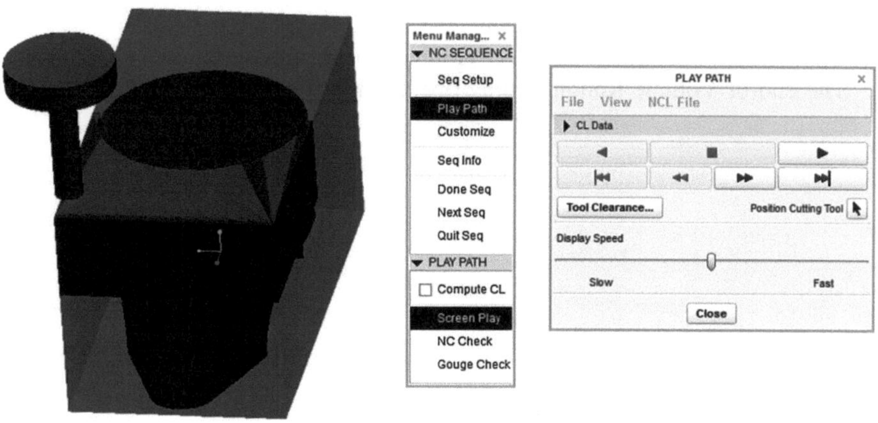

Fig. 12.42 Activated cutting Tool and PLAY PATH dialogue box

Click on the Display View Style icon  to activate its drop-down menu ≫ Click on Wireframe on the drop-down list, the cutting Tool and Workpiece changes to Wireframe display as shown in Fig. 12.43.

Note: The Display Speed dial on the PLAY PATH dialogue box can be adjusted to individual preference depending on the speed of your computer processor.

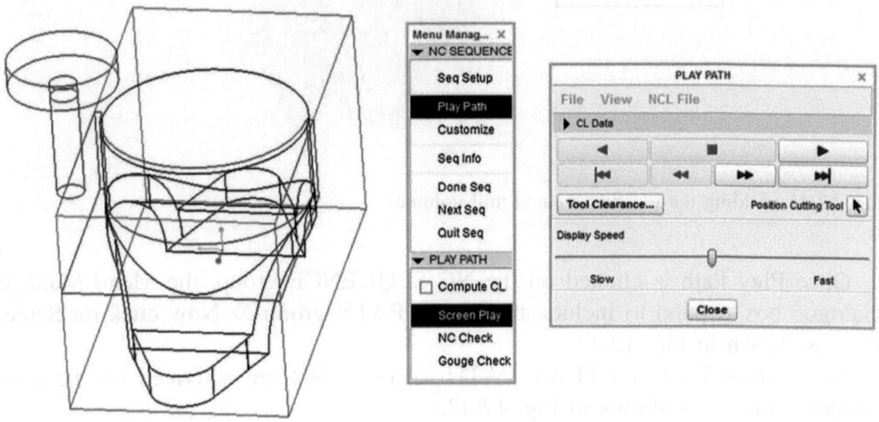

Fig. 12.43 Cutting Tool and Workpiece in Wireframe display

Click on the Play tab to start the Volume Rough milling process.
The end of Volume Rough milling operation is shown in Fig. 12.44.

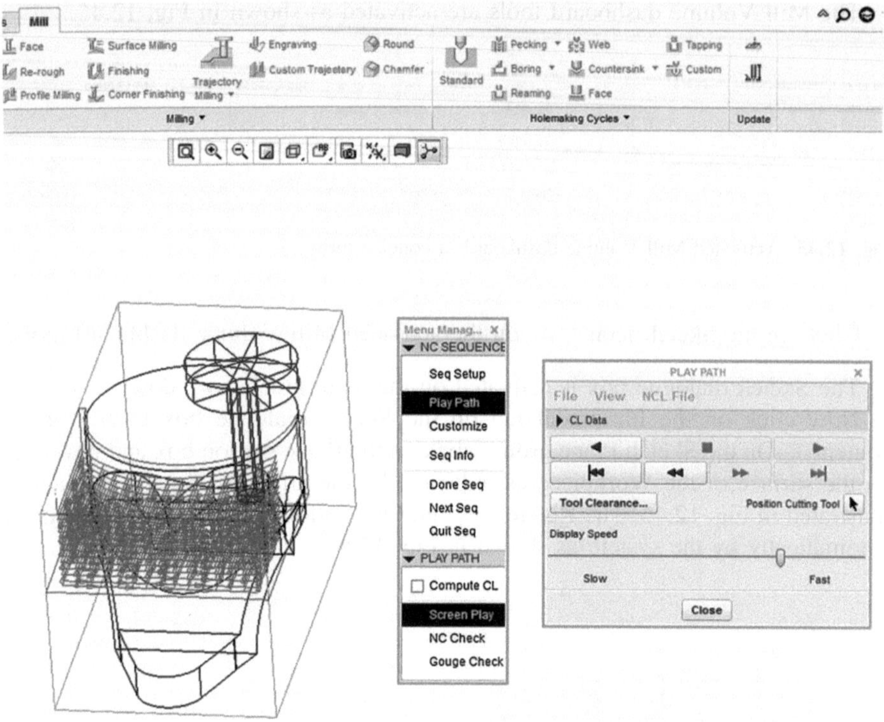

Fig. 12.44 End of Volume Rough milling operation in Wireframe display

Click on Close tab to exit the PLAY PATH dialogue box.

Click on Done Seq on the NC SEQUENCE group on the Menu Manager dialogue box to exit the Volume Rough milling operation.

12.6.3 Create Second Mill Volume

Click on Mill Volume icon tab as indicated by the arrow as shown below.

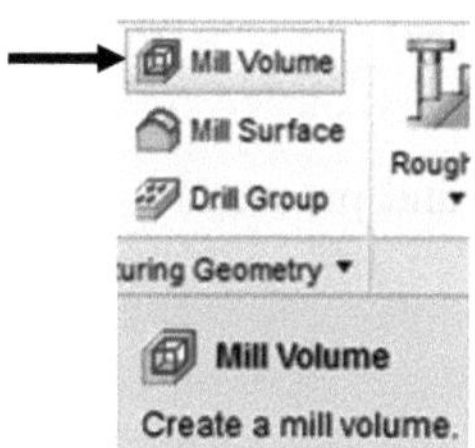

The Mill Volume dashboard tools are activated as shown in Fig. 12.45.

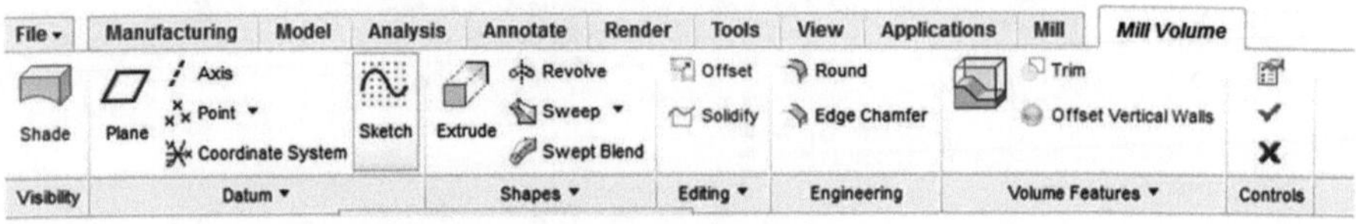

Fig. 12.45 Activated Mill Volume dashboard in concise form

Click on the Sketch icon  on the activated Mill Volume dashboard tools.

The Sketch dialogue box is activated on the main graphic window.

Now click on the Placement tab on the Sketch dialogue box to activate its content ≫ On the Sketch Plane group, click on the Plane section box and now click on the surface of the Workpiece as indicated by the arrow as the sketch plane as illustrated in Fig. 12.46. The References and Orientation section boxes are updated automatically by the system as shown in Fig. 12.46.

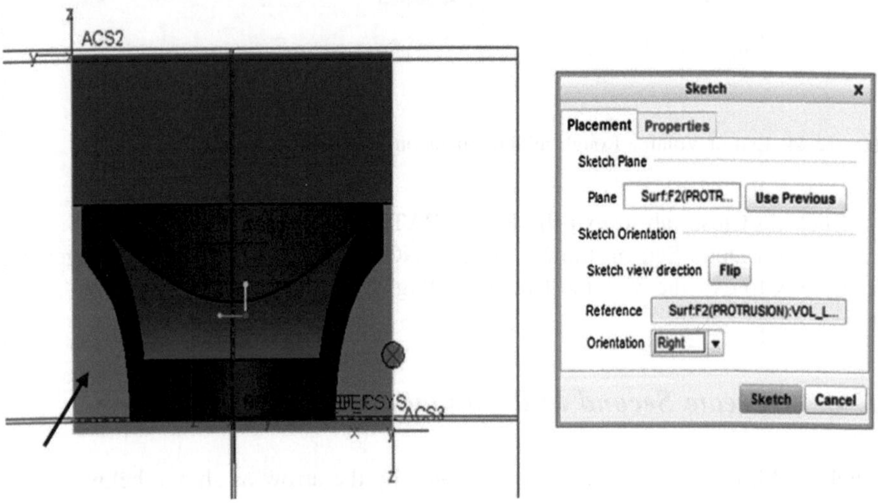

Fig. 12.46 Activating the sketch plane

Click on Sketch tab to exit and proceed.

The Sketch tools are activated as shown in Fig. 12.47.

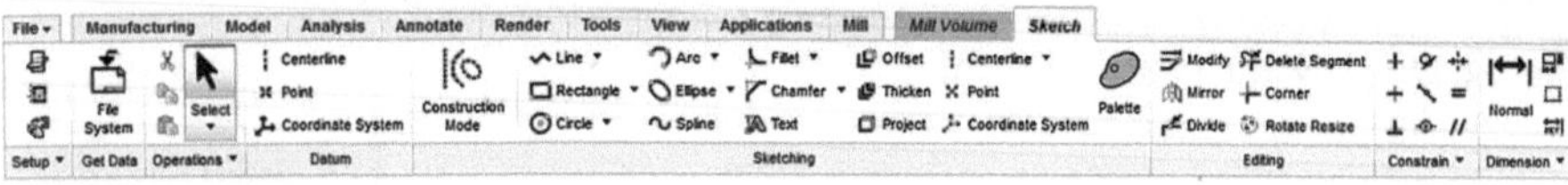

Fig. 12.47 Activated Sketch tools in concise form

Click on Sketch view icon  to orient the Workpiece to the correct sketch view/plane.

The Workpiece is now orientated on the correct sketch plane as shown in Fig. 12.48.

Fig. 12.48 Workpiece in correct orientation on the sketch plane

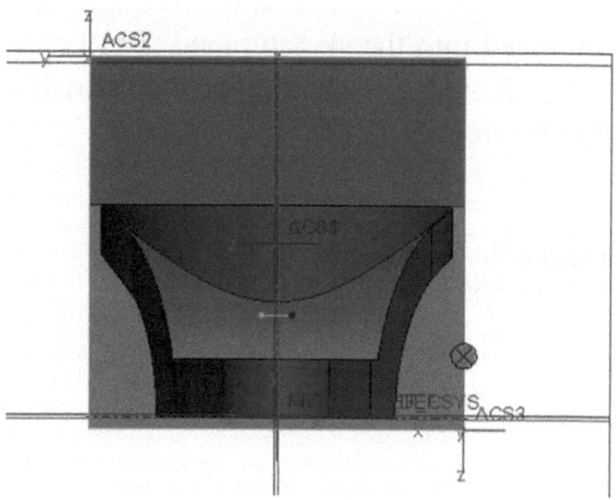

Project Sketch

Click on the Project icon tab as shown below.

The Type dialogue box is activated on the main graphic window as shown in Fig. 12.49.

Fig. 12.49 Activated Type dialogue box

Click on the Lines on the Workpiece as indicated by the arrows. The Lines are projected into the sketch plane. The Line, Spline, Arc and Delete Segment tools on the activated Sketch toolbar can also be used to accomplish the Sketch as highlighted in Fig. 12.50.

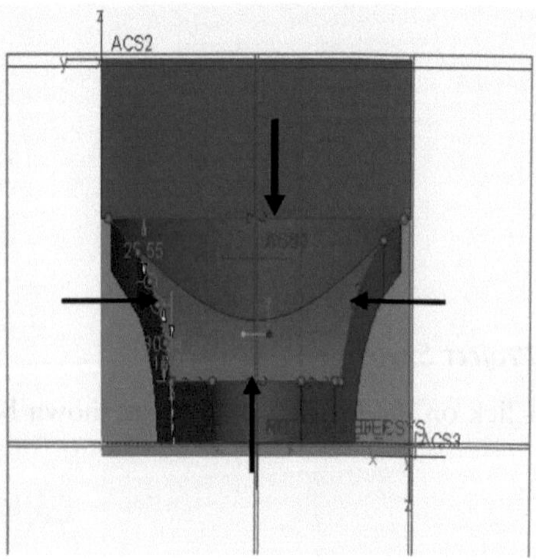

Fig. 12.50 Sketched Mill Volume highlighted

Click on Close tab on the Type dialogue box.
Click on the Check Mark icon to exit the Sketch application.

Click on the Extrude icon (Extrude) and the Extrude dashboard tools are activated » Click on Exclude type icon to activate its drop-down menu list » Click on

"Extrude to selected point, curve, plane or surface"  as indicated by the arrow in Fig. 12.51.

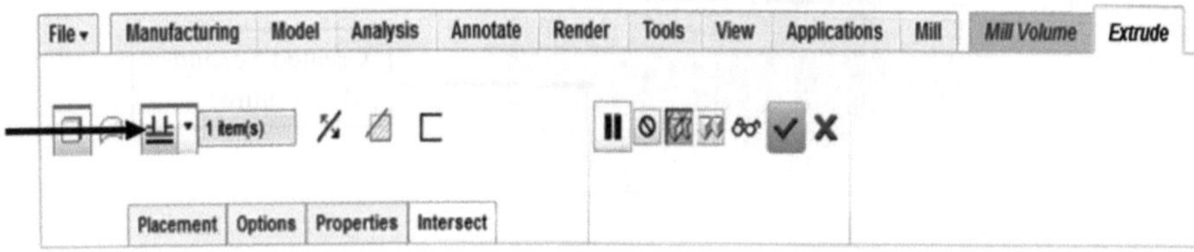

Fig. 12.51 Selecting Extrude to selected point, curve, plane or surface extrusion option type

Now click on the next surface to be Extrude to and if required, flip the direction of extrusion as shown in Fig. 12.52.

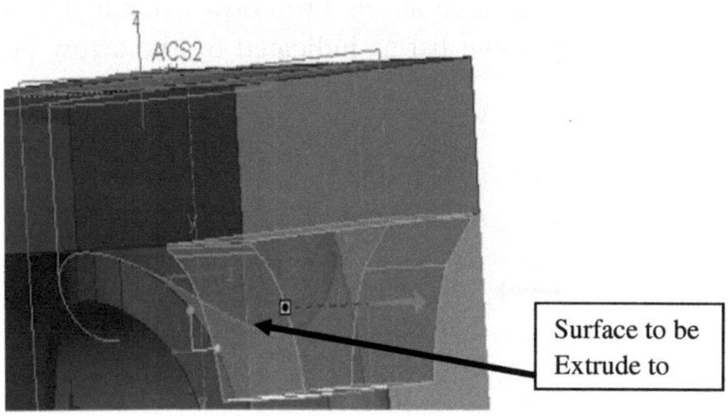

Fig. 12.52 Extruding the sketched Mill Volume to the next surface

Click on the Check Mark icon to exit Extrude application.
Click on Check Mark icon to exit Mill Volume application.
The arrow in Fig. 12.53 indicates the created Second Mill Volume.

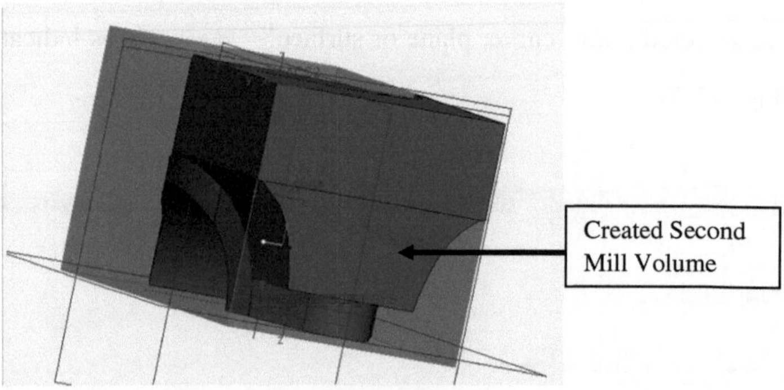

Fig. 12.53 Second created Mill Volume

12.6.4 Volume Rough Milling Operation for the Second Mill Volume

Click on Roughing icon tab to activate its drop-down menu list ≫ Now click on Volume Rough on the drop-down list as indicated by the arrow below.

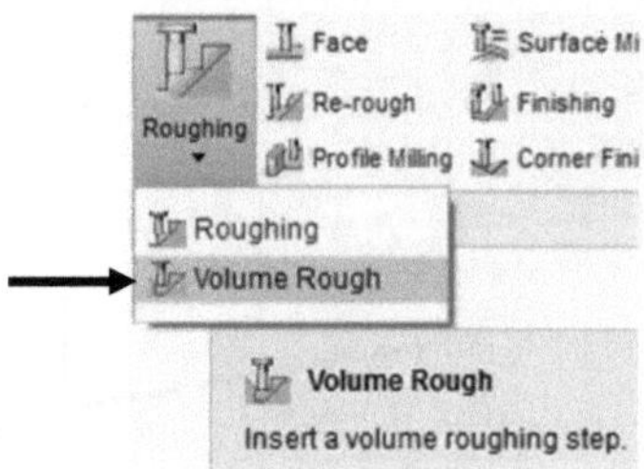

The NC SEQUENCE Menu Manager dialogue box is activated on the main graphic window ≫ Check Mark the Tool, Parameters, Retract Surf and Volume square boxes if they are not automatically activated by the system as shown in Fig. 12.54.

Fig. 12.54 Activated NC SEQUENCE Menu Manager dialogue box

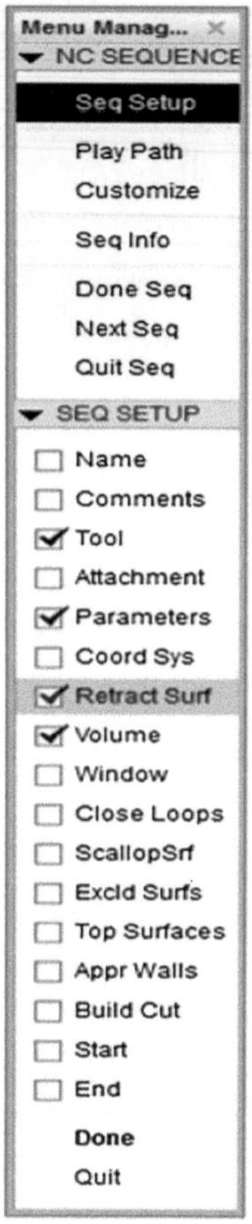

Click on Done.

The Tools Setup dialogue window is activated again on the main graphic window ≫ Click on T02 as the cutting Tool for the second Volume Rough milling operation as shown in concise form in Fig. 12.55.

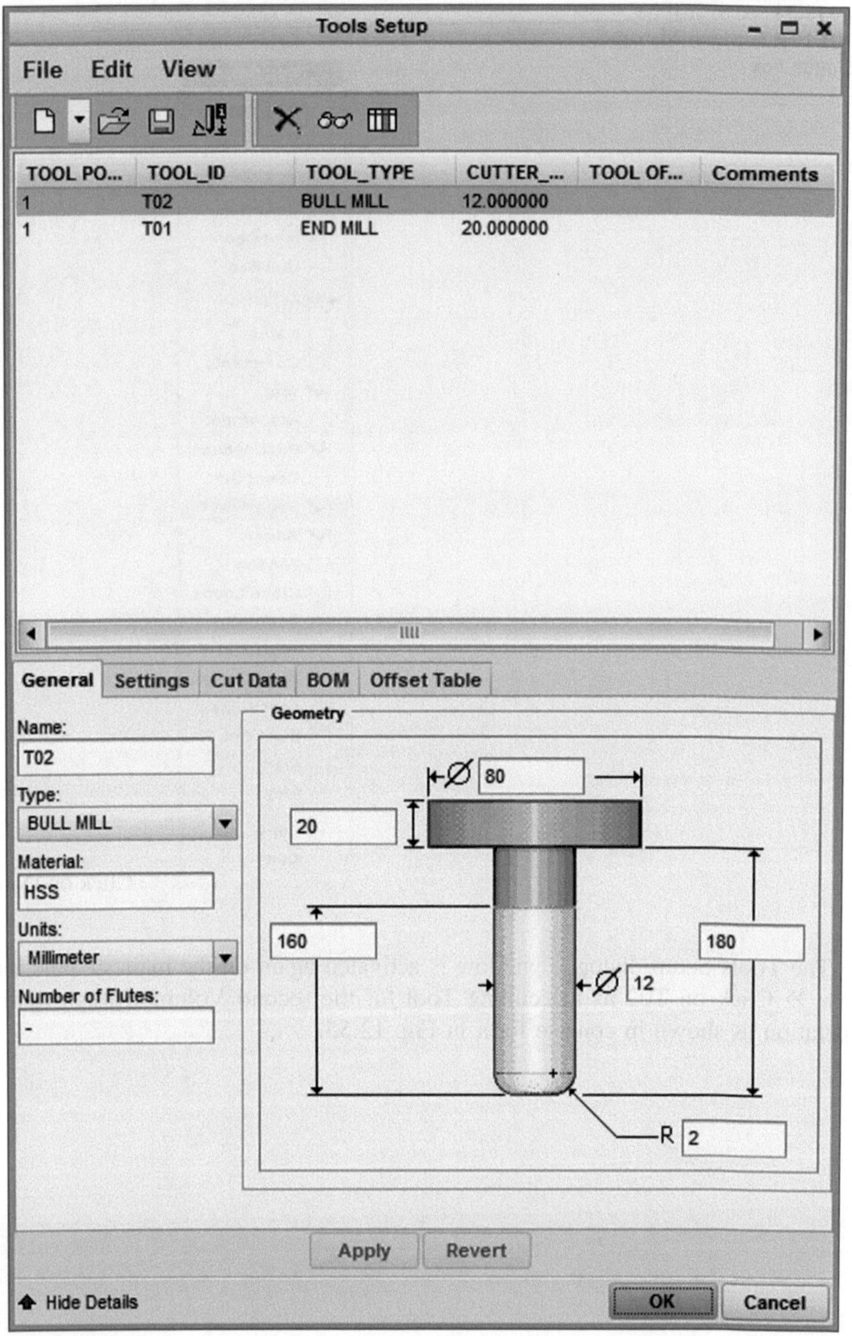

Fig. 12.55 Activated Tools Setup dialogue window with T02 selected

Click on the OK tab to add T02 and exit the Tools Setup application.

The Edit Parameters of Sequence "Volume Milling" dialogue window is activated ≫ Input the parameter values for the Volume Rough milling operation (see Fig. 12.56).

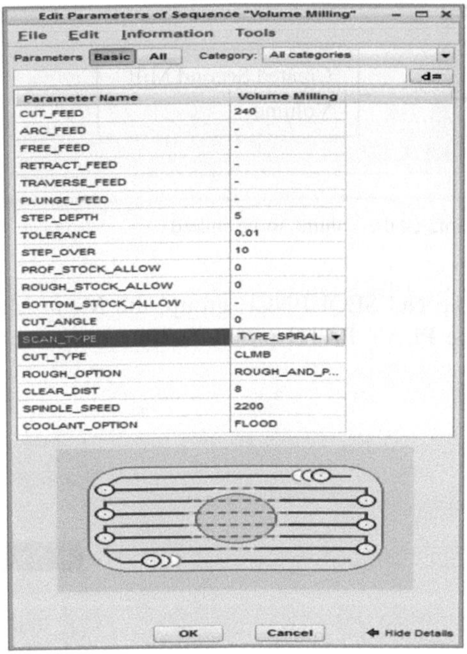

Click on the OK tab to exit.

Fig. 12.56 Volume Rough milling parameters for the second mill volume

The Retract Setup dialogue box is activated on the main graphic window as shown in Fig. 12.57.

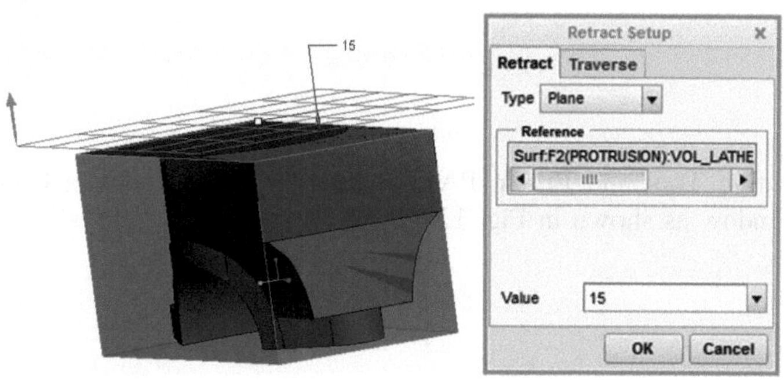

Fig. 12.57 Activated Retract plane and Retract Setup dialogue box

Click on Ok tab to accept and exit.

System is asking for the created Mill Volume to be selected/added on the message bar information » Now click on the created Second Mill Volume as indicated by the arrow in Fig. 12.58.

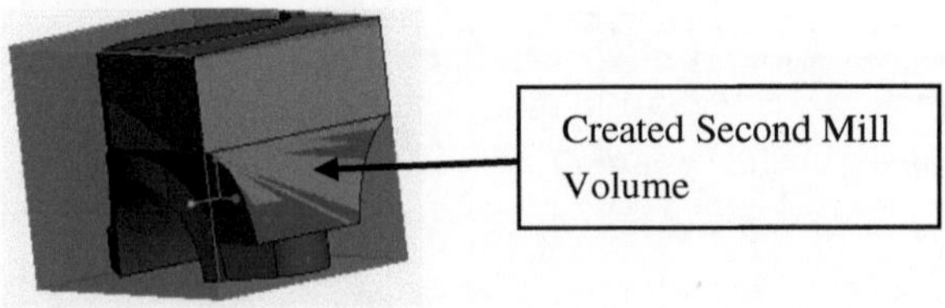

Fig. 12.58 Adding the second mill volume as the volume to be milled

Once Play Path is clicked on the NC SEQUENCE group, the Menu Manager dialogue box expand to include the PLAY PATH group » Now click on Screen Play as higlighted in Fig. 12.59.

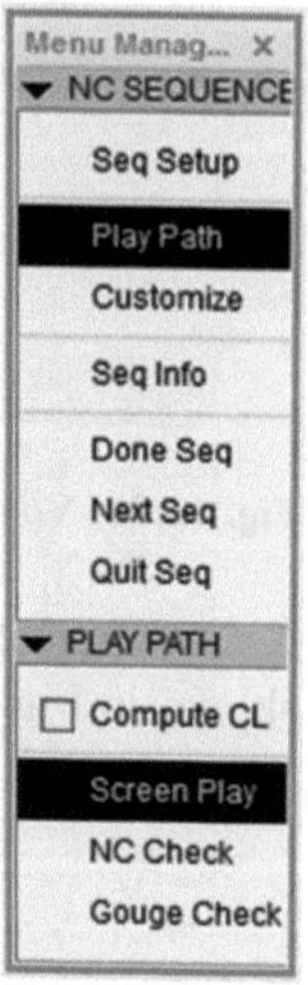

Fig. 12.59 Activating Play Path and on Screen Play

The cutting Tool and PLAY PATH dialogue box are activated on the main graphic window as shown in Fig. 12.60.

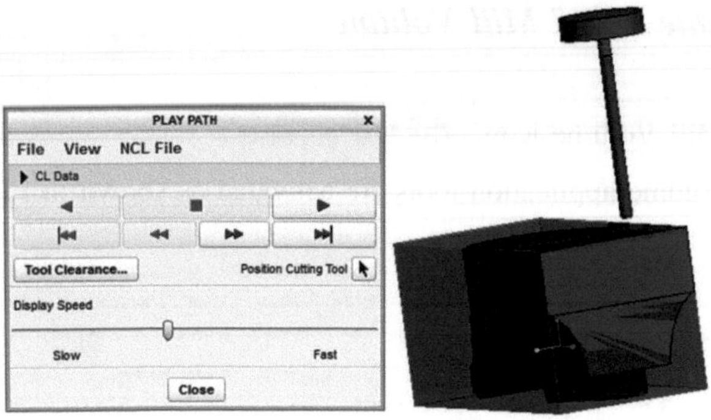

Fig. 12.60 Activated cutting Tool and PLAY PATH dialogue box

Click on the Display View Style icon  on the View toolbar to activate its drop-down menu list ≫ Click on Wireframe on the drop-down list, the cutting Tool and Workpiece changes to Wireframe display as shown in Fig. 12.61.

Click on the Play tab to start the Volume Rough milling operation.

The end of Volume Rough milling operation is shown in Fig. 12.61.

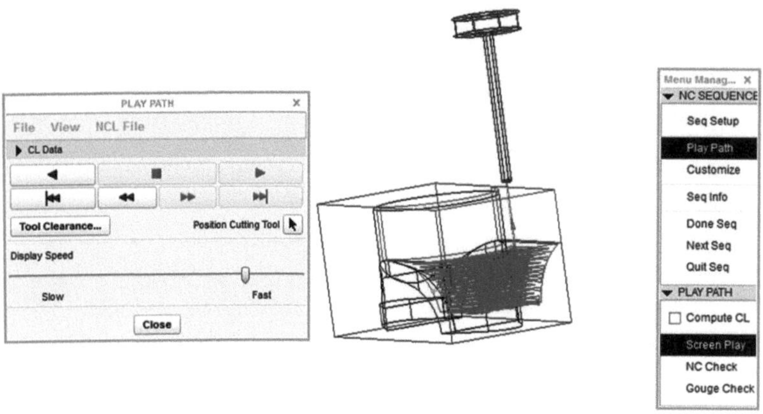

Fig. 12.61 End of Volume Rough milling operation in Wireframe display

Click on Close tab to exit the PLAY PATH dialogue box.

Click on Done Seq on the NC SEQUENCE group on the Menu Manager dialogue box to exit the Volume Rough milling operation.

12.6.5 *Create Third Mill Volume*

Click on the Mill Volume icon 

.

The Mill Volume application tools are activated as shown in Fig. 12.62.

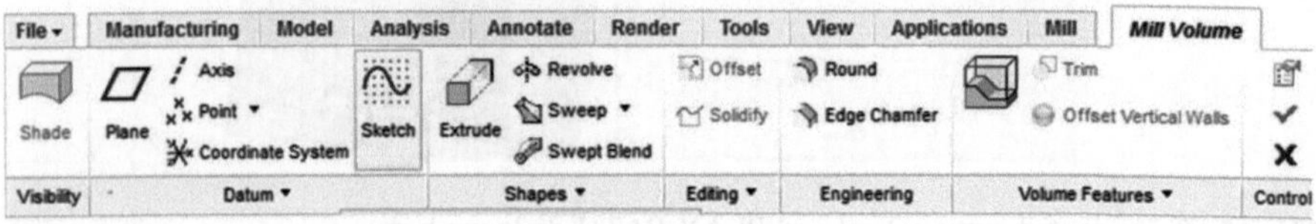

Fig. 12.62 Activated Mill Volume ribbon toolbar

Click on the Sketch icon on the activated Mill Volume ribbon.

The Sketch dialogue box is activated on the main graphic window ≫ Click on the Placement tab ≫ On the Sketch Plane group, click on the Plane section box and now click on the Top of the Workpiece as the sketch plane. The Sketch dialogue box is updated automatically as shown in Fig. 12.63.

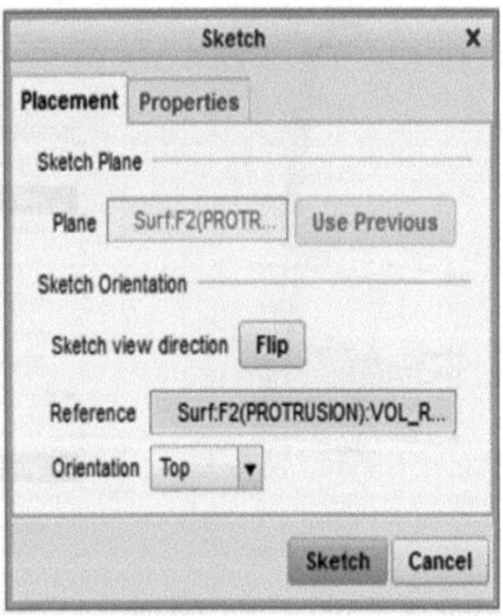

Click on Sketch tab to proceed.

Fig. 12.63 Updated Sketch dialogue box

The Sketch tools are activated. See Fig. 12.64.

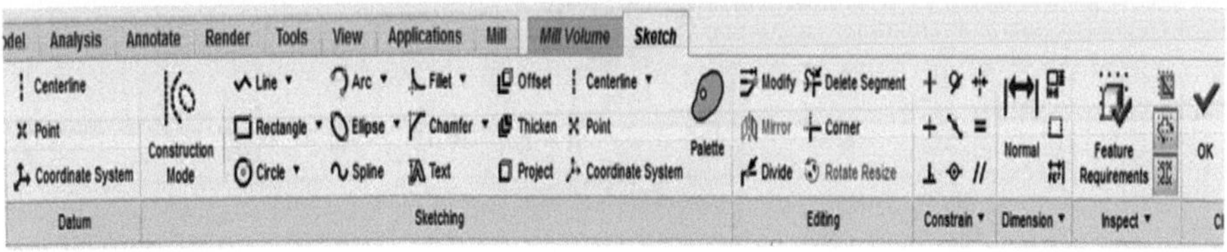

Fig. 12.64 Activated Sketch tools in concise form

Click on Sketch view icon  to orient the Workpiece to the correct sketch view/plane.

The Workpiece is now orientated on the correct sketch plane as shown in Fig. 12.65.

Fig. 12.65 Workpiece in correct orientation

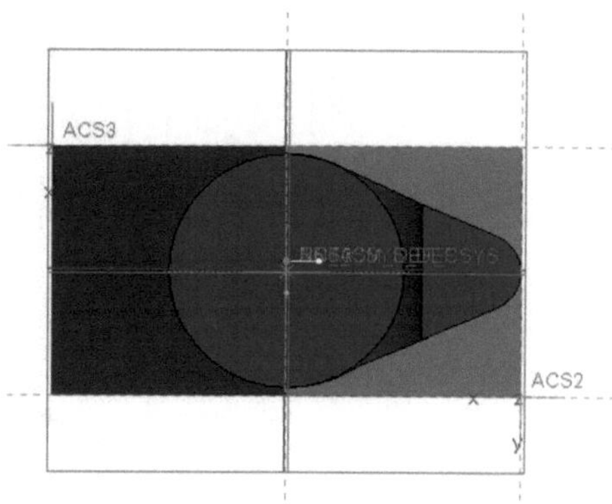

Create Reference

No new References will be created as all need references have been created earlier.

Project Sketch

Click on the Project icon tab .

The Type dialogue box is activated on the main graphic window as shown in Fig. 12.66.

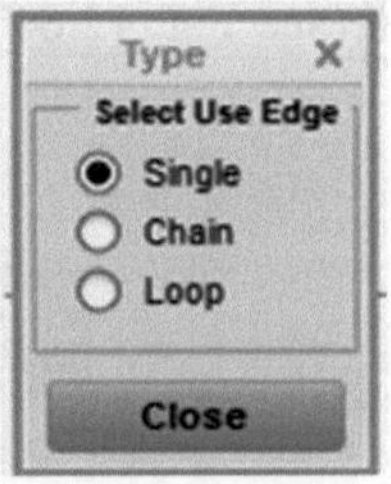

Fig. 12.66 Activated Type dialogue box

Click on the Lines and Curves on the Workpiece as indicated by the arrows. The Lines and Curves are projected into the sketch plane. The Line, Spline, Arc and Delete Segment tools on the activated Sketching tools can also be used to accomplish the Sketch as highlighted in Fig. 12.67.

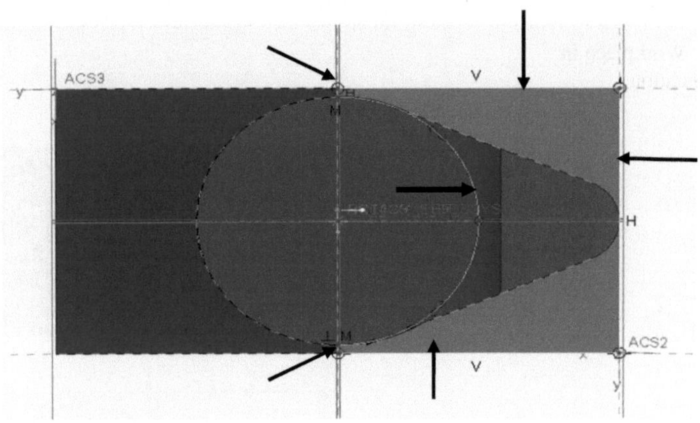

Fig. 12.67 Projecting Lines and Curves into the sketch plane

Click on Close tab on the Type dialogue box to exit.
Click on the Check Mark icon to exit the Sketch application.

Click on Exclude icon 

» The Extrude dashboard tools are activated » Click on the Extrusion type icon to activate its drop-down menu list » Click on "Extrude to selected point, curve, plane or surface" icon 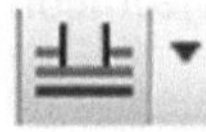 as indicated by the arrow in Fig. 12.68.

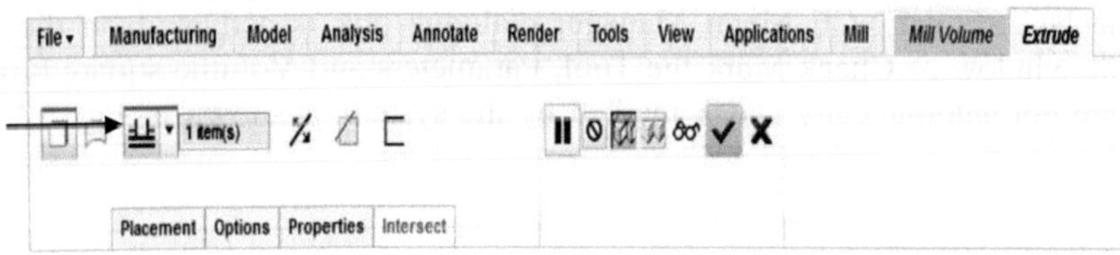

Fig. 12.68 Activated Extrude dashboard

Now click on the next surface to "Extrude to" on the Workpiece as indicated by the arrow in Fig. 12.69 and if required, flip the direction of Extrusion.

Fig. 12.69 Extruding the created third mill volume

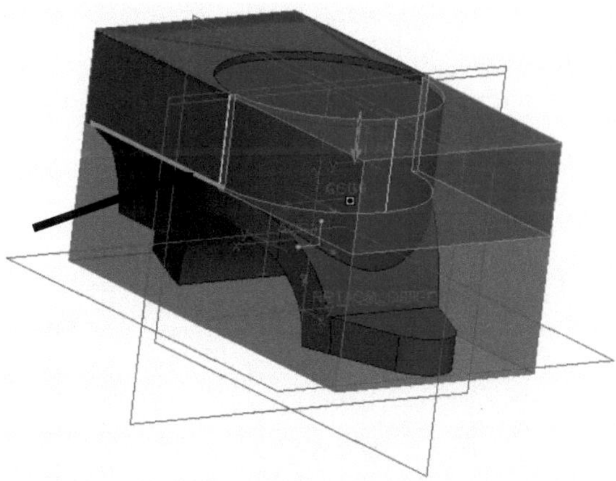

Click on the Check Mark icon to exit the Extrude application.
Click on the Check Mark icon to exit the Mill Volume application.
The third Mill Volume is now created and added to the Workpiece.

12.6.6 *Volume Rough Milling Operation for the Third Mill Volume*

Click on Roughing icon tab ≫ Now click on Volume Rough on the drop-down menu list as indicated by the arrow below.

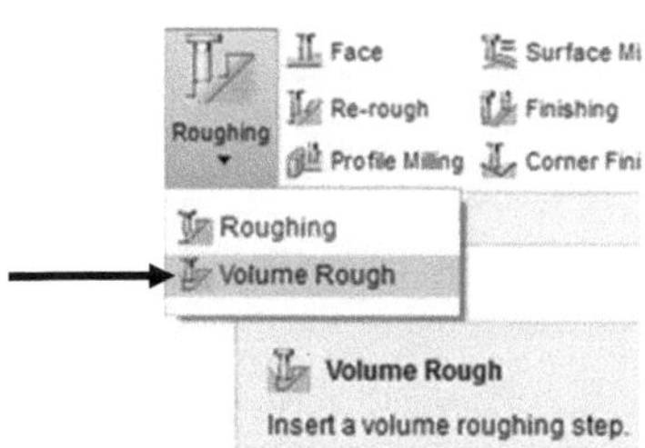

The NC SEQUENCE Menu Manager dialogue box is activated on the main graphic window ≫ Check Mark the Tool, Parameters and Volume square boxes if they are not automatically check marked by the system as shown in Fig. 12.70.

Fig. 12.70 Tool, Parameters and Volume check marked on the SEQ SETUP group

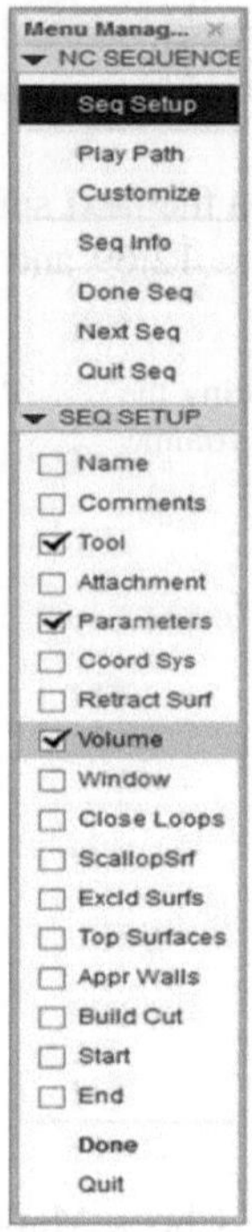

Click on Done.

The Tools Setup dialogue window is activated on the graphic window ≫ Select T01 as the cutting Tool for the third Volume Rough milling operation ≫ Click on the OK tab, accept the cutting Tool and exit the Tools Setup dialogue window.

The Edit Parameters of Sequence "Volume Milling" dialogue window is activated again on the main graphic window ≫ Input parameter values for the Volume Rough milling operation as shown in Fig. 12.71.

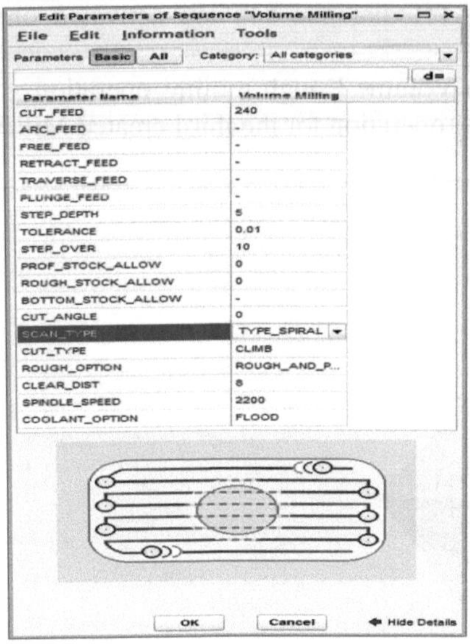

Click on OK tab to proceed.

Fig. 12.71 Parameters for Volume Rough milling operation

> Note: System is asking for the created Mill Volume to be selected/added on the message bar information.

Now click on the created third Mill Volume, if the Mill Volume has not been added automatically by the system.

Click on Play Path on the NC SEQUENCE group and the Menu Manager dialogue box expand to include the PLAY PATH group ≫ Now click on Screen Play. See Fig. 12.72 as highlighted.

Fig. 12.72 Activating Play Path and on Screen Play

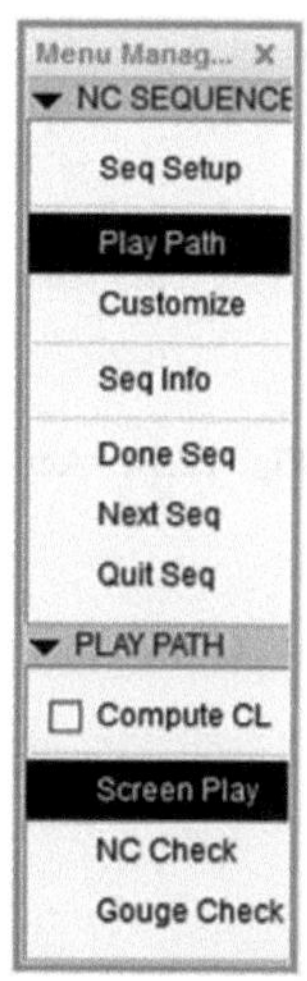

The cutting Tool and PLAY PATH dialogue box are activated on the main graphic window.

Click on the Play tab to start the Volume Rough milling operation.

The end of Volume Rough milling operation for the third created Mill Volume is shown in Fig. 12.73.

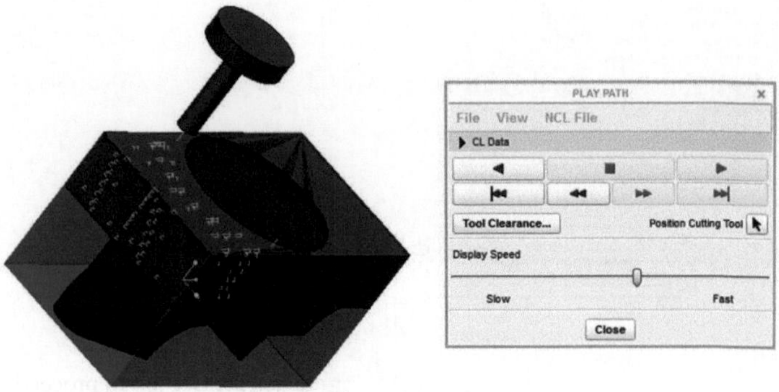

Fig. 12.73 End of Volume Rough milling operation

Click on Close tab to exit the PLAY PATH dialogue box.

Click on Done Seq on the NC SEQUENCE group on the Menu Manager dialogue box to exit the Volume Rough milling operation.

12.6.7 *Create Fourth Mill Volume*

Click on the Mill Volume icon .

The Mill Volume dashboard tools are activated as shown in Fig. 12.74.

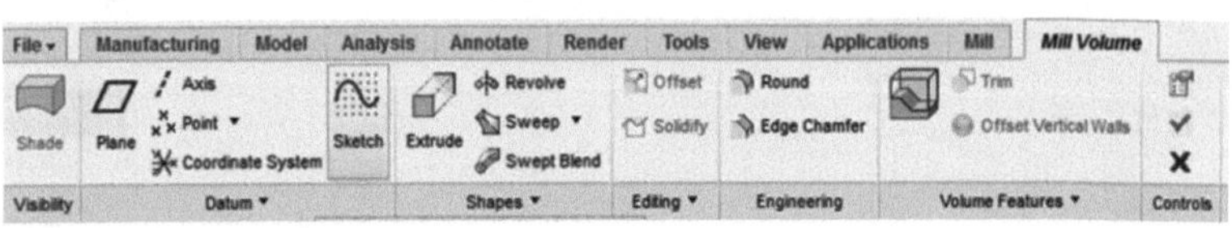

Fig. 12.74 Activated Mill Volume dashboard in concise form

Click on the Sketch icon  on the activated Mill Volume toolbar.

The Sketch dialogue box is activated on the main graphic window.

Click on the Placement tab to activate its content ≫ On the Sketch Plane group, click on the Plane section box and click on the surface of the Workpiece as indicated by the arrow below as the sketch plane as shown in Fig. 12.75.

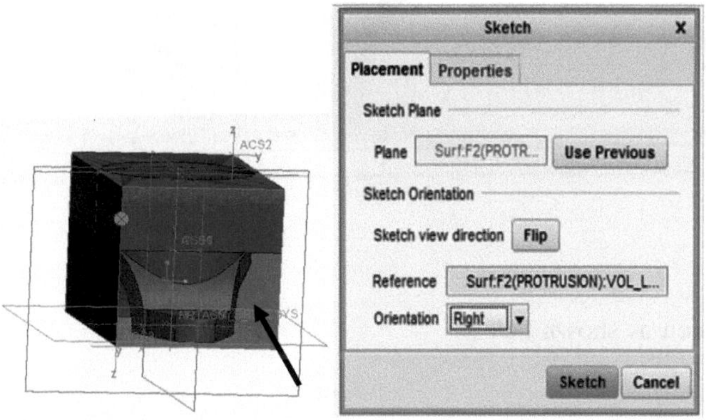

Click on Sketch tab.

Fig. 12.75 Creating the Sketch plane reference

The Sketch tools are activated as shown in Fig. 12.76 in concise form.

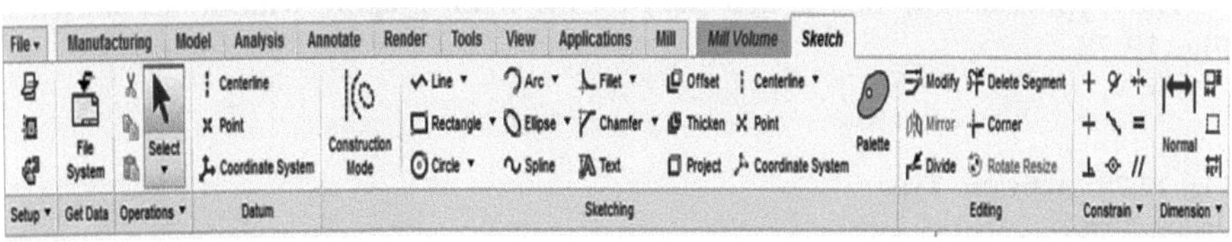

Fig. 12.76 Activated Sketch tools

Click on Sketch view icon  on the Sketch ribbon toolbar to orient the Workpiece to the correct sketch view/plane.

The Workpiece is now oriented into the correct sketch plane as shown in Fig. 12.77.

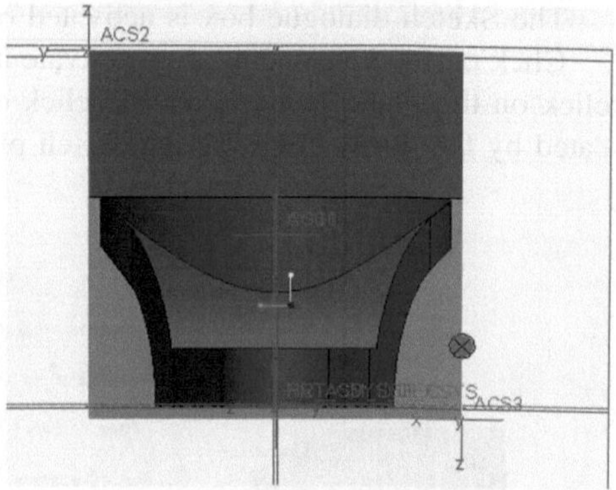

Fig. 12.77 Workpiece on correct orientation

Project Sketch

Click on the Project icon as shown below.

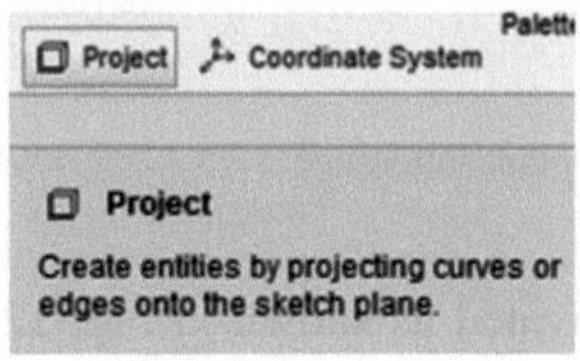

The Type dialogue box is activated on the main graphic window as shown in Fig. 12.78.

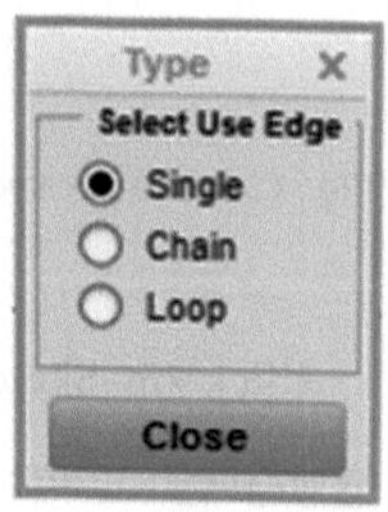

Fig. 12.78 Activated Type dialogue box

Click on the Lines on the Workpiece as indicated by the arrows. The Lines are projected into the sketch plane. The Line, Spline, Arc and Delete Segment tools on the activated Sketch tools can also be used to accomplish the Sketch as highlighted in Fig. 12.79.

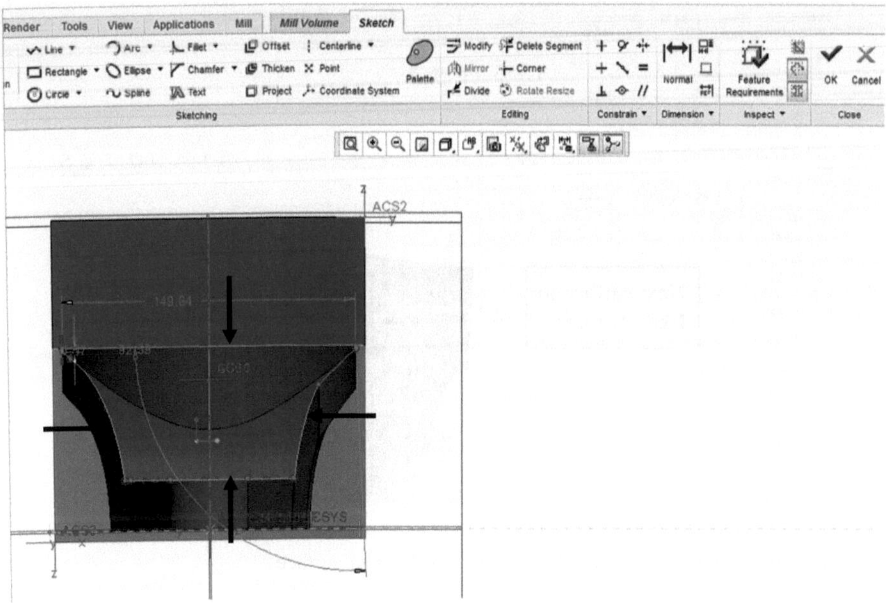

Fig. 12.79 Projecting the lines into the sketch plane

Click on Close tab on the Type dialogue box to exit.
Click on the Check Mark icon to exit the Sketch application.

Click on Exclude type icon 

>> The Extrude dashboard tools are activated >> Click on the Extrusion type icon to activate its drop-down menu list >> Click on "Extrude to selected point, curve, plane or surface" icon 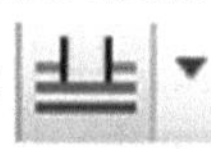 as indicated by the arrow in Fig. 12.80.

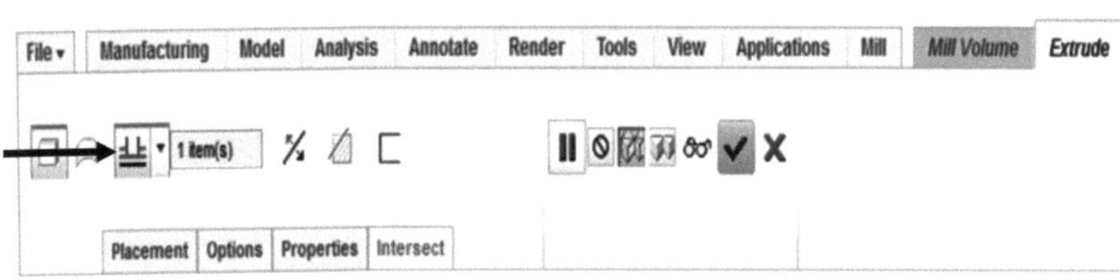

Fig. 12.80 "Extrude to selected point, curve, plane or surface" option type

Now click on the surface to "Extrude to" as indicated by the arrow in Fig. 12.81 and if required, flip the direction of extrusion.

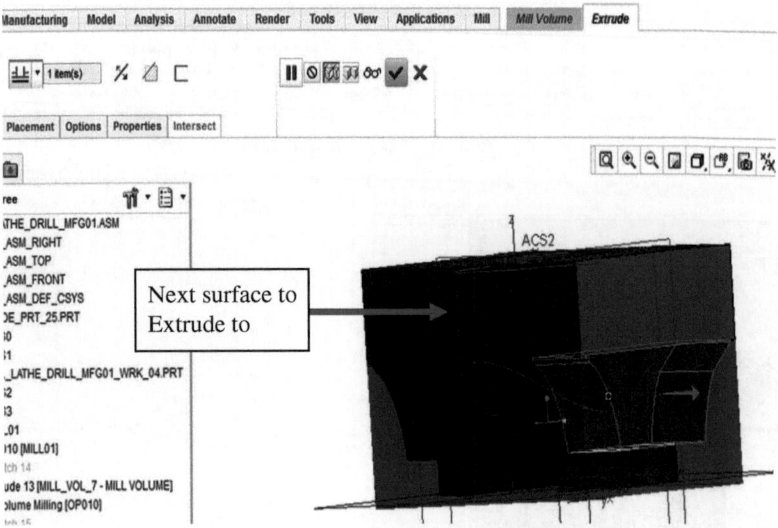

Fig. 12.81 Extruding the created Mill Volume to the next surface

Click on the Check Mark icon to exit Extrude application.
Click on Check Mark icon to exit Mill Volume application.

12.6.8 Volume Rough Milling Operation for the Fourth Mill Volume

Click on Roughing icon tab to activate its drop-down menu list ≫ Now click on Volume Rough on the drop-down list as indicated by the arrow below.

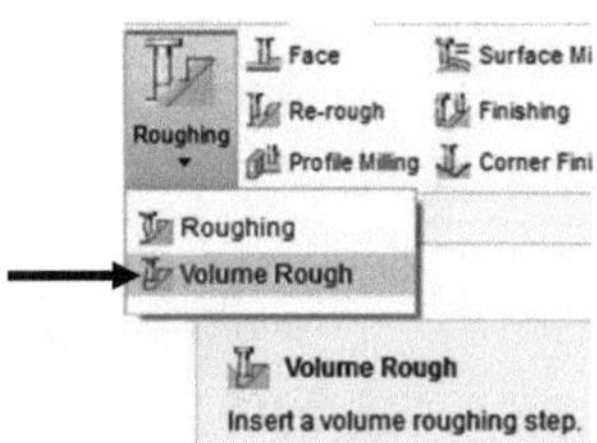

The NC SEQUENCE Menu Manager dialogue box is activated on the main graphic window » Check Mark the Tool, Parameters and Volume square boxes as shown in Fig. 12.82.

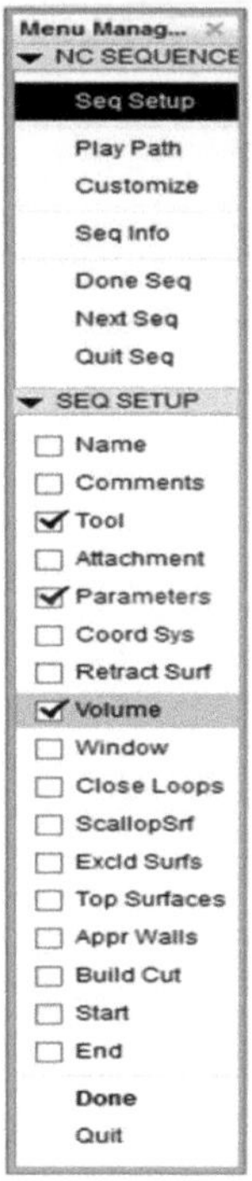

Fig. 12.82 NC SEQUENCE Menu Manager dialogue box with Tool, Parameters and Volume checked marked

Click on Done.

The Tools Setup dialogue window is activated again on the main graphic window » Click on T02 as the cutting Tool for the fourth Volume Rough milling operation as shown in Fig. 12.83.

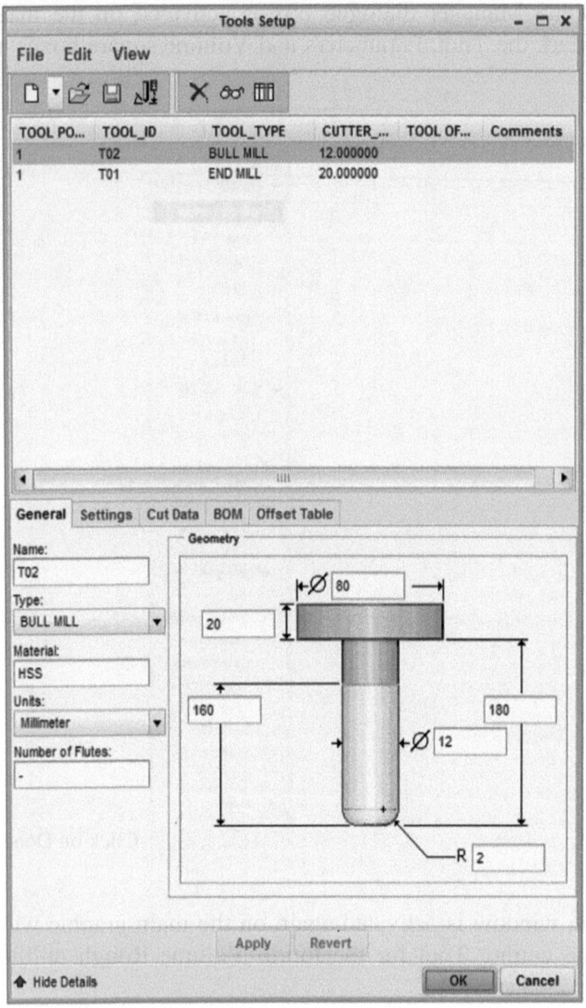

Click on the OK tab to exit.

Fig. 12.83 Activating Tool T02 for the fourth Mill Volume

Click on the OK tab to accept and proceed.

The Edit Parameters of Sequence "Volume Milling" dialogue window is activated again on the main graphic window ≫ Input the parameter values for the Volume Rough milling operation as shown in Fig. 12.84.

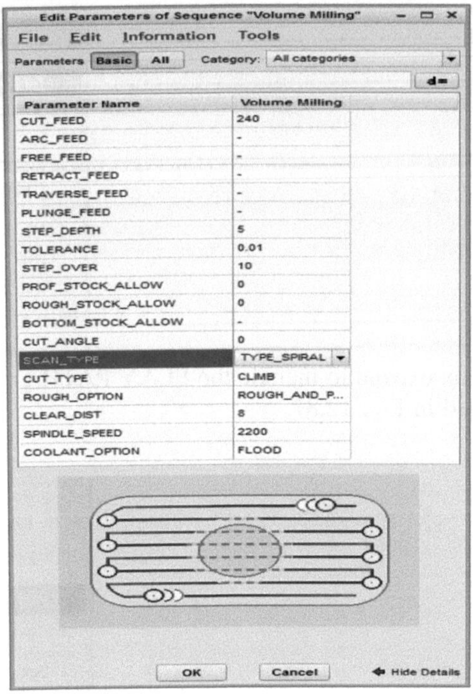

Click on the OK tab to exit.

Fig. 12.84 Parameters for Volume Rough milling operation

On the message bar information, system is asking for the created Mill Volume to be selected/added ≫ Now click on the created fourth Mill Volume as indicated by the arrow in Fig. 12.85.

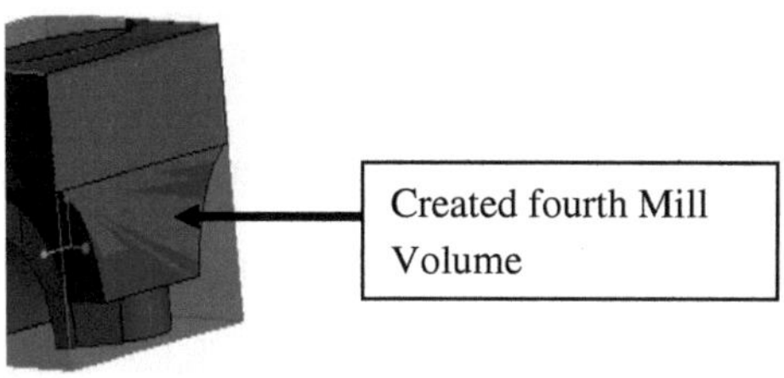

Fig. 12.85 Fourth Mill Volume for the milling operation

Click on Play Path on the NC SEQUENCE group in Fig. 12.86.

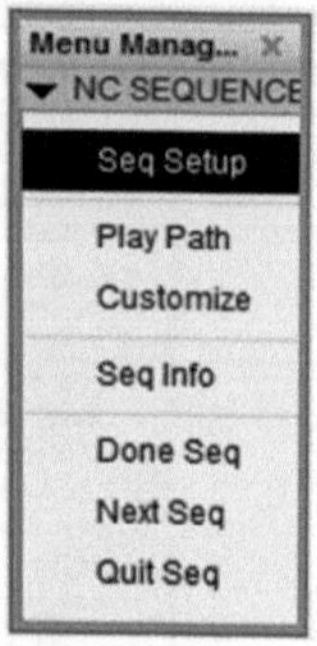

Fig. 12.86 Activating Play Path

The Menu Manager dialogue box expand to include the PLAY PATH group ≫ Click on Screen Play as highlighted in Fig. 12.87.

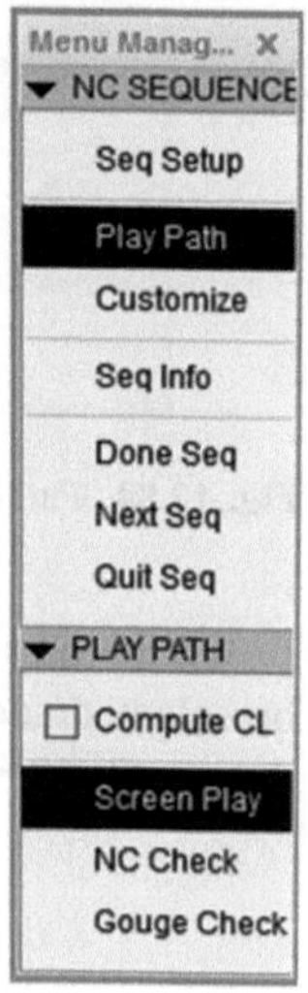

Fig. 12.87 Activating on Screen Play

The cutting Tool and the PLAY PATH dialogue box are activated on the main graphic window.

Click on the Play tab on the PLAY PATH dialogue box to start the Volume Rough milling operation.

The end of Volume Rough milling operation is shown in Fig. 12.88.

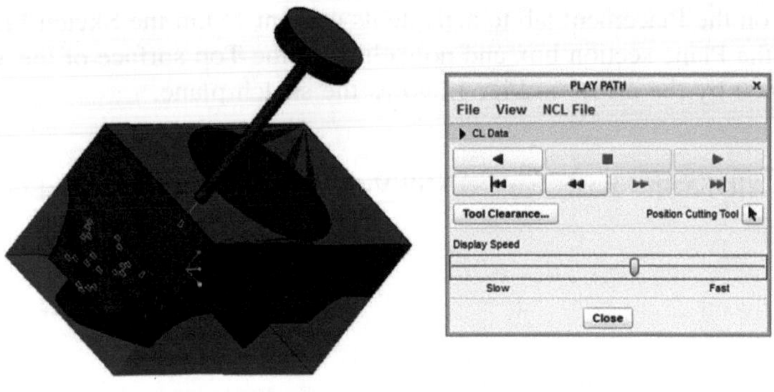

Fig. 12.88 End of Volume Rough milling operation

Click on Close tab to exit the PLAY PATH dialogue box.
Click on the Done Seq on the NC SEQUENCE group to exit the operation.

12.6.9 Create the Fifth Mill Volume

Click on the Mill Volume icon as indicated by the arrow below.

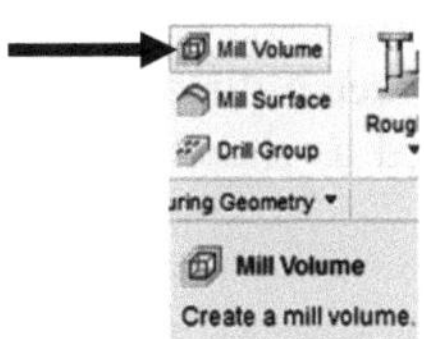

The Mill Volume ribbon dashboard tools are activated as shown in Fig. 12.89.

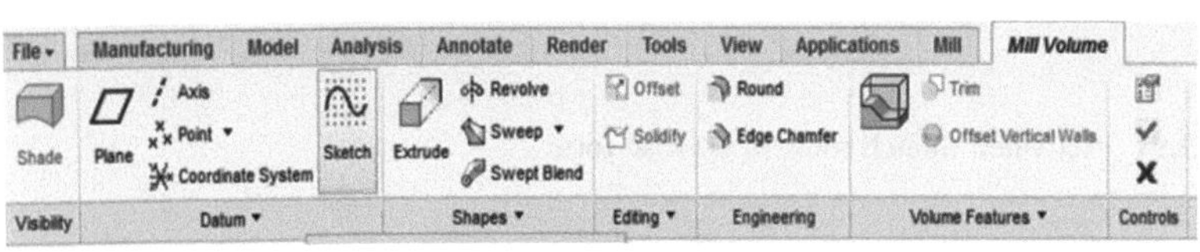

Fig. 12.89 Activated Mill Volume dashboard in concise form

Click on the Sketch icon  on the activated Mill Volume dashboard.

The Sketch dialogue box is activated on the main graphic window.

Click on the Placement tab to activate its content ≫ On the Sketch Plane group, click on the Plane section box and now click on the Top surface of the Workpiece as indicated by the arrow in Fig. 12.90 as the sketch plane.

Note: All the previously created Mill Volumes are all hidden. See Fig. 12.90.

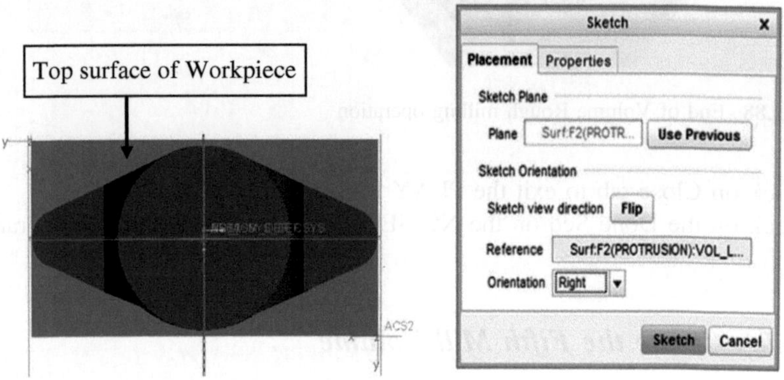

Fig. 12.90 Activating the sketch plane

Click on Sketch tab on the Sketch dialogue box.
The Sketch tools are activated as shown in Fig. 12.91.

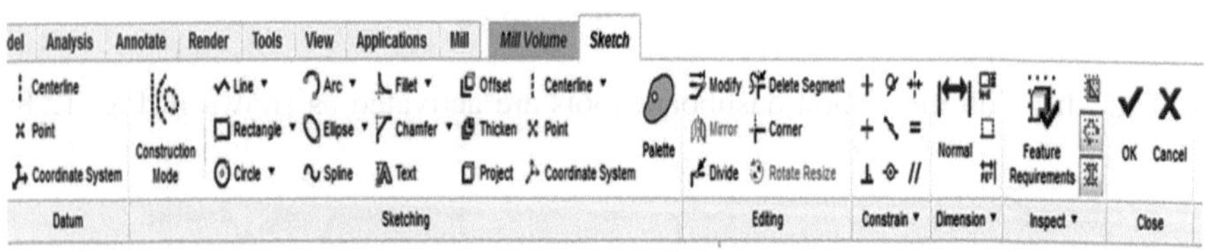

Fig. 12.91 Activated Sketch tools in concise form

Click on Sketch view icon  on the Sketch tools to orient the Workpiece to the correct sketch view/plane.

The Workpiece is now oriented on the correct sketch plane as shown in Fig. 12.92.

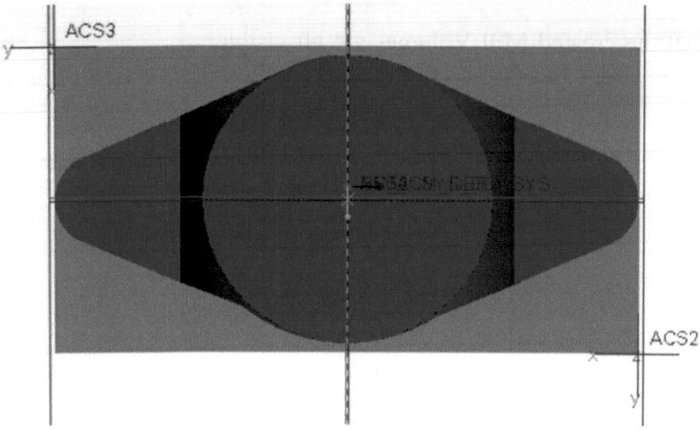

Fig. 12.92 Correct orientation of Workpiece Stock

Project Sketch

Click on the Project icon as shown below.

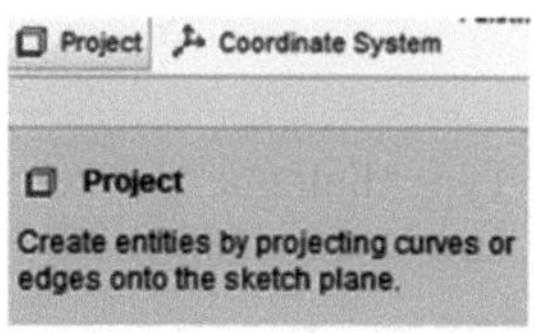

The Type dialogue box is activated on the main graphic window as shown in Fig. 12.93.

Fig. 12.93 Activated Type dialogue box

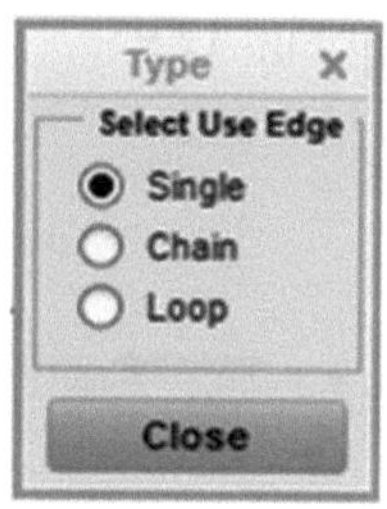

Click on circular curve on the Workpiece as indicated by the arrow. The circular Curve is now projected into the sketch plane as highlighted in Fig. 12.94.

Note: All the created Mill Volumes are all visible.

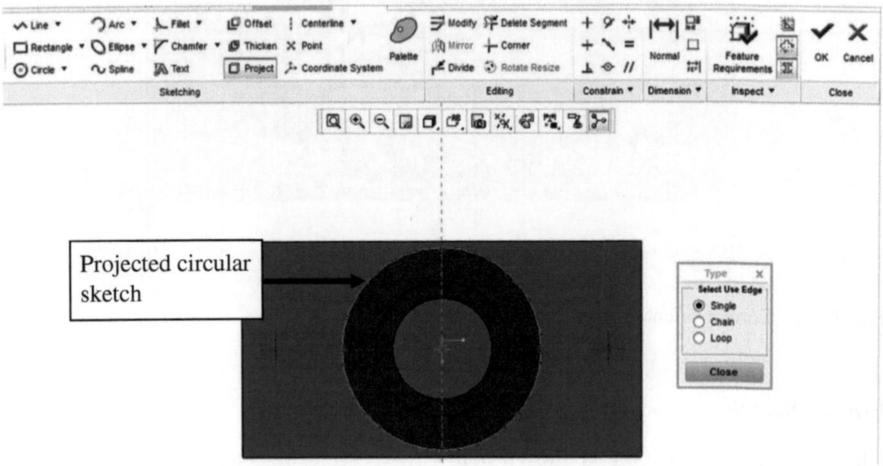

Fig. 12.94 Projected circular sketch

Click on Close tab on the Type dialogue box to exit.
Click on the Check Mark icon to exit the Sketch application.

Click on Exclude icon  >> The Extrude ribbon toolbar is activated >> Click on the Extrusion type icon to activate its drop-down menu list >> Click on "Extrude to selected point, curve, plane or surface" icon 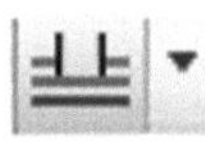 as indicated by the arrow in Fig. 12.95.

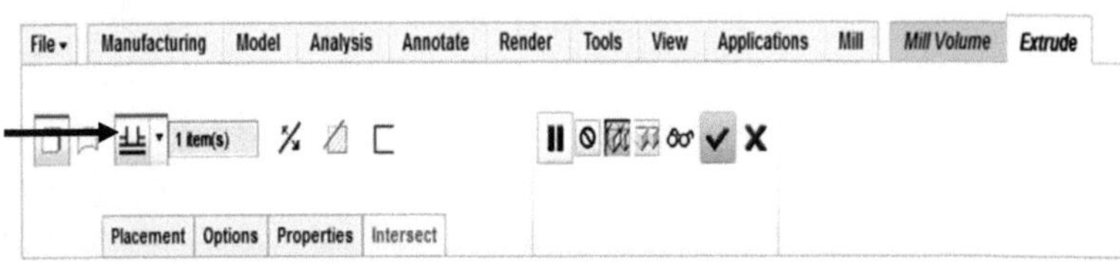

Fig. 12.95 Activating Extrude to selected point, curve, plane or surface extrusion option type

Now click on the next surface to "Extrude to" (Top surface of Part) on the Workpiece as indicated by the arrow in Fig. 12.96 and if required, flip the direction of Extrusion. See Fig. 12.96.

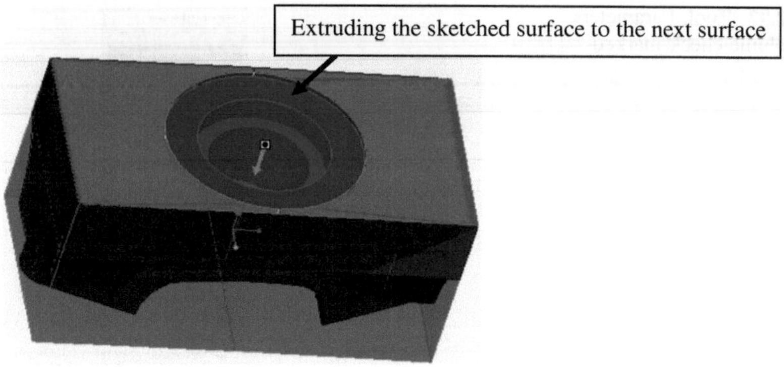

Fig. 12.96 Extruding the sketched mill volume

Click on the Check Mark icon to exit the Extrude application.
Click on the Check Mark icon to exit the Mill Volume application.
The fifth Mill Volume is now created and added to the Workpiece.

12.6.10 Volume Rough Milling Operation for the Fifth Mill Volume

Click on Roughing icon tab to activate its drop-down menu list ≫ Now click on Volume Rough as indicated by the arrow below.

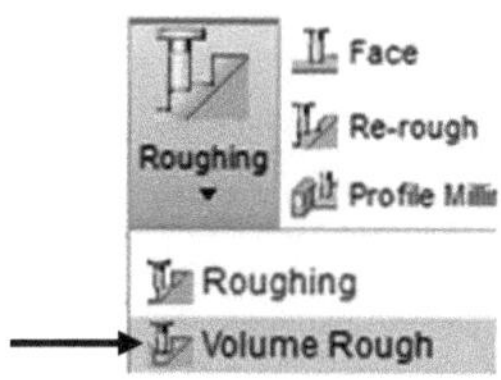

The NC SEQUENCE Menu Manager dialogue box is activated on the main window ≫ Check Mark the Tool, Parameters and Volume square boxes as shown in Fig. 12.97.

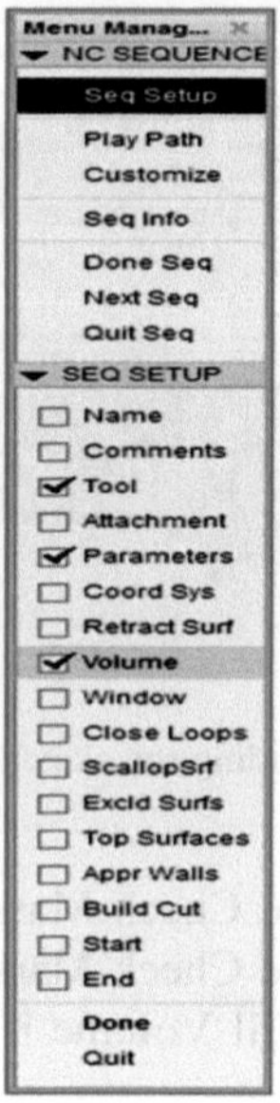

Fig. 12.97 Tool, Parameters and Volume check marked

Click on Done.

The Tools Setup dialogue window is activated on the graphic window ≫ Click on T03 as the cutting Tool for the fifth Volume Rough milling operation as shown in Fig. 12.98.

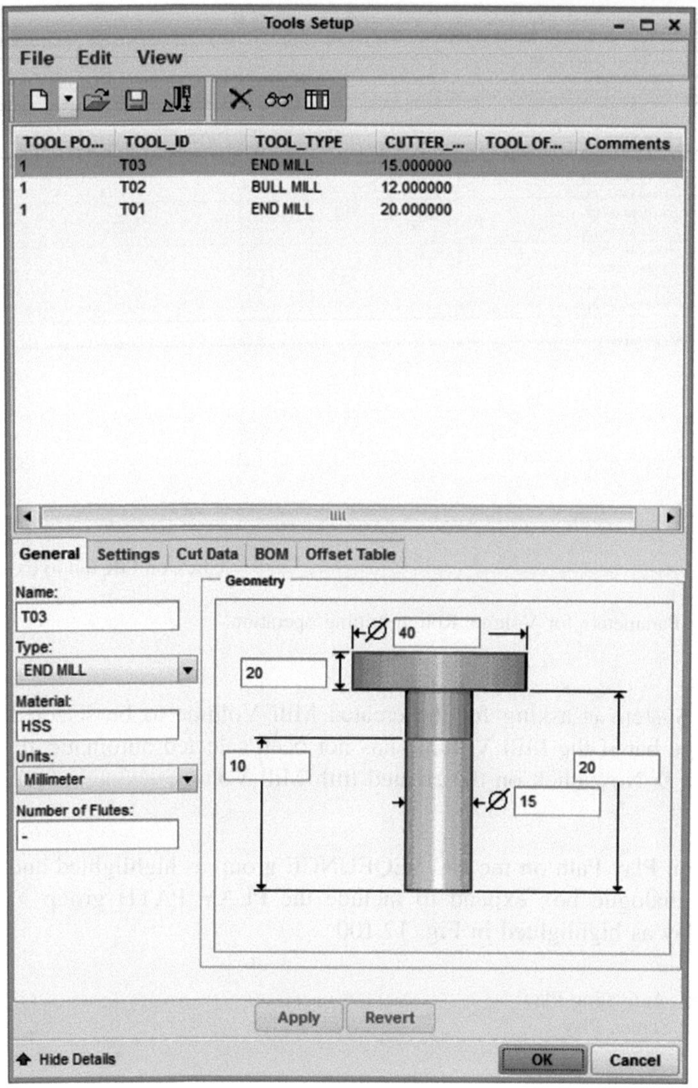

Fig. 12.98 Selected fifth Volume Rough mill Tool (T03)

Click on OK tab to exit.

The Edit Parameters of Sequence "Volume Milling" dialogue window is activated on the main graphic window ≫ Add values to the parameters as shown in Fig. 12.99.

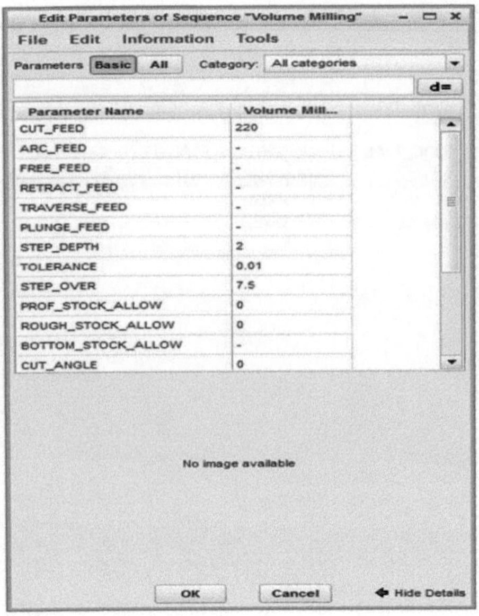

Click on OK tab to exit.

Fig. 12.99 Parameters for Volume Rough milling operation

Note: System is asking for the created Mill Volume to be selected on the message bar if the Mill Volume has not been selected automatically by the system ≫ Now click on the created fifth Mill Volume

Click on Play Path on the NC SEQEUNCE group as highlighted and the Menu Manager dialogue box expand to include the PLAY PATH group ≫ Click on Screen Play as highlighted in Fig. 12.100.

Fig. 12.100 Activating Play Path and on Screen Play

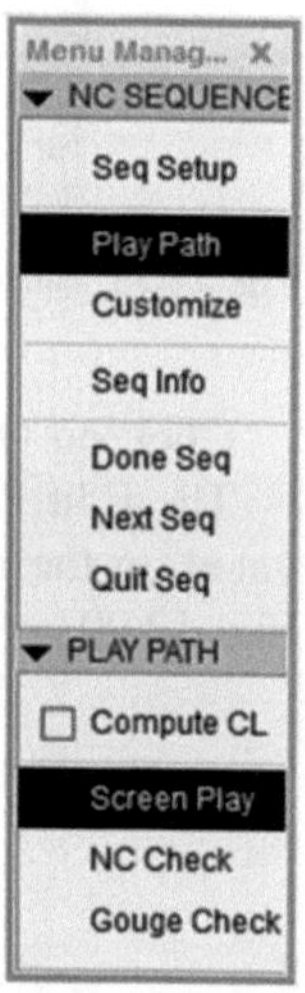

The cutting Tool and the PLAY PATH dialogue box are activated on the main graphic window as shown in Fig. 12.101.

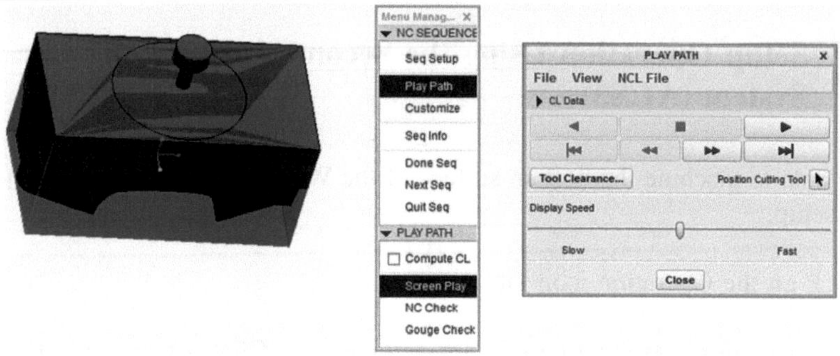

Fig. 12.101 Activated cutting Tool and the PLAY PATH dialogue box

Click on the Display View Style icon  to activate its drop-down menu list

≫ Now click on Wireframe on the drop-down list, the cutting Tool and Workpiece changes to Wireframe display.

Click on Play tab on PLAY PATH dialogue box to start the Volume Rough milling process.

The end of Volume Rough milling operation for the fifth created Mill Volume is shown in Fig. 12.102.

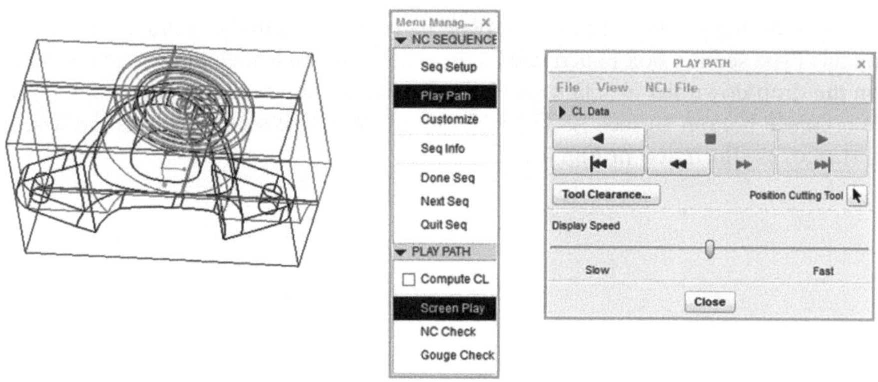

Fig. 12.102 End of Volume Rough milling operation in Wireframe display

Click on Close tab to exit the PLAY PATH dialogue box.

Click on the Done Seq on the NC SEQEUNCE group to exit the Volume Rough milling operation.

12.7 Setup Operation Using the Second Coordinate System (ACS3)

To be able to machine the bottom surface of the Workpiece, another Operation has to be setup.

Click on the Operation icon tab .

The Operation dashboard tools are activated ≫ Click on the Machine icon section box to activate its drop-down list, and now click on MILL01 ≫ Click on the Coordinate System icon section box, and now go to the Model Tree and click on ACS3 (Fig. 12.103).

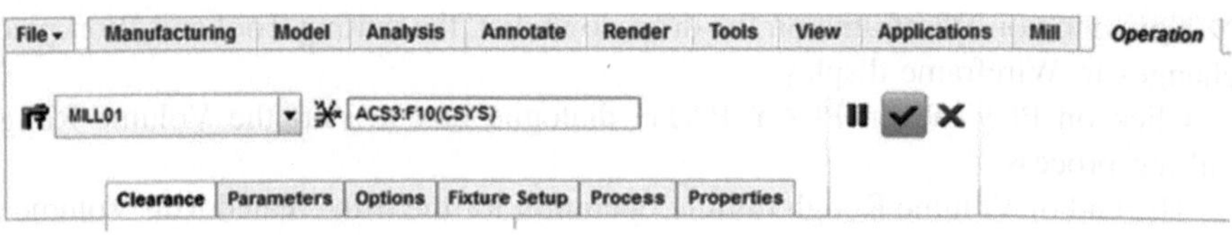

Fig. 12.103 Operation parameters for the bottom Volume Rough milling operations

Click on the Clearance tab to activate its content ≫ On the Retract group, click on the Type section box to activate its drop-down menu list, and now click on Plane on the drop-down list ≫ Click on the Reference section box, and now click on the Workpiece bottom surface ≫ Type 15 mm in the Value section box as the Retract distance, as illustrated in Fig. 12.104.

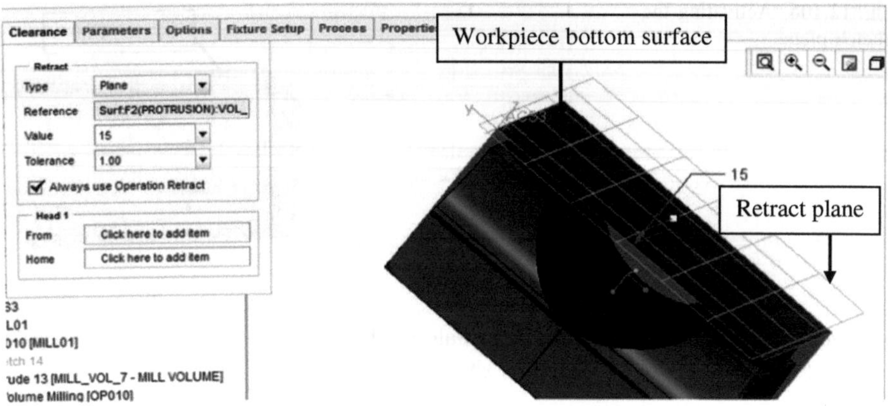

Fig. 12.104 Creating the Retract plane for the bottom milling operations

Click on the Check Mark icon to exit Operation setup.

12.7.1 *Create the First Bottom Mill Volume*

Click on the Mill Volume icon tab as shown below.

The Mill Volume dashboard tools are activated. See Fig. 12.105.

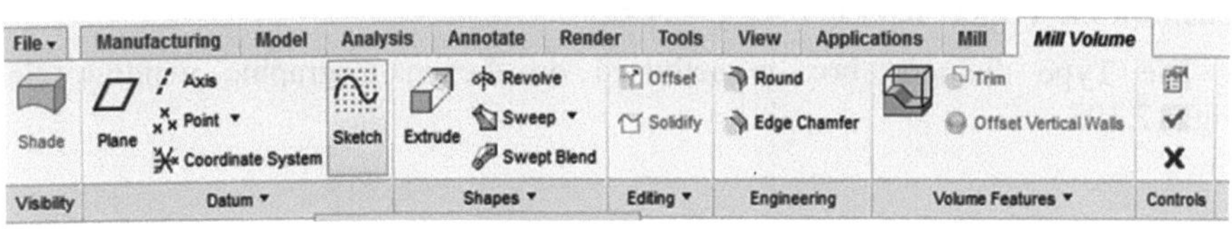

Fig. 12.105 Activated Mill Volume dashboar in concise form

Click on the Sketch icon  on the activated Mill Volume dashboard.

The Sketch dialogue box is activated on the main graphic window ≫ Now click on the Placement tab to activate its content ≫ On the Sketch Plane group, click on the Plane section box and select the Bottom surface of the Workpiece as indicated by the arrow in Fig. 12.106 as the sketch plane.

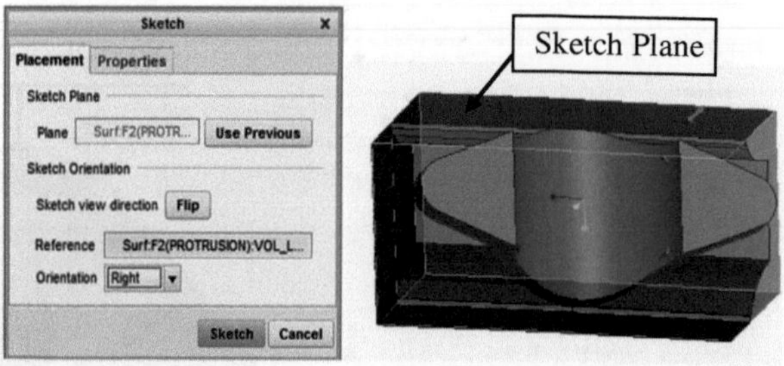

Fig. 12.106 Activating the Sketch plane

Click on Sketch tab on the Sketch dialogue box to exit.

The Sketch tools are activated as shown in Fig. 12.107.

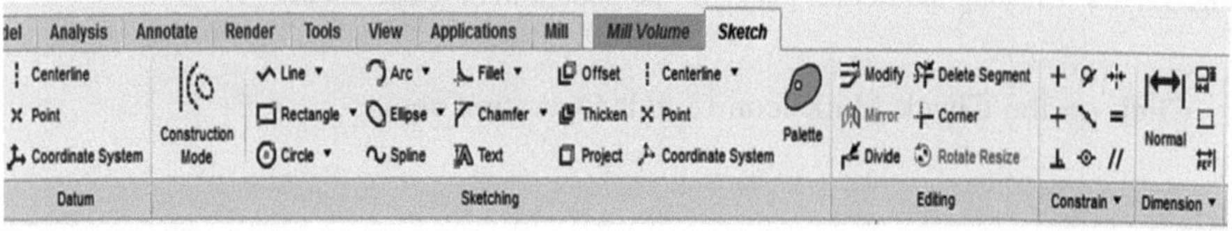

Fig. 12.107 Activated Sketch tools in concise form

Click on Sketch view icon  on the Sketch toolbar to orient the Workpiece to the correct sketch view/plane.

The Workpiece is now orientated on the correct sketch plane.

Project Sketch

Click on the Project icon tab 

.

The Type dialogue box is activated on the main graphic window. See Fig. 12.108.

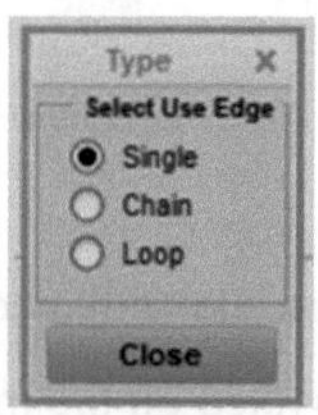

Fig. 12.108 Activated Type dialogue box

Click on the Lines and Curves on the Workpiece as indicated by the arrows in Fig. 12.109. The Lines and Curves are projected into the sketch plane. The Line,

Spline, Arc and Delete Segment tools on the activated Sketch ribbon can also be used to accomplish the Sketch as highlighted in Fig. 12.109.

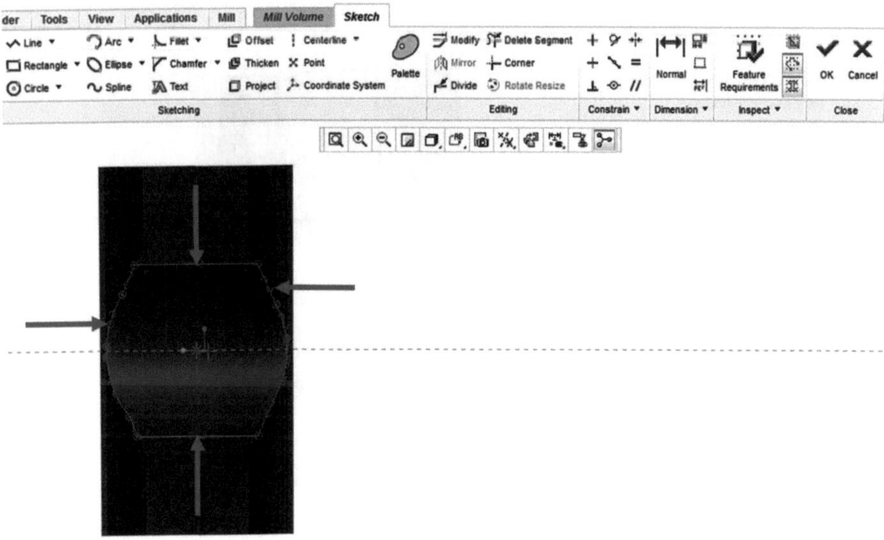

Fig. 12.109 Projecting Lines and Curves into the Sketch plane

Click on Close tab to exit the Type dialogue box
Click on the Check Mark icon to exit the Sketch application.

Click on Extrude icon 

>> The Extrude dashboard tools are activated >> Click on the Extrusion type icon to activate its drop-down menu list >> Click on "Extrude to selected point, curve, plane or surface" icon 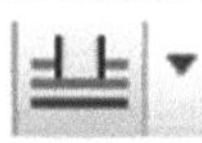 as indicated by the arrow below (Fig. 12.110).

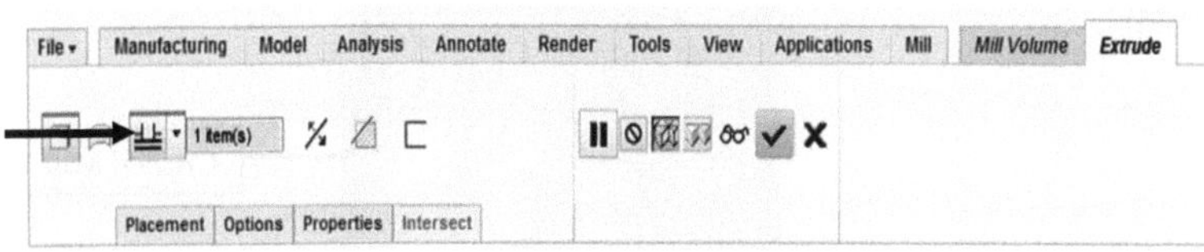

Fig. 12.110 Activating the Extrude to selected point, curve, plane or surface option

Now click on the next surface to "Extrude to" on the Workpiece as indicated by the arrow in Fig. 12.111 and if required, flip the direction of Extrusion.

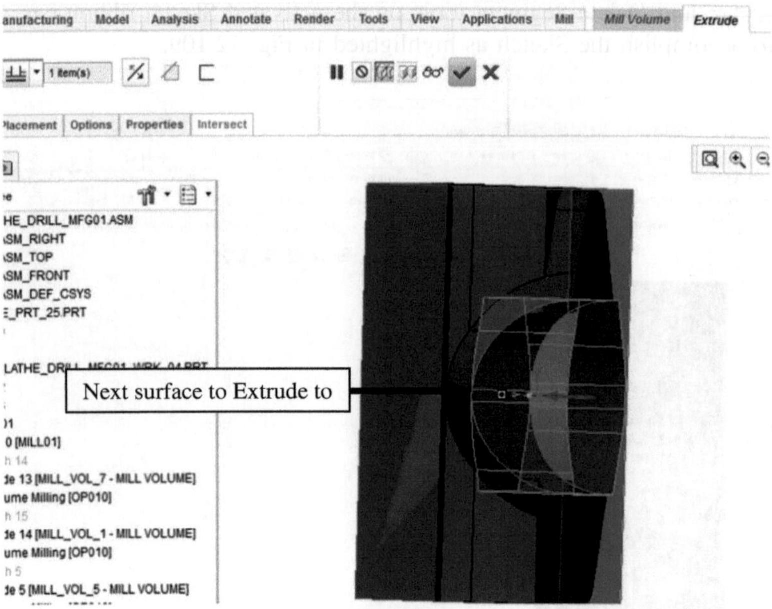

Fig. 12.111 Extruding the sketch to the next surafce

Click on the Check Mark icon to exit Extrude application.
Click on Check Mark icon again to exit Mill Volume application.
The arrow as shown in Fig. 12.112 indicates the Extruded sketch (Mill Volume).

Note: All the created Mill Volumes are visible.

Fig. 12.112 Created first bottom mill volume

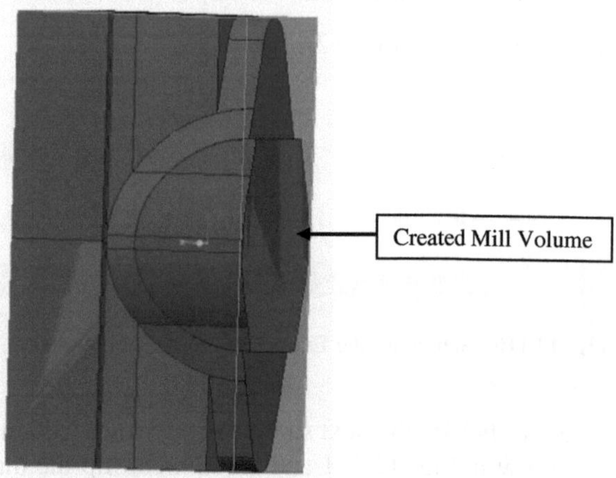

12.7.2 *Volume Rough Milling Operation for the First Bottom Mill Volume*

Volume Rough is accessed as illustrated below.

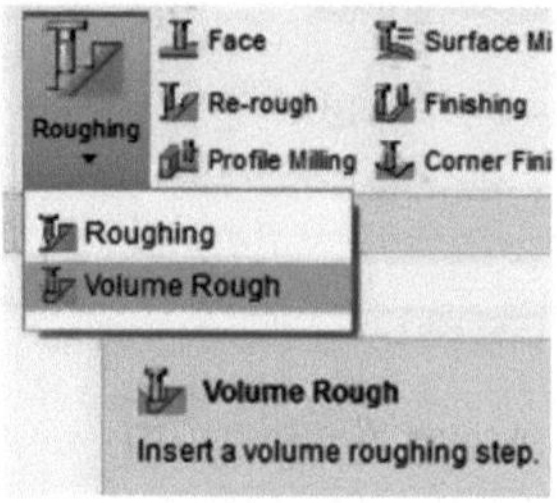

Once Volume Rough is clicked ≫ NC SEQUUENCE Menu Manager dialogue box is activated on the main graphic window ≫ Check Mark Tool, Parameters, Retract Surf and Volume on the SEQ SETUP group as shown in Fig. 12.113.

Fig. 12.113 Tool, Parameters, Retract Surf and Volume checked marked on SEQ SETUP group

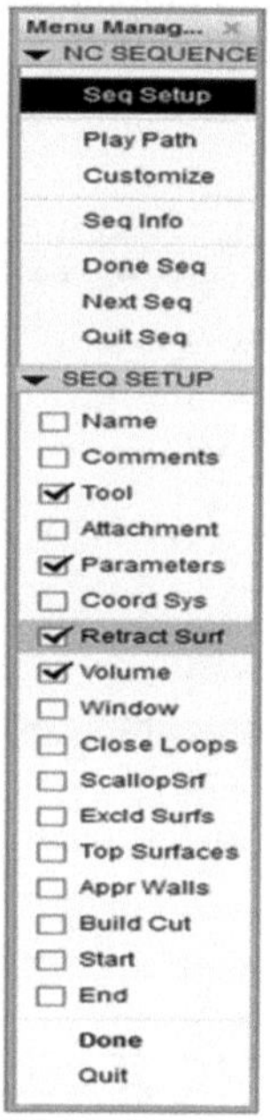

Click on Done.

The Tools Setup dialogue window is activated on the main graphic window ≫ Click on T01 as the cutting Tool for the first bottom Volume Rough milling operation as shown in Fig. 12.114 in concise form.

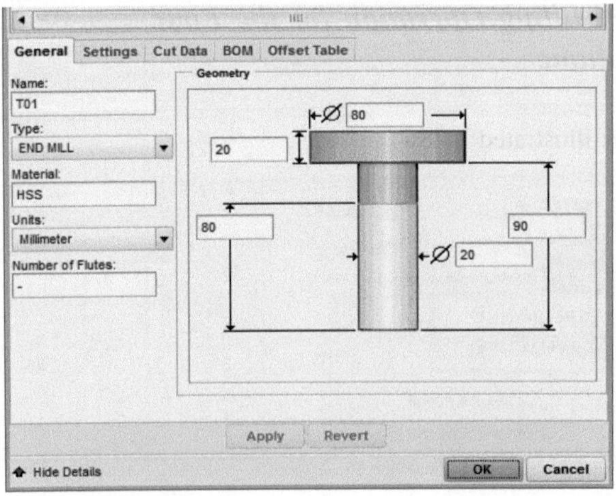

Click on the OK tab to exit.

Fig. 12.114 Correct Tool for the first bottom Volume Rough milling operation

The Edit Parameters of Sequence "Volume Milling" dialogue window is activated on the main graphic window ≫ Input parameter values for the Volume Rough milling operation as shown in Fig. 12.115.

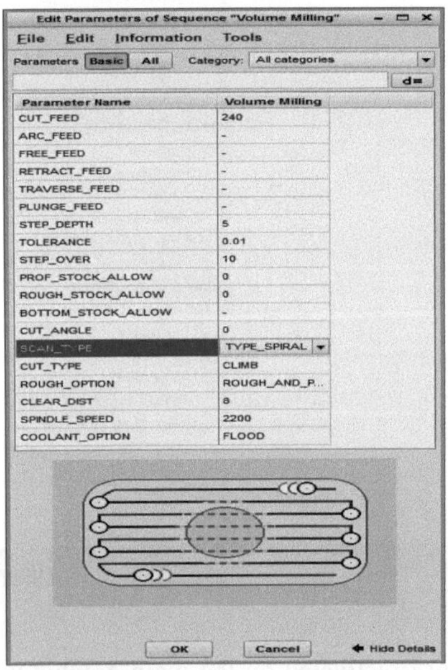

Click on the OK tab to exit.

Fig. 12.115 Volume Rough milling parameters

The Retract Setup dialogue box is activated on the graphic window. See Fig. 12.116.

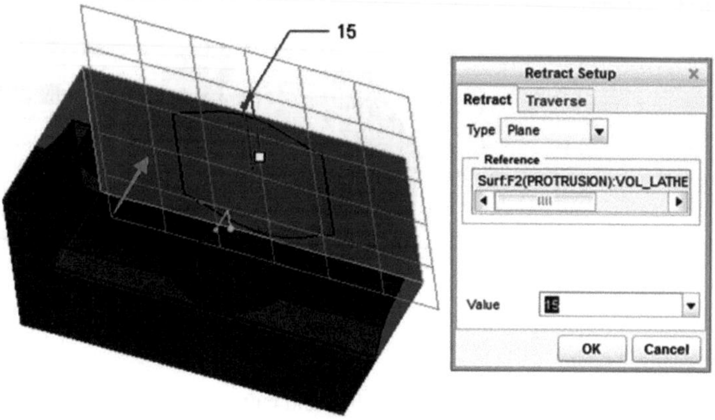

Fig. 12.116 Activated Retract Setup dialogue box and plane

Click on the OK tab to exit the Retract Setup dialogue box.

Note: System is asking for the created Mill Volume to be selected on the message bar if the Mill Volume has not been selected automatically by the system

Now click on the created Mill Volume.

Click on Play Path on the NC SEQUENCE group and the Menu Manager dialogue box expand to include the PLAY PATH group ≫ Now click on Screen Play as highlighted in Fig. 12.117.

Fig. 12.117 Activating Play Path and Screen Play

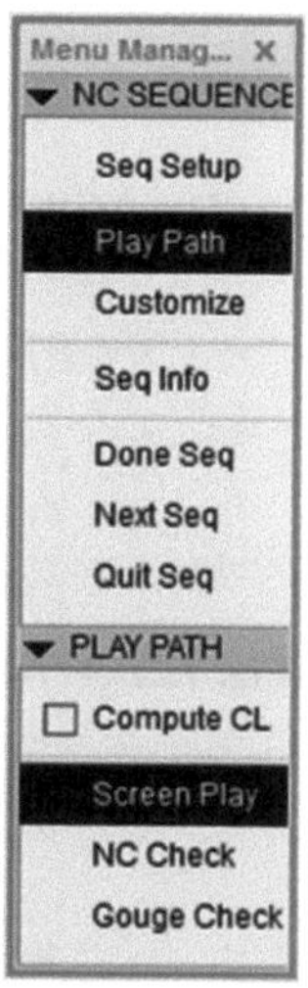

Once Screen Play is clicked, the cutting Tool and the PLAY PATH dialogue box are activated on the main graphic window as shown in Fig. 12.118.

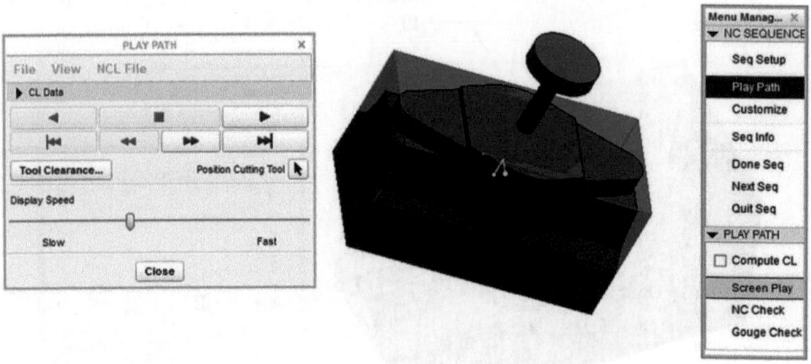

Fig. 12.118 Activated PLAY PATH dialogue box and cutting Tool

Click on the Play tab to start the Volume Rough milling process.

At the end of the on Screen Volume Rough milling operation, click on the Close tab to exit PLAY PATH dialogue box.

Click on Done Seq on NC SEQUENCE group to exit the Volume Rough milling operation.

12.7.3 Create the Second Mill Volume on the Bottom Side of the Workpiece

Click on the Mill menu bar to activate the Mill ribbon toolbar ≫ Now click on the Mill Volume icon as shown below.

The Mill Volume dashboard tools are activated. See Fig. 12.119.

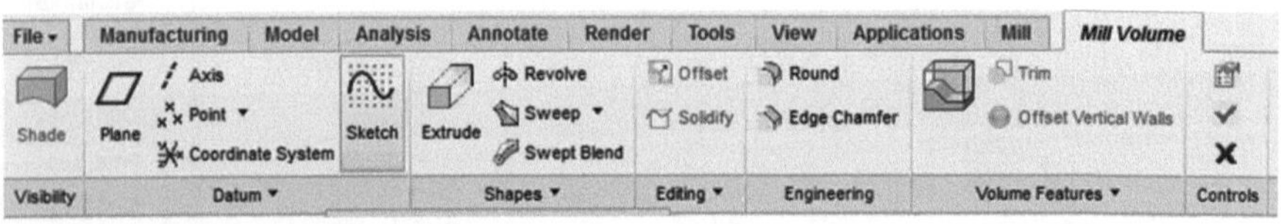

Fig. 12.119 Activated Mill Volume dashboard in concise form

Click on the Sketch icon .

The Sketch dialogue box is activated on the main graphic window ≫ Now click on the Placement tab to activate its content ≫ On the Sketch Plane group, click on the Plane section box and now click on the Bottom surface of the Workpiece as indicated by the arrow below as the sketch plane. The Reference and Orientation section boxes are automatically updated by the system on the Sketch dialogue box. See Fig. 12.120.

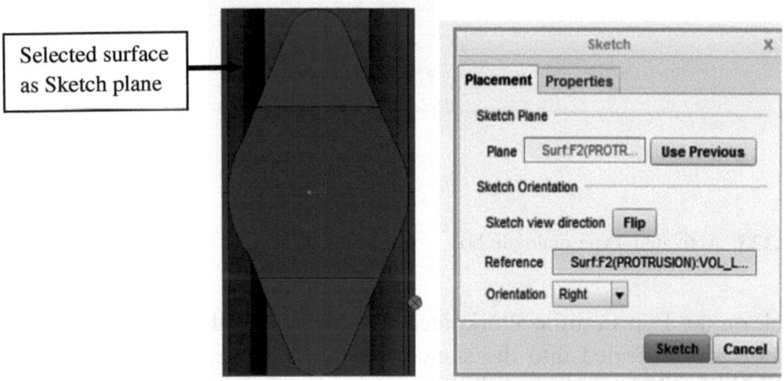

Fig. 12.120 Activating the Sketch plane

Click on Sketch tab to exit the Sketch dialogue box.

The Sketch tools are activated. See Fig. 12.121.

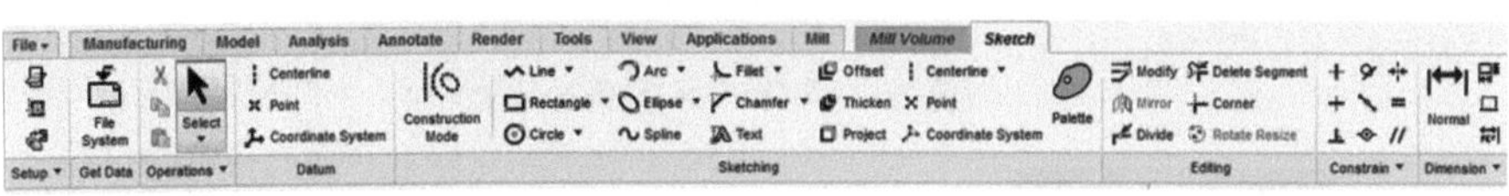

Fig. 12.121 Activated Sketch tools in concise form

Click on Sketch view icon  on the Sketch ribbon to orient the Workpiece to the correct sketch view/plane.

Workpiece is oriented on the correct sketch plane as shown in Fig. 12.122.

Fig. 12.122 Correct orientation of Workpiece

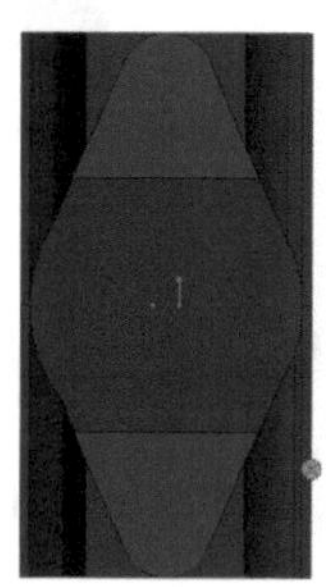

Project Sketch

Click on the Project icon tab 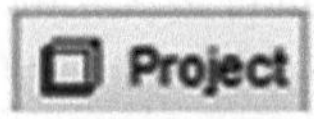

.

The Type dialogue box is activated on the main graphic window. See Fig. 12.123.

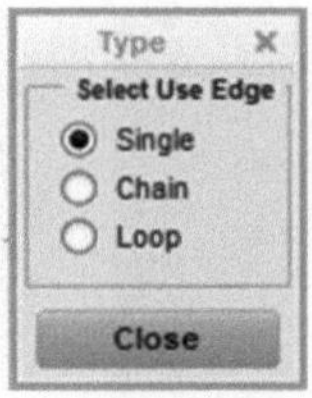

Fig. 12.123 Activated Type dialogue box

Click on the Lines on the Workpiece as indicated by the arrows in Fig. 12.124. The Lines are projected into the sketch plane. The Line, Spline, Arc and Delete Segment tools on the activated Sketch ribbon can also be used to accomplish the Sketch as highlighted in Fig. 12.124.

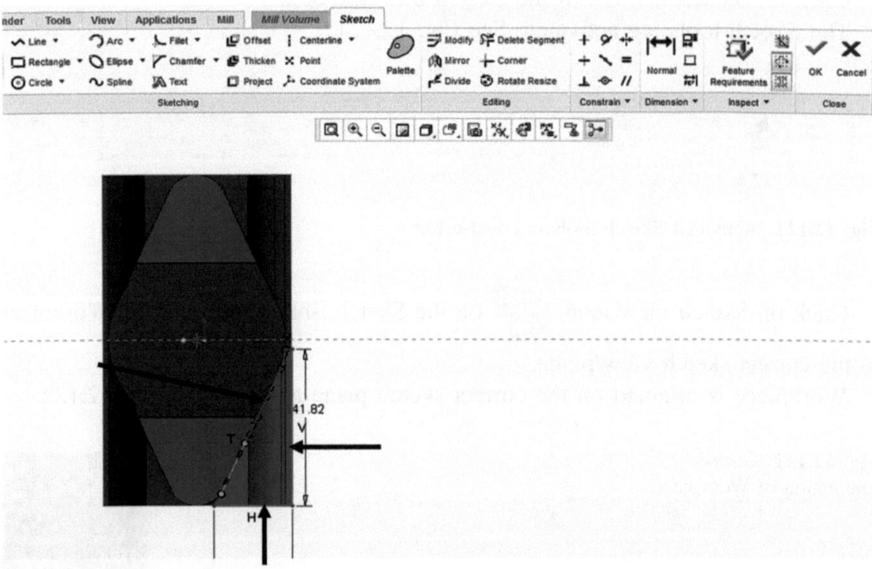

Fig. 12.124 Projecting Lines on the Workpiece into the Sketch plane

Click on Close tab to exit the Type dialogue box.

Click on the Check Mark icon/OK to exit the Sketch application.

Click on Exclude icon 

» The Extrude dashboard tools are activated » Click on the Extrusion type icon to activate its drop-down menu list » Click on "Extrude to selected point, curve, plane or surface" icon 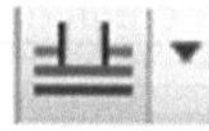 as indicated by the arrow in Fig. 12.125.

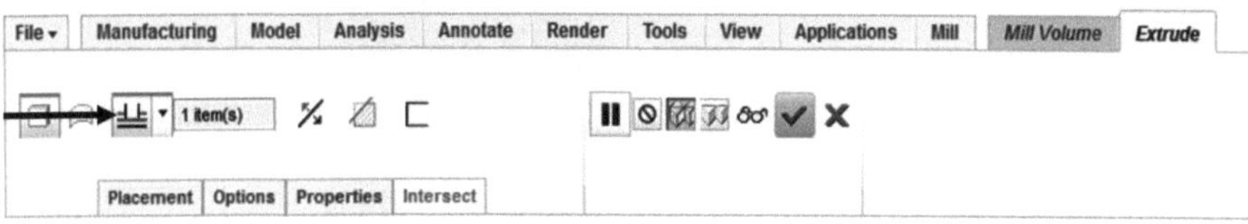

Fig. 12.125 Activating Extrude to selected point, curve, plane or surface extrusion option type

Now click on the next surface to "Extrude to" on the Workpiece as indicated by the arrow in Fig. 12.126 and if required, flip the direction of Extrusion.

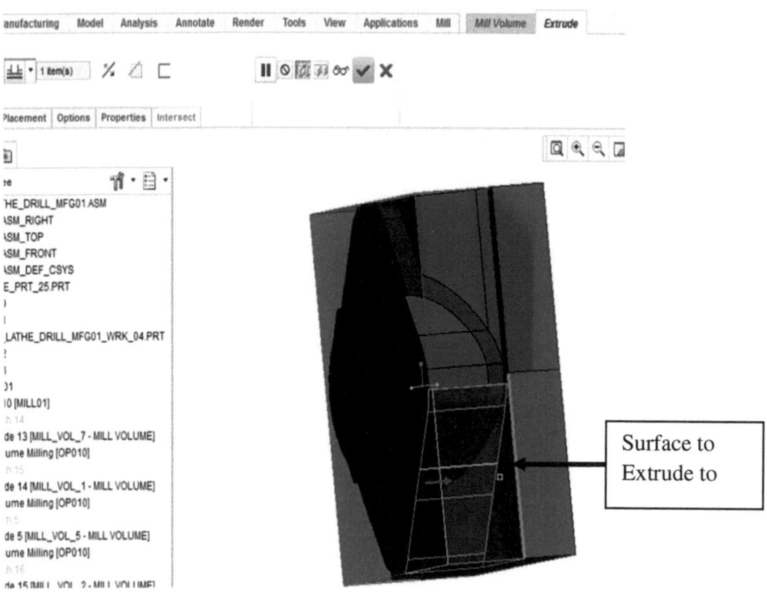

Fig. 12.126 Extruding the sketch to the next surafce

Click on the Check Mark icon to exit Extrude application.

Click on Check Mark icon again to exit Mill Volume application.
The arrow as shown in Fig. 12.127 indicates the new created Mill Volume.

Note: All the created Mill Volumes are visible.

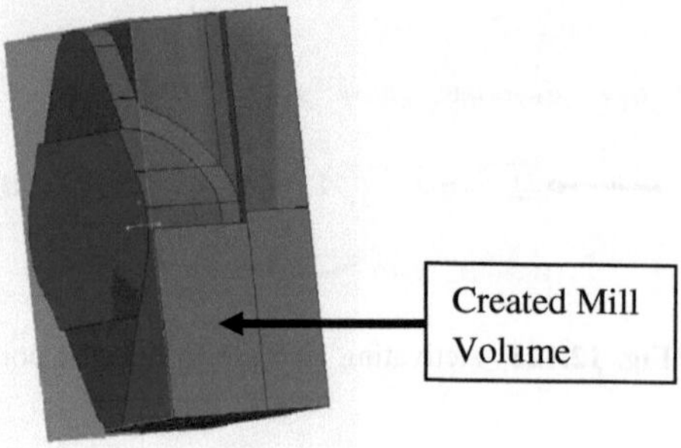

Fig. 12.127 Created second right bottom Mill Volume

12.7.4 Volume Rough Milling Operation for the Second Bottom Mill Volume

Click on the Mill menu bar to activate the Mill ribbon ≫ Click on Roughing icon tab to activate its content ≫ Now click on Volume Rough as shown below.

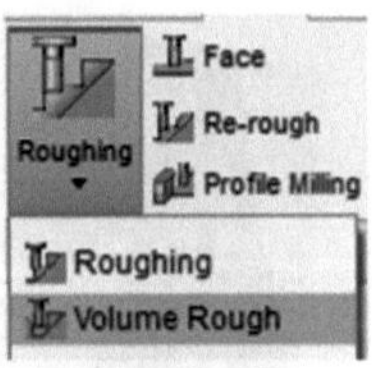

NC SEQUUENCE Menu Manager dialogue box is activated on the main graphic window ≫ Check Mark Tool, Parameters, Retract Surf and Volume on the SEQ SETUP group as shown in Fig. 12.128.

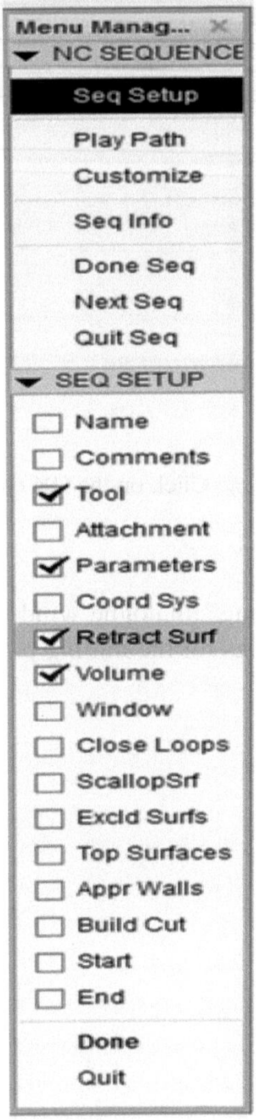

Click on Done on the SEQ SETUP group.

Fig. 12.128 Tool, Parameters, Retract Surf and Volume check marked

The Tools Setup dialogue window is activated on the main graphic window ≫ Click on T02 as the cutting Tool for the second Volume Rough milling operation for the bottom right Mill Volume as shown in Fig. 12.129 in concise form.

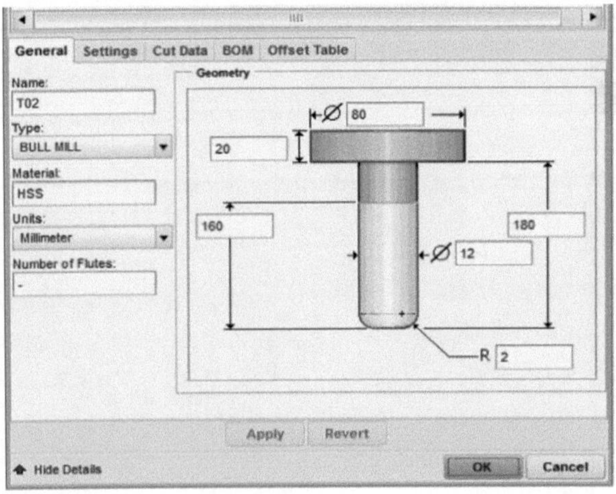

Click on the OK tab to exit.

Fig. 12.129 Activating Tool (T02) as the correct cutting Tool

The Edit Parameters of Sequence “Volume Milling” dialogue window is activated on the graphic window ≫ Input parameter values for the milling process. See Fig. 12.130.

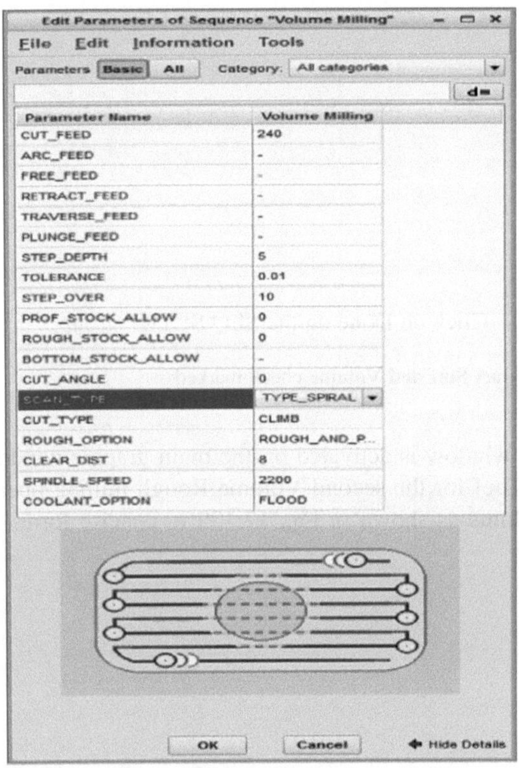

Click on the OK tab to exit.

Fig. 12.130 Volume Rough milling parameters

The Retract Setup dialogue box is activated on the main graphic window as illustrated in Fig. 12.131.

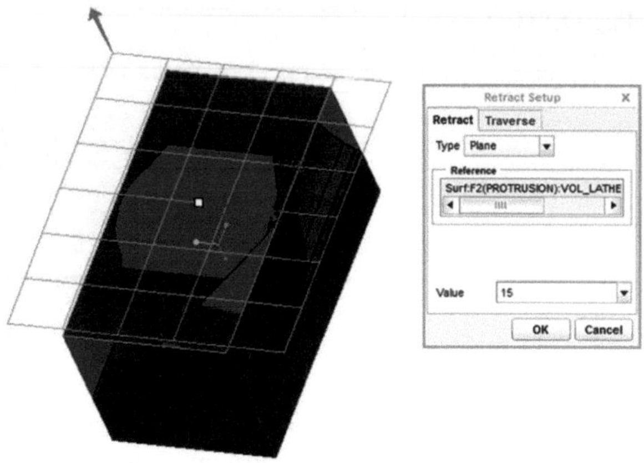

Fig. 12.131 Activated Retract Setup dialogue box

Click on the OK tab to exit.

Note: System is asking for the created Mill Volume to be selected on the message bar if the Mill Volume has not been selected automatically by the system

Now click on the created Mill Volume.

Click on Play Path on the NC SEQUENCE group and the Menu Manager dialogue box expand to include the PLAY PATH group' ≫Now click on Screen Play as shown in Fig. 12.132.

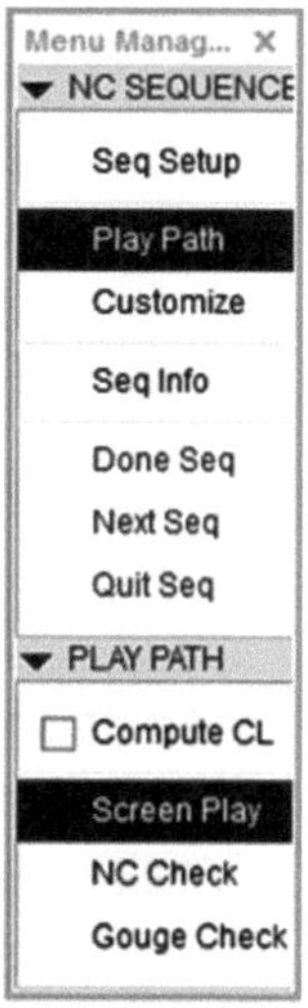

Fig. 12.132 Activating Play Path and on Screen Play

The cutting Tool and the PLAY PATH dialogue box are activated on the main graphic window as shown in Fig. 12.133.

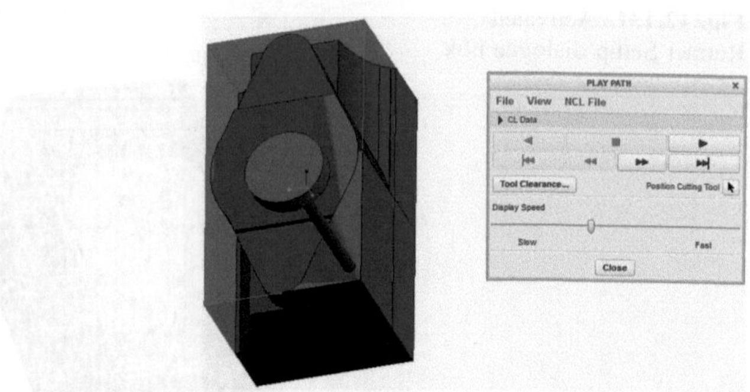

Fig. 12.133 Activated cutting Tool and the PLAY PATH dialogue box

Click on the Play tab to start the on Screen Volume Rough milling process.
At the end of the on Screen Volume Rough milling operation, click on Close tab.
Now click on Done Seq to exit the Volume Rough milling operation.

12.7.5 Create the Third Mill Volume on the Bottom Side of the Workpiece

Note: The step-by-step guide on how to create the second Mill Volume is the same as the steps used in creating the second Mill Volume on Sect. 12.7.3.

Click on the Mill Volume icon as shown below.

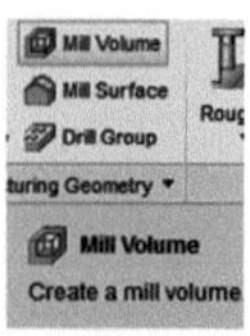

The Mill Volume dashboard tools are activated as shown in Fig. 12.134.

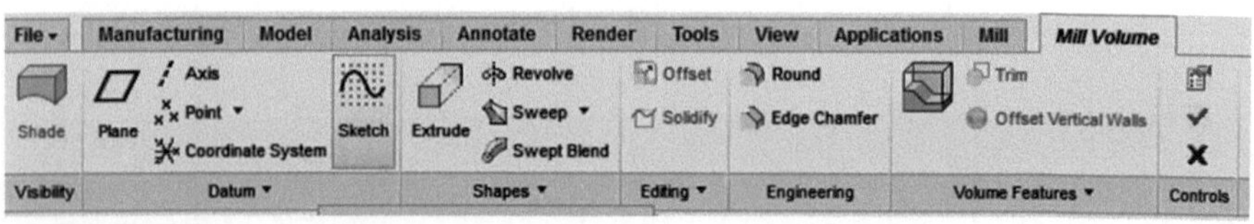

Fig. 12.134 Activated Mill Volume dashboard tools in concise form

Click on the Sketch icon .

The Sketch dialogue box is activated on the main graphic window ≫ Now click on the Placement tab ≫ On the Sketch Plane group, click on the Plane section box and now click on the Bottom surface of the Workpiece. The Reference and Orientation section boxes are updated automatically by the system. Make sure that Orientation is set to the right.

Now click on Sketch tab to exit the Sketch dialogue box.

The Sketch tools are activated. See Fig. 12.135.

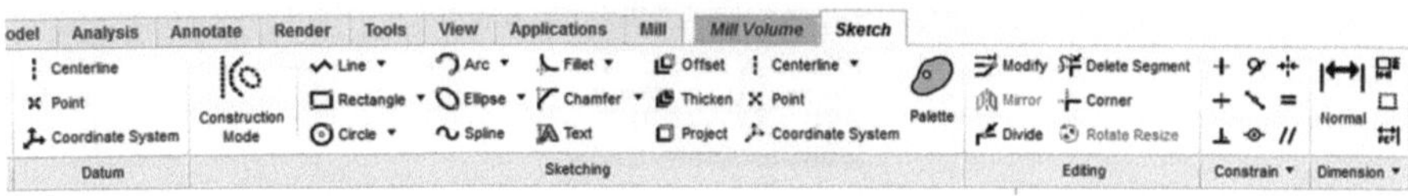

Fig. 12.135 Activated Sketch tools in concise form

Click on Sketch view icon  on the Sketch ribbon toolbar to orient the Workpiece to the correct sketch view/plane.

Workpiece is now oriented on the correct sketch plane.

Project Sketch

Click on the Project icon tab .

The Type dialogue box is activated on the main graphic window (see Fig. 12.136).

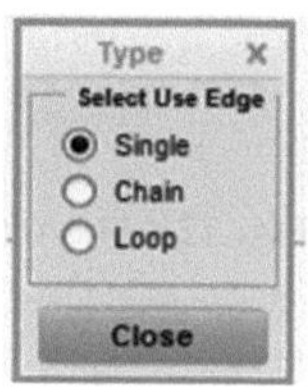

Fig. 12.136 Activated Type dialogue box

Click on the Lines on the Workpiece as indicated by the arrows in Fig. 12.137. The Lines are projected into the sketch plane. The Line, Spline, Arc and Delete Segment tools on the activated Sketch ribbon can also be used to accomplish the Sketch as highlighted in Fig. 12.137.

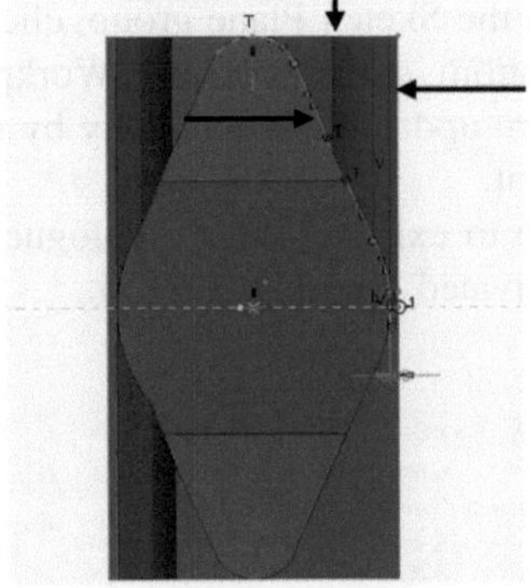

Fig. 12.137 Projecting sketch Lines on the Workpiece into the Sketch plane

Click on Close tab on the Type dialogue box to exit.
Click on the Check Mark icon/OK to exit the Sketch application.

Click on Exclude icon 

≫ The Extrude dashboard tools are activated ≫ Click on the Extrusion type icon to activate its drop-down menu list ≫ Click on "Extrude to selected point, curve, plane or surface" icon 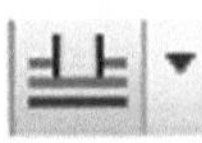 as indicated by the arrow in Fig. 12.138.

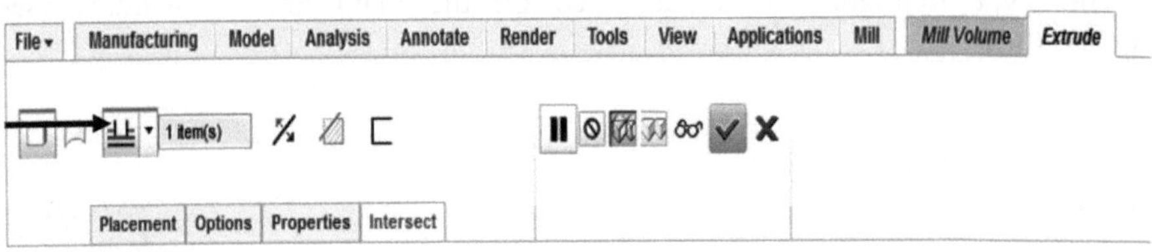

Fig. 12.138 Activating Extrude to selected point, curve, plane or surface extrusion type type

Now click on the next surface to "Extrude to" on the Workpiece and if required, flip the direction of Extrusion.

The arrow as shown in Fig. 12.139 indicates the new created Mill Volume.

Note: All the created Mill Volumes are visible.

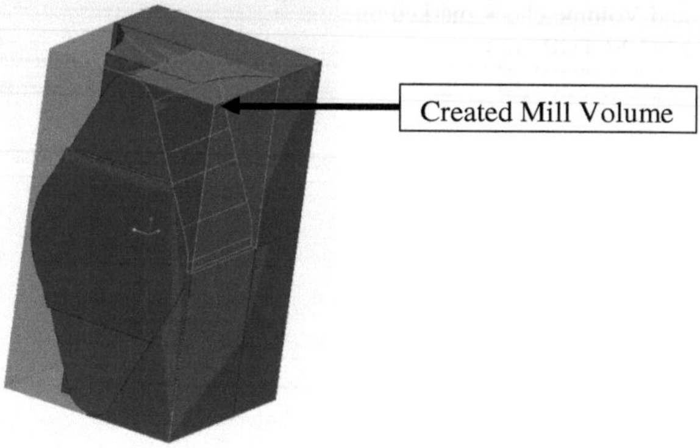

Fig. 12.139 Created third bottom Mill Volume

Click on the Check Mark icon to exit Extrude application.
Click on Check Mark icon again to exit Mill Volume application.

12.7.6 Volume Rough Milling Operation for the Third Bottom Mill Volume

Click on Volume Rough icon tab to activate its drop-down menu list ≫ Now click on Volume Rough as indicated by the arrow below.

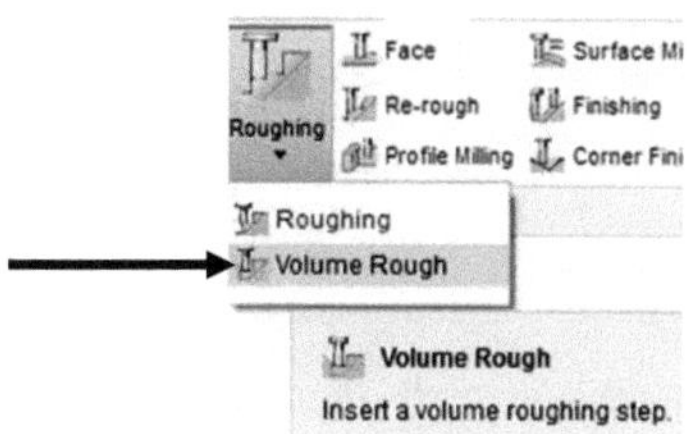

NC SEQUUENCE Menu Manager dialogue box is activated on the main graphic window ≫ Check Mark Tool, Parameters and Volume on the SEQ SETUP group as shown in Fig. 12.140.

Fig. 12.140 Tool, Parameters and Volume check marked on SEQ SETUP group

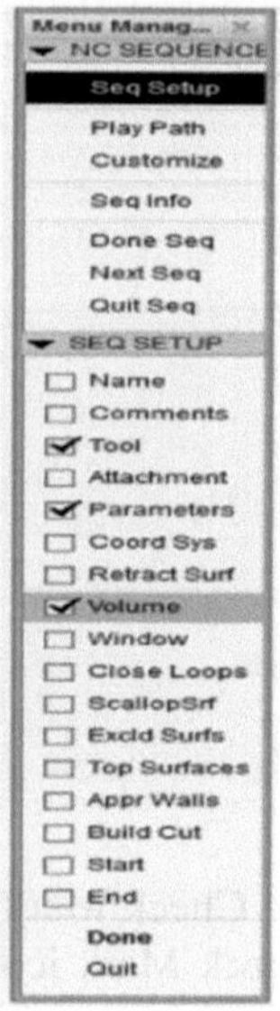

Click on Done.

The Tools Setup dialogue window is activated on the main graphic window ≫ Click on T02 as the cutting Tool for the third Volume Rough milling operation.

Click on OK tab on the Tools Setup window to exit.

The Edit Parameters of Sequence "Volume Milling" dialogue window is activated on the main graphic window ≫ Input parameter values as shown in Fig. 12.141.

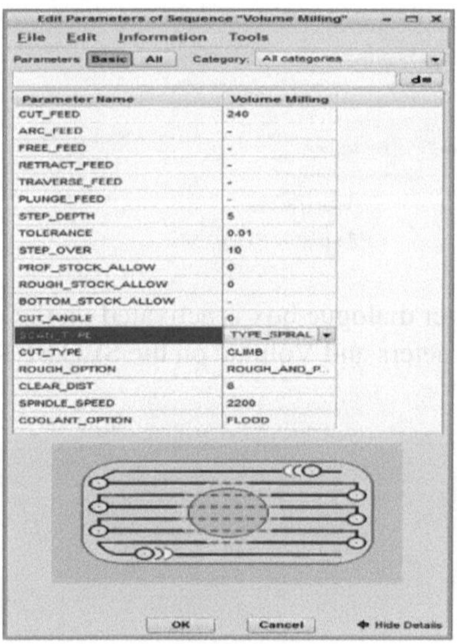

Click on OK tab to exit.

Fig. 12.141 Parameters for Volume Rough milling

Note: System is asking for the created Mill Volume to be selected on the message bar if the Mill Volume has not been selected automatically by the system.

Now click on the newly created Mill Volume.

Once Play Path is clicked on the NC SEQUENCE group, the Menu Manager dialogue box expand to include the PLAY PATH group ≫ Now click on Screen Play as shown in Fig. 12.142.

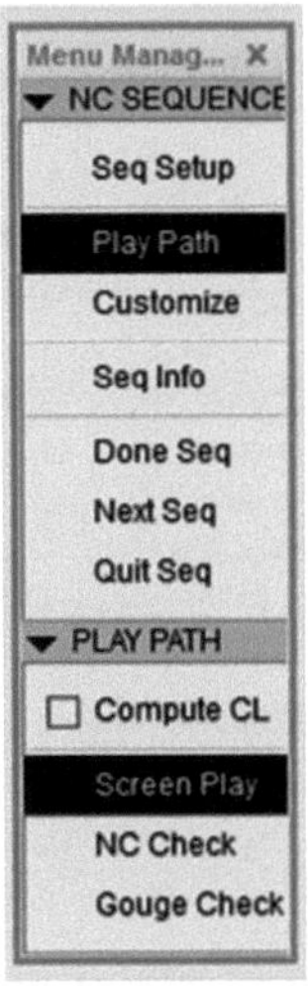

Fig. 12.142 Activating on Screen Play

The cutting Tool and the PLAY PATH dialogue box are activated on the main graphic window as shown in Fig. 12.143.

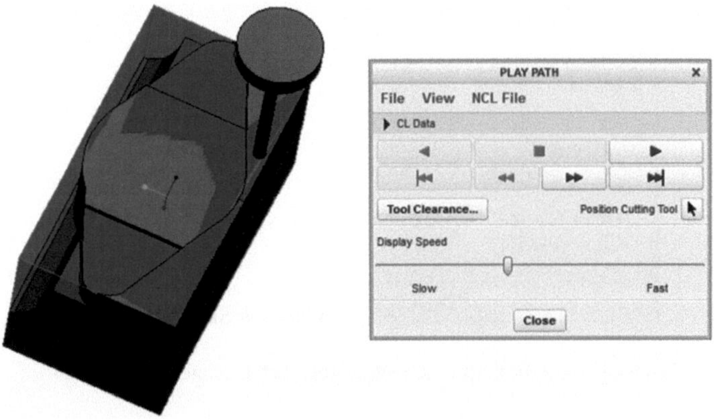

Fig. 12.143 Activated cutting Tool and PLAY PATH dialogue box

Click on the Play tab to start the Volume Rough milling process.
At the end of the on-screen Volume Rough milling operation, click on Close tab.
Click on Done Seq to exit the Volume Rough Milling operation.

12.7.7 Create the Fourth Mill Volume on the Bottom Side of the Workpiece

Click on the Mill Volume icon tab as shown below.

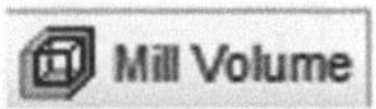

The Mill Volume ribbon is activated as shown in concise form in Fig. 12.144.

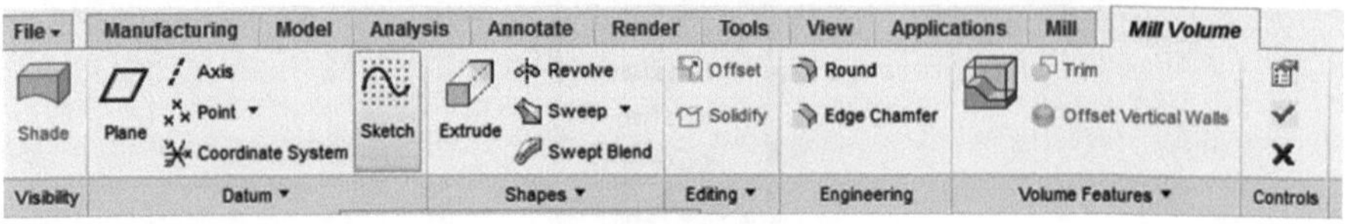

Fig. 12.144 Activated Mill Volume ribbon toolbar

Click on the Sketch icon 

.

The Sketch dialogue box is activated on the main graphic window ≫ Now click on the Placement tab to activate its content ≫ On the Sketch Plane group, click on the Plane section box and now click on the Bottom surface of the Workpiece as the sketch plane. The Reference and Orientation section boxes are automatically updated by the system as shown in Fig. 12.145.

Click on Sketch tab to exit.

Fig. 12.145 Sketch dialogue box after activating the sketch plane

The Sketch tools are activated as shown in Fig. 12.146 in a concise form.

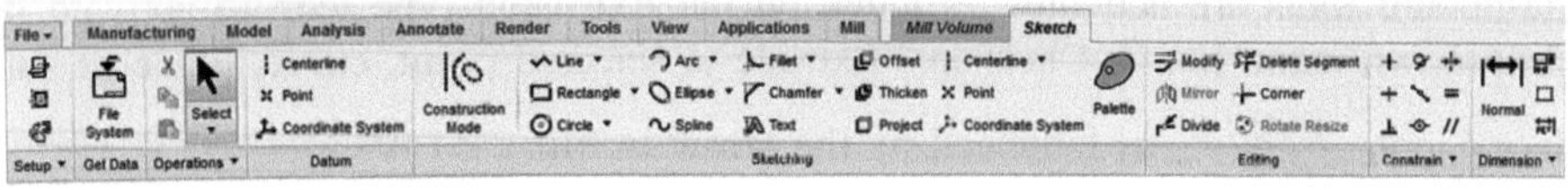

Fig. 12.146 Activated Sketch ribbon/tools

Click on Sketch view icon  on the Sketch ribbon toolbar to orient the Workpiece to the correct sketch view/plane.

Workpiece is oriented on the correct sketch plane.

Project Sketch

Click on the Project icon tab 

.

The Type dialogue box is activated on the main graphic window (See Fig. 12.147).

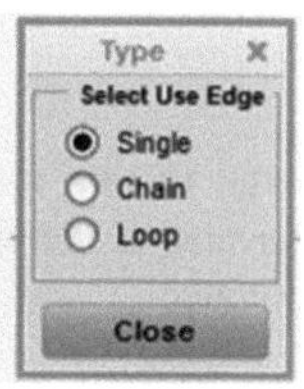

Fig. 12.147 Activated Type dialogue box

Click on Lines on the Workpiece as indicated by the arrows in Fig. 12.148. The Lines are projected into the sketch plane. The Line, Spline, Arc and Delete Segment tools on the activated Sketch ribbon can also be used to accomplish the Sketch as highlighted in Fig. 12.148.

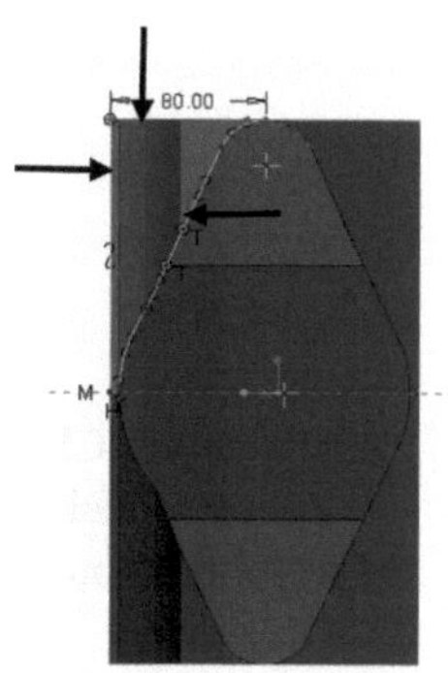

Fig. 12.148 Projecting Lines on Workpiece into the Sketch plane

Click on Close tab to exit the Type dialogue box.

Click on the Check Mark icon/OK to exit the Sketch application.

Click on Exclude icon  on the Mill Volume dashboard tools ≫ The Extrude dashboard tools are activated ≫ Click on the Extrusion type icon to activate its drop-down menu list ≫ Click on "Extrude to selected point, curve, plane or surface" icon 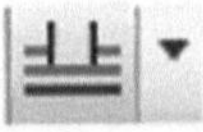 as indicated by the arrow in Fig. 12.149.

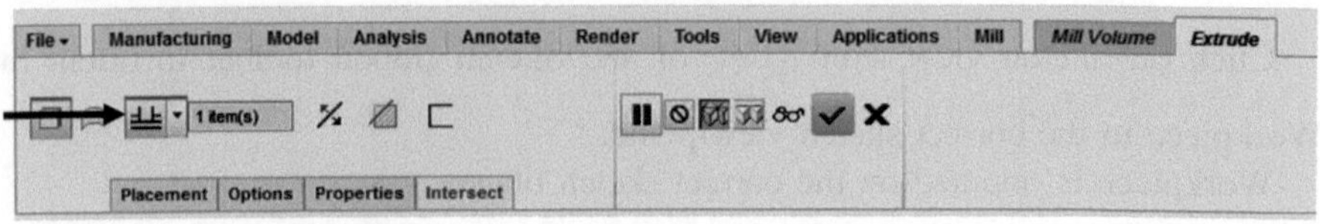

Fig. 12.149 Activating Extrude to selected point, curve, plane or surface extrusion option type

Now click on the next surface to "Extrude to" on the Workpiece as indicated by the arrow in Fig. 12.150 and if required, flip the direction of Extrusion.

Note: All the created Mill Volumes are visible.

Fig. 12.150 Extruding created sketch to the next surafce

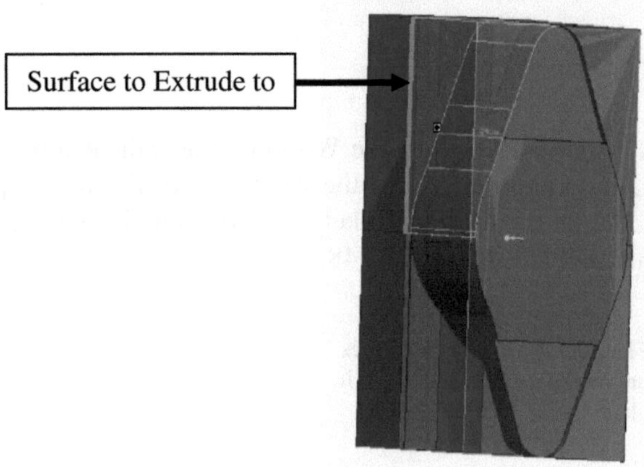

Click on the Check Mark icon to exit Extrude application.
Click on Check Mark icon again to exit Mill Volume application.
The fourth bottom Mill Volume is now created.

12.7.8 *Volume Rough Milling Operation for the Fourth Bottom Mill Volume*

Click on Roughing icon to activate its drop-down menu list ≫ Now click on Volume Rough on the drop-down menu list as shown below.

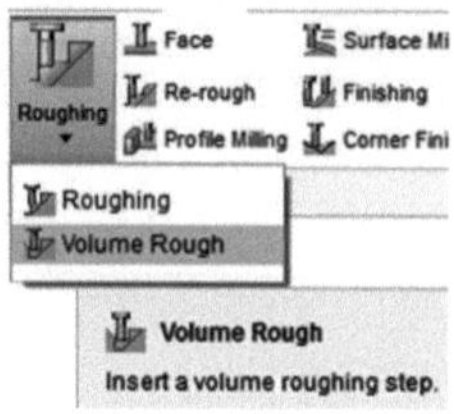

NC SEQUUENCE Menu Manager dialogue box is activated on the main graphic window ≫ Check Mark Tool, Parameters and Volume on the SEQ SETUP group as shown in Fig. 12.151.

Fig. 12.151 Tool, Parameters and Volume check marked on the SEQ SETUP group

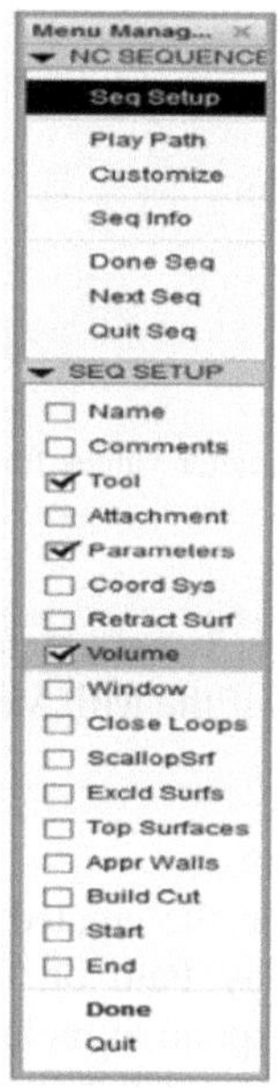

Click on Done.

The Tools Setup dialogue window is activated on the main graphic window ≫ Click on T02 as the cutting Tool for the fourth Volume Rough milling operation.

Now click on OK tab to exit Tools Setup dialogue window.

The Edit Parameters of Sequence "Volume Milling" dialogue window is activated on the main graphic window ≫ Input parameter values for the Volume Rough milling operation as shown in Fig. 12.152.

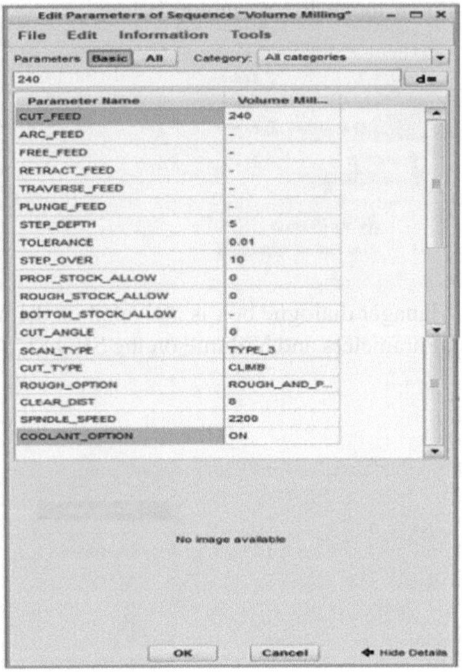

Click on OK tab.

Fig. 12.152 Parameter values for the Volume Rough milling operation

> Note: System is asking for the created Mill Volume to be selected on the message bar if the Mill Volume has not been selected/added automatically by the system

Now click on the created Mill Volume.

Click on Play Path on the NC SEQUENCE group and the Menu Manager dialogue box expand to include the PLAY PATH group ≫ Click on Screen Play as highlighted in Fig. 12.153.

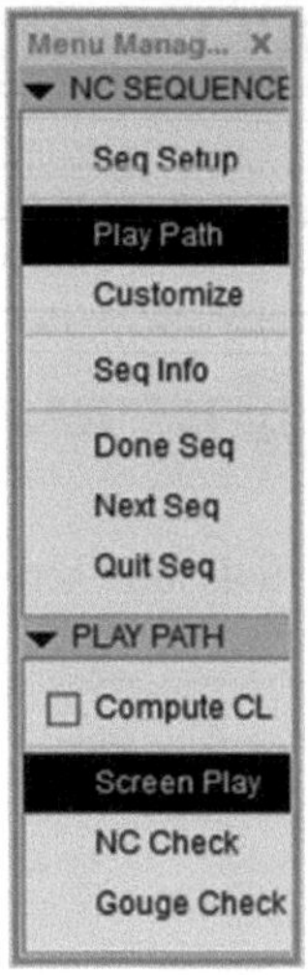

Fig. 12.153 Activating on Screen Play

The cutting Tool and the PLAY PATH dialogue box are activated on the main graphic window as shown in Fig. 12.154.

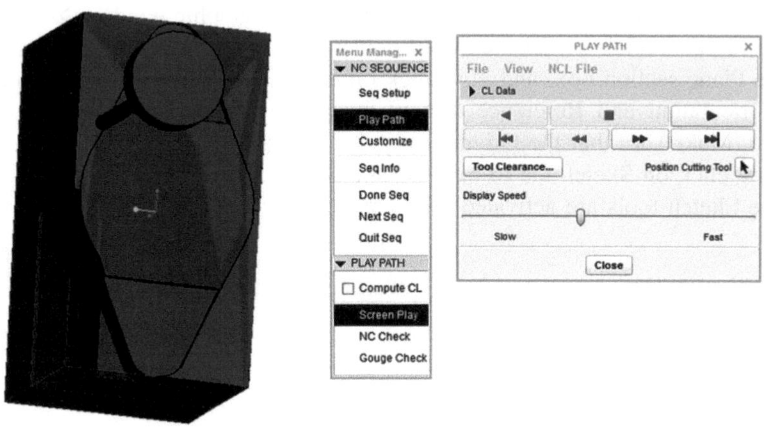

Fig. 12.154 Activated cutting Tool and the PLAY PATH dialogue box

Click on the Play tab to start the Volume Rough milling process.

At the end of the on-screen Volume Rough milling operation, click on the Close tab.

Click on Done Seq on the NC SEQUECE Menu Manager to exit the Volume Rough milling operation.

12.7.9 Create the Fifth Mill Volume on the Bottom Side of the Workpiece

Click on the Mill Volume icon as shown below.

The Mill Volume dashboard tools are activated (See Fig. 12.155 in concise form).

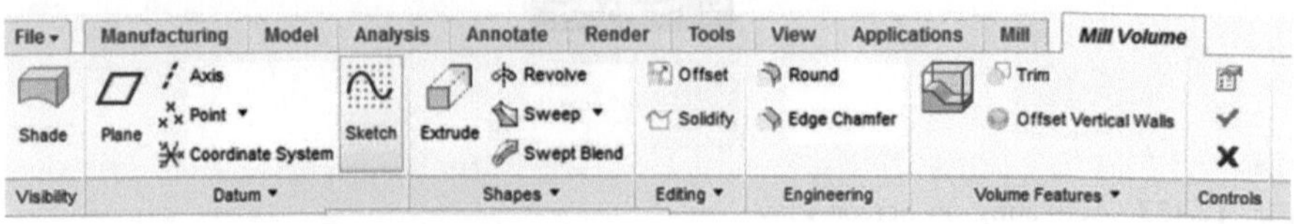

Fig. 12.155 Activated Mill Volume dashboard

Click on the Sketch icon  to activate Sketch dialogue box.

The Sketch dialogue box is activated on the main graphic window ≫ Now click on the Placement tab to make active its content ≫ On the Sketch Plane group, click on the Plane section box, and now click on the Bottom surface of the Workpiece. The Orientation and Reference section boxes are automatically updated by the system. Make sure that the Orientation is set to right.

Now click on Sketch tab to exit the Sketch dialogue box.

The Sketch tools are activated. See Fig. 12.156.

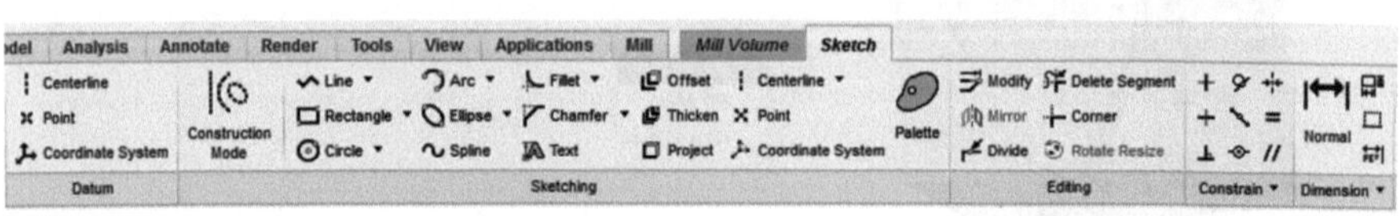

Fig. 12.156 Activated Sketch tools in concise form

Click on Sketch view icon  on the Sketch ribbon to orient the Workpiece to the correct sketch view/plane.

Project Sketch

Click on the Project icon tab 

.

The Type dialogue box is activated on the main graphic window (See Fig. 12.157).

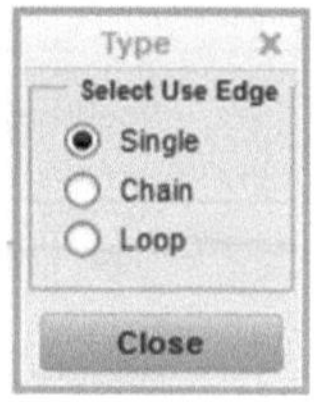

Fig. 12.157 Activated Type dialogue box

Click on Lines on the Workpiece as indicated by the arrows in Fig. 12.158. The Lines are projected into the sketch plane. The Line, Spline, Arc and Delete Segment tools on the activated Sketch toolbar can also be used to accomplish the Sketch as highlighted in Fig. 12.158.

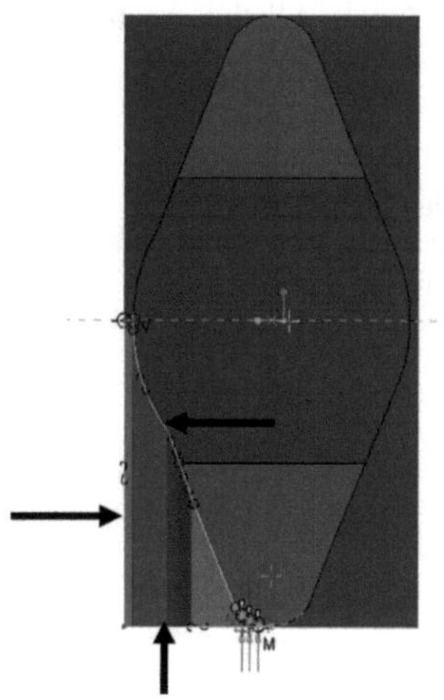

Fig. 12.158 Projecting Lines on Workpiece into the Sketch plane

Click on Close tab on the Type dialogue box to exit.
Click on the Check Mark icon/OK to exit the Sketch application.

Click on Exclude icon 

on the Mill Volume dashboard tools ≫ The Extrude dashboard tools are activated ≫ Click on the Extrusion type icon to activate its drop-down menu list ≫ Click on "Extrude to selected point, curve, plane or surface" icon 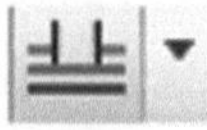 as indicated by the arrow in Fig. 12.159.

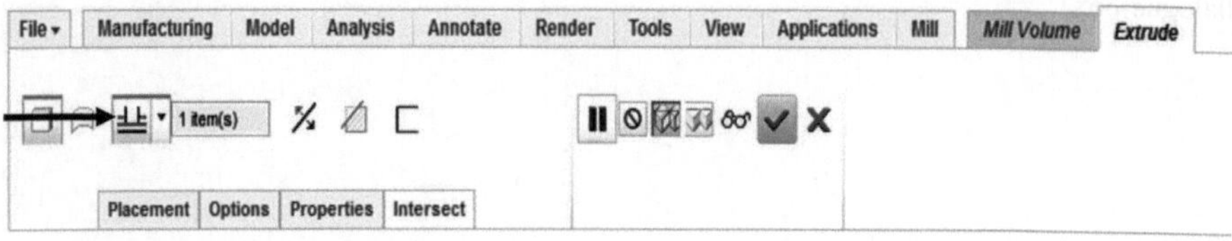

Fig. 12.159 Activating Extrude to selected point, curve, plane or surface extrusion option type

Now click on the next surface to "Extrude to" on the Workpiece as indicated by the arrow in Fig. 12.160 and if required, flip the direction of Extrusion.

Note: All the created Mill Volumes are visible.

Fig. 12.160 Extruding created sketch to the next surafce

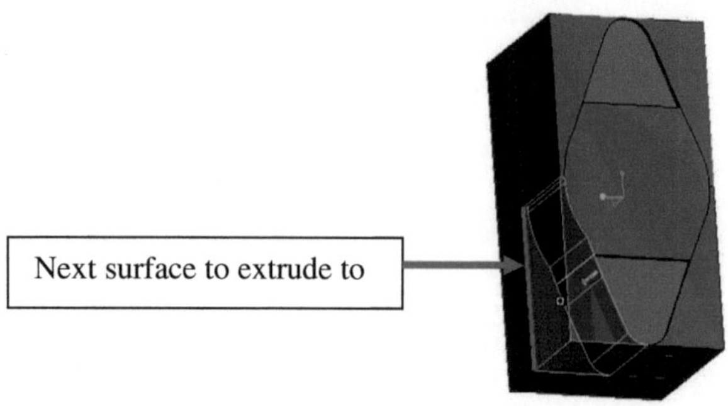

Click on the Check Mark icon to exit Extrude application.
Click on Check Mark icon again to exit Mill Volume application.
The fifth Mill Volume is now created on the Workpiece.

12.7.10 *Volume Rough Milling Operation for the Fifth Mill Volume*

Click on Roughing icon to activate its drop-down menu » Now click on Volume Rough on the activated drop-down menu list as shown below.

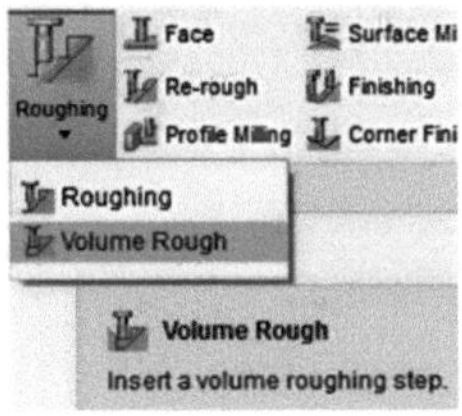

NC SEQUUENCE Menu Manager dialogue box is activated on the main graphic window » Click on Tool, Parameters and Volume on the SEQ SETUP group square boxes to check mark them if they are not checked marked automatically by the system as shown in Fig. 12.161.

Fig. 12.161 SEQ SETUP parameters check marked

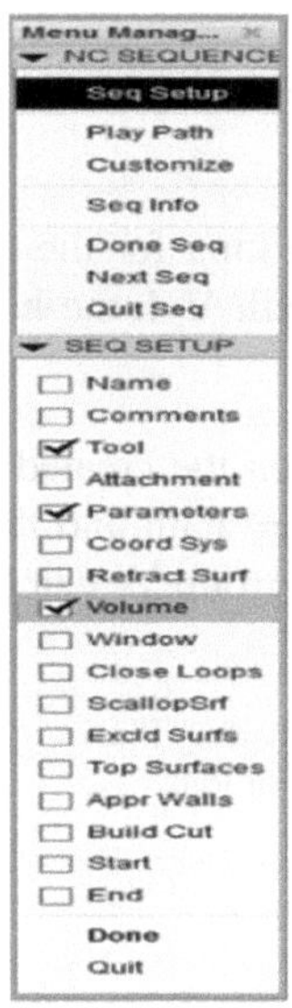

Click on Done.

The Tools Setup dialogue window is activated on the main graphic window » Click on T02 as the cutting Tool for the fifth Volume Rough milling operation.

The Edit Parameters of Sequence "Volume Milling" dialogue window is activated on the main graphic window » Input parameter values as shown in Fig. 12.162.

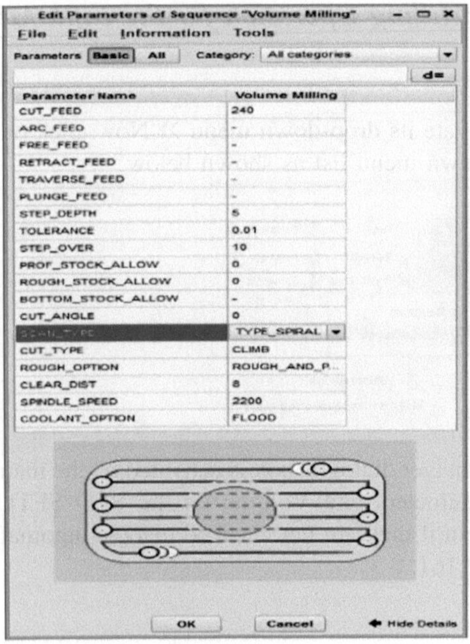

Click on OK tab to exit.

Fig. 12.162 Volume Rough operation parameters

System is asking for the created Mill Volume to be selected on the message bar if the Mill Volume has not been selected automatically by the system.

Now click on the created Mill Volume.

Click on Play Path and the Menu Manager dialogue box expand to include the PLAY PATH group ≫ Click on Screen Play as highlighted in Fig. 12.163.

Fig. 12.163 Activating Play Path and on Screen Play

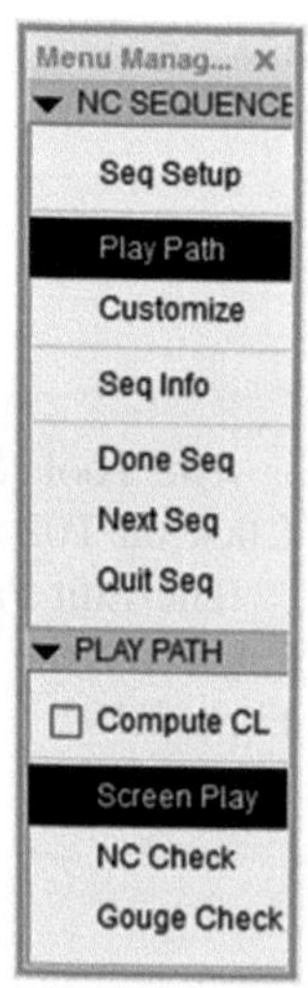

The cutting Tool and the PLAY PATH dialogue box are activated on the main graphic window as shown in Fig. 12.164.

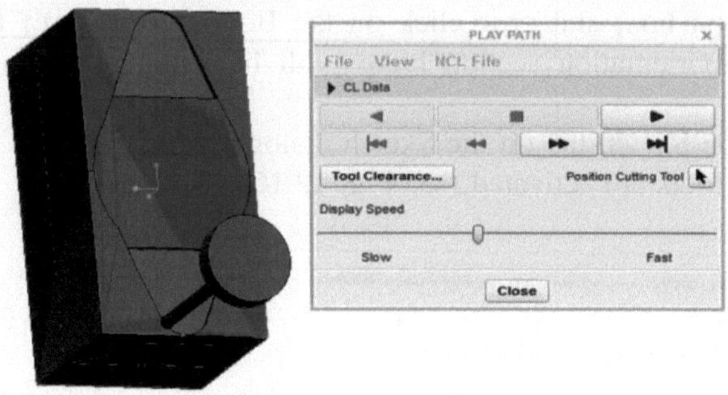

Fig. 12.164 Acivated cutting Tool and PLAY PATH dialogue box

Click on the Play tab to start the Volume Rough milling process.

At the end of the on Screen Volume Rough milling operation, click on the Close tab.

Click on Done Seq to exit the Volume Rough milling process.

12.7.11 Create the Sixth Mill Volume on the Workpiece

Click on Mill menu bar to activate its ribbon/tools ≫ Now click on Mill Volume icon tab as indicated by the arrow below.

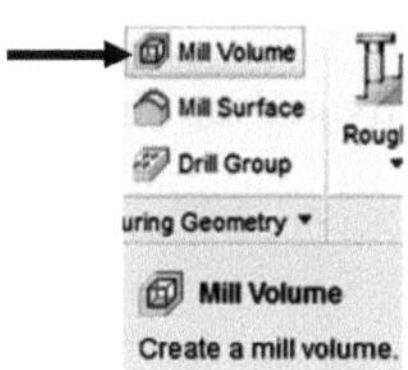

The Mill Volume dashboard tools are activated. See Fig. 12.165.

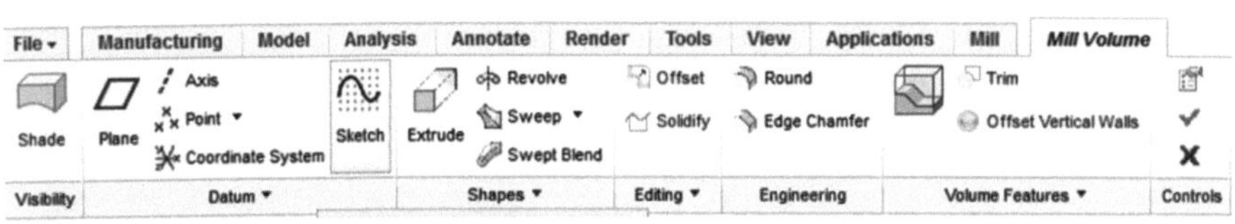

Fig. 12.165 Acivated Mill Volume dashboard in concise form

Click on the Sketch icon 

.

The Sketch dialogue box is activated on the main graphic window » Now click on the Placement tab to activate its content » On the Sketch Plane group, click on the Plane section box, and now click on the Bottom surface of the Workpiece. Reference and Orientation section boxes will be automatically updated by the system.

Now click on Sketch tab on the Sketch dialogue box.

The Sketch tools are activated (see Fig. 12.166).

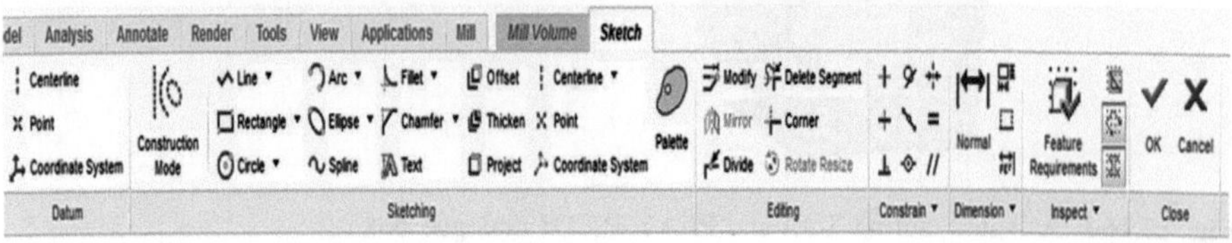

Fig. 12.166 Acivated Sketch tools in concise form

Click on Sketch view icon  to orient the Workpiece in the correct sketch plane.

Project Sketch

Click on the Project icon as shown below.

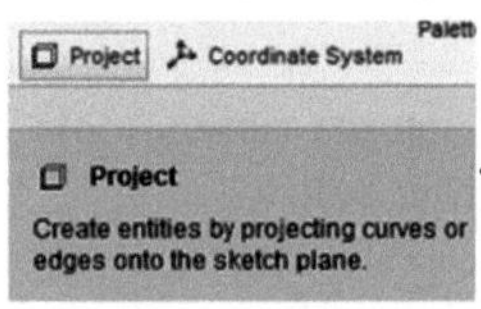

.

The Type dialogue box is activated on the main graphic window as shown in Fig. 12.167.

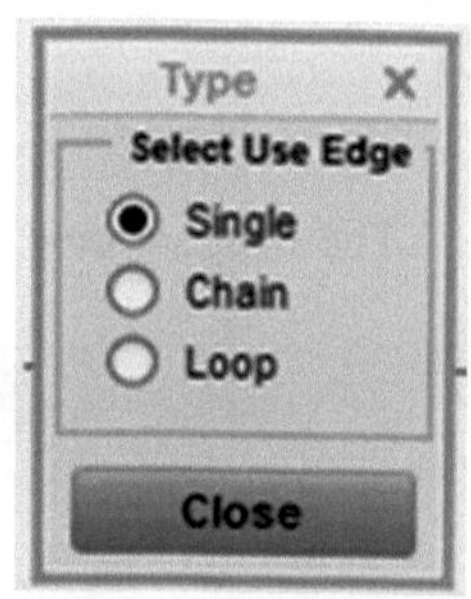

Fig. 12.167 Activated Type dialogue box

Click on the Lines and Curves on Workpiece as indicated by the arrows in Fig. 12.168. The Lines are projected into the sketch plane. The Line, Spline, Arc and Delete Segment tools on the activated Sketch toolbar can also be used to accomplish the Sketch as highlighted in Fig. 12.168.

Fig. 12.168 Projecting Lines and Curves on Workpiece into the Sketch plane

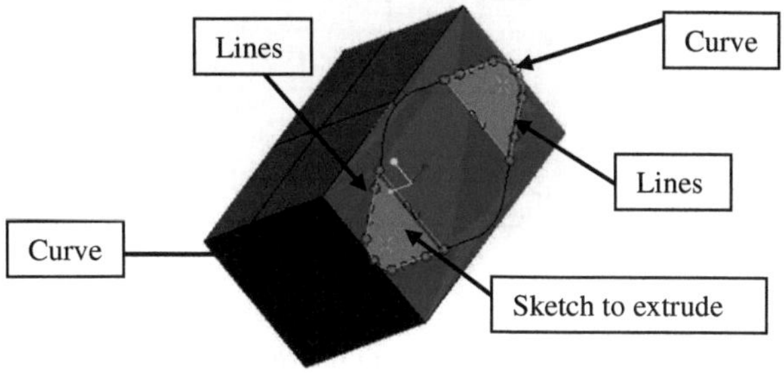

Click on Close tab on the Type dialogue box to exit.
Click on the Check Mark icon/OK to exit the Sketch application.

Click on Exclude icon 

on the activated Mill Volume ribbon » The Extrude ribbon/tools are activated » Click on the Extrusion type icon to activate its drop-down menu list » Click on "Extrude to selected point, curve, plane or surface" icon  as indicated by the arrow in Fig. 12.169.

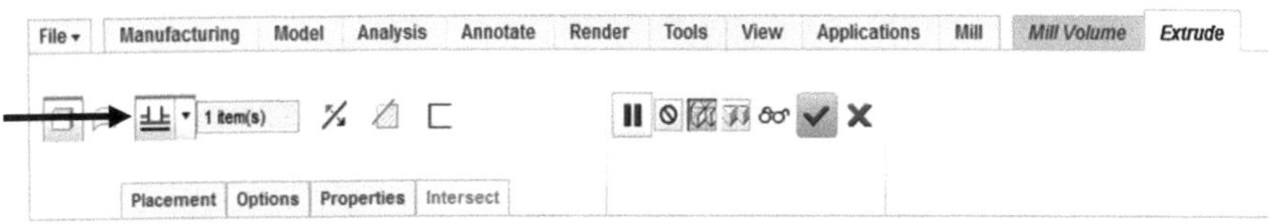

Fig. 12.169 Activating Extrude to selected point, curve, plane or surface extrusion option type

Now click on the next surface to "Extrude to" on the Workpiece as indicated by the arrow in Fig. 12.170 and if required, flip the direction of Extrusion.

Note: All the created Mill Volumes are visible.

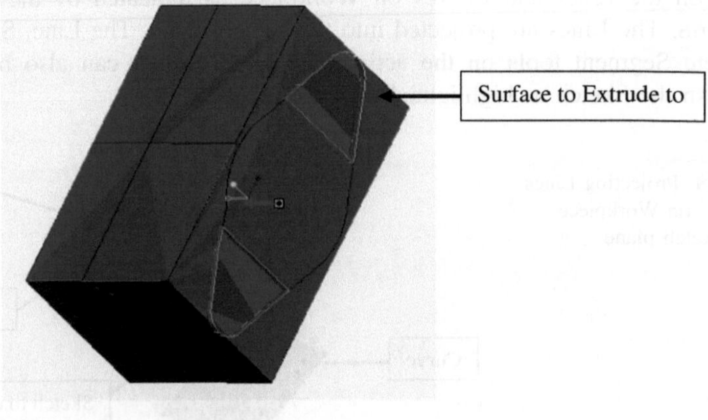

Fig. 12.170 Extruding Sketch to the Next surface

Click on the Check Mark icon to exit Extrude application.
Click on Check Mark icon again to exit Mill Volume application.

12.7.12 Volume Rough Milling Operation for the Sixth Mill Volume

Click on Roughing icon tab to activate its drop-down menu list ≫ Now click on Volume Rough on the activated drop-down menu list as shown below.

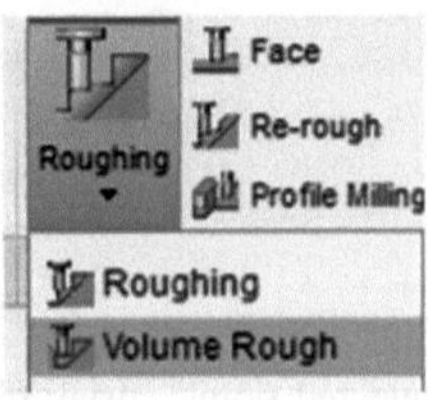

The NC SEQUUENCE Menu Manager dialogue box is activated on the main graphic window ≫ Check Mark Tool, Parameters and Volume on the SEQ SETUP group as shown in Fig. 12.171.

Fig. 12.171 Tool, Parameters and Volume check marked

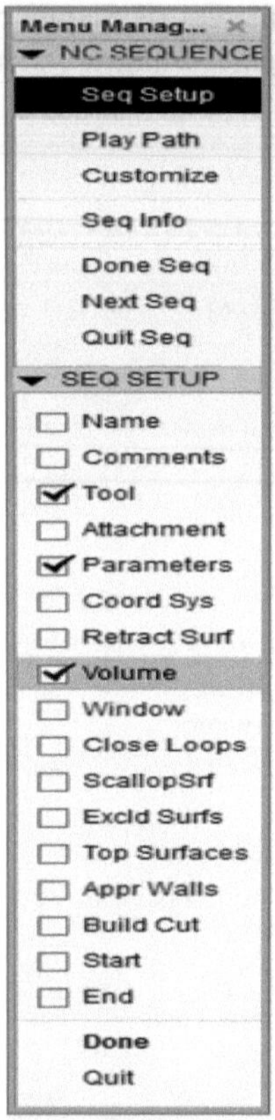

Click on Done.

The Tools Setup dialogue window is activated on the main graphic window » Click on T03 as the cutting Tool for the sixth Volume Rough milling as shown in Fig. 12.172.

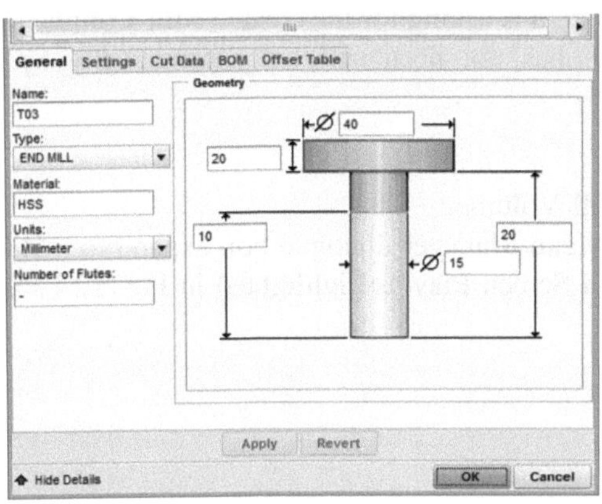

Click on OK tab to exit.

Fig. 12.172 Activating Tool (T03) as the cutting Tool in concise form

The Edit Parameters of Sequence "Volume Milling" dialogue window is activated on the graphic window ≫ Input parameter values for the Volume Rough process as shown in Fig. 12.173.

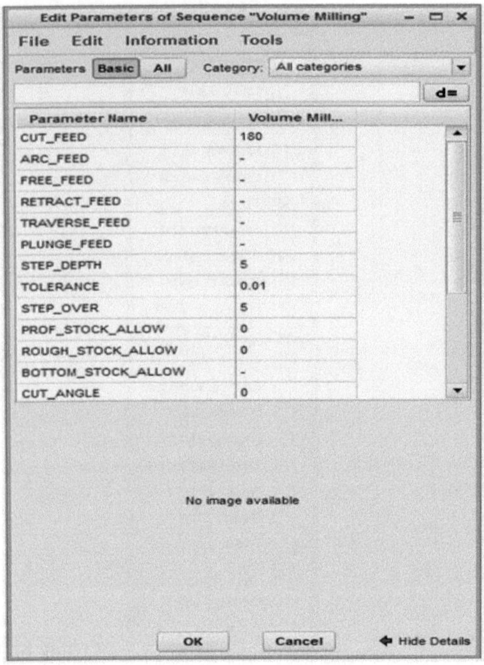

Click on OK tab to exit.

Fig. 12.173 Volume Rough milling parameters

> Note: On the message bar, system is asking for the created Mill Volume to be selected if the Mill Volume has not been selected automatically by the system.

Now click on the created Mill Volume.

Click on Play Path and the Menu Manager dialogue box expand to include the PLAY PATH group ≫ Click on Screen Play as highlighted in Fig. 12.174.

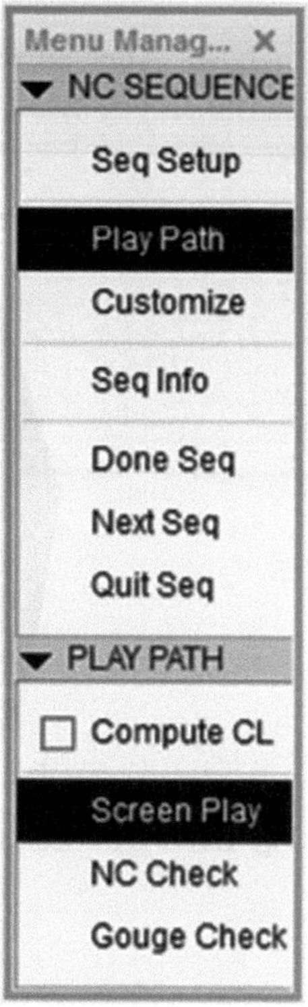

Fig. 12.174 Activating on Screen Play

The cutting Tool and the PLAY PATH dialogue box are activated on the main graphic window as shown in Fig. 12.175.

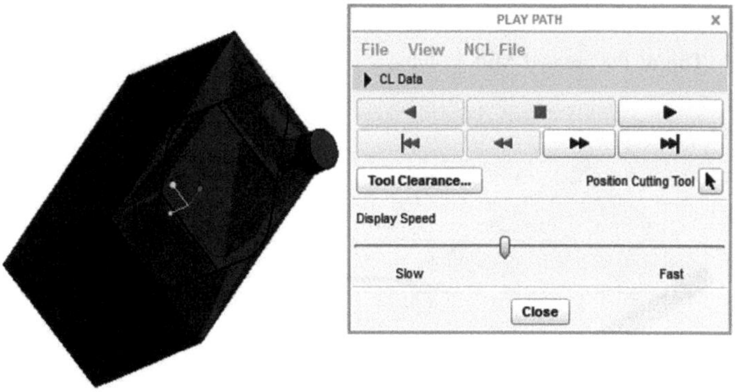

Fig. 12.175 Activated cutting Tool and PLAY PATH dialogue box

Click on the Play tab to start the Volume Rough milling process.

At the end of the on-screen Volume Rough milling operation, click on the Close tab.

Click on Done Seq to exit the Volume Rough milling operation.

Note: After all the Mill Volumes are created, the Workpiece will look as shown in Fig. 12.176.

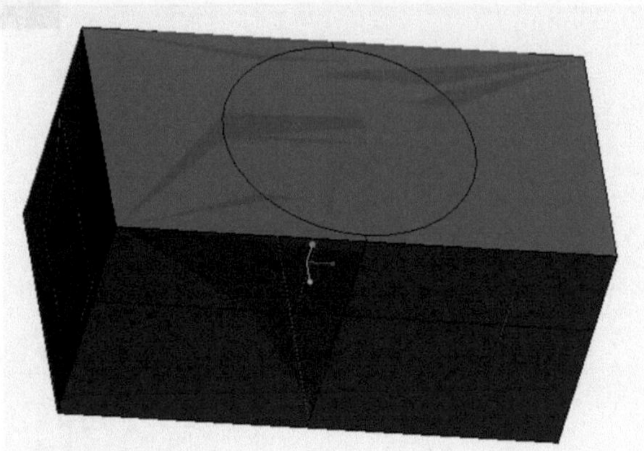

Fig. 12.176 All created Mill Volumes

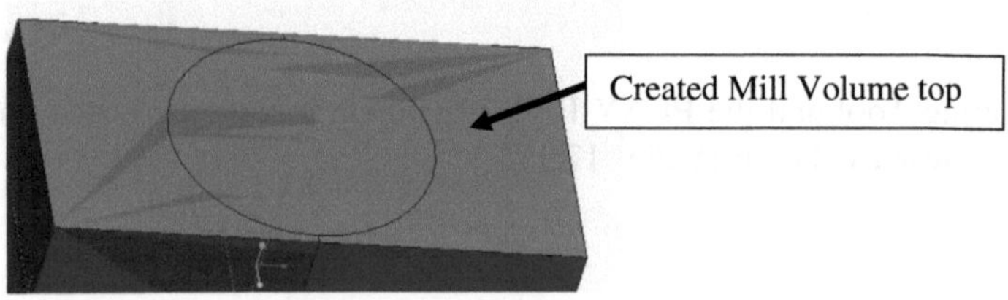

Fig. 12.177 Top of the created Mill Volume

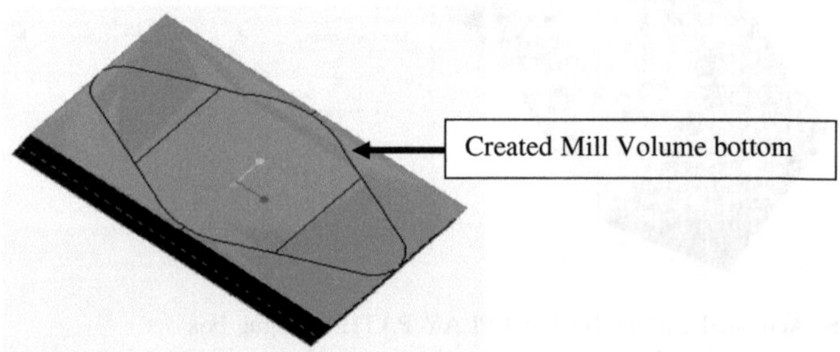

Fig. 12.178 Bottom of the created Mill Volume

12.8 Surface Milling

Surface Milling is used to refine or smoothen the surface of the Workpiece after Roughing operation.

12.8.1 Create Surface Milling for the Outer Surface of the Workpiece

Click on Mill menu bar to activate its ribbon ≫ Now click on the Surface Milling icon as indicated by the arrow below.

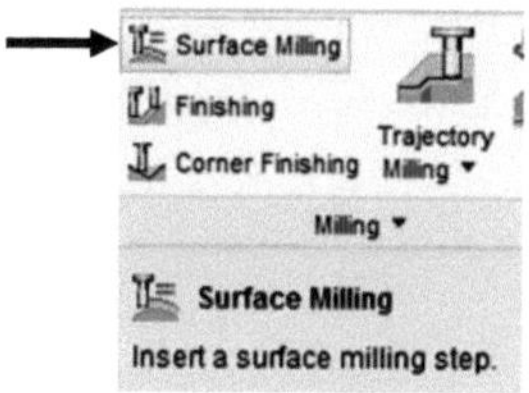

The MACH AXES Menu Manager dialogue box is activated on the main graphic window ≫ Click on 5 Axis as highlighted in Fig. 12.179.

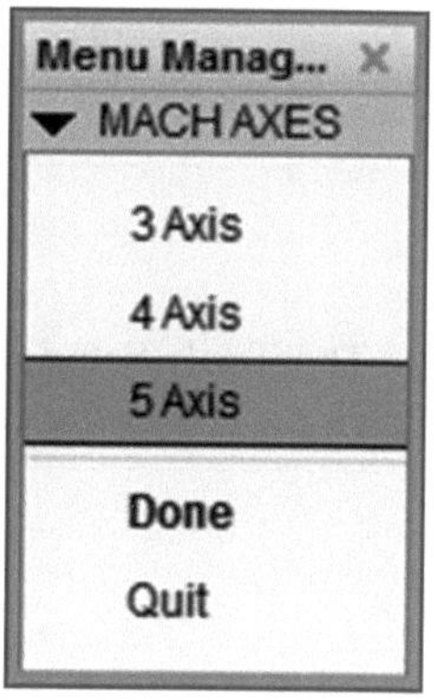

Fig. 12.179 Activating five-axis milling application

Click on Done.

The NC SEQUENCE Menu Manager dialogue box is activated on the main graphic window ≫ Check Mark Tool, Parameters, Surfaces and Define Cut on the SEQ SETUP group as shown in Fig. 12.180.

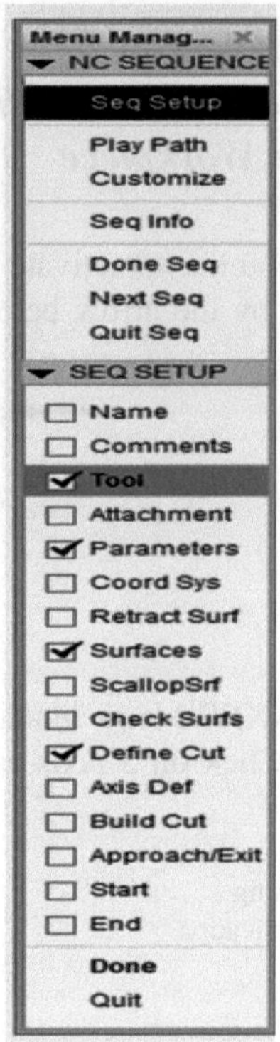

Fig. 12.180 Tool, Parameters, Surfaces and Define Cut check marked

Click on Done.

The Tools Setup dialogue window is activated on the main graphic window ≫ Click on T02 as the cutting Tool for the Surface Milling operation. See Fig. 12.181.

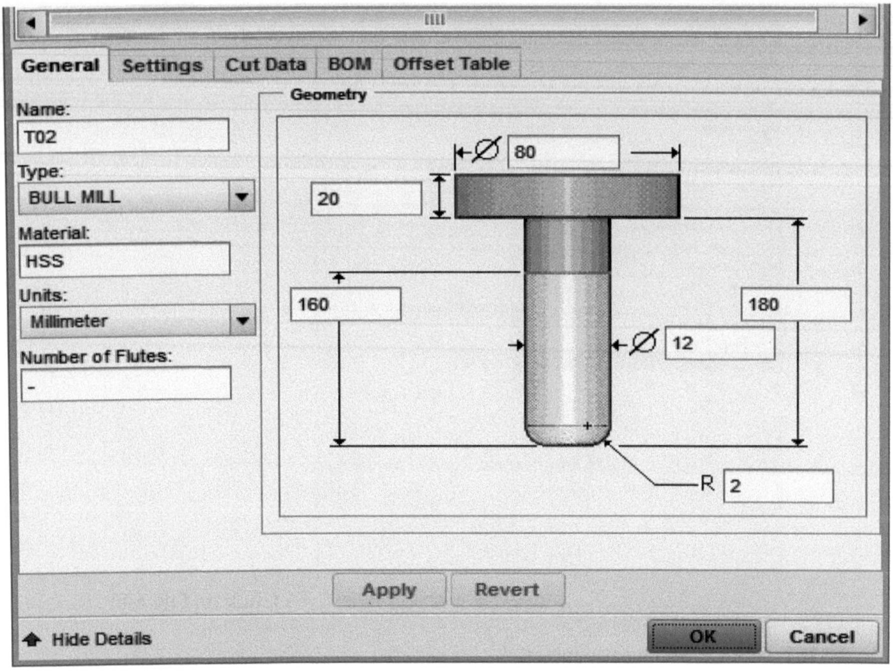

Fig. 12.181 Activating Tool (T02) for the Surface Milling operation in concise form

Click on OK tab to exit.

The Edit Parameters Sequence "Surface Milling" dialogue window is activated on the main graphic window ≫ Input values for the Surface Milling parameters as shown in Fig. 12.182.

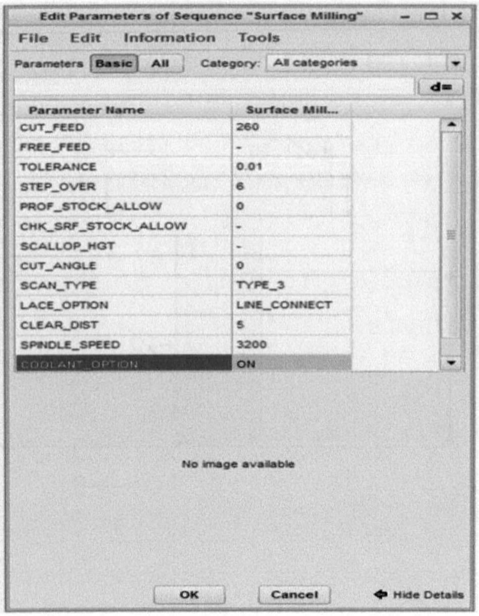

Click on OK tab.

Fig. 12.182 Surface Milling parameters

The NCSEQ SURFS and SURF PICK groups are automatically added to the NC SEQUENCE Menu Manager dialogue box as shown in Fig. 12.183.

Fig. 12.183 Activated NCSEQ SURFS and SURF PICK groups

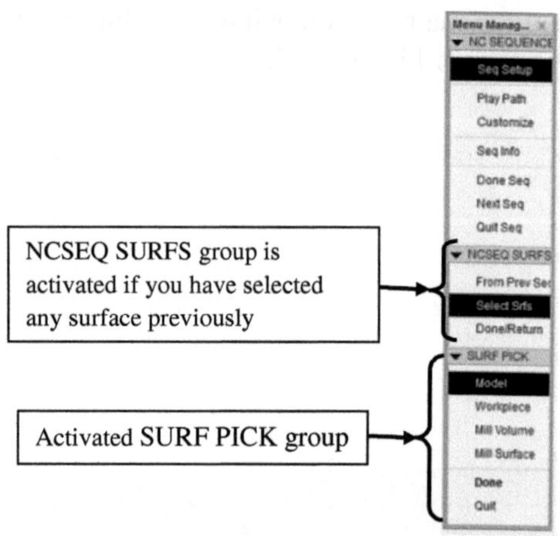

Note: NCSEQ SURFS group in Fig. 12.183 is activated if you have selected any surface previously for Surface milling.

Alternatively, Fig. 12.184 will be activated by the system.

The SELECT SRFS group is automatically activated as indicated by the arrow

» Click on Add as highlighted in Fig. 12.184.

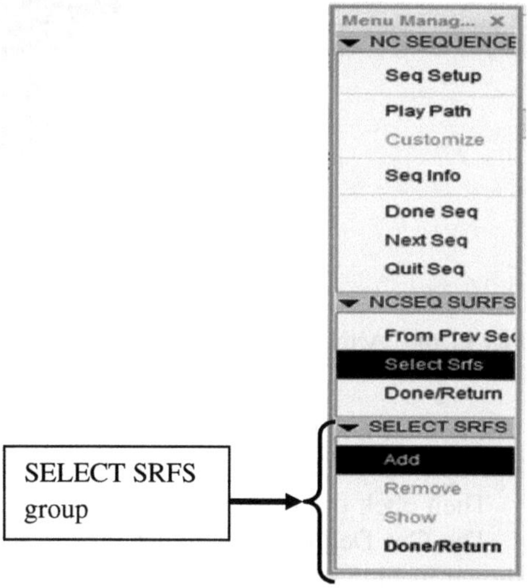

Fig. 12.184 Select Srfs and Add highlighted

The Select dialogue box is activated on the main graphic window as shown in Fig. 12.185.

Fig. 12.185 Activated Select dialogue box

Note: At this stage, it is advisable to hide/suppress all the created Mill Volumes to avoid confusion. In this tutorial, not all the created Mill Volumes will be hidden as shown in the proceeding figures.

Now click on the Outside Surfaces of the Reference Model/Part by holding down the Ctrl key while clicking on the Outer Surfaces of Part as indicated by the arrow as shown in Fig. 12.186.

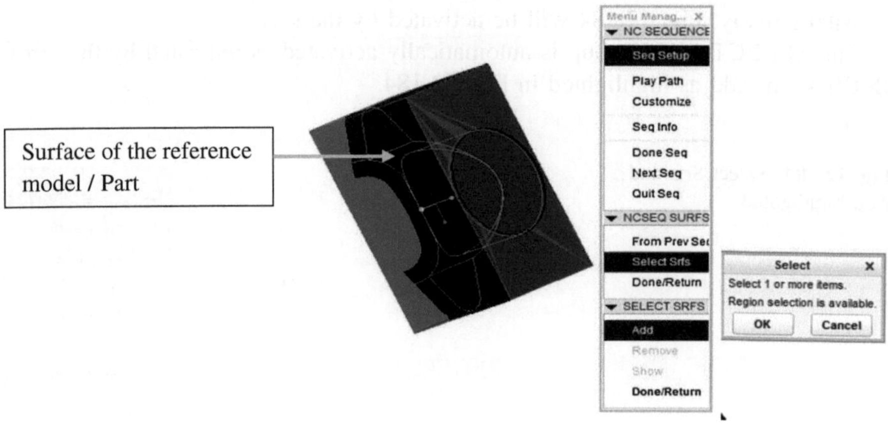

Fig. 12.186 Activating Part surface for Surface Milling

Click on OK tab to exit the Select dialogue box.
Click on the Done/Return on the SELECT SRFS group.
Then click on the Done/Return on the NCSEQ SURFS if activated.
The Cut Definition dialogue box is activated on the main graphic window as shown in Fig. 12.187.

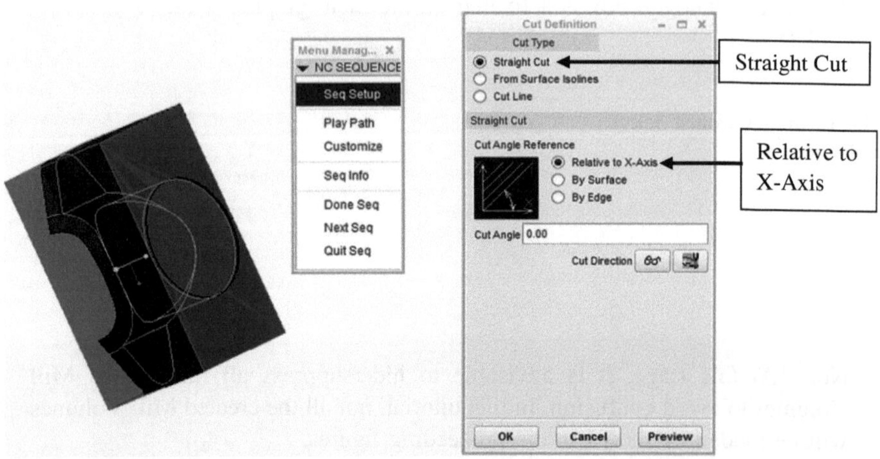

Fig. 12.187 Activated Cut Definition dialogue box on the main graphic window

Note: In this tutorial, click on the Relative to *X* axis radio button to make it active on the Cut Angle References group » Click on the Straight Cut radio button to make it active on the Cut Type group as indicated by the arrows in Fig. 12.187.

Click on the Preview tab on the Cut Definition dialogue box to preview the operation on the Reference Model/Part as shown in Fig. 12.188.

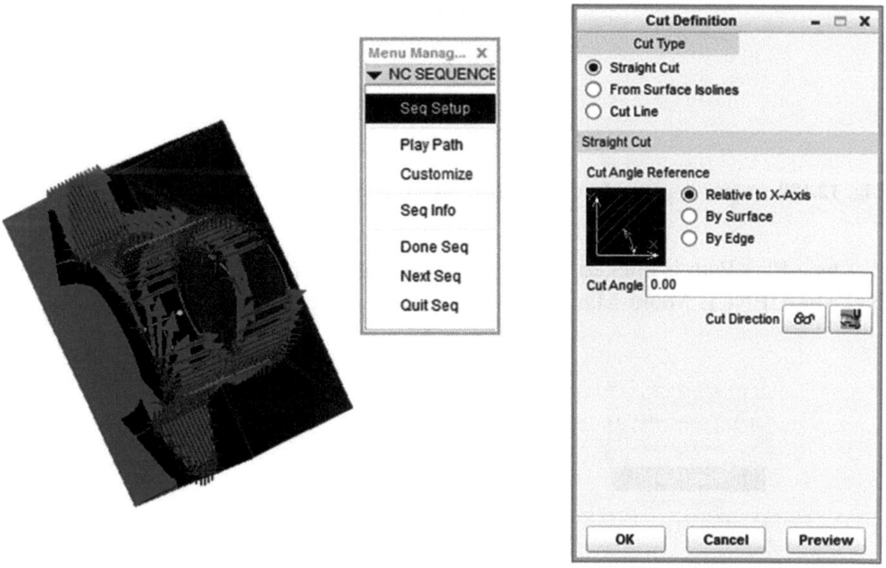

Fig. 12.188 Previw of the Surface Milling operation on Part/Reference Model

Click on OK tab to exit the Cut Definition dialogue box.

Click on the Play Path on the NC SEQUENCE Menu Manager dialogue box as highlighted in Fig. 12.189.

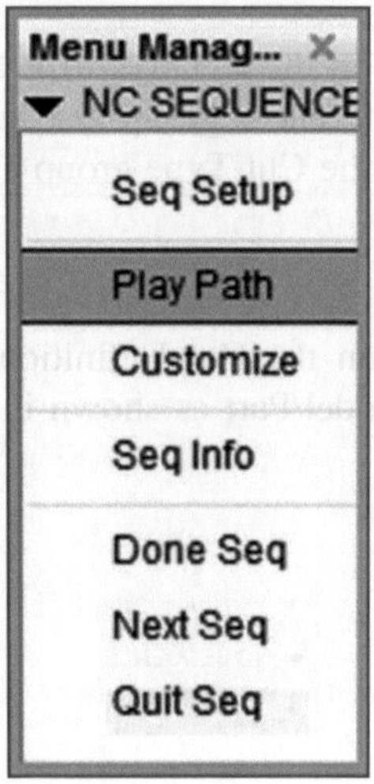

Fig. 12.189 Activating Play Path

Once Play Path is clicked, the PLAY PATH group is automatically added to the NC SEQUENCE Menu Manager dialogue box as shown in Fig. 12.190.

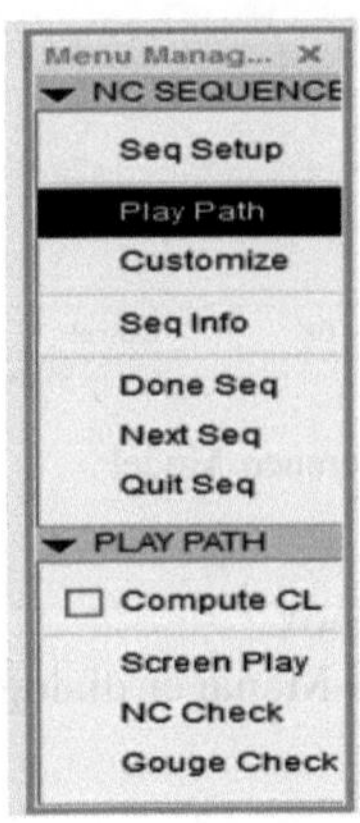

Click on Screen Play on the PLAY PATH group.

Fig. 12.190 Activating Screen Play

The cutting Tool and the PLAY PATH dialogue box are activated. See Fig. 12.191.

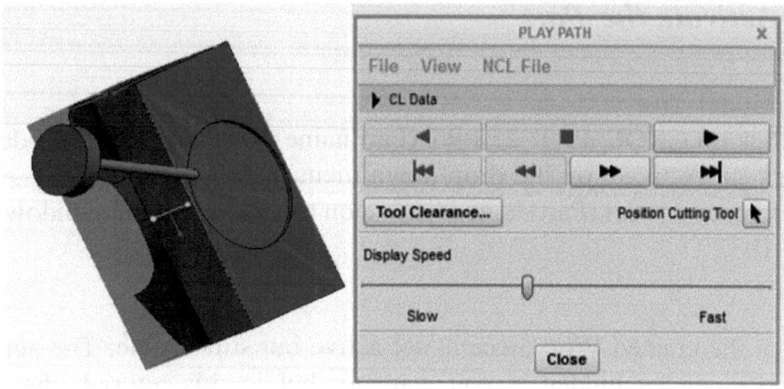

Fig. 12.191 Activated cutting Tool and the PLAY PATH dialogue box

Activate the Wireframe display style
Click on Play tab to start the Surface Milling process.
End of Surface Milling operation is shown in Fig. 12.192.

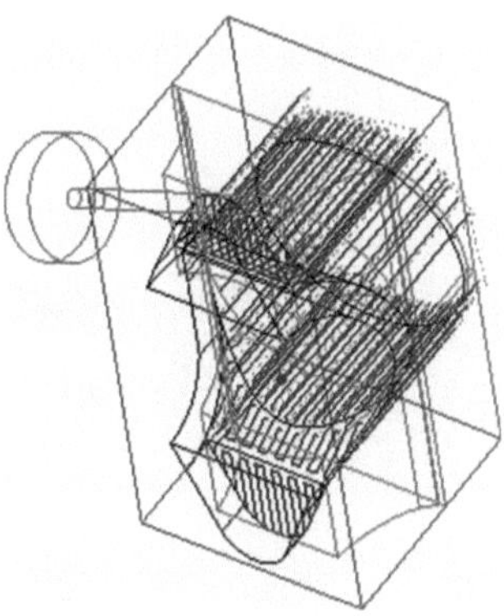

Fig. 12.192 End of Surface Milling operation in Wireframe display

Click on Close tab to exit the PLAY PATH dialogue box.
Click on Done Seq to exit the Surface Milling operation.

12.9 Create Drill Cycle on Workpiece

Before carrying out the drilling operation of the Holes, the suppressed Holes have to be un-suppressed.

12.9.1 Activate the Part

Go to the Model Tree.

Right click on CROE_PRT_25.PRT (Part name) to activate its drop-down menu list ≫ Click on Activate on the drop-down menu list.

The Reference Model (Part) is now active on the main graphic window as shown in Fig. 12.193.

> Note that the created Workpiece is not active but still visible. The automatic Workpiece can be hidden if you want to, but in this tutorial, the created Workpiece will not be hidden

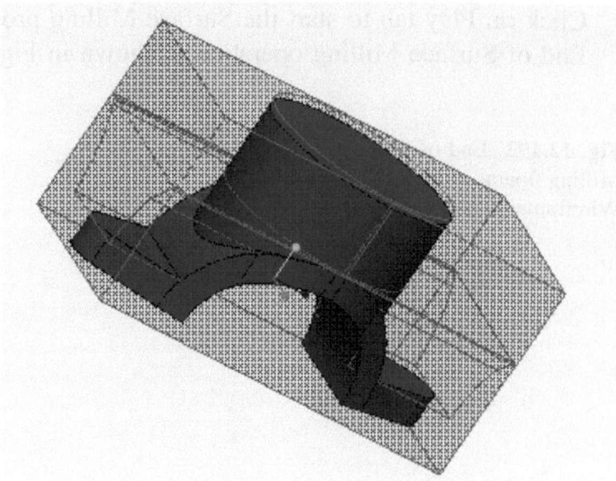

Fig. 12.193 Active Part/Reference Model

Click on Operations to activate its drop-down menu list ≫ Click on Resume ≫ Now click on Resume All on the drop-down list as shown in Fig. 12.194.

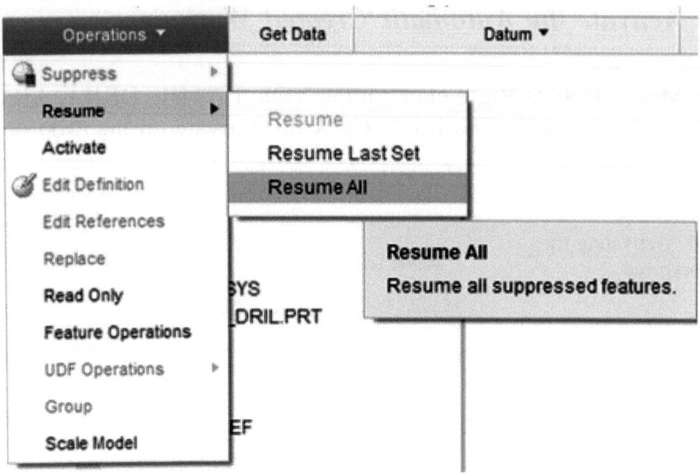

Fig. 12.194 Activating Resume All via Operations

Once Resume all is clicked, the Part changes with all the hidden/suppressed Holes activated as shown in Fig. 12.195.

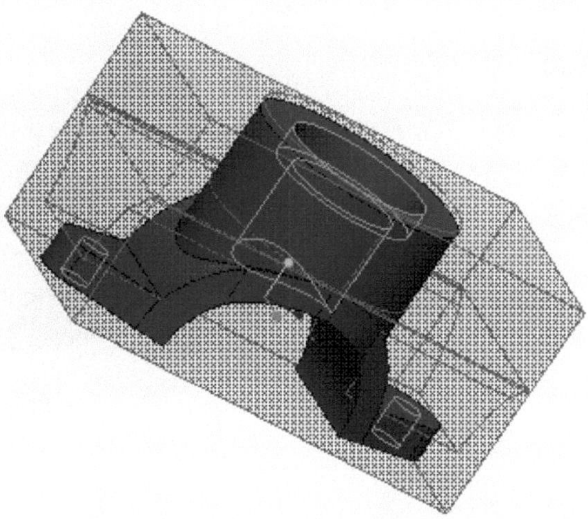

Fig. 12.195 All sppressed Holes now active

12.9.2 Activate the Automatic/Created Workpiece

Go to the Model Tree >> Right click on the VOL_LATHE_DRILL_MFG01.ASM to activate its drop-down menu list >> Click on Activate on the drop-down list as highlighted in Fig. 12.196.

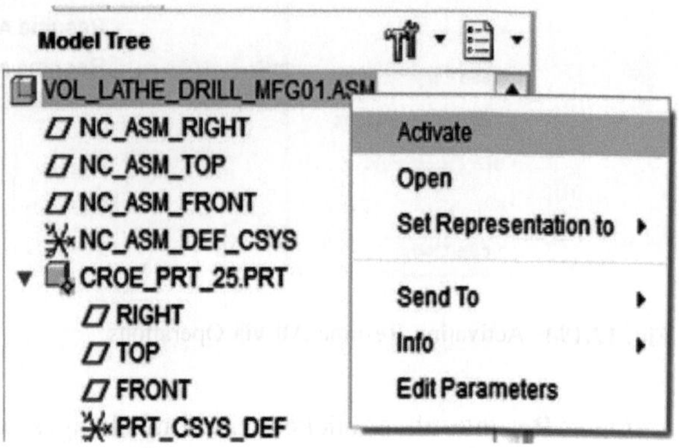

Fig. 12.196 Activating the created Workpiece

Once Activate is clicked, the automatic/created Workpiece is active again with all the Holes visible on the Part as shown in Fig. 12.197.

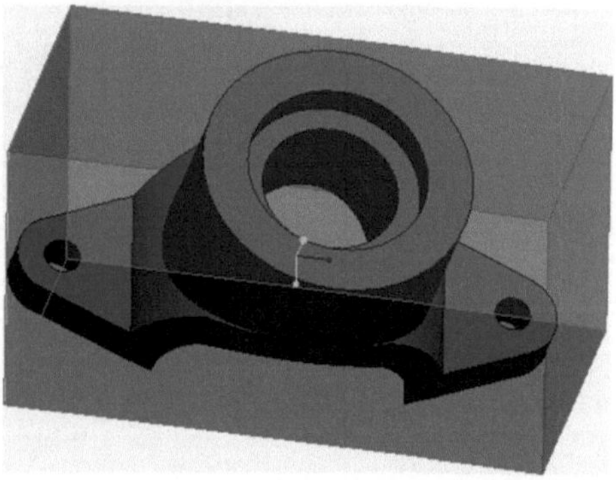

Fig. 12.197 Activated created Workpiece

Note: In Fig. 12.197, all the created Mill Volumes are all hidden, so that the Holes and axes can be easily seen during the drilling operation.

12.9.3 *Standard Drilling Operation for 100 mm Hole*

Click on the Mill menu bar to activate the Mill ribbon.

Click on Standard icon tab as indicated by the arrow in Fig. 12.198.

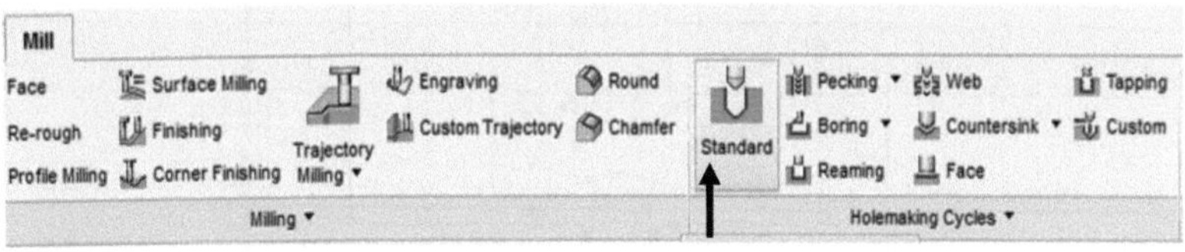

Fig. 12.198 Activating Standard drill operation

The Drilling dashboard tools are activated. See Fig. 12.199.

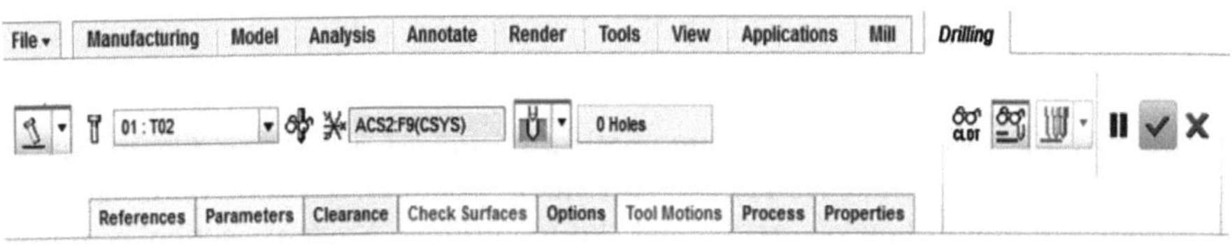

Fig. 12.199 Activated Drilling ribbon toolbar

Now Click on Tool icon  on the activated Drilling dashboard ≫ The Tools Setup dialogue window is activated on the main graphic window (See Fig. 12.200) ≫ Click on T05 as the Drilling Tool for this operation as shown in concise form in Fig. 12.200.

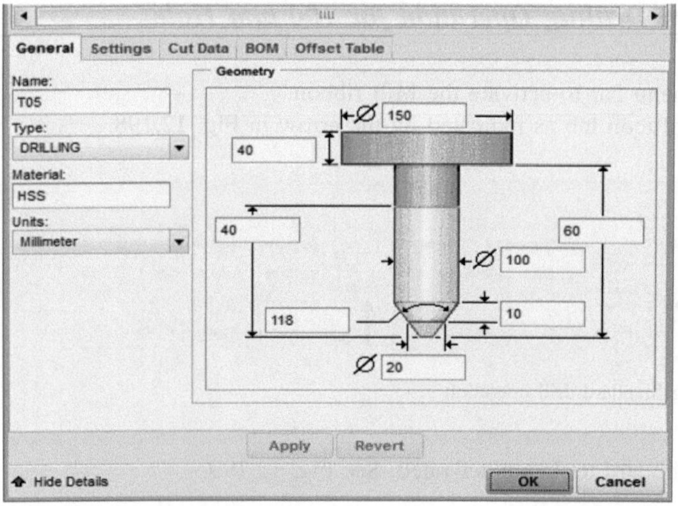

Click on OK tab.

Fig. 12.200 Activating Tool (T05) for the first drill operation

> Note: If the system does not automatically add ACS2 in the Coordinate System section box in Fig. 12.201, add ACS2 yourself. To add ACS2, go to Model Tree and click on ACS2 and the Coordinate System section box is automatically updated by the system as shown in Fig. 12.201

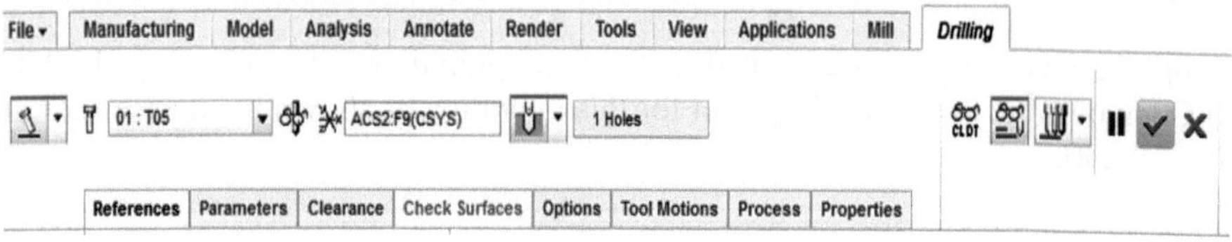

Fig. 12.201 Tool and Coordinate System parameters updated

Click on References tab to activate its panel ≫ Click on the Type section box to activate its drop-down menu list, and now click on Axes on the activated drop-down list ≫ Click on the Holes section box, and now click on the axis of the Hole to be Drilled on the Part as illustrated in Fig. 12.202.

> Note: The Start and End of the Drilling operation are set to Auto as illustrated in Fig. 12.202

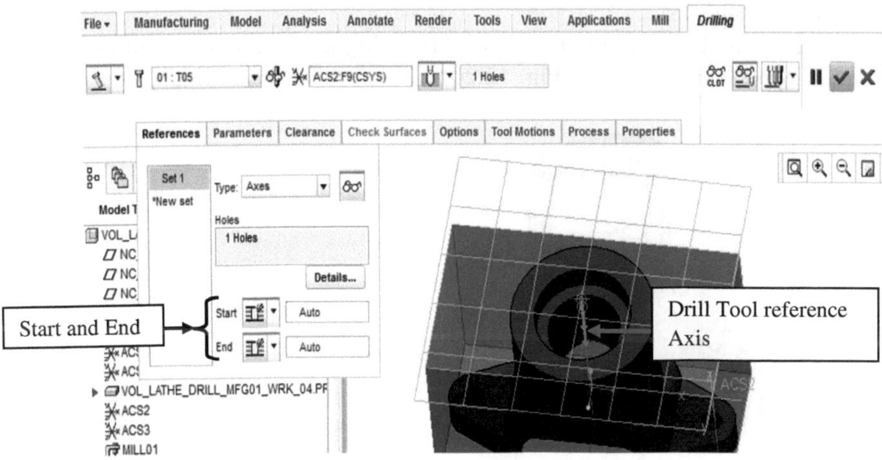

Fig. 12.202 Activating the Reference parameters

Click on Parameters tab to activate its panel >> Now input values for the Drilling parameters as shown in Fig. 12.203.

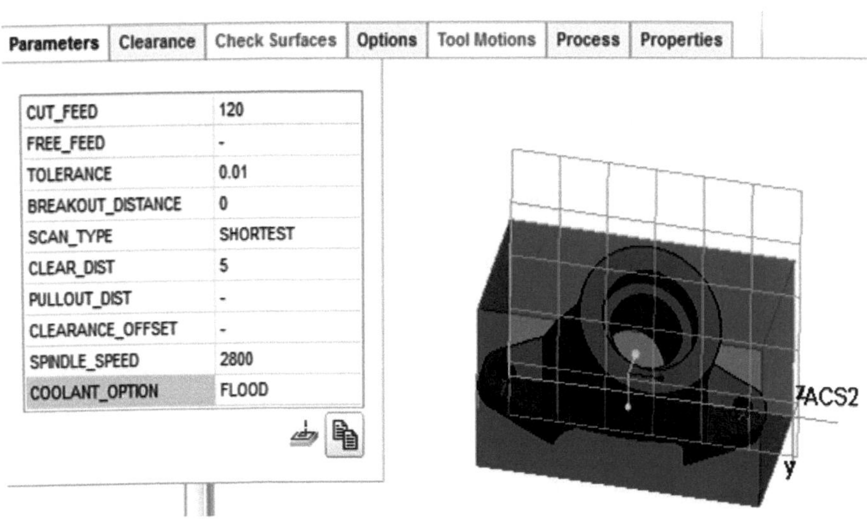

Fig. 12.203 Parameters for drilling operation

Click on the Check Mark icon ✔ to exit.

Activate the Drilling Operation for 100 mm Hole

Click on the Manufacturing menu bar to activate the Manufacturing ribbon/tools ≫ Click on Play Path icon.

Alternatively, go to the Model Tree ≫ Right click the mouse on Drilling 1 (OP010) to activate its drop-down menu list ≫ Click on Play Path on the drop-down menu list.

The Drilling Tool and the PLAY PATH dialogue box are activated on the main graphic window as shown in Fig. 12.204.

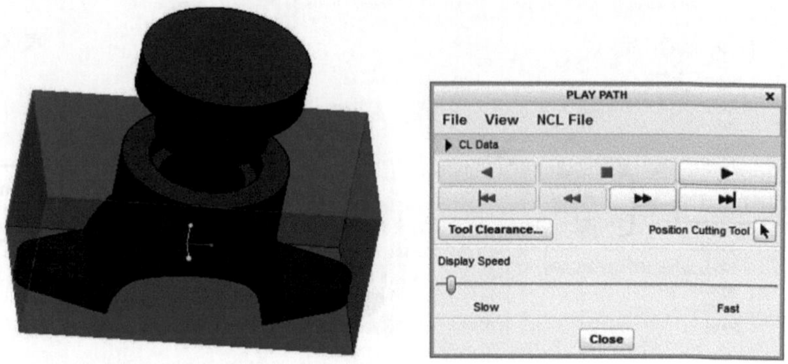

Fig. 12.204 Activated Drilling Tool and the PLAY PATH dialogue box

Click on Play tab on the PLAY PATH dialogue box to start the on-screen Drilling operation.

Click on Close tab on the PLAY PATH dialogue box at the end of the Drilling operation to exit.

12.9.4 Standard Drilling Operation for 70 mm Hole

Click on the Mill menu bar to activate the Mill ribbon.

Click on Standard icon 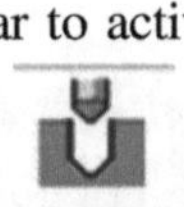 on the Mill ribbon.

The Drilling dashboard tools are activated. See Fig. 12.205.

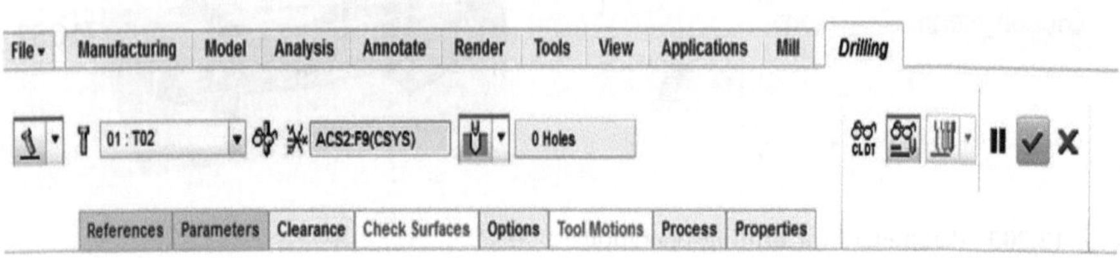

Fig. 12.205 Activated Drilling dashboard

Now Click on Tool icon  on the activated Drilling ribbon toolbar ≫ The Tools Setup dialogue window is activated on the main graphic window as shown ≫ Click on T06 as the Drilling Tool for this operation. See Fig. 12.206.

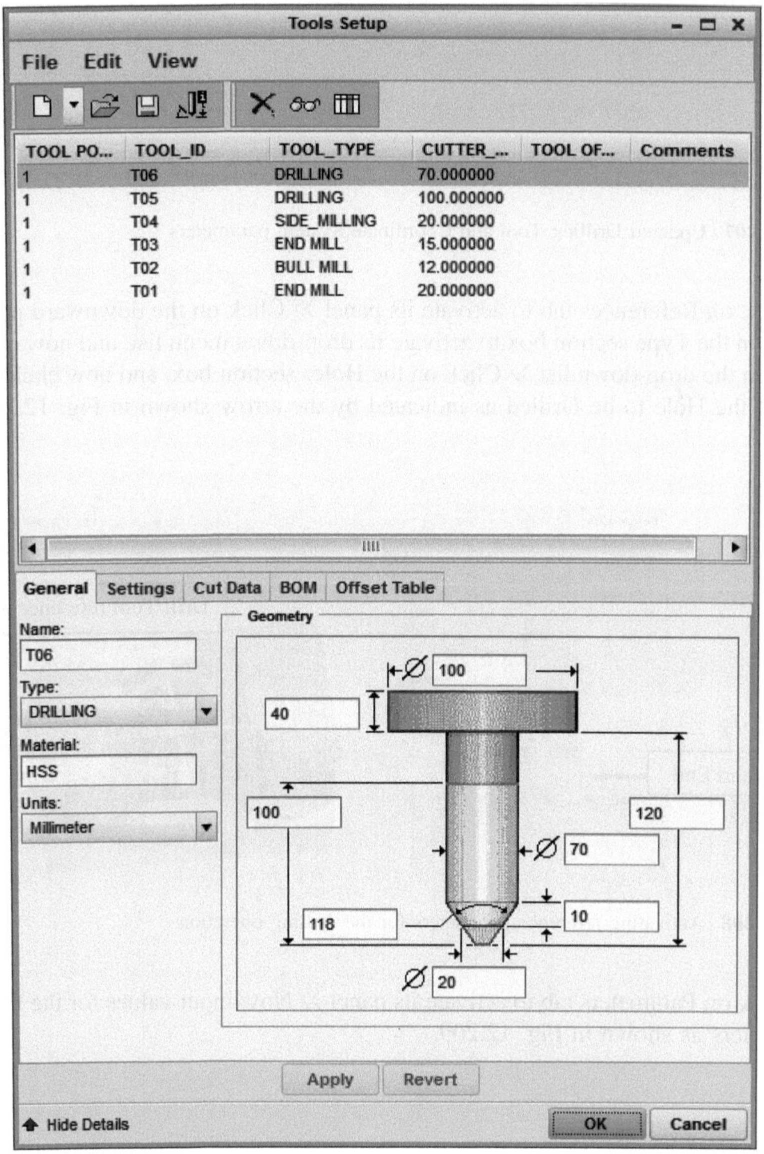

Fig. 12.206 Activating Drilling Tool (T06)

Click on OK tab to exit.

If the system does not automatically add ACS2 in the Coordinate System section box in Fig. 12.207, add ACS2 yourself. To add ACS2, go to Model Tree and click on ACS2 and the Coordinate System section box is automatically updated by the system as shown in Fig. 12.207.

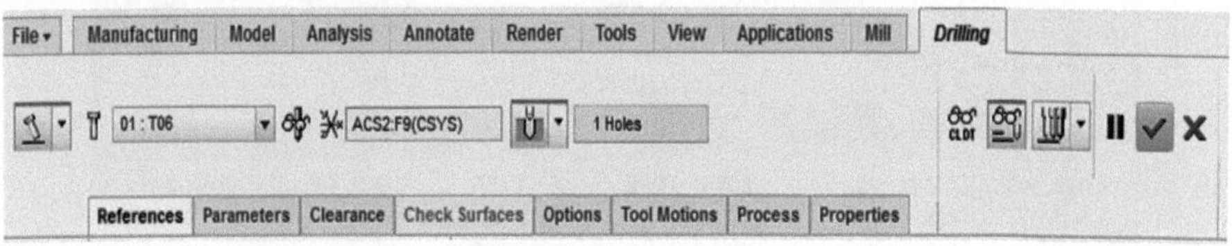

Fig. 12.207 Updated Drilling Tool and Coordinate System parameters

Click on References tab to activate its panel ≫ Click on the downward pointing arrow on the Type section box to activate its drop-down menu list, and now click on Axes on the drop-down list ≫ Click on the Holes section box, and now click on the axis of the Hole to be Drilled as indicated by the arrow shown in Fig. 12.208.

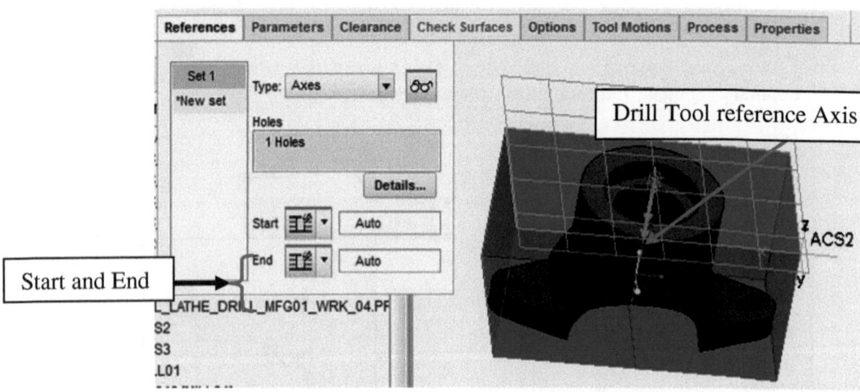

Fig. 12.208 Activating reference parameters for the drilling operation

Click on Parameters tab to activate its panel ≫ Now input values for the Drilling parameters as shown in Fig. 12.209.

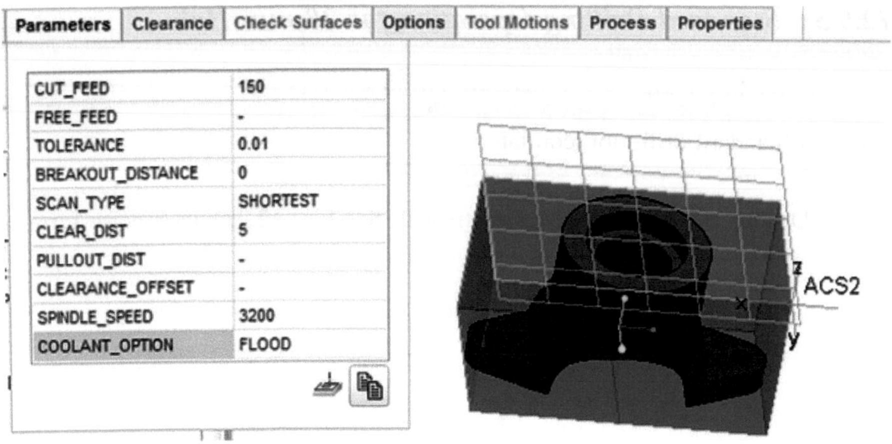

Fig. 12.209 Parameter values for the drilling operation

Click on the Check Mark icon ✔ to exit.

Activate the Drilling Operation for the 70 mm Hole

Go to the Model Tree ≫ Right click the mouse on Drilling 2 (OP010) to activate its drop-down menu list ≫ Now click on Play Path on the drop-down list.

The Drilling Tool and the PLAY PATH dialogue box are activated on the main graphic window as shown in Fig. 12.210.

Fig. 12.210 Drilling Tool and PLAY PATH dialogue box activated

Click on Play tab on the PLAY PATH dialogue box to start the on-screen Drilling operation.

Click on Close tab at the end of the Drilling operation to exit.

12.9.5 *Standard Drilling Operation for 20 mm Hole*

Click on Standard drill tool icon tab 

.

The Drilling dashboard tools are activated. See Fig. 12.211 in concise form.

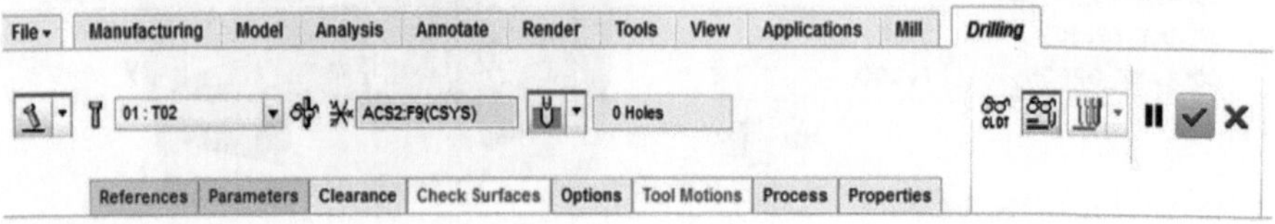

Fig. 12.211 Activated Drilling dashboard

Note: If the system has not automatically added ACS2 in the Coordinate System section box, add ACS2 yourself. To add ACS2 click on the Coordinate System icon section box, now go to the Model Tree and click on ACS2.

Click on the Tool icon  on the activated Drilling dashboard.

The Tools Setup dialogue window is activated on the main graphic window >> Click on T08 as the Drilling Tool for the operation. See Fig. 12.212.

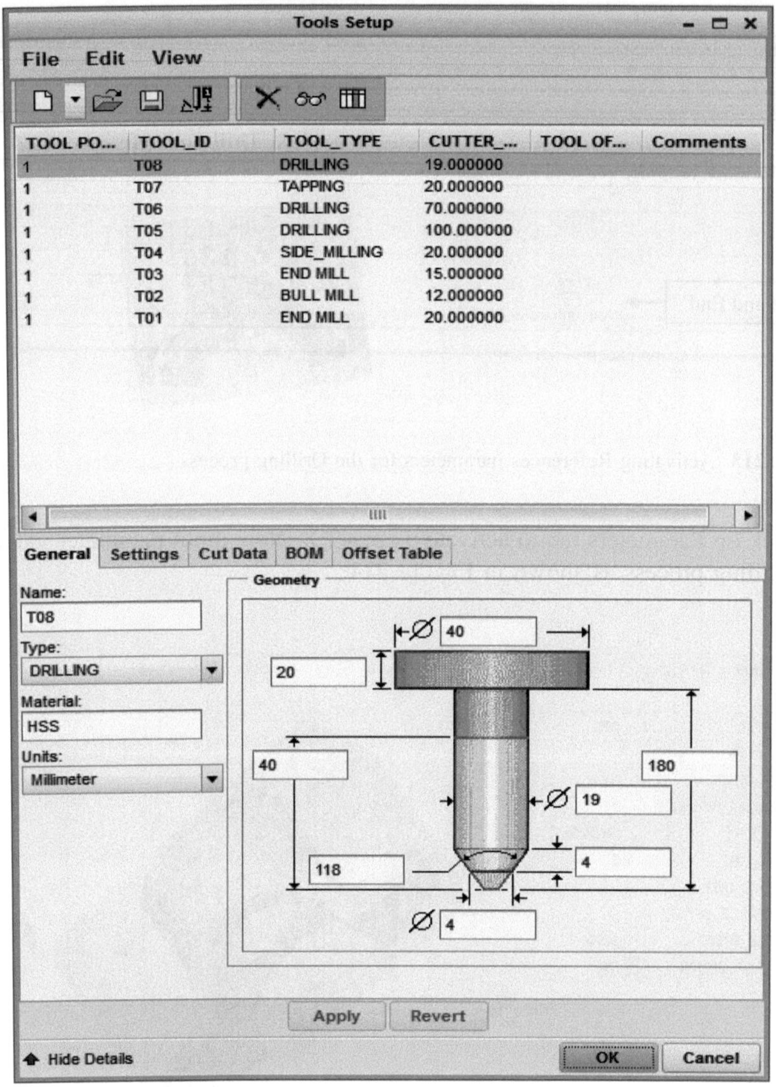

Fig. 12.212 Activating the Drilling Tool (T09) in concise form

Click on OK tab to exit.

Click on References tab to activate its panel ≫ Click on the Type section box to activate its drop-down menu list, and now click on Axes on the drop-down list ≫ Click on the Holes section box, and now click on the axes of the Holes to be Drilled on the Part as illustrated in Fig. 12.213.

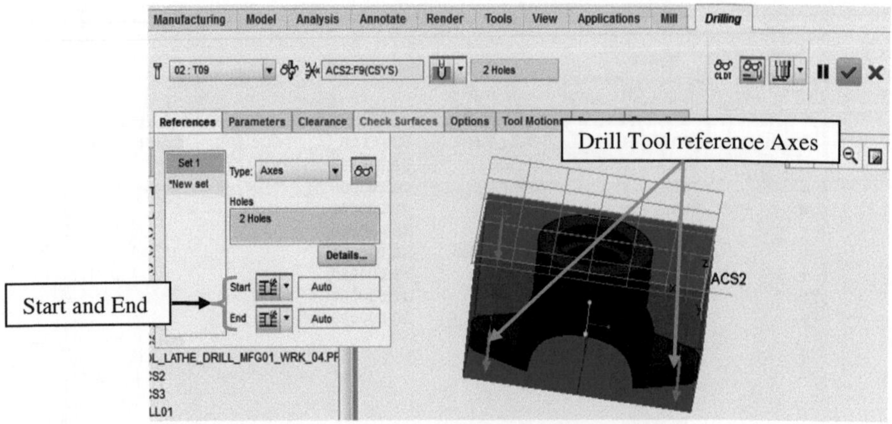

Fig. 12.213 Activating References parameters for the Drilling process

Click on Parameters tab to activate its panel ≫ Now input parameter values for the Drilling process as shown in Fig. 12.214.

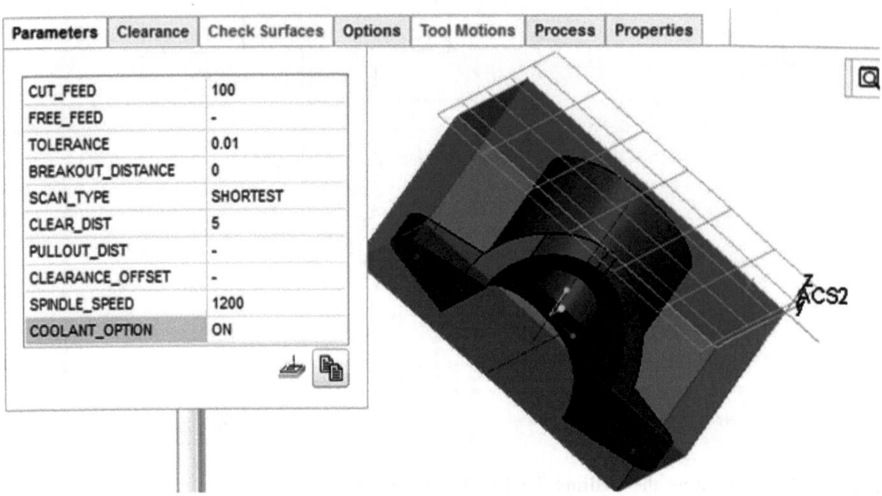

Fig. 12.214 Parameters for the Drilling process

Click on the Check Mark icon ✓ to accept the inputs and exit the Drilling ribbon toolbar.

Activate the Drilling Operation for 20 mm Hole

Go to the Model Tree ≫ Right click the mouse on Drilling 3 (OP010) to activate its drop-down menu list ≫ Click on Play Path on the drop-down list.

The Drilling Tool and the PLAY PATH dialogue box are activated on the main graphic window as shown in Fig. 12.215.

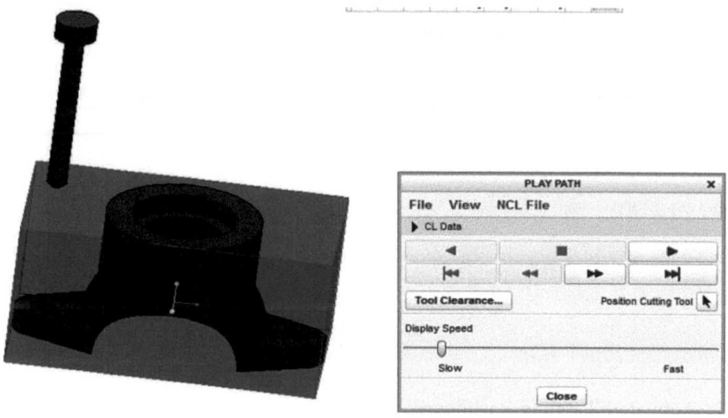

Fig. 12.215 Activated Drilling Tool and the PLAY PATH dialogue box

Click on Play tab on the PLAY PATH dialogue box.
Click on Close tab at the end of the Drilling operation to exit.

12.9.6 Create Tapping Operation for 20 mm Hole

Click on the Mill menu bar to activate the Mill ribbon.

Click on the Tapping icon tab on the activated Mill ribbon as indicated by the arrow as shown in Fig. 12.216.

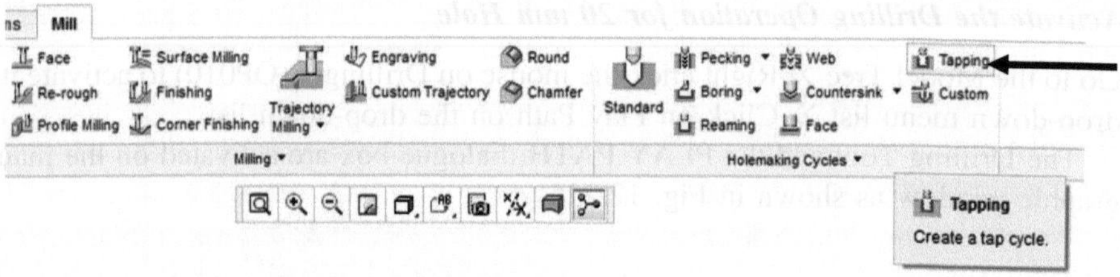

Fig. 12.216 Activating the Tapping cycle

The Tapping dashboard tools are activated. See Fig. 12.217.

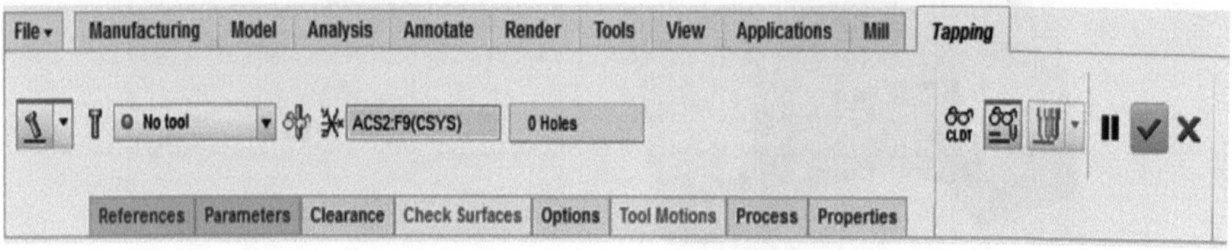

Fig. 12.217 Activated Tapping dashboard

Note: If the system did not add ACS2 automatically in the Coordinate System section box, add ACS2 yourself. To add ACS2 click on the Coordinate System icon section box, now go to the Model Tree and click on ACS2.

Click on the Tool icon .

The Tools Setup dialogue window is activated ≫ Click on T07 as the Tapping Tool for this operation as shown in Fig. 12.218.

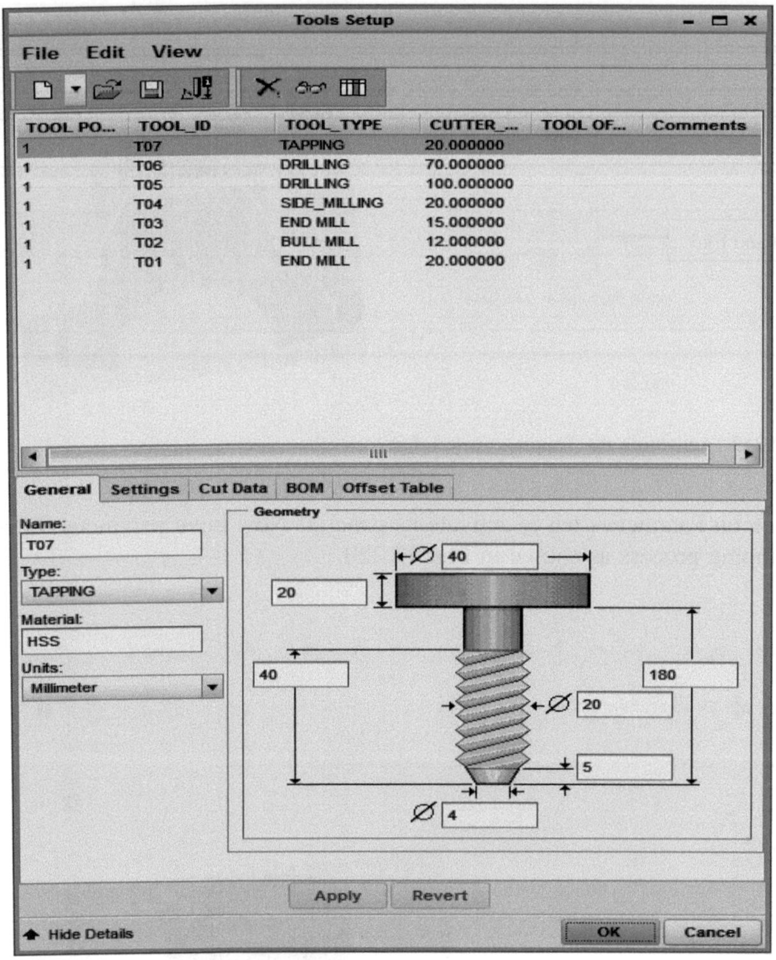

Fig. 12.218 Activating the Tapping Tool

Click on the OK tab to exit.

Click on References tab to activate its panel >> Click on the Type section box to activate its drop-down menu list, and now click on Axes on the drop-down list >> Click on the Holes section box, and now click on the axis of the Holes to be Tapped as illustrated in Fig. 12.219.

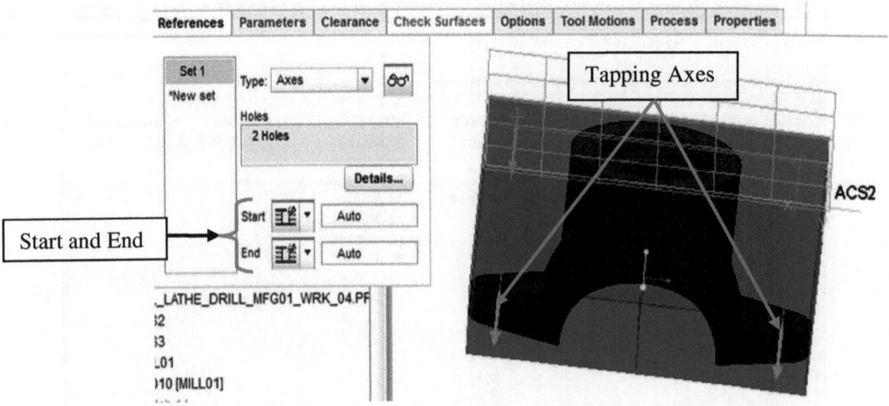

Fig. 12.219 Activating the Tapping cycle references

Click on Parameters tab to activate its panel ≫ Now input parameter values for the Tapping process as shown in Fig. 12.220.

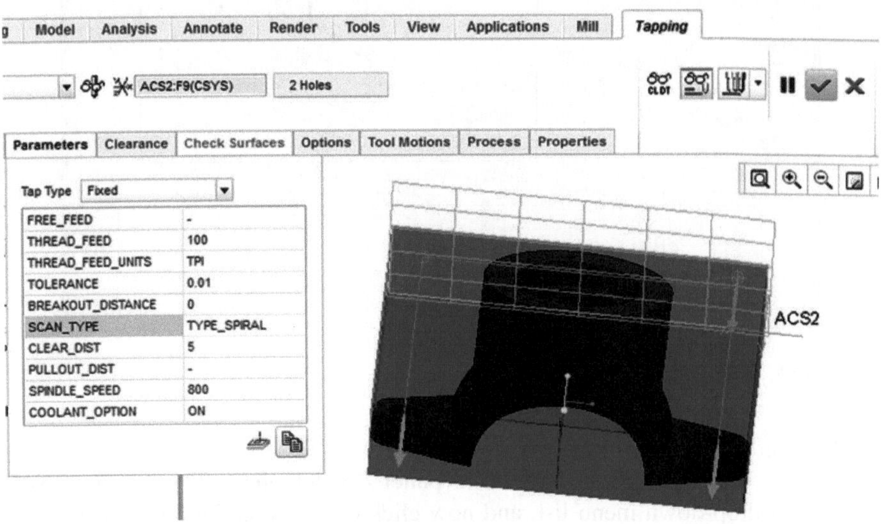

Fig. 12.220 Prameters for Tapping cycle

Click on the Check Mark icon  to exit.

Activate Tapping Operation for 20 mm Hole

Go to the Model Tree.

Right click the mouse on Tapping 1 (OP010) to activate its drop-down menu list ≫ Click on Play Path on the drop-down list.

The Tapping Tool and the PLAY PATH dialogue box are activated on the main graphic window as shown in Fig. 12.221.

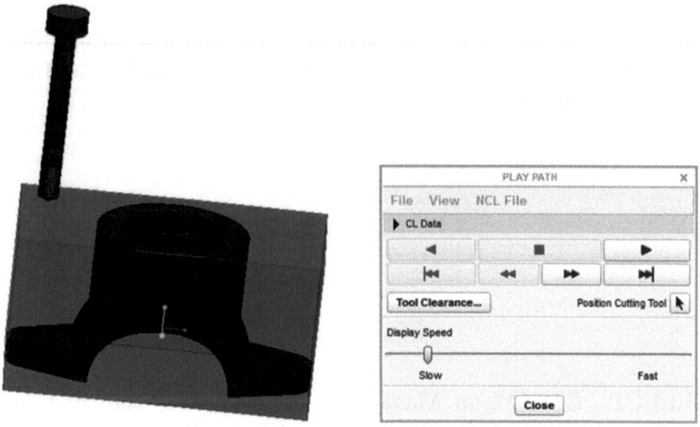

Fig. 12.221 Activated Drilling Tool and PLAY PATH dialogue box

Click on the Display View Style icon  to activate its drop-down menu ≫ Click on Wireframe on the drop-down menu list, the cutting Tool and Workpiece changes to Wireframe display as shown in Fig. 12.222.

Click on Play tab on the PLAY PATH dialogue box to start the on-screen Tapping operation.

The end of the Tapping cycle is shown in Fig. 12.222.

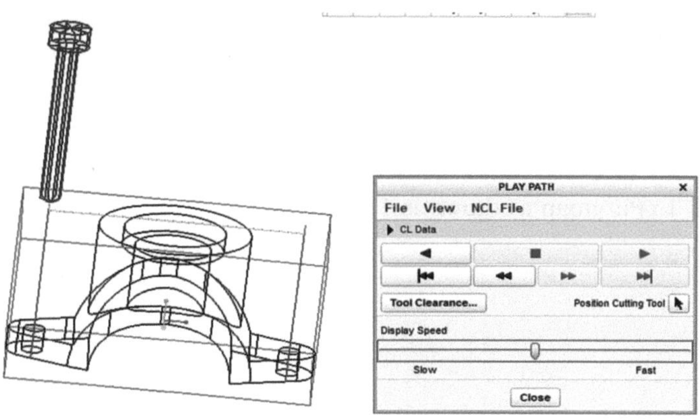

Fig. 12.222 End of the Tapping cycle in Wireframe display

Click on Close tab on the PLAY PATH dialogue box to exit the Tapping operation.

12.10 Create the Cutter Location (CL) Data

12.10.1 Create Cutter Location for OP010

Click on the Manufacturing menu bar to activate the Manufacturing ribbon ≫ Click on Save a CL File icon tab to activate its drop-down menu list, and now click on Save a CL File on the menu list as shown below.

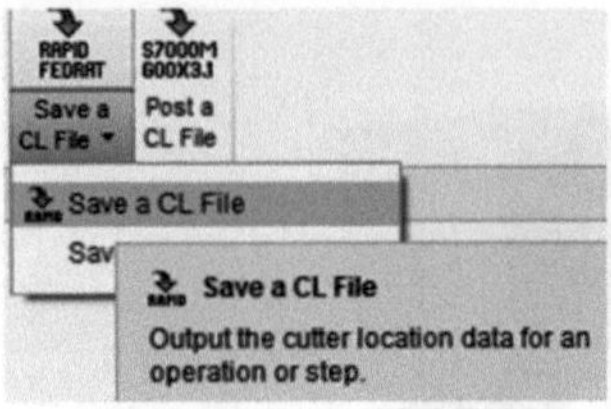

The SELECT FEAT Menu Manager dialogue box is activated on the main graphic window ≫ Click on Operation on the SELECT FEAT group and the SEL MENU group is activated as shown in Fig. 12.223.

Fig. 12.223 Activating Operation and OP010 on Menu Manager dialogue box

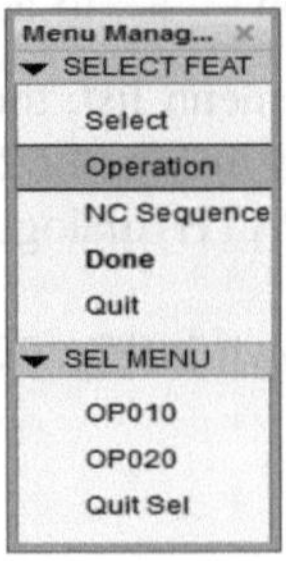

Click on OP010 on the SEL MENU group.

Once OP010 is clicked, the PATH Menu Manager dialogue box is activated on the main graphic window ≫ File is automatically highlighted by the system on the PATH group ≫ Check Mark the CL File and Interactive square boxes on the OUTPUT TYPE group as shown in Fig. 12.224.

Fig. 12.224 Activated PATH Menu Manager dialogue box

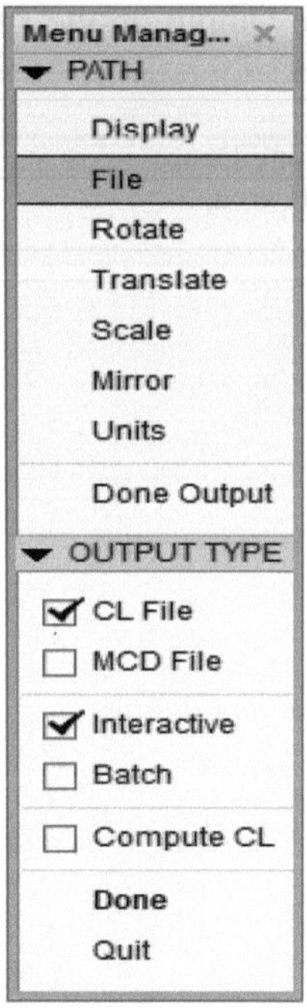

Click on Done.

The Save a Copy dialogue window is activated on the main graphic window. See Fig. 12.225 in concise form. Make sure that the name on the New Name section box is correct. You can now save the CL file in your already chosen directory folder. You can also change the file name and directory where you want to save the CL data.

Fig. 12.225 Activated Save a Copy dialogue window

Click on OK tab to exit.

Note: The Cutter Location Data is now created and saved in the chosen directory.

Click on Done Output on the PATH group as highlighted in Fig. 12.226.

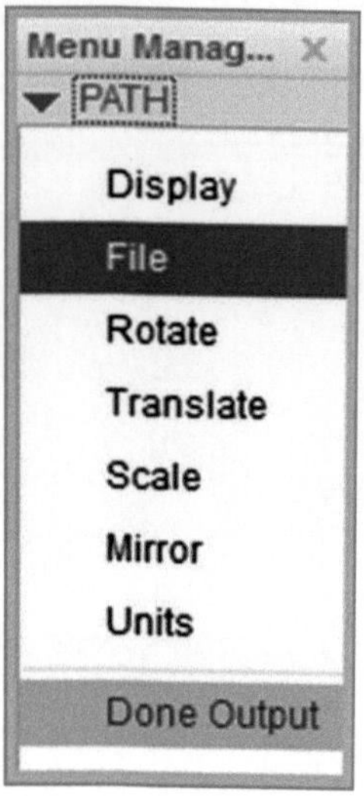

Fig. 12.226 Activating Done Output on the PATH group

12.10.2 Create Cutter Location for OP020

Click on the Save a CL File icon tab to activate its menu list ≫ Now click on Save a CL File on the drop-down menu list as shown below.

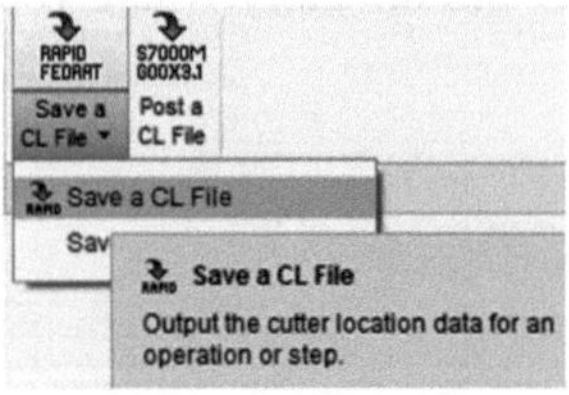

The SELECT FEAT Menu Manager dialogue box is activated on the main graphic window ≫ Operation is automatically highlighted by the system, if it is not, click on Operation on the SELECT FEAT group as shown in Fig. 12.227.

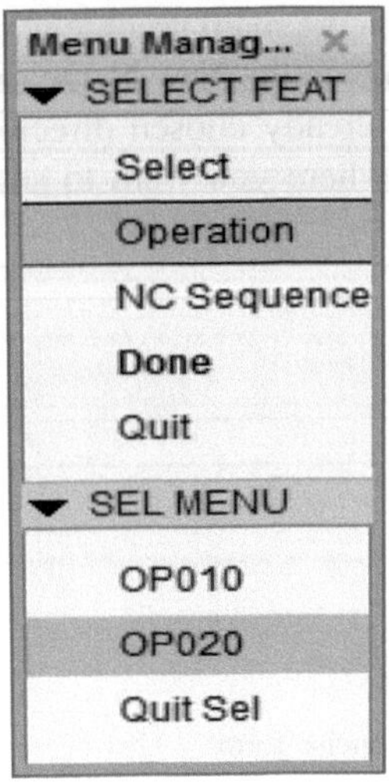

Click on OP020 on the SEL MENU group.

Fig. 12.227 Activating Operation and OP020 on Menu Manager dialogue box

Once OP020 is clicked, the PATH Menu Manager dialogue box is activated on the main graphic window ≫ File is automatically selected by the system on the PATH group ≫ Check Mark the CL File and Interactive square boxes on the OUTPUT TYPE group as shown in Fig. 12.228.

Fig. 12.228 CL File and Interactive check marked on the OUTPUT TYPE group

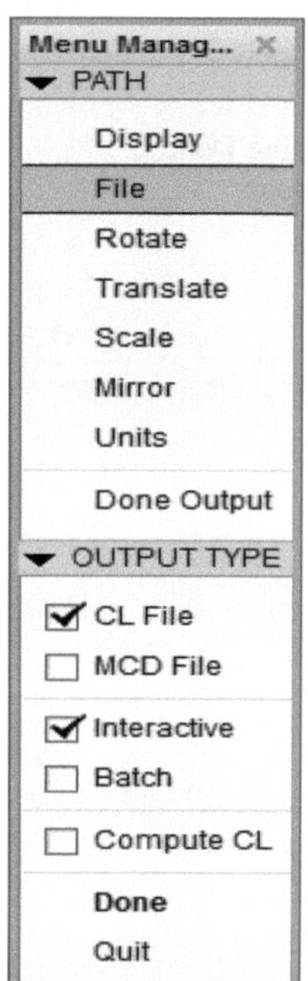

Click on Done.

The Save a Copy dialogue window is activated on the main graphic window. See Fig. 12.229 in concise form. Make sure that the name on the New Name section box is correct. You can now save the CL file in your already chosen directory folder. You can also change the file name and directory where you want to save the CL data.

Fig. 12.229 Activated Save a Copy dialogue window in concise form

Click on OK tab to exit.

Note: The Cutter Location Data is now created and saved in the chosen directory.

Click on Done Output on the PATH group as highlighted in Fig. 12.230.

Fig. 12.230 Activating Done Output on PATH group

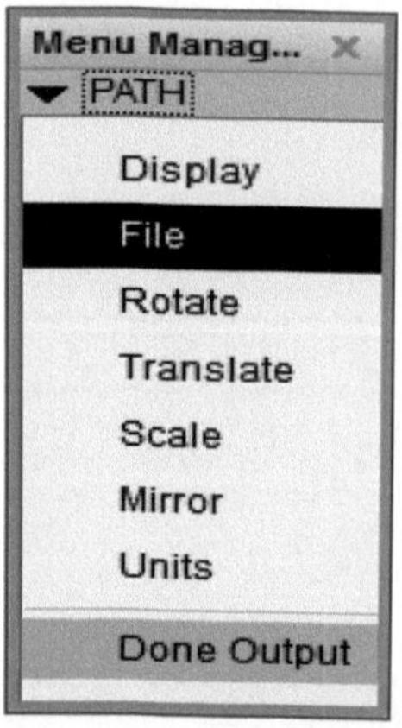

12.11 Create the G-Code Data

12.11.1 Generate the G-Code Data for OP010.ncl

Click on the Post a CL File icon tab as shown below.

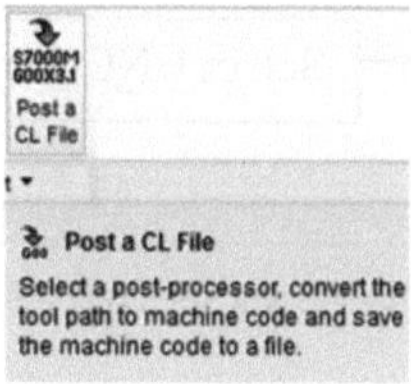

The Open dialogue window is activated on the main graphic window. See Fig. 12.231.

Fig. 12.231 Activated Open dialogue window in concise form

Click on Open tab to exit.

The PP OPTIONS Menu Manager dialogue box is activated on the main graphic window ≫ Check Mark Verbose and Trace as shown in Fig. 12.232, only when it is not automatically activated by the system itself.

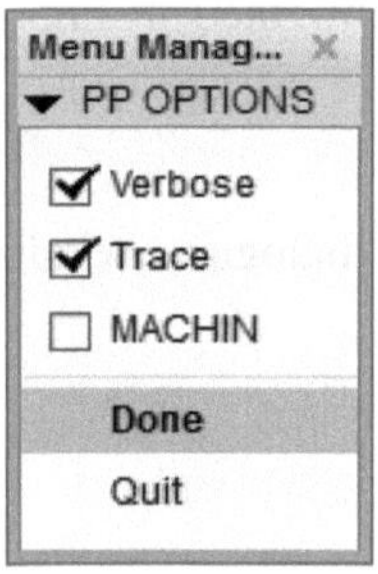

Click on Done.

Fig. 12.232 Activated PP OPTIONS Menu Manager dialogue box Verbose and Trace Check Marked

The PP LIST dialogue box is activated on the main graphic window as shown in Fig. 12.233.

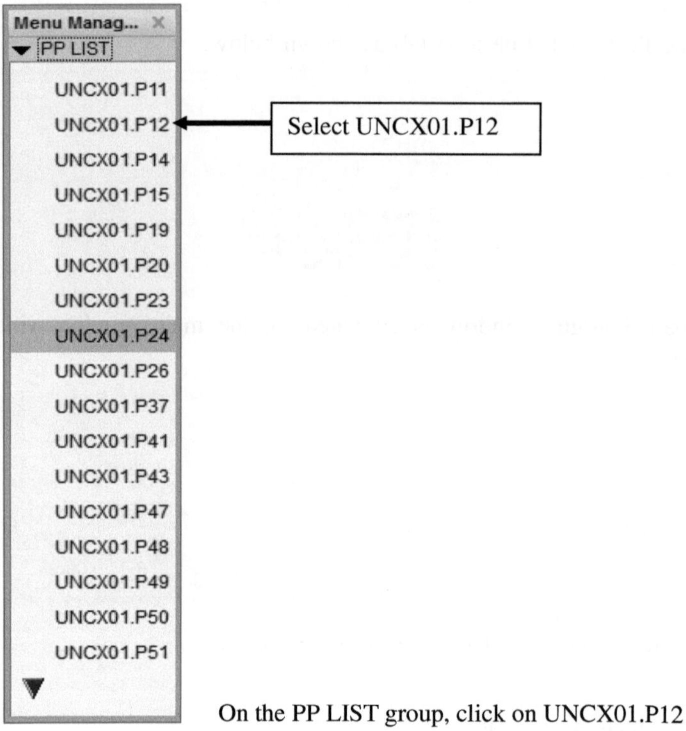

Fig. 12.233 Activated PP LIST dialogue box

Note: The UNCX01.P24 P24 (MAHO MH—1000S VMC/HMC—PHILLIPS CNC 432–5 AXES) as highlighted in Fig. 12.233 is a five-axis CNC milling machine

The INFORMATION WINDOW dialogue window is activated. See Fig. 12.234.

Fig. 12.234 Activated INFORMATION WINDOW in concise form

12.11.2 Generate the G-Code Data for OP020.ncl

Click on the Post a CL File icon tab as shown below.

The Open dialogue window is activated on the main graphic window. See Fig. 12.235.

Fig. 12.235 Activated Open dialogue window in concise form

Click on Open tab to exit.

The PP OPTIONS Menu Manager dialogue box is activated on the graphic window ≫ Check Mark Verbose and Trace as shown in Fig. 12.236.

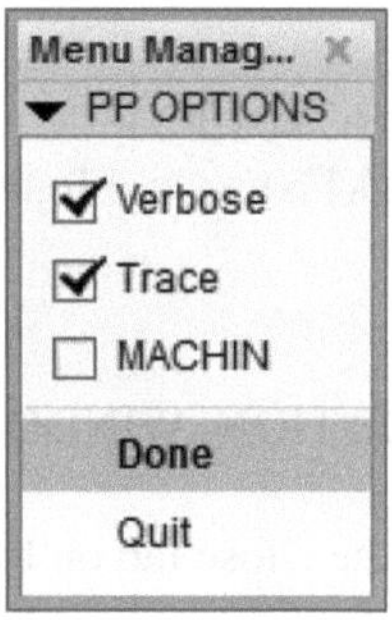

Fig. 12.236 Activated PP OPTIONS Menu Manager dialogue box

Click on Done.

The PP LIST dialogue box is activated on the main graphic window. See Fig. 12.237.

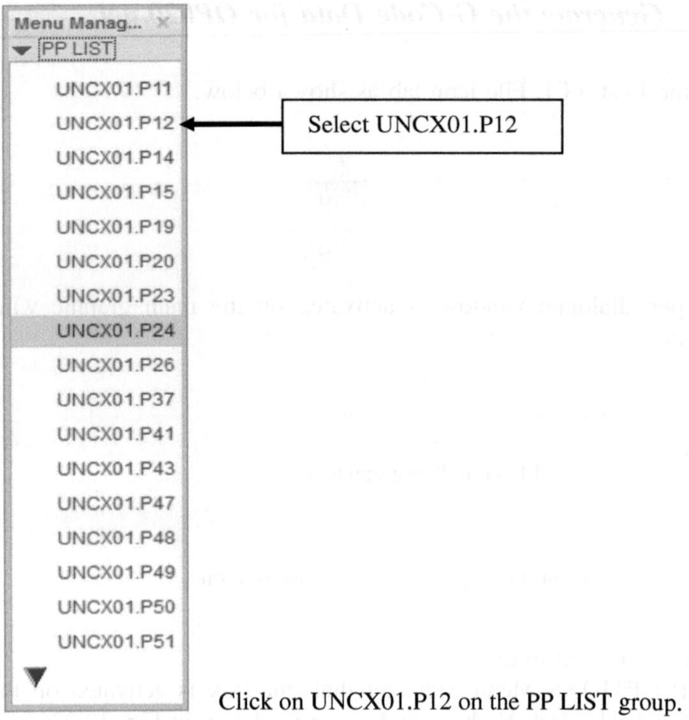

Fig. 12.237 Activated PP LIST dialogue box

Note: The UNCX01.P24 (MAHO MH—1000S VMC/HMC—PHILLIPS CNC 432–5 AXES) as highlighted in Fig. 12.237 is a five-axis CNC milling machine

The INFORMATION WINDOW dialogue window is activated on the main graphic window.

Now click on the Close tab on the activated INFORMATION WINDOW to exit.

Note: The generated G-codes are stored as a TAP file in the chosen work directory. Open the TAP file using Notepad to view the generated G-codes for both OP010.ncl and OP020.ncl.

Chapter 13
Surface Milling of Intricate Cast Part

13.1 Five Axes Surface Milling of Cast Part

In this tutorial, the following step-by-step guide will be covered.

- Start the Manufacturing application
- Reference Model will be Imported into the Manufacturing GUI window
- Reference Model will be Constrained
- No stock will be created because this is a Cast Part and also only Surface Milling operation will be carried out
- Create Programme Zeroes or Coordinate Systems
- Tools will be created
- Surface Milling will be performed using Cut Lines

13.1.1 Start New Manufacturing Application

Start Creo Parametric either on your Computer Desktop or from Programme ≫ Click on File to activate its drop-down menu list ≫ Click on New icon as indicated by the arrow in Fig. 13.1.

P.O. Kanife, *Computer Aided Virtual Manufacturing Using Creo Parametric*,
DOI 10.1007/978-3-319-23359-8_13

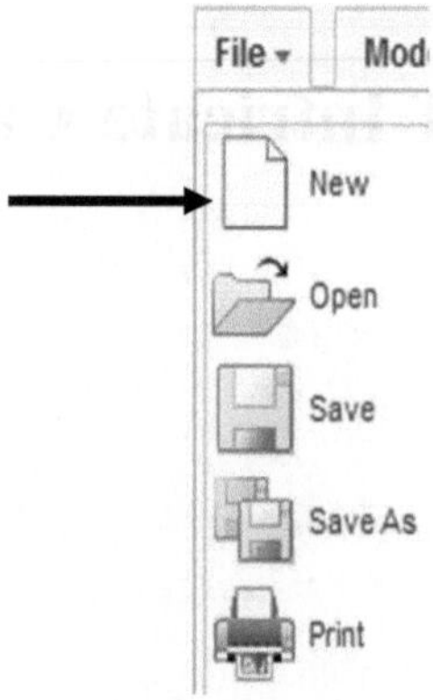

Fig. 13.1 Activating the New dialogue box via File

The New dialogue box is activated on the main graphic window ≫ Click on the Manufacturing radio button on the Type group ≫ On the Sub-type group, click on the NC Assembly radio button ≫ Click on the Name section box and type Cast_mfg01 ≫ Click on the Use default template square box to clear the Check Mark icon as shown in Fig. 13.2.

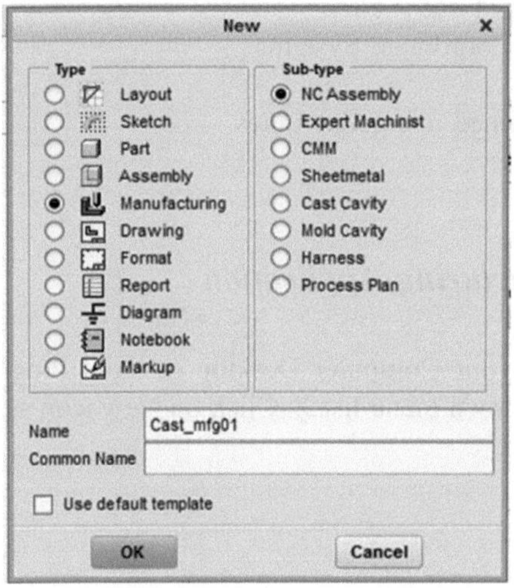

Click on OK tab to exit.

Fig. 13.2 Activating the Manufacturing application

The New File Options dialogue window is activated on the main graphic window ≫ Click on “mmns_mfg_nc” on the Template group ≫ On the Parameters group, type your name in the MODELLED BY section box and in the

DESCRIPTION section box, type whatever you want to type as description. In this tutorial, the section boxes will be left blank. See Fig. 13.3.

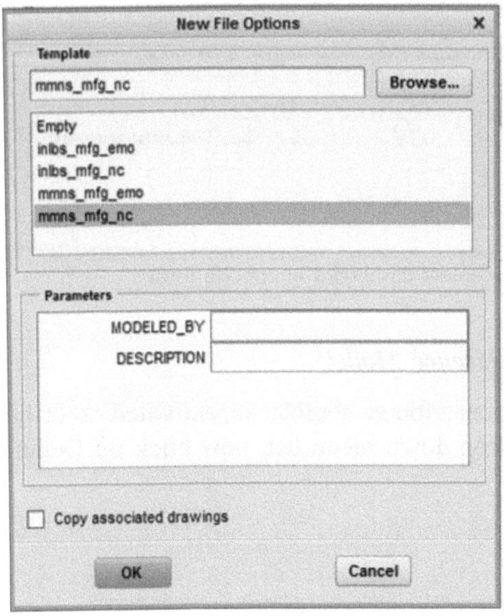

Click on OK tab to exit.

Fig. 13.3 Activating the New File Options dialogue window

Creo Parametric Manufacturing application GUI window is activated.

13.1.2 Import and Constrain the Reference Model

To import the Part into the Manufacturing graphic window, click on the Reference Model icon tab to activate its drop down menu list ≫ Now click on Assemble Reference Model on the drop down list as shown below.

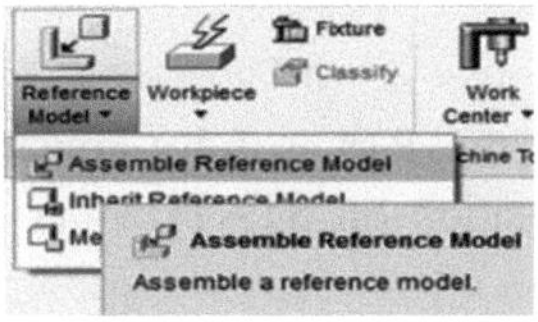

The Open dialogue window is activated on the Manufacturing GUI window.

Make sure that the File name that you want to open is in the File name section box.

Click on the Open tab and the Reference Model is imported into the Manufacturing graphic window.

The Component Placement dashboard tools are activated. See Fig. 13.4.

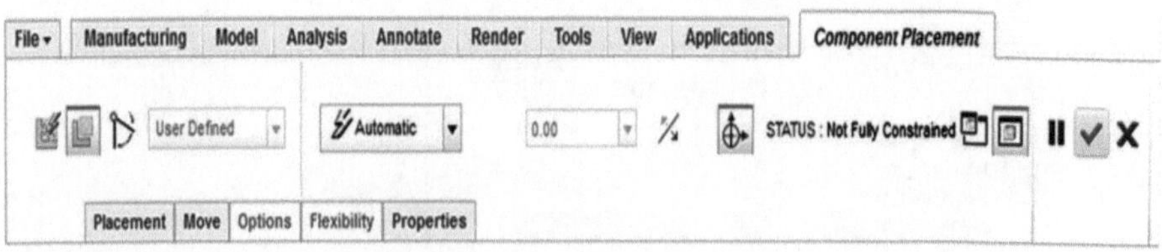

Fig. 13.4 Activated Component Placement dashboard in concise form

Constraining the imported Reference Model

Once the Component Placement ribbon toolbar is activated » Click on the Automatic tab to activate its drop down menu list, now click on Default as highlighted in Fig. 13.5.

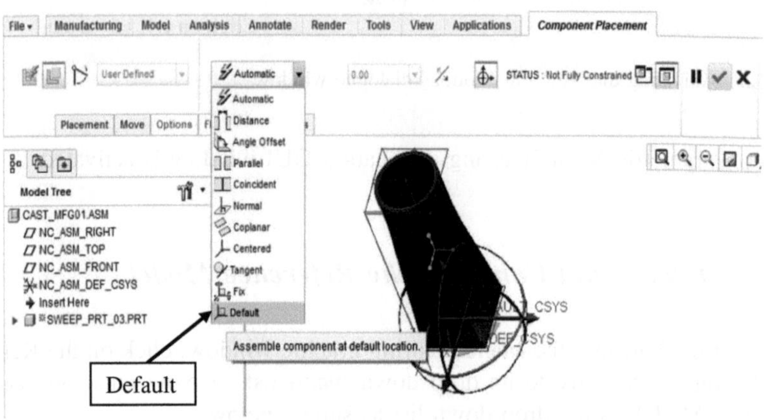

Fig. 13.5 Activating Default to constrain the reference model

Once Default is clicked, the Part is fully constrained on the Manufacturing graphic window. The information on STATUS bar reads Fully Constrained. See Fig. 13.6.

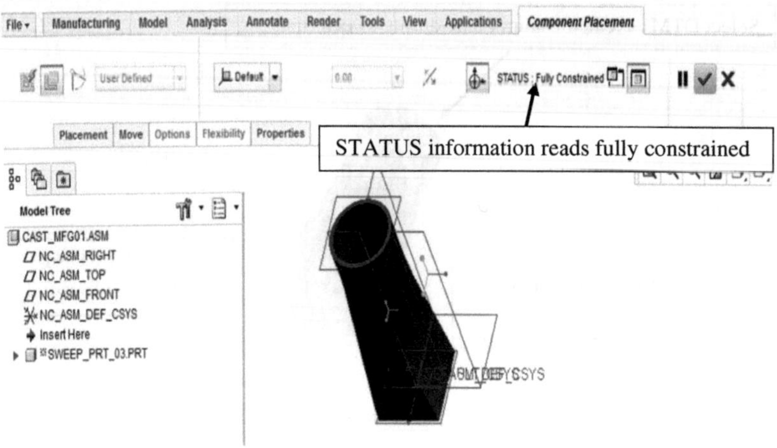

Fig. 13.6 Reference model fully constrained

Click on the Check Mark icon to proceed.

Add Stock to Reference Model

No stock will be added to the Reference Model.

13.2 Create Programme Zero or Coordinate System

Click on the Coordinate System icon as shown below.

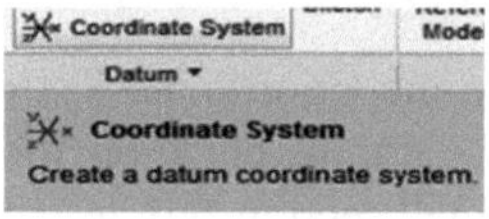

The Coordinate System dialogue box is activated on the main graphic window ≫ Now click on DTM1, NC_ASM_FRONT and NC_ASM_RIGHT as the reference Datum Planes as indicated by the arrows in Fig. 13.7.

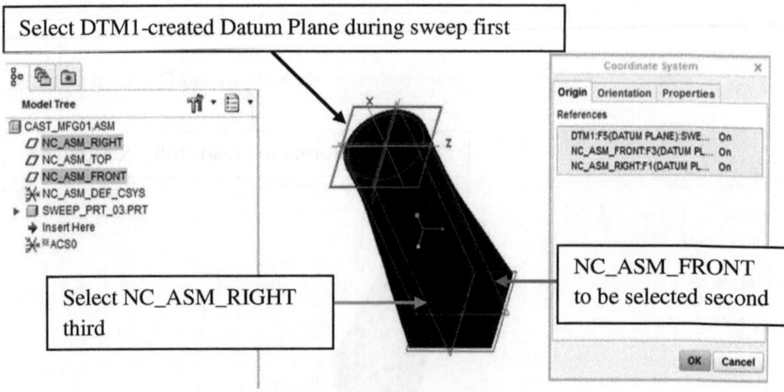

Fig. 13.7 Activating the reference datum planes for the Coordinate System

The *X*, *Y* and *Z* axes of the Coordinate System are not pointing in the correct orientation as shown in Fig. 13.8.

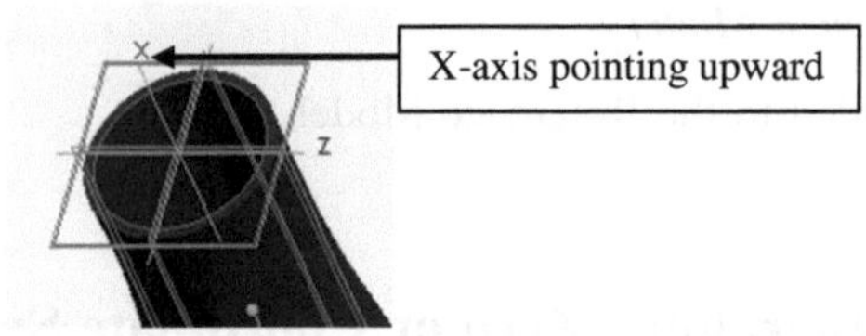

Fig. 13.8 Wrong orientation of *X*, *Y* and *Z* axes of the Coordinate System

To orientate the X, Y and Z coordinate axes to the correct orientation

Click on the Orientation tab on the Coordinate System dialogue box to activate its content ≫ On the "Orient by" group, click on the "References selection" radio button ≫ Click on "to determine" section box to activate its drop down menu list, now click on *Z*-axis on the activated drop down list ≫ Click on the "to project" section box to activate its drop down list, now click on *Y*-axis on the activated drop down list. Click on the Flip tab, to flip *Y*-axis orientation. See Fig. 13.9.

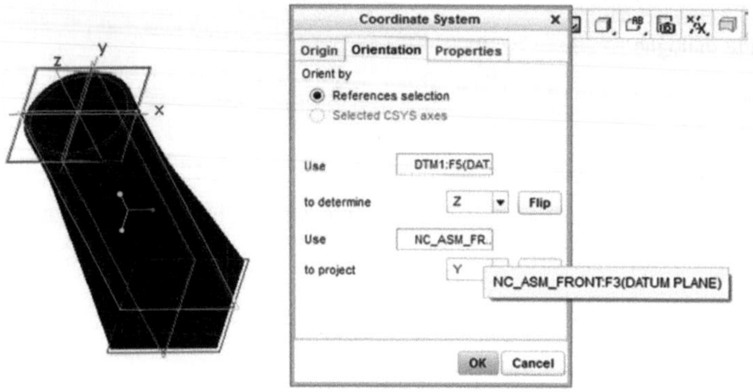

Fig. 13.9 Correct orientations of *X*, *Y* and *Z* axes of the Coordinate System

Click on OK tab to exit the Coordinate System dialogue box.

The new Coordinate System named ACS0 is created as indicated by the arrow as shown in Fig. 13.10.

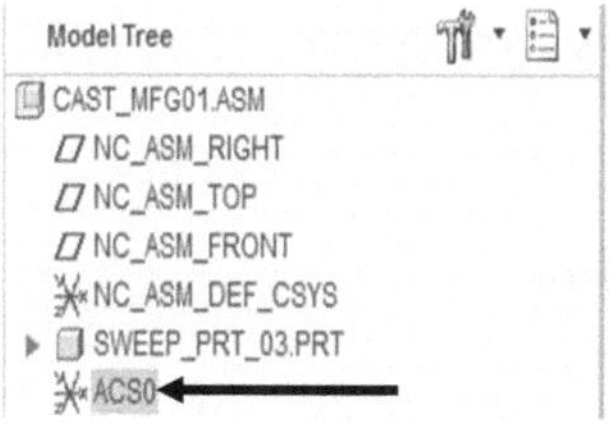

Fig. 13.10 New created Coordinate System (ACS0) on the Model Tree

13.3 Create Work Centre

Click on the Work Centre icon tab to activate its drop-down menu list ≫ Now click on Mill on the activated drop down list as shown below.

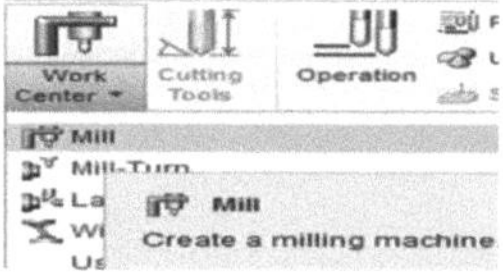

The Milling Work Centre dialogue window is activated on the main graphic window ≫ Click on the Number of Axes section box to activate its drop-down menu list ≫ Click on 5 Axis on the drop-down menu. See Fig. 13.11.

Fig. 13.11 Activated Milling Work Centre dialogue window

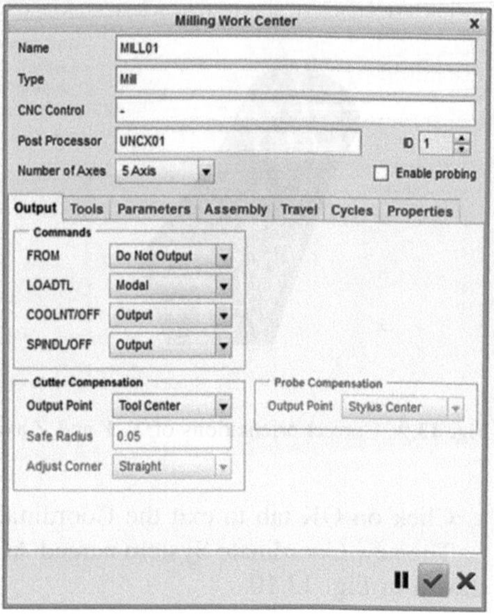

Click on the Check Mark icon to exit.

13.4 Create Operation

Click on the Operation icon tab as shown below.

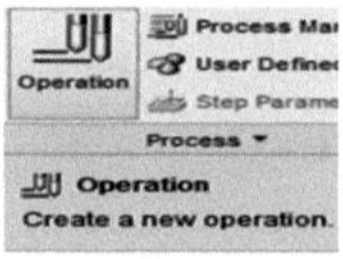

The Operation dashboard tools are activated. See Fig. 13.12.

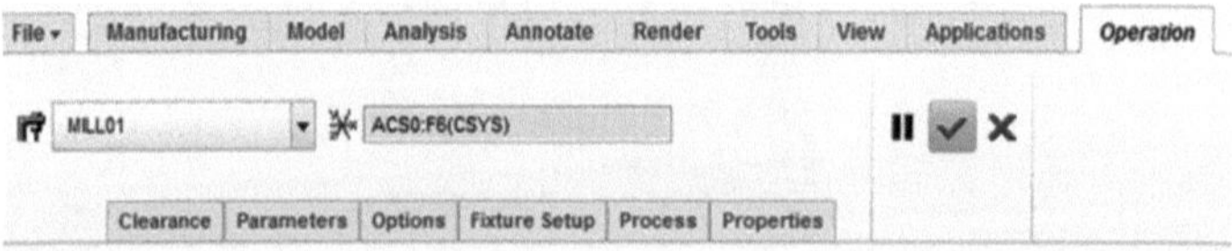

Fig. 13.12 Activated Operation dashboard in concise form

If MILL01 and ACS0 are not automatically added in their respective section boxes by the system, add them yourself.

To add them, click on the Coordinate System icon section box, now go to the Model Tree and click on the created Coordinate System, that is ACS0 » Click on the Work Centre icon section box, and click on the created MILL01 on the drop down list. See Fig. 13.13.

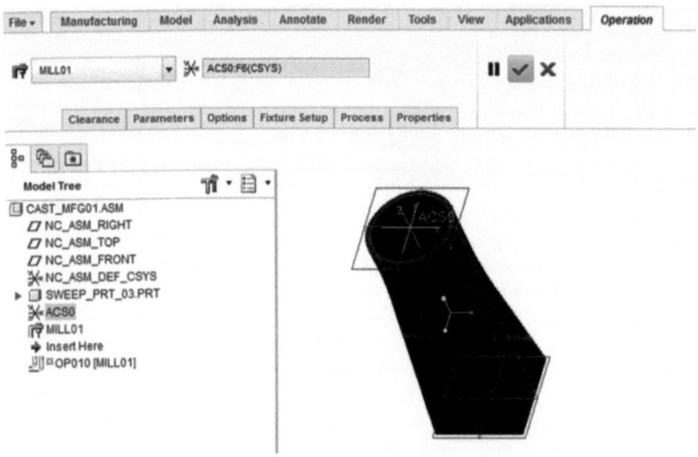

Fig. 13.13 MILL01 and ACS0 added into their respective section boxes

Click on Check Mark icon to exit Operation setup.

13.5 Create Cutting Tool

Click on the Cutting Tools icon tab as shown below.

The Tools Setup dialogue window is activated on the main graphic window.

To add new LOLIPOP Tool » Click on the General tab if not active, to activate its content » Click on the Name section box and type T0001 » Click on the Type section box to activate its drop down menu list, now click on LOLIPOP Tool on the activated drop down menu list » Add dimensions to the LOLIPOP Tool » Now click on the Apply. See Fig. 13.14.

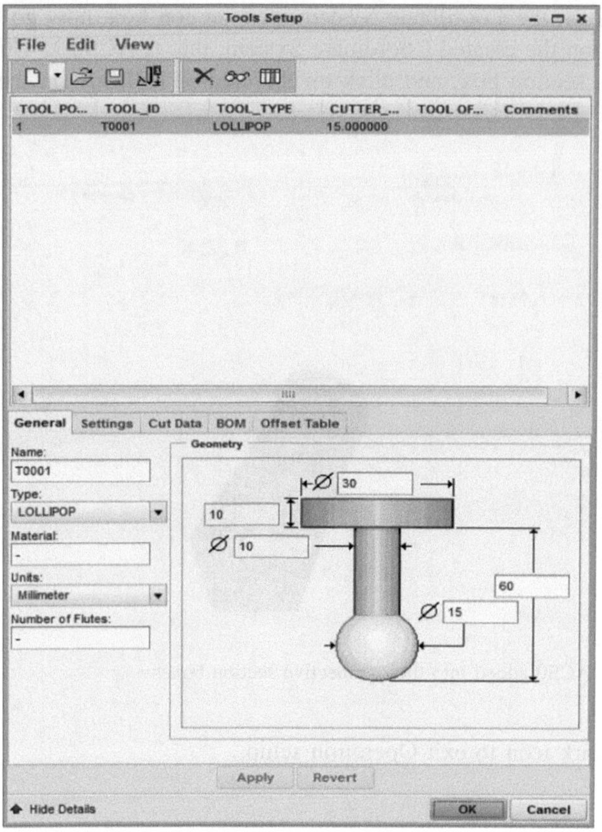

Click on OK tab.

Fig. 13.14 Created Surface Milling Tool (LOLIPOP)

Add the 5-Axis application Tools (Using Add Method)

Activate 5-Axis side mill as it will add more Tools options to Surface Milling and axis control.

Click on File to activate its content ≫ Click on Options on the drop down menu list ≫ The Creo Parametric Options dialogue window is activated on the main graphic window ≫ On the list on the left side of the dialogue window, click on Configuration Editor to activate its content. See Fig. 13.15.

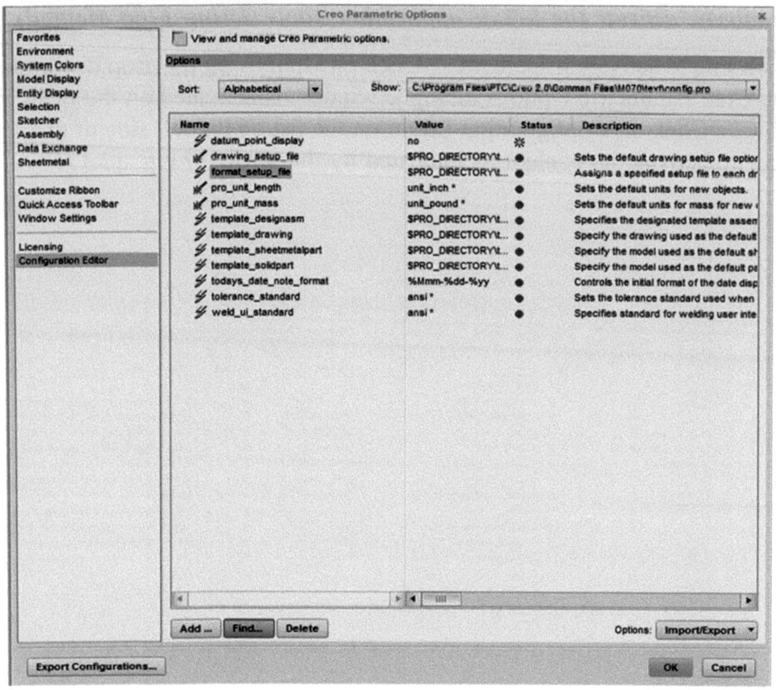

Fig. 13.15 Activated Configuration Editor Content is activated on the Creo Parametric Options dialogue window

Now click on Add tab.

The Options dialogue box is activated on the main graphic window ≫ Now type "5_axis_side_mill" in the Options name section box ≫ Type "yes" in the Option value section box. See Fig. 13.16.

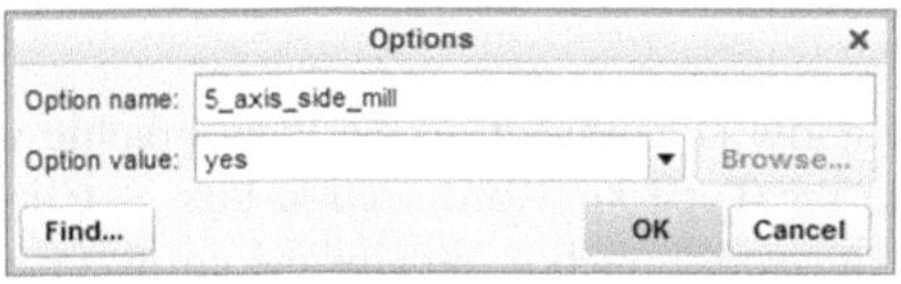

Click on OK tab to exit.

Fig. 13.16 Activated Options dialogue box

Click on the OK tab on the Creo Parametric Options dialogue window to exit. Click on OK tab to exit.

Alternatively, activate the 5-Axis application Tools (Using Find Method)

Click on File to activate its content ≫ Click on Options on the drop down menu list ≫ The Creo Parametric Options dialogue window is activated on the main graphic window ≫ Click on Configuration Editor on the list on the left side of the dialogue window to activate its content as indicated by the arrow in Fig. 13.17.

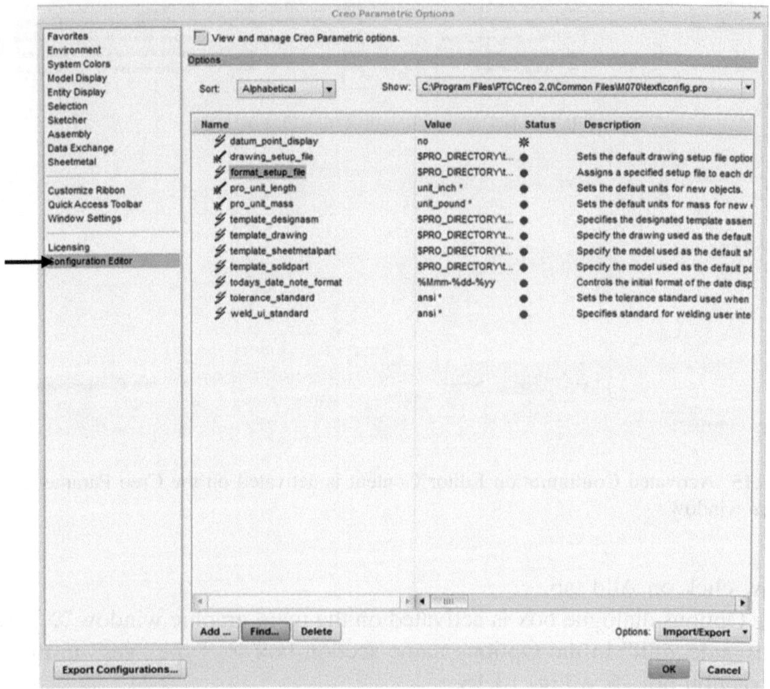

Fig. 13.17 Activated Configuration Editor content on Creo Parametric Options dialogue window

Click on the Find tab.

The Find Option dialogue window is activated on the main graphic window ≫ Now type "5_axis_side_mill" in the Type keyword section box ≫ Click on Find Now tab ≫ Click on the Set value section box downward arrow to activate its content, now click on 5-axis side mill on the drop down menu list or type it if you cant find it ≫ Click on Add/Change tab to add the 5-axis_side_mill to the system as shown in Fig. 13.18.

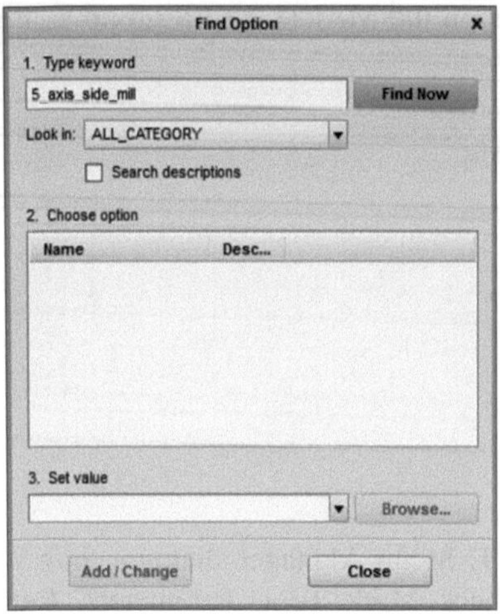

Fig. 13.18 Activating "5_axis_side_mill" on Find Option dialogue window

Save into the configuration file if prompted to do so.
Click on the Close tab to exit the Find Option dialogue window.
Click on the OK tab on the Creo Parametric Options dialogue window to exit.

13.6 Create Surface Milling Sequences

13.6.1 Create Surface Milling for Internal Surface of the Cast Part

Click on the Mill menu bar to activate its ribbon ≫ Now click on the Surface Milling icon tab as shown below.

The MACH AXES Menu Manager dialogue box is activated on the main graphic window as shown. See Fig. 13.19.

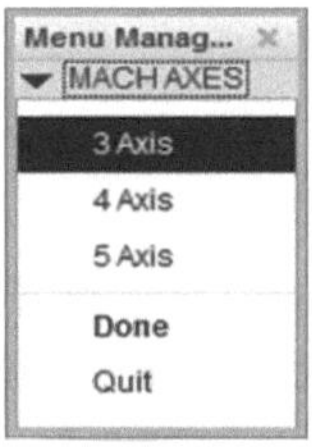

Fig. 13.19 Activated MACH AXES Menu Manager dialogue box

Now click on 5-Axis on the MACH AXES group as highlighted on Fig. 13.20.

Fig. 13.20 Activating 5 Axis on MACH AXES group

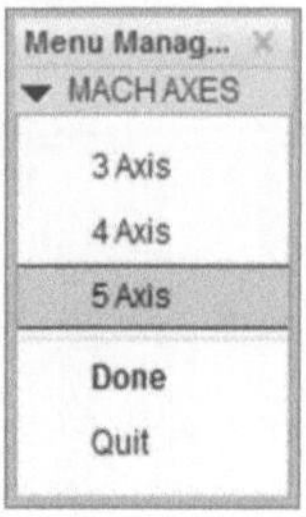

Click on Done.

The NC SEQUENCE Menu Manager dialogue box is activated on the main graphic window ≫ Check Mark Tool, Parameters, Retract Surf, Surfaces and Define Cut on the SEQ SETUP group as shown in Fig. 13.21.

Fig. 13.21 Tool, Parameters, Retract Surf, Surfaces and Define Cut check marked

Click on Done.

The Tools Setup dialogue window is activated again. Make sure that the Tool activated and its dimensions are correct as shown in Fig. 13.22.

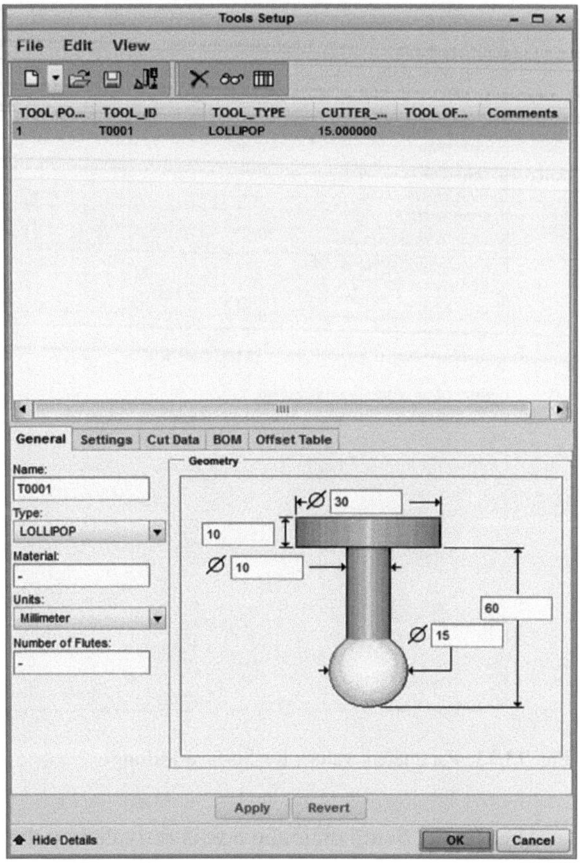

Fig. 13.22 Surface Milling Tool activated

Click on OK tab to exit.

The Edit Parameters Sequence "Surface Milling" dialogue window is activated on the main graphic window ≫ Now input parameters values for the Surface Milling process as shown in Fig. 13.23.

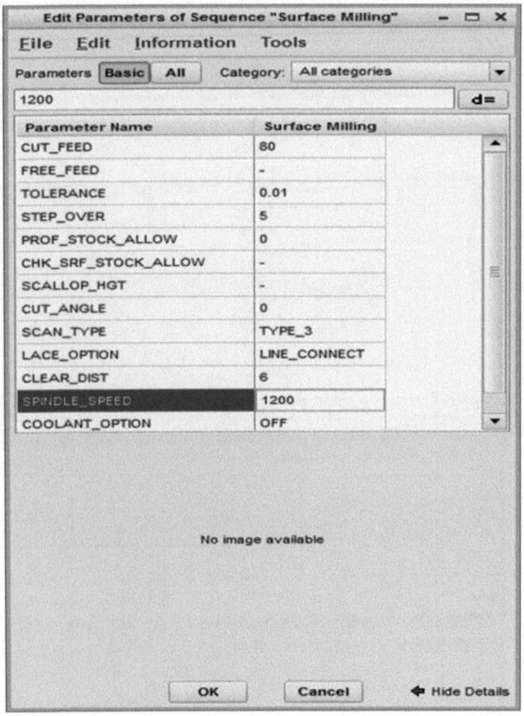

Click on OK tab to exit.

Fig. 13.23 Parameters values for Surface Milling

The Retract Setup dialogue box is activated on the main graphic window ≫ Now input 15 mm as the retract value. See Fig. 13.24.

Fig. 13.24 Retract Setup Plane parameter

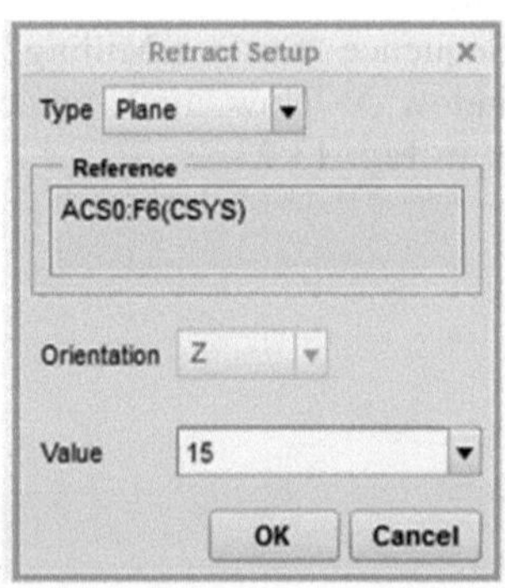

Click on OK tab to exit.

The SURF PICK group is automatically added to the Menu Manager dialogue box as shown in Fig. 13.25. Now select Part as the Model.

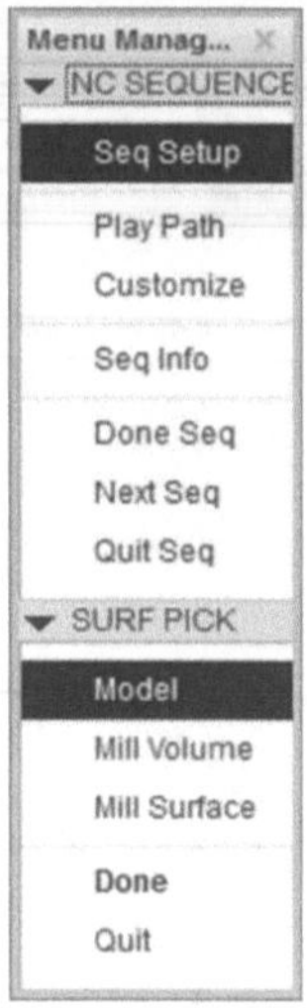

Click on Done.

Fig. 13.25 Menu Manager dialogue box with SURF PICK group activated

The Select dialogue box is activated on the main graphic window as shown (Fig. 13.26).

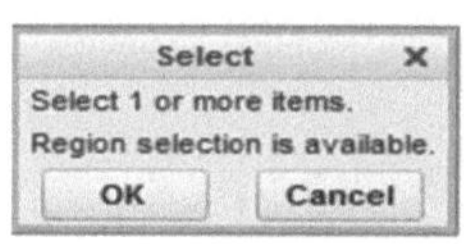

Fig. 13.26 Activated Select dialogue box

The SELECT SRFS group is automatically added by the system on the Menu Manager dialogue box as shown in Fig. 13.27.

Fig. 13.27 Activated SELECT SRFS groups with Add highlighted

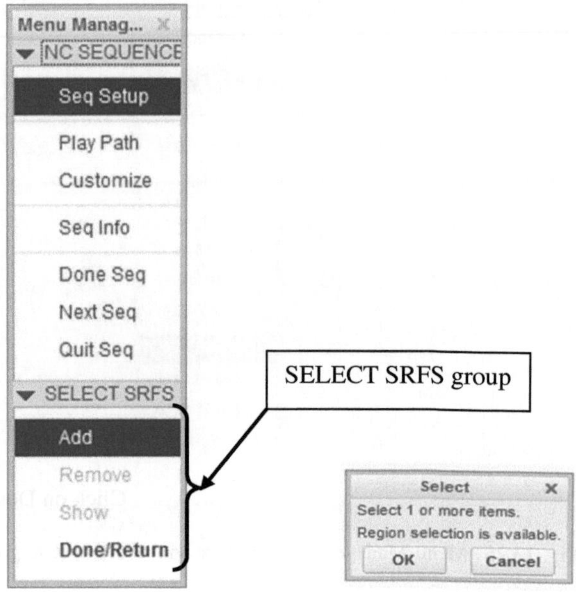

Now click on the outside Surfaces of the Reference Model while holding down the Ctrl key when selecting each of the four outside surfaces as indicated by the arrows in Fig. 13.28 ≫ The selected outside surfaces are shown in Fig. 13.28 in wireframe display.

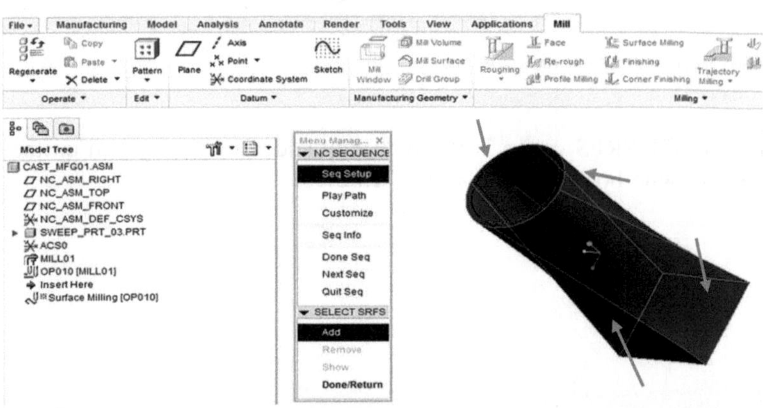

Fig. 13.28 Activating the outside surfaces for the surface milling operation

Click on OK tab on the Select dialogue box to exit if it is still active.
Click on Done/Return on the SELECT SRFS group.

The Cut Definition dialogue box is activated on the main graphic window as shown in Fig. 13.29.

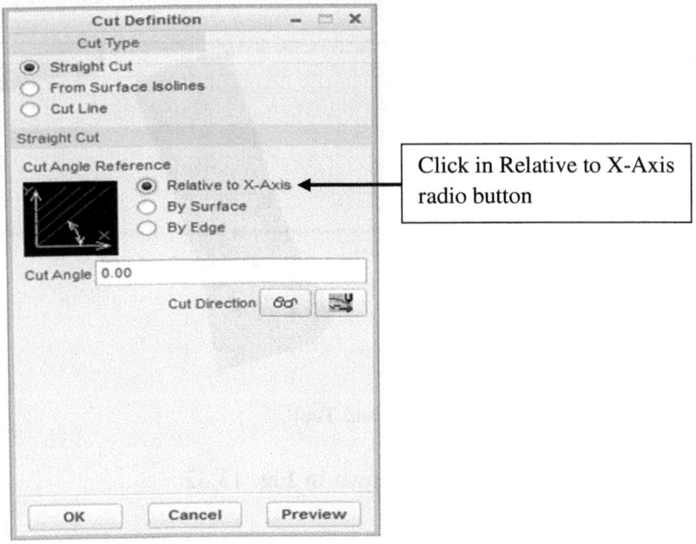

Fig. 13.29 Activating Cut Line and Relative to *X*-Axis radio buttons

Change the Cut Angle to 15° as shown in Fig. 13.30.

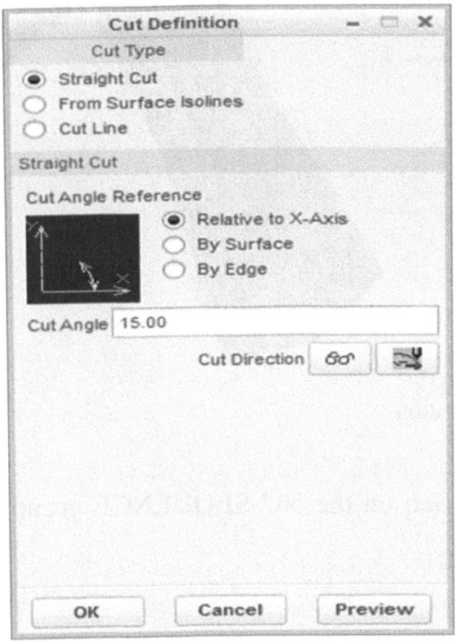

Click on the Ok tab to proceed.

Fig. 13.30 Activating Cut Lines parameters

The PLAY PATH dialogue window and Tool are activated as shown in Fig. 13.31.

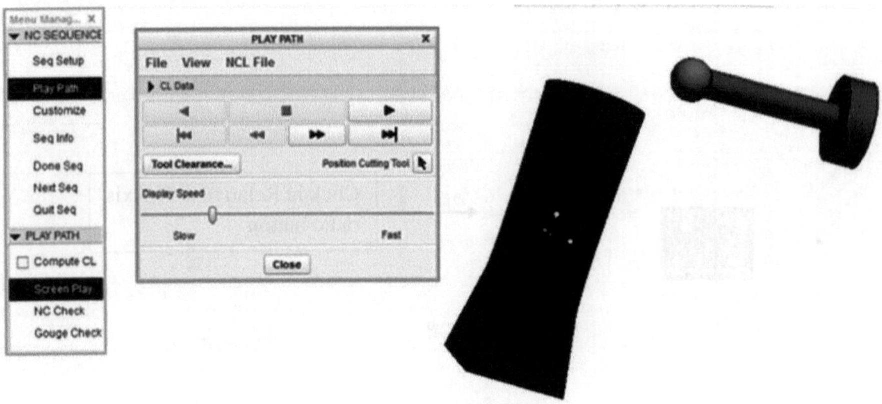

Fig. 13.31 PLAY PATH dialogue window and Tool

End of the surface milling is as shown in Fig. 13.32.

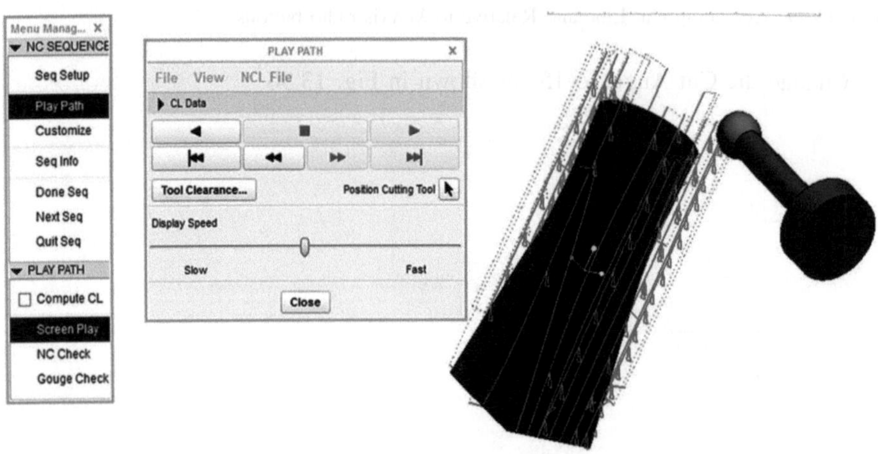

Fig. 13.32 End of the surface milling operation

Click on Close tab and the Done Seq on the NC SEQUENCE group to exit.

13.6.2 *Create Edge Surface Milling for the Outer Circular Surface of the Cast Part*

Click on the Surface Milling icon as shown below.

The MACH AXES Menu Manager dialogue box is activated on the main graphic window » Click on 5-Axis as highlighted in Fig. 13.33.

Fig. 13.33 Activating 5-Axis machine type

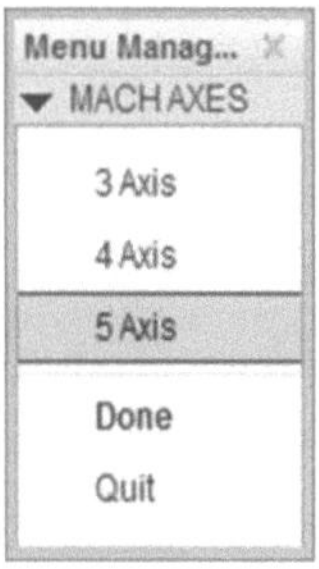

Click on Done on the MACH AXES group.

The NC SEQUENCE Menu Manager dialogue window is activated on the main graphic window » Check Mark Tool, Parameters, Surfaces and Define Cut as shown in Fig. 13.34.

Fig. 13.34 Tool, Parameters, Surfaces and Define Cut are check marked

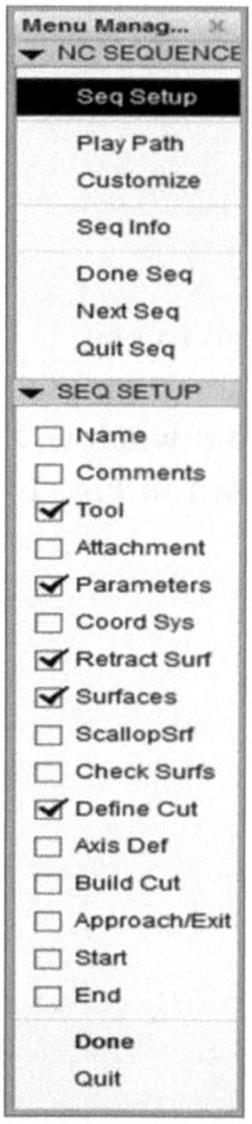

Click on Done.

The Tools Setup dialogue window is activated again on the main graphic window. Make sure that the Tool activated and its dimensions are correct. See Fig. 13.35.

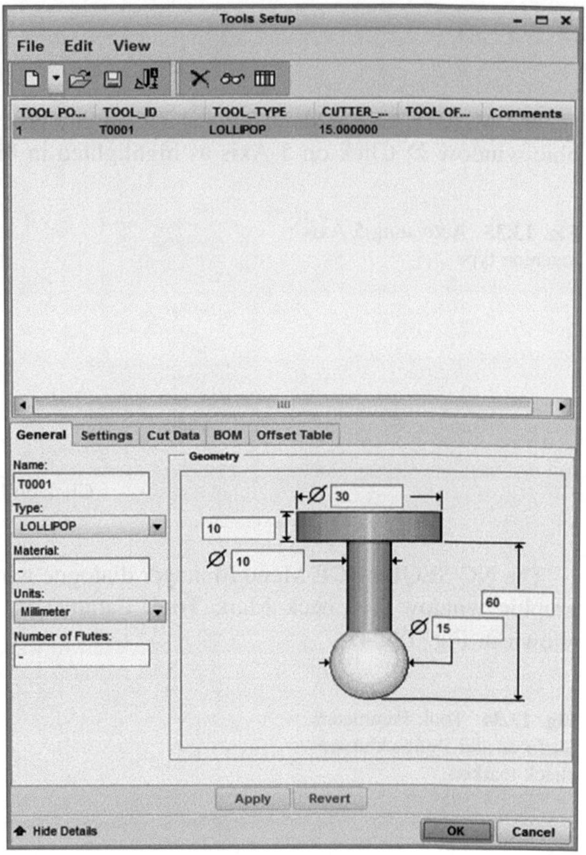

Fig. 13.35 Surface Milling Tool activated

Click OK tab to exit.

The Edit Parameters Sequence "Surface Milling" dialogue box is activated on the main graphic window ≫ Now input parameter values for the Surface Milling process as shown in Fig. 13.36.

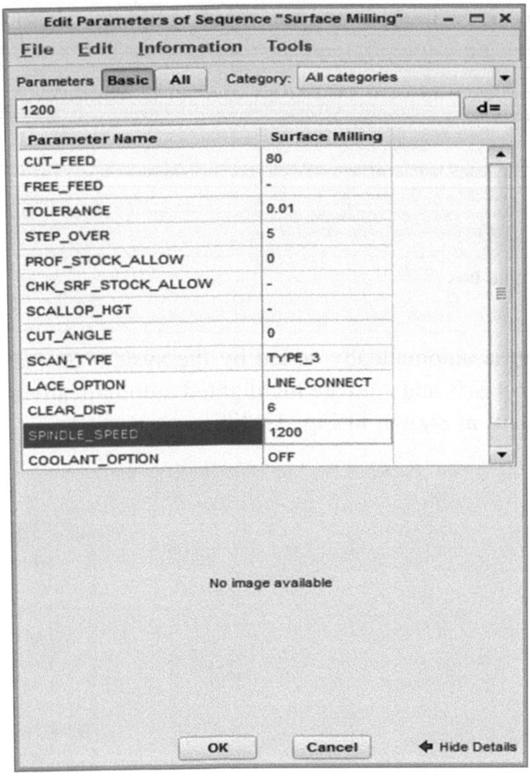

Click on OK tab to exit.

Fig. 13.36 Parameters values for Surface Milling on Edit parameters Sequence "Surface Milling" dialogue window

The Retract dialogue box is activated ≫ Use previous defined and then click Ok.

The SURF PICK group is automatically added into the Menu Manager dialogue box as shown in Fig. 13.37.

Fig. 13.37 SURF PICK group auto added into the Menu Manager dialogue box

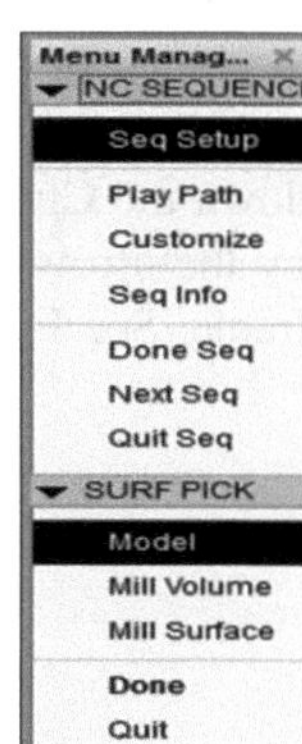

Select the Part as the Model ≫ Now click on Done.

The Select dialogue box is activated on the main graphic window. See Fig. 13.38.

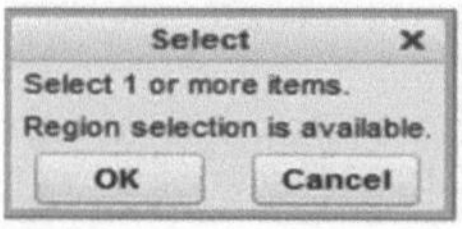

Fig. 13.38 Activated Select dialogue box

The SELECT SRFS group is automatically added by the system on the Menu Manager dialogue box ≫ Select Srfs and Add are highlighted automatically on their respective groups by the system as shown in Fig. 13.39.

Fig. 13.39 Add are highlighted on the SELECT SRFS groups

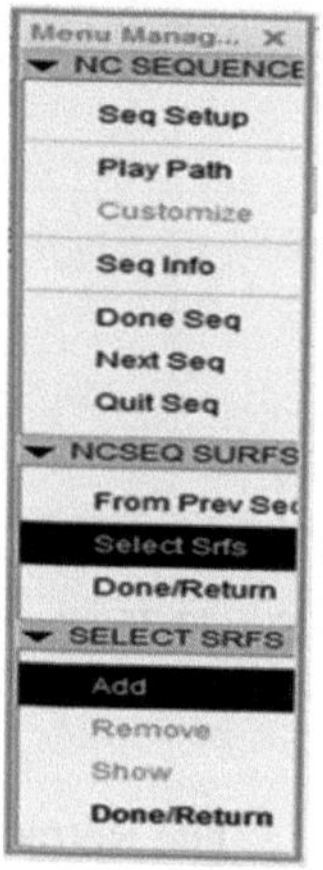

Now click on the Display View Style icon  to activate its drop-down menu ≫ Click on Wireframe on the drop-down list, the Part changes to Wireframe display as shown in Fig. 13.40.

Now click on the four outside Surfaces of the Reference Model, while holding down the Ctrl key when clicking on each of the four outside Surfaces as indicated by the arrows in Fig. 13.40 ≫ The selected outside Surfaces are shown in wireframe. See Fig. 13.40.

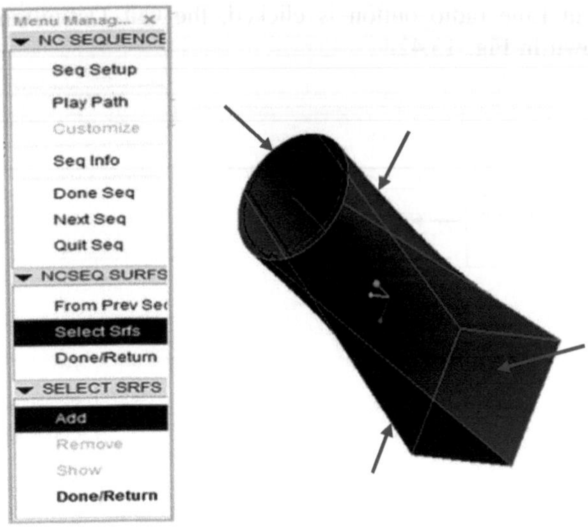

Fig. 13.40 Activating the outside surfaces of the cast Part

Click on OK tab on the Select dialogue box to exit if it's still active.

Click on Done/Return on the SELECT SRFS group.

Click on Done/Return on the activated NCSEQ SURFS group if activated and added by system.

The Cut Definition dialogue box is activated on the main window. See Fig. 13.41.

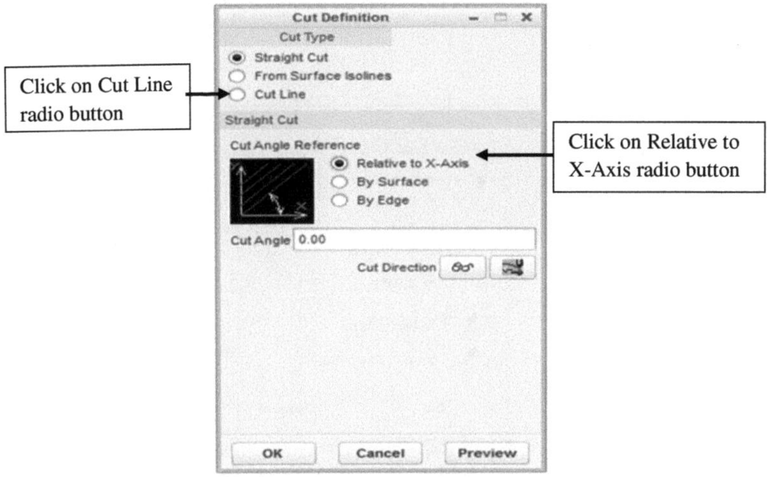

Fig. 13.41 Activated Cut Definition dialogue box

Once the Cut Line radio button is clicked; the Cut Definition dialogue box changes as shown in Fig. 13.42.

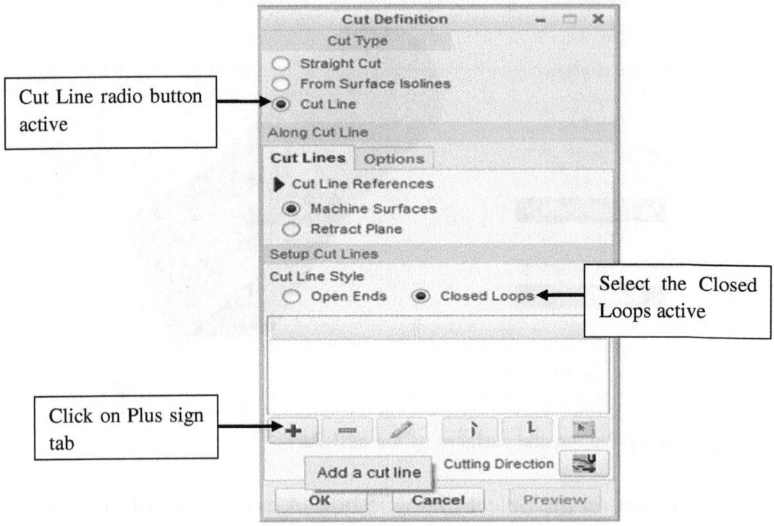

Fig. 13.42 Activating Cut Lines parameters

Click on Plus sign tab.

Once the Plus sign tab is clicked, the Add/Redefine Cutline, CHAIN and CHOOSE dialogue boxes are activated on the main graphic window, respectively, as shown in Fig. 13.43.

Fig. 13.43 Activated Add/Redefine Cutline, CHAIN and CHOOSE dialogue boxes

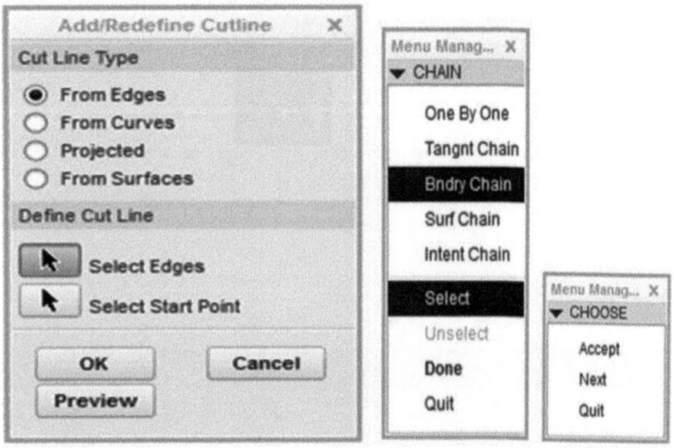

Hold down the Alt key first to enable selection of the circular edge ≫ Hold down both Ctrl and Alt keys to complete the selection as indicated by the arrow in Fig. 13.44.

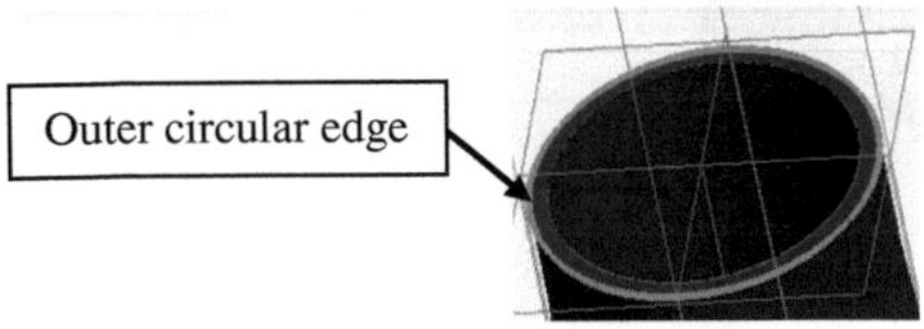

Fig. 13.44 Activated outer circular edge of the part

Click on Accept on the CHOOSE Menu Manager dialogue box ≫ Click on Done on the CHAIN Menu Manager dialogue box.

Click on the OK tab on the Add/Redefine Cutline dialogue box.

Cutline 1 From Edges is added as indicated by the arrow in Fig. 13.45.

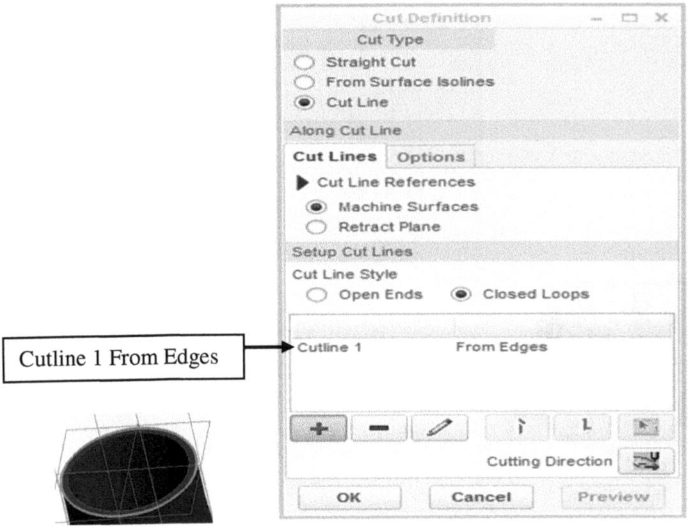

Fig. 13.45 Cutline 1 From Edges added into the Setup Cut Lines section box

Click the Plus sign tab again, the Add/Redefine Cutline, CHAIN and the CHOOSE dialogue boxes are activated again on the main graphic window, respectively ≫ Now click on the edge of the four outer lines while holding down the Ctrl key as indicated by the arrow in Fig. 13.46.

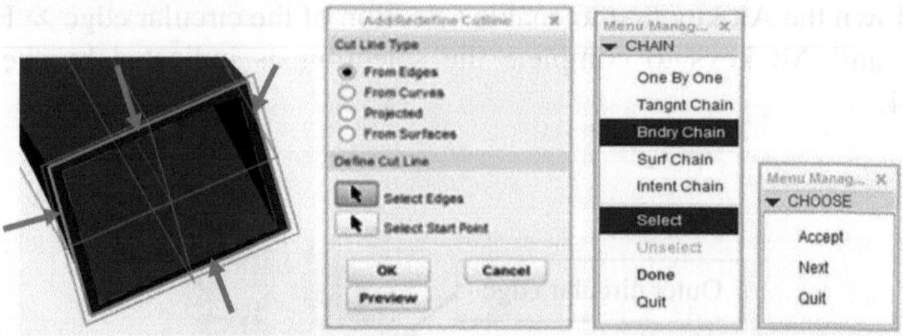

Fig. 13.46 Activated Add/Redefine Cutline, CHAIN and CHOOSE dialogue boxes with outer square edges active

Click on Accept on the CHOOSE Menu Manager dialogue box ≫ Click on Done on the CHAIN Menu Manager dialogue box.

Click on OK tab on the Add/Redefine Cutline dialogue box.

Cutline 2 From Edges is added as indicated by the arrow in Fig. 13.47.

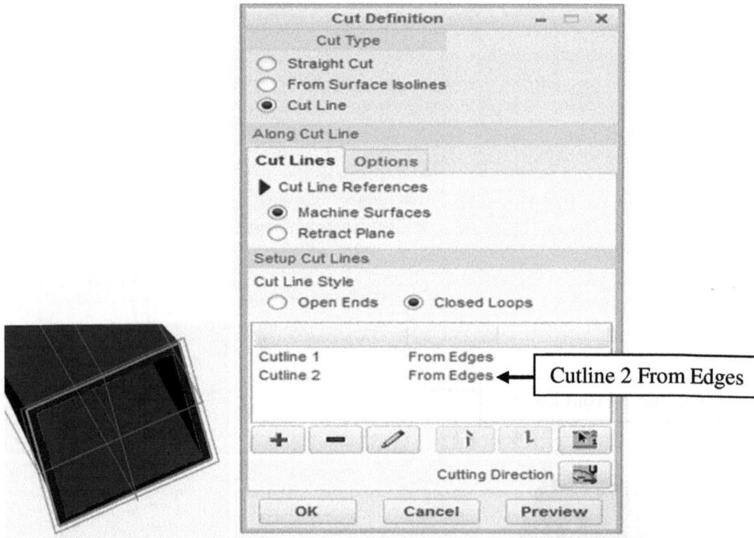

Fig. 13.47 Cutline 2 From Edges added into the Setup Cut Lines section box

Click on OK tab to exit.

13.6.3 Activate on Screen Play for the Outer Surface Milling Process

Click on Play Path on NC SEQUENCE group as highlighted in Fig. 13.48.

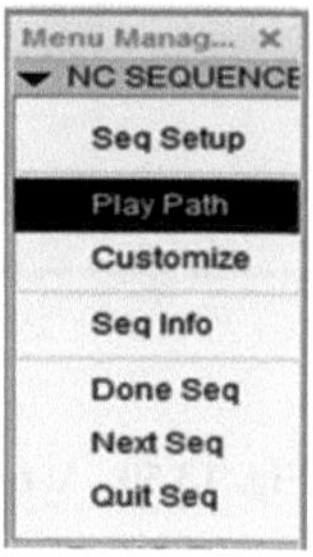

Fig. 13.48 Activating Play Path on NC SEQUENCE group

Once Play Path is clicked, the PLAY PATH group is automatically added by the system ≫ Now click on Screen Play as highlighted on the PLAY PATH group in Fig. 13.49.

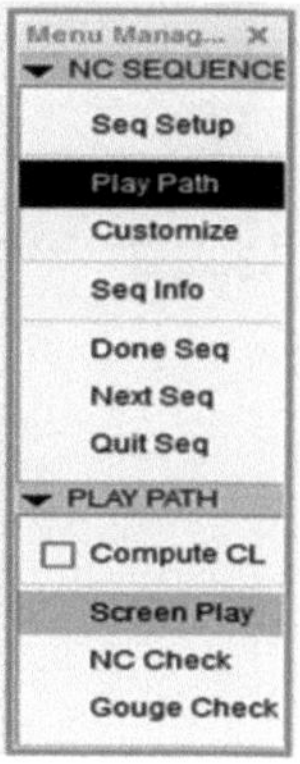

Fig. 13.49 Activating on Screen Play

The cutting Tool and the PLAY PATH dialogue box are activated on the main graphic window as shown in Fig. 13.50.

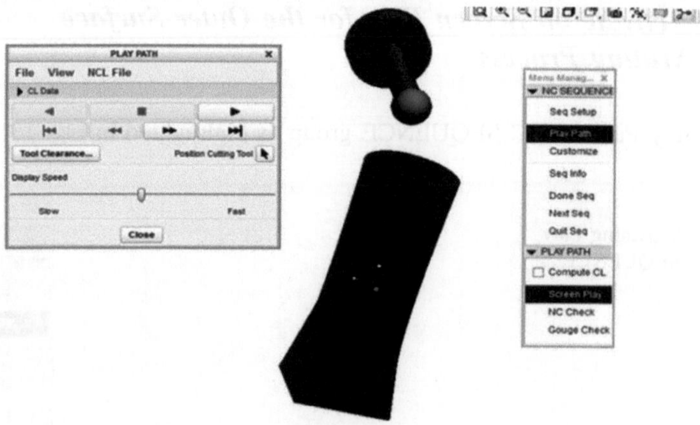

Fig. 13.50 Activated cutting Tool and PLAY PATH dialogue box

End of the circular edge surface milling process is as shown in Fig. 13.51.

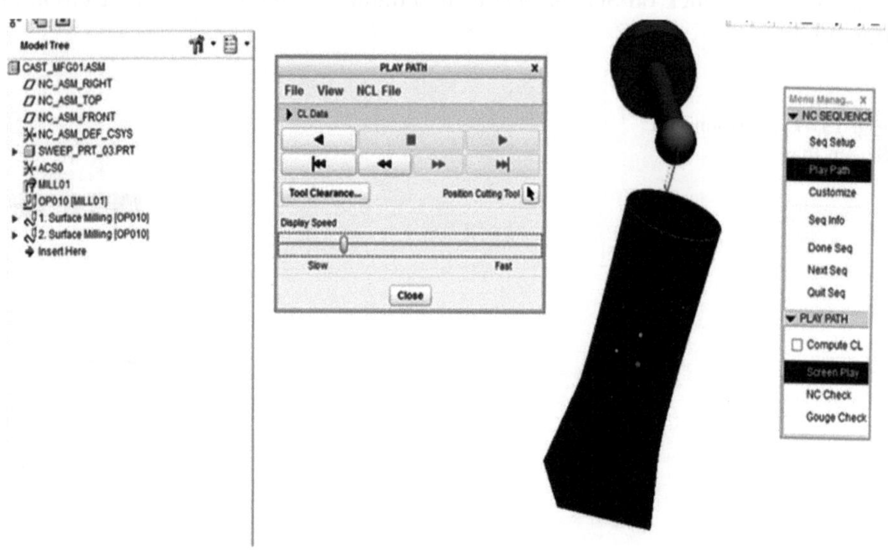

Fig. 13.51 End of circular edge milling process

Edit the Scan Type

Click on Seq Setup on NC SEQUENCE Menu Manager dialogue box (in Fig. 13.51) ≫ Check Mark Parameters ≫ Now click on Done ≫ The Edit Parameters Sequence "Surface Milling" dialogue box is activated ≫ Change the Scan Type to Type_Helical ≫ Click on the OK tab to exit.

Now click on Play tab on the PLAY PATH dialogue box to start external Surface Milling of the cast Part.

At the end of the Surface Milling operation, click on Close tab on the PLAY PATH dialogue box to exit.

Click on Done Seq on NC SEQUENCE group on the Menu Manager dialogue box to exit.

Further Reading

Creo Parametric 2.0 software (2012) PTC learning exchange tutorials. Copy right of Parametric Technology Inc.
PTC Learning Exchange Resource. Available at http://learningexchange.ptc.com/tutorials/by_sub_products.id:1
http://www.ptc.com

P.O. Kanife, *Computer Aided Virtual Manufacturing Using Creo Parametric*,
DOI 10.1007/978-3-319-23359-8

Index

P.O. Kanife, *Computer Aided Virtual Manufacturing Using Creo Parametric*,
DOI 10.1007/978-3-319-23359-8

MIX
Papier aus verantwortungsvollen Quellen
Paper from responsible sources
FSC® C105338

If you have any concerns about our products,
you can contact us on
ProductSafety@springernature.com

In case Publisher is established outside the EU,
the EU authorized representative is:
Springer Nature Customer Service Center GmbH
Europaplatz 3, 69115 Heidelberg, Germany

Printed by Libri Plureos GmbH
in Hamburg, Germany